Biodegradation of Nitroaromatic Compounds and Explosives

Biodegradation of
Nitroaromatic Compounds and Explosives

Edited by
Jim C. Spain
Joseph B. Hughes
Hans-Joachim Knackmuss

Sponsored by the
U.S. Air Force Office of Scientific Research and the
U.S. Defense Threat Reduction Agency

CRC Press
Taylor & Francis Group
Boca Raton London New York

CRC Press is an imprint of the
Taylor & Francis Group, an **informa** business

CRC Press
Taylor & Francis Group
6000 Broken Sound Parkway NW, Suite 300
Boca Raton, FL 33487-2742

First issued in paperback 2019

ISBN-13: 978-1-56670-522-6 (hbk)
ISBN-13: 978-0-367-39849-1 (pbk)

Library of Congress Card Number 00-026824

Library of Congress Cataloging-in-Publication Data

Biodegradation of nitroaromatic compounds and explosives / edited by Jim C. Spain, Joseph B. Hughes, Hans-Joachim Knackmuss.
    p. cm.
    Includes bibliographical references and index.
    ISBN 1-56670-522-3 (alk. paper)
    1. Nitroaromatic compounds—Biodegradation. 2. Explosives—Biodegradation. I. Spain, Jim C. II. Hughes, Joseph B. III. Knackmuss, Hans-Joachim. IV. United States Air Force. Office of Scientific Research. V. United States. Defense Threat Reduction Agency.

QP801.N55 B543 2000
662'.2—dc21

00-026824
CIP

Visit the Taylor & Francis Web site at
http://www.taylorandfrancis.com

and the CRC Press Web site at
http://www.crcpress.com

Preface

During the past 10 years, there has been an increasing amount of research on the biodegradation of nitroaromatic compounds because of the growing awareness of environmental contamination by pesticides, synthetic intermediates, and explosives. The resulting discoveries have dramatically extended our knowledge about the strategies used by biological systems for dealing with such compounds. In addition, the basic research has provided a foundation for a number of practical applications of biological systems for destruction of nitroaromatic contaminants.

In 1995 the advances in basic understanding of microbial degradation of nitroaromatic compounds were summarized in a book of reviews written by a group of experts in biochemistry, microbiology, and soil chemistry. Since that time the fundamental understanding has led to several practical applications involving biodegradation of nitroaromatic compounds previously thought to be recalcitrant to biological transformation. In September 1999, the Second International Symposium on Biodegradation of Nitroaromatic Compounds and Explosives was held in Leesburg, VA. The symposium brought together a wide range of scientists and engineers to discuss not only basic advances, but also field applications based on biodegradation and biotransformation of nitroaromatic compounds and explosives. The invited speakers were asked to critically review recent advances in their areas of expertise and to describe how the basic work has led to practical applications. They were also asked to discuss the questions that remain to be answered by future research. Some of the areas still involve mostly basic research on biochemistry and microbiology, whereas in other areas there is a considerable amount of experience in field applications.

The first goal of this book is to provide the reader with a timely synthesis of ongoing recent research and an appreciation for the remarkable range of biochemical strategies available for the transformation of nitroaromatic compounds. The second goal is to give a realistic evaluation of the current and potential field applications of the various strategies. Each chapter is designed to stand alone, thus there is occasionally some overlap in the introductory material and some experiments have been used as examples in more than one chapter. The book should be of interest to microbiologists, biochemists, engineers, and anyone concerned about the environmental behavior and destruction of nitroaromatic compounds and explosives.

Chapter 1 provides a brief introduction and overview of the scope of the problem with environmental contamination by nitroaromatic compounds. Most of the incentive for research in the U.S. comes from concerns about contamination with explosives. The extent of such contamination in much of the rest of the world is not known. The second chapter describes the strategies used by aerobic bacteria for converting nitroaromatic compounds into sources of carbon and energy. Most of the catabolic pathways are well understood, and there is a growing body of knowledge on the molecular biology of such systems. In addition, bacteria that grow on nitroaromatic compounds have been used in a variety of pilot- and field-scale applications. Chapter 3 examines the biochemistry and molecular biology of nitroarene dioxygenases, key enzymes that initiate the productive metabolism of a variety of substrates.

The insight about both the catalytic mechanism and the evolutionary origin of the enzymes provides a sense of how productive pathways might have evolved for organic compounds that have only been in the biosphere for a brief time.

Chapter 4 introduces more complex molecules that cannot serve as growth substrates for microorganisms, but, nonetheless, can be extensively transformed and detoxified by binding to soil. The exciting advances in the biochemistry of the initial enzymatic attack on polynitroaromatic compounds are described in detail along with the mechanisms of binding of the intermediates to soil humic material. Enzymes that attack the aromatic ring of picric acid can lead to the elimination of nitrite and subsequent mineralization. In contrast, enzymes that reduce the nitro groups of nitrotoluenes can produce intermediates that bind to soil under appropriate conditions. Both strategies can lead to effective solutions for environmental contamination. Very recent discoveries, described in Chapter 5, reveal the sequences and arrangement of the genes involved in the bacterial degradation of picric acid. The work also illustrates the remarkable effectiveness of mRNA differential display in the discovery of genes involved in the degradation pathway and how the information has led to elucidation of the pathway. Simple nitrophenols and nitrobenzoates were the first nitroaromatic compounds discovered to be degraded by bacteria. Chapter 6 describes the recent advances in understanding the molecular biology of the degradative pathways of simple polar nitroaromatic compounds and their relationships to other catabolic sequences.

Chapter 7 provides an overview of the current understanding of the biotransformation of nitroglycerin. Because the molecule is a nitrate ester rather than a nitroaromatic compound, the reductase enzymes that catalyze the initial attack on the molecule can cause the elimination of nitrite. This chapter also describes the remarkable discovery that the bacterial nitrate ester reductase can catalyze the elimination of nitrite from a wide range of nitro-substituted compounds, including TNT.

Although many workers have reported the reduction of nitro groups of explosives by anaerobic bacteria, most of the work has been done with complex mixed cultures where it is impossible to study the details of the biochemistry of the transformation pathways. In contrast, Chapter 8 describes the details of TNT transformation by *Clostridium* and the central, previously unexpected, role of hydroxylamino derivatives as key intermediates.

Fungi are the only organisms capable of significant amounts of TNT mineralization. The ligninolytic enzymes of *Phanerochaete chrysosporium* have been studied extensively because of their ability to transform TNT. Unfortunately, *Phanerochaete* does not grow well in soil and is inhibited by high concentrations of TNT. The authors of Chapter 9 have evaluated the TNT degradation abilities of a wide range of fungi adapted to a variety of habitats and discovered several that have considerable potential for use in practical applications. Higher plants have recently been discovered to transform both TNT and RDX extensively. Phytoremediation is becoming widely accepted as a strategy for remediation of contaminated water and soil. Chapter 10 explores the recent advances in understanding transformation pathways based on studies with plant cell cultures, intact plants, and phytoremediation systems in the field.

RDX and HMX have been used extensively by the military since World War II and are found widely as contaminants at military sites, yet their biodegradation has only been studied sporadically. In Chapter 11, the author summarizes the previous understanding of the degradation mechanisms and then outlines some very recent discoveries that provide new insight about the reactions leading to mineralization by bacteria. Some of the newly proposed reactions differ considerably from the ones that have been accepted for many years.

Most of the interest in biodegradation of nitroaromatic compounds stems from concerns about their fate and transport in the environment. Studies of biodegradation in complex natural systems are meaningless without a good understanding of the many abiotic reactions that affect the transformation and behavior of nitroaromatic molecules. Chapter 12 explains the abiotic transformations of the nitro substituents as well as the geomicrobiology involved in generation of the materials that interact with nitro compounds.

Chapters 13 and 14 illustrate how the basic research in the past has led to practical treatment strategies based on composting and other large-scale systems. Composting has been the most extensively used approach to field-scale treatment of excavated soil. It is widely accepted in the U.S. and has been optimized by the U.S. Army. Chapter 13 explains the basic understanding of the composting process and the state of the art in application. Chapter 14 describes the scope of the explosives contamination problem in the U.S. and compares the cost and effectiveness of a variety of commercially available treatments based on biodegradation. The comparison reveals that several technologies can be effective and relatively inexpensive for destruction of TNT in excavated soil.

The authors would like to thank Alice Giraitis for outstanding editorial assistance and Loreen Kollar of the Florida State University Institute for International Cooperative Environmental Research for organizing the symposium. The authors also thank Shirley Nishino for taking on all the extra tasks required to make this book possible. Much of the work described in this book would not have been possible without the unfailing support of Walter Kozumbo of the U.S. Air Force Office of Scientific Research. He developed an outstanding basic research program to shed light on the strategies used by biological systems to detoxify and destroy nitroaromatic compounds. He provided funding for a number of outstanding investigators to work in critical areas and is responsible to a large extent for the recent exciting advances in the field.

Editors

Jim C. Spain has a B.S. in biology from the University of Texas at Arlington and a Ph.D. in microbiology from the University of Texas at Austin. For the past 20 years he has studied the mechanisms of biodegradation of synthetic organic compounds. He worked on the biodegradation of pesticides for five years at the U.S. Environmental Protection Agency Marine Environmental Research Laboratory before joining the Air Force Research Laboratory where he currently directs the Environmental Biotechnology in-house research program.

Dr. Spain is responsible for both basic and applied research on the application of biotechnology for degradation and synthesis of organic compounds. He has organized several international symposia and edited a previous book on the biodegradation of nitroaromatic compounds and explosives. He has published extensively on the pathways involved in the biodegradation of a variety of nitroaromatic compounds and has also conducted a considerable amount of field work in the area.

Joseph B. Hughes is an Associate Professor at Rice University and Chair of the Department of Environmental Science and Engineering. He received his B.A. in chemistry (1987) from Cornell College. Dr. Hughes attended graduate school at the University of Iowa. His M.S. and Ph.D. were awarded in 1992 by the Department of Civil Engineering and Environmental Engineering, where he was a graduate fellow in biocatalysis and bioprocessing. He joined the faculty of Rice University immediately thereafter.

Dr. Hughes' research and teaching focus on the metabolism of hazardous environmental pollutants by living organisms — bacteria and plants primarily. Much of his most recent research has dealt with the metabolism of nitroaromatic compounds, which have complex environmental chemistry. Specifically, his research group is trying to understand the pathways of metabolism and to identify the intermediates and products of metabolism. His work has relied heavily on the use of modern isolation and structural identification methods to define intermediates and products from complex mixtures of organic compounds.

Dr. Hughes has received the 1997 Rice University ASCE outstanding teaching award, and was the National Academy of Engiqneering's invited speaker on Frontiers of Environmental Engineering (1995).

Hans-Joachim Knackmuss obtained his education as a diploma chemist (1963) and his degree of doctor rerum naturarum (1964) from the University of Heidelberg. After working (1964–1971) as a research scientist with Richard Kuhn (Heidelberg) and Motimer Starr (Davis) he qualified as a university lecturer (habilitation) by submitting a thesis on "Azaquinones and Glycorylazaquinones from Bacteria." As a university lecturer and associate professor at the University of Göttingen (1971–1979), he initiated his work on the biodegradation of xenobiotic compounds. In 1979 he was appointed full professor for microbiology at the University of Wuppertal and at the University of Stuttgart (since 1986).

Besides being a university professor, he was a collaborator and advisor of the Bayer AG, Leverkusen (1979–1991), and since 1986 he has been head of the Department of Chemical Microbiology at the Fraunhofer Institute for Interfacial Engineering and Biotechnology in Stuttgart. He published about 180 papers, largely on the biodegradation of xenobiotic compounds. His current research interest focuses on the elimination of xenobiotics from waste water, soil and exhaust gases, the design of new biodegradable synthetic products, and the development of new synthons by regio-, stereo-, and enantioselective biotransformation.

Contributors

Christof Achtnich
Fraunhofer Institute für Grenzflächen-
und Bioverfahrenstechnik
Stuttgart, Germany
ach@igb.fhg.de

Farrukh Ahmad
Department of Environmental Science
and Engineering
Rice University
Houston, Texas
ahmadf@rice.edu

Sang-Weon Bang
Biotechnology Center for Agriculture
and the Environment
Rutgers–The State University
New Brunswick, New Jersey
swbang@socrates.berkeley.edu

Neil C. Bruce
Institute of Biotechnology
University of Cambridge
Cambridge, United Kingdom
n.bruce@biotech.cam.ac.uk

Dirk Bruns-Nagel
Institute of Immunology and
Environmental Hygiene
Philipps University Marburg
Marburg, Germany
dirk.bruns-nagel@aventis.com

Joel G. Burken
Environmental Research Center
Department of Civil Engineering
University of Missouri-Rolla
Rolla, Missouri
burken@umr.edu

Wolfgang Fritsche
Institute of Microbiology
Friedrich-Schiller University Jena
Jena, Germany
schein@merlin.biologie.uni-jena.de

Diethard Gemsa
Institute of Immunology and
Environmental Hygiene
Philipps University Marburg
Marburg, Germany

Stefan B. Haderlein
Swiss Federal Institute of
Environmental Science and
Technology (EAWAG)
Dübendorf, Switzerland
Stefan.haderlein@eawag.ch

Jalal Hawari
Biotechnology Research Institute
National Research Council of Canada
Montreal, Quebec, Canada
jalal.hawari@nrc.ca

Zhongqi He
Air Force Research Laboratory, MLQ
Tyndall Air Force Base, Florida
zhe@maine.edu

Angela Herre
Institute of Microbiology
Friedrich-Schiller University Jena
Jena, Germany

Martin Hofrichter
Institute of Microbiology
Friedrich-Schiller University Jena
Jena, Germany
mhofrich@LadyBird.helsinki.fi

Thomas B. Hofstetter
Swiss Federal Institute of Technology
 (ETH)
Dübendorf, Switzerland
hofstetter@tech.chem.ethz.ch

Joseph B. Hughes
Department of Environmental Science
 and Engineering
Rice University
Houston, Texas
hughes@owlnet.rice.edu

Douglas E. Jerger
IT Corporation
Knoxville, Tennessee
djerger@itcrp.com

Hans-Joachim Knackmuss
Fraunhofer Institute für Grenzflächen-
 und Bioverfahrenstechnik
and Institut für Mikrobiologie,
 Universität Stuttgart
Stuttgart, Germany
imbhjk@po.uni-stuttgart.de

Hiltrud Lenke
Fraunhofer Institute für Grenzflächen-
 und Bioverfahrenstechnik
Stuttgart, Germany
len@igb.fhg.de

Lisa M. Newman
Biotechnology Center for Agriculture
 and the Environment
Rutgers–The State University
New Brunswick, New Jersey
lisa_newman@maxygen.com

Shirley F. Nishino
Air Force Research Laboratory, MLQ
Tyndall Air Force Base, Florida
shirley.nishino@tyndall.af.mil

Rebecca E. Parales
Department of Microbiology
University of Iowa
Iowa City, Iowa
rebecca-parales@uiowa.edu

Lynda Perry
Biotechnology Center for Agriculture
 and the Environment
Rutgers–The State University
New Brunswick, New Jersey
callicot@eden.rutgers.edu

Pierre E. Rouvière
E.I. DuPont de Nemours Company
Wilmington, Delaware
pierre.e.rouviere@usa.dupont.com

Rainer Russ
Institut für Mikrobiologie, Universität
 Stuttgart
Stuttgart, Germany
russ@rhrk.uni-kl.de

Katrin Scheibner
Institute of Microbiology
Friedrich-Schiller University Jena
Jena, Germany

René P. Schwarzenbach
Swiss Federal Institute of
 Environmental Science and
 Technology (EAWAG)
Dübendorf, Switzerland
rene.schwarzenbach@eawag.ch

Jacqueline V. Shanks
Department of Chemical Engineering
Iowa State University
Ames, Iowa
jshanks@iastate.edu

Jim C. Spain
Air Force Research Laboratory, MLQ
Tyndall Air Force Base, Florida
jim.spain@tyndall.af.mil

Klaus Steinbach
Department of Chemistry
Philipps University Marburg
Marburg, Germany

Phillip L. Thompson
Department of Civil and Environmental
  Engineering
Seattle University
Seattle, Washington
thompson@seattleu.edu

Eberhard von Löw
Institute of Immunology and
  Environmental Hygiene
Philipps University Marburg
Marburg, Germany
vonloewe@post.med.uni-marburg.de

Dana M. Walters
E.I. DuPont de Nemours Company
Wilmington, Delaware

Richard E. Williams
Institute of Biotechnology
University of Cambridge
Cambridge, United Kingdom
r.williams@biotech.cam.ac.uk

Patrick M. Woodhull
IT Corporation
Findlay, Ohio

Gerben J. Zylstra
Biotechnology Center for Agriculture
  and the Environment
Rutgers–The State University
New Brunswick, New Jersey
zylstra@aesop.rutgers.edu

Contents

CHAPTER 1

Introduction

Jim C. Spain

Explosives and other nitrated compounds (Figure 1.1) are widely distributed environmental contaminants. Nitroaromatic pesticides such as dinoseb, dinitrocresol, parathion, and methylparathion are intentionally released in soil and water worldwide. They are also spilled accidentally at loading facilities and during agricultural use. Nitrophenols and nitrotoluenes are used extensively as feedstocks in industry and are often released to surface water in waste streams. The pesticides and simple nitroaromatic compounds are biodegradable by soil bacteria and do not accumulate in the environment unless concentrations exceed the assimilative capacity of the ecosystem. In contrast, explosives such as 2,4,6-trinitrotoluene (TNT), hexahydro-1,3,5-trinitro-1,3,5-triazine (RDX), and octahydro-1,3,5,7-tetranitro-1,3,5,7-tetrazine (HMX) are less biodegradable and often persist for extended periods in soil or groundwater.

The major explosives manufacturing, handling, and storage sites in the U.S. (Figure 1.2) have been identified by the U.S. Army Environmental Center.[3] The most heavily contaminated sites are the army ammunition plants where explosives were manufactured for much of this century. Fortunately, most of the sites are owned by the U.S. government and are therefore not accessible to the public. A number of the sites have been cleaned up, and many are in the process of cleanup. Major exceptions include Volunteer and Ravenna Army Ammunition Plants where contamination is extensive and cleanup of contaminated soil and groundwater has not started. A substantial amount of dinitrotoluene-contaminated soil also remains at the Badger Army Ammunition Plant. The most recent public document on the scope of the problem[1] describes a timeline for the cleanup of contaminated soil at many of the sites, but does not provide information about contaminated groundwater. More detailed information about the scope of the problem and the current cleanup strategies is provided in Chapter 14.

In Germany, the situation is more problematic because many of the explosives manufacturing facilities (Figure 1.3) were demolished at the end of World War II

1

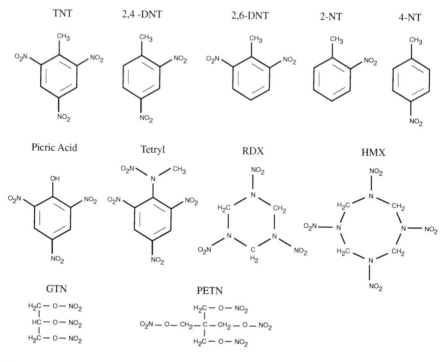

Figure 1.1   Major military explosives and production intermediates: TNT, 2,4,6-trinitrotoluene; 2,4-DNT, 2,4-dinitrotoluene; 2,6-DNT, 2,6-dinitrotoluene; 2-NT, 2-nitrotoluene; 4-NT, 4-nitrotoluene; picric acid, 2,4,6-trinitrophenol; tetryl, 2,4,6-trinitrophenylmethylnitramine; RDX, hexahydro-1,3,5-trinitro-1,3,5-triazine; HMX, octahydro-1,3,5,7-tetranitro-1,3,5,7-tetrazocine; GTN, nitroglycerin; and PETN, pentaerythritol tetranitrate.

with little regard for environmental consequences. Most of the sites have since been used for industry and housing, and only a few of the larger sites have remained undeveloped. To date only a limited amount of site characterization has been done, and few cleanup projects have been started. The German government has established a joint research group funded by the Federal Ministry of Education and Research to develop and test processes for the bioremediation of contaminated soils.[4] The results will be published in a handbook and will provide the basis for the use of biotreatment for cleanup of contaminated soil at the sites.

In the U.K., Canada, and Australia some site characterization has been done, but very little cleanup is in progress. In the rest of the world the extent of contamination with explosives is either undetermined or not available to the public. Anecdotal information suggests that the scope of the problem is significant.

Research during the past 10 years has yielded a wealth of information about the biodegradation of nitroaromatic compounds. All of the nitroaromatic compounds and explosives listed in Figure 1.1 are subject to transformation by microorganisms, and the processes will be discussed in detail in subsequent chapters. Some, including picric acid, 2,4- and 2,6-dinitrotoluene, mononitrotoluenes, dinitrocresol, nitrobenzoates, and

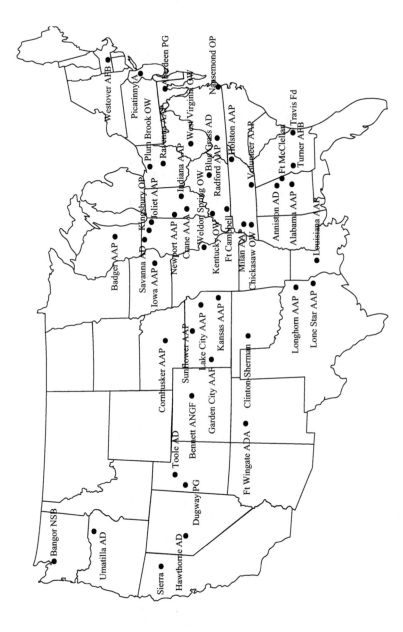

Figure 1.2 Explosive-contaminated manufacturing, processing, and storage sites in the U.S. A, Arsenal; AAP, Army Ammunition Plant; AD, Army Depot; AFB, Air Force Base; ANGF, Air National Guard Field; NSB, Naval Submarine Base; OW, Ordnance Works; and PG, Proving Ground. (Adapted from U.S. Army Environmental Center. 1996).

Figure 1.3 Explosives-contaminated manufacturing, processing, and storage sites in the Federal Republic of Germany. Sites shown processed more than 1700 tons of explosives per month. Numbers in parentheses are tons of TNT per month. (Data from Preuss, J. and R. Haas. 1987. *Geogr. Umschau.* 39:578.)

nitrophenols, can serve as growth substrates for bacteria. Therefore, such compounds are prime candidates for bioremediation because the process can be self-sustaining and inexpensive. A major conundrum and an area of active research is the question of why the biodegradable compounds persist in the environment.

Other explosives including TNT, RDX, and HMX do not serve as growth substrates for bacteria. They can, however, be transformed and detoxified by cometabolic processes. Several alternative, cost-effective strategies are now available for the cometabolic treatment of explosives in contaminated soil (Chapter 14), and phytoremediation shows considerable promise (Chapter 10). A variety of such cometabolic strategies will be discussed in subsequent chapters. The disadvantage of such processes is that they are more expensive because they require a second growth substrate for the microorganisms. In addition, the products are often difficult to characterize and the processes are difficult to control.

Much of the current research on biodegradation of explosives is focused on discovery and development of strategies that would allow the more recalcitrant explosives to serve as growth substrates for microorganisms. Recent advances in molecular biology pave the way for pathway engineering that would allow complete degradation in a variety of organisms. Success would reduce the cost of bioremediation considerably for major explosives, including TNT.

In the U.S., cleanup of soils heavily contaminated with explosives is being conducted primarily by excavation followed by composting or incineration. Large volumes of lightly contaminated soils would be more appropriately treated *in situ* if a suitable technology were available. Treatment of explosives-contaminated groundwater is done almost exclusively by pump and treat with carbon sorption. In other countries the cleanup of explosives-contaminated sites is just beginning or has not been considered to date. Thus, although new strategies for *ex situ* bioremediation of explosives-contaminated soil may not be fielded in time to help solve the problem in the U.S., they could be very useful in international applications. Novel *in situ* soil and groundwater remediation strategies could be fielded in the U.S. in time to provide considerable cost savings.

REFERENCES

1. Broder, M. F. and R. A. Westmoreland. 1998. An estimate of soils contaminated with secondary explosives. Final Report. SFIM-AEC-ET-CR-98002. U.S. Army Environmental Center.
2. Preuss, J. and R. Haas. 1987. Die Standorte der Pulver-, Sprengstoff-, Kampf-und Nebelstofferzeugung im ehemaligen Deutschen Reich. *Geogr. Umschau.* 39:578–584.
3. U.S. Army Environmental Center. 1996. WWW site. U.S. Army Environmental Center.
4. Michels, J. 1999. Second International Symposium on Biodegradation of Nitroaromatic Compounds and Explosives. Leesburg, VA.

CHAPTER 2

Strategies for Aerobic Degradation of Nitroaromatic Compounds by Bacteria: Process Discovery to Field Application

Shirley F. Nishino, Jim C. Spain, and Zhongqi He

CONTENTS

2.1 INTRODUCTION

2.1.1 Rationale for Bioremediation of Nitroaromatic Compounds

Nitroaromatic compounds constitute a major class of environmental contaminants. They are important as industrial feedstocks due to the versatile chemistry of the nitro group. For example, 97% of nitrobenzene (NB) produced worldwide is used for aniline production,[53] which increased 50% between 1988 and 1998 to 1.5 billion pounds.[36] In 1982 the U.S. accounted for about 31% of the worldwide dinitrotoluene (DNT) production of 2.3 billion pounds,[152] and in 1999 the largest DNT producer in the U.S. expanded DNT production capacity by 50% to 1.5 billion pounds.[37] NB, chloronitrobenzene, and chloroaniline have all been found in the North Sea.[17] Nitrophenolic compounds are widely and regularly dispersed as agricultural pesticides. Military explosives production, testing, and distribution has led to contaminated sites worldwide. Costs for conventional cleanup are enormous; estimated costs for incineration are $400 to $600/yd³ of soil,[71] and for composting only slightly less at $404 to $467/yd³ of soil,[116] although economies of scale may reduce those costs considerably (see Chapter 14). Estimated costs for cometabolic bioslurry treatment of trinitrotoluene (TNT)-contaminated soil are $290 to $350/yd³.[131] If sustainable strategies that rely on mineralization of nitroaromatic compounds can be developed, costs might be reduced dramatically. Nitroaromatic compounds in the environment create intense selection pressure which has led to the evolution of microorganisms able to degrade a growing number of nitroaromatic compounds. The catabolic pathways that have been discovered can be exploited for bioremediation, waste minimization, and the rapidly expanding field of biocatalysis.

2.1.1.1 Toxicity, Recalcitrance, and Regulatory Requirements

The stability, persistence, and toxicity that make nitroaromatic compounds valuable to industry render them hazardous when released into the environment. The electron-withdrawing character of the nitro group makes oxidative attack, a major mode of catalysis of aromatic compounds,[64] increasingly difficult as the number of nitro substituents on the aromatic ring increases.[174] The same electron-withdrawing properties lead to facile reduction of the nitro group under both anaerobic and aerobic conditions.[174] Nitroaromatic compounds are readily reduced to more reactive and, potentially, more carcinogenic or mutagenic derivatives when introduced into mammalian systems.[132,205] Intestinal microflora, as well as mammalian organ systems, possess nonspecific nitroreductases that catalyze the conversion of nitro groups to physiologically more harmful nitroso and hydroxylamino groups.[34] Epidemiological studies suggest that both amino- and nitroaromatic compounds are powerful carcinogens.[8] As a result, a number of nitroaromatic compounds including nitrobenzenes, dinitrotoluenes, and mono- and dinitrophenols are regulated priority pollutants.[112] Waste streams generated from production of NB, aniline, DNT, diaminotoluene, and explosives are regulated by the U.S. Environmental Protection Agency as toxic wastes from specific sources (40 CFR 261.32).

2.1.1.2 Mineralization vs. Cometabolism

Early surveys found few nitroaromatic compounds that were amenable to microbial attack.[5,83] Many of the studies used activated sludges to test the potential treatability of nitroaromatic compounds and came to the conclusion that nitroaromatic compounds were, in general, much more recalcitrant to biological breakdown than the nonnitrated analogs.[5] Researchers looking for a biological solution to the problem of widespread munitions contamination came to the conclusion that polynitroaromatic compounds could only be transformed to dead end aminonitro compounds and not mineralized.[110]

It is important to distinguish between biodegradation of a contaminant and the less desirable transformation, disappearance, or immobilization. Mineralization (biodegradation) — the complete catabolism of a compound to its inorganic components — is the preferred goal of bioremediation systems. When a compound is mineralized, it is consumed to yield energy and/or incorporated into biomass.[4] Because energy is derived from the catabolic process, degradation is self-sustaining as long as the compound (contaminant) is present and provides a strong selective advantage to the degradative organism. The mineralization process is in contrast to cometabolism, the nonspecific transformation by enzymes involved in the catabolism of some other growth substrate and also induced by some other growth substrate.[4] In some cases the cometabolized substrate can provide carbon and energy, but only as long as a primary substrate is present. The requirement for a primary substrate and the absence of a selective advantage renders cometabolic systems more expensive and difficult to control than systems which rely on mineralization. Another key difference between degradation and transformation is the difference between the endpoints. Degradation produces harmless minerals and biomass, whereas transformation produces organic derivatives of the parent molecule whose identity and toxicity must be established for each individual situation.

2.1.2   Biodegradable Nitroaromatic Compounds

In recent years, a number of nitroaromatic compounds have been discovered to be susceptible to aerobic microbial degradation (Table 2.1). In most instances where mineralization has been demonstrated, the catabolic pathway has also been determined. In a few instances the genes that encode the degradative enzymes have been cloned and sequenced. Understanding the pathway by which a compound is degraded is a key element in the development of an effective bioremediation system. Given the knowledge of the degradative pathway and the physiological requirements of the degradative microorganisms, systems can be designed to meet the growth requirements of the organisms and to control or monitor the catabolic process.

A considerable body of knowledge has been developed about the biodegradation of many of the compounds listed in Table 2.1. Bacteria have evolved a variety of aerobic strategies for removal of the nitro group during conversion of the nitroaromatic compounds to central metabolites (Figure 2.1). The nitro group is released as nitrite following dioxygenation of the aromatic ring to a dihydroxy intermediate, monooxygenation to an epoxide, or formation of a hydride–Meisenheimer complex. Ammonia is released by pathways that proceed via partial reduction to hydroxylaminobenzenes. The hydroxylaminobenzenes can further undergo a mutase-mediated rearrangement to *ortho*-aminophenols in some strains. Alternatively, the hydroxylamino compound can be converted to the corresponding catechol with the elimination of ammonia. To date there is little evidence for an aerobic pathway that involves complete reduction to the amine prior to ring fission by the same bacterium. Some general observations can be made about the metabolism of the various classes of nitroaromatic compounds.

Mononitrophenols including 2-nitrophenol,[215] 4-nitrophenol,[189,191] and 4-chloro-2-nitrophenol[28] are all hydroxylated at the nitro group with release of nitrite to form an *ortho*- or *para*-dihydroxybenzene (Figure 2.1, pathway I). Some compounds, including parathion[183] and methyl parathion,[169] are first hydrolyzed to 4-nitrophenol, which then undergoes hydroxylation at the nitro group. Similarly, 4-nitroanisole[177] is *O*-demethylated to 4-nitrophenol prior to hydroxlyation at the nitro group. In some bacteria 4-nitrophenol is degraded via an initial hydroxylation to produce 4-nitrocatechol;[100,106,107,171] then the nitro group is replaced by a hydroxyl group. In some instances the monooxygenation reactions are catalyzed by flavoprotein monooxygenases,[107] and all the monooxygenation reactions probably involve a quinone intermediate. The intermediate had been postulated based on the stoichiometry of the cofactor requirements for the reaction,[191] but the

Table 2.1 Biodegradable Nitroaromatic Compounds

Compound	Pathway Ref.	Bioremediation Ref.
Nitrobenzene	Nishino and Spain 1993 Nishino and Spain 1995 Jung et al. 1995	Livingston 1993 Dickel et al. 1993 Swindoll et al. 1993 Oh and Bartha 1997 Greene et al. 1999 Peres et al. 1998
1,3-Dinitrobenzene	Dickel and Knackmuss 1991 Nishino and Spain 1992	Bringmann and Kühn 1971 Mitchell and Dennis 1982 Dey et al. 1986 Dey and Godbole 1986
2-Nitrotoluene	Haigler et al. 1994	Bringmann and Kühn 1971 Struijs and Stoltenkamp 1986 Lendenmann et al. 1998 Spain et al. 1999
3-Nitrotoluene	Ali-Sadat et al. 1995	Hallas and Alexander 1983
4-Nitrotoluene	Haigler and Spain 1993 Rhys-Williams et al. 1993 Michán et al. 1997 Spiess et al. 1998	Bringmann and Kühn 1971 Hallas and Alexander 1983 Lendenmann et al. 1998 Spain et al. 1999
2,4-Dinitrotoluene	Spanggord et al. 1991 Haigler et al. 1994 Haigler et al. 1999	Bausum et al. 1992 Berchtold et al. 1995 Cheng et al. 1996 Bradley et al. 1997 Lendenmann et al. 1998 Nishino et al. 1999 Spain et al. 1999

continued

Table 2.1   (continued) Biodegradable Nitroaromatic Compounds

Compound	Pathway Ref.	Bioremediation Ref.
2,6-Dinitrotoluene	Nishino et al. In Press	Bradley et al. 1995 Lendenmann et al. 1998 Nishino et al. 1999 Spain et al. 1999
2-Nitrophenol	Zeyer and Kearney 1984	Donlon et al. 1996
3-Nitrophenol	Zeyer and Kearney 1984 Meulenberg et al. 1996 Schenzle et al. 1997	
4-Nitrophenol	Sudhakar-Barik et al. 1978 Barkay and Pritchard 1988 Spain and Gibson 1991 Jain et al. 1994 Kadiyala and Spain 1998	Donlon et al. 1996 Heitkamp and Orth 1998 Ray et al. 1999
2,4-Dinitrophenol	Schmidt and Gier 1989 Hess et al. 1990 Lenke et al. 1992	Gisi et al. 1997
2,6-Dinitrophenol	Ecker et al. 1992	
4-Chloro-2-nitrophenol (constructed strain)	Bruhn et al. 1988	

Compound	References	
2,4,6-Trinitrophenol	Rajan et al. 1996 Behrend and Heesche-Wagner 1999 Rieger et al. 1999 Ebert et al. 1999	Perkins et al. 1995 Rajan and Sariaslani 1996
2-Nitrobenzoic acid	Cain 1958 Mironov et al. 1992	
3-Nitrobenzoic acid	Cain 1958 Nadeau and Spain 1995	Hallas and Alexander 1983 Goodall et al. 1998
4-Nitrobenzoic acid	Cain 1958 Groenewegen and de Bont 1992 Groenewegen et al. 1992	Hallas and Alexander 1983 Goodall et al. 1998 Peres et al. 1999
4-Nitroanisole	Schäfer et al. 1996	
Pesticides (Parathion, DNOC, Dinoseb)	Tewfik and Evans 1966 Jensen and Lautrup-Larsen 1967 Serdar et al. 1982	Kaake et al. 1992 Gisi et al. 1997
Chloronitrobenzenes	Livingston 1993 Katsivela et al. 1999	Livingston and Willacy 1991 Livingston 1993
Dyes (nitrobenzenesulfonic acid)	Takeo et al. 1997	

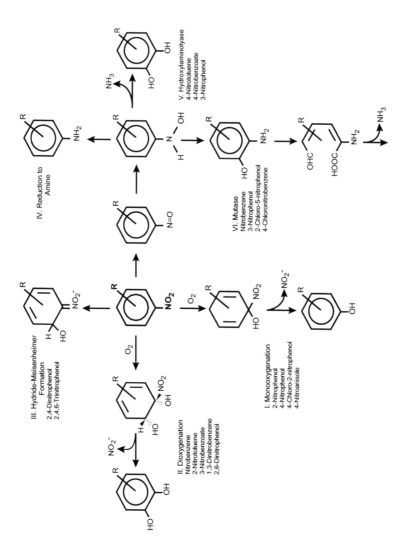

Figure 2.1   Mechanisms for degradation of nitroaromatic compounds. Convincing evidence for pathway IV is only known from anaerobic systems.

benzoquinone intermediates in most instances are too unstable to allow isolation. The first stable quinone intermediate to be identified and quantitatively analyzed was 2-hydroxy-5-methylquinone,[77,81] the monooxygenation product of 4-methyl-5-nitrocatechol in the 2,4-dinitrotoluene degradation pathway (see below).

Some bacteria eliminate a nitro group following initial dioxygenation to a dihydroxy intermediate (Figure 2.1, pathway II). Compounds susceptible to dioxygenase attack include 2,4-dinitrotoluene (2,4-DNT),[192] 2,6-dinitrotoluene (2,6-DNT),[143] 2-nitrotoluene (2-NT),[82] NB,[147] 2,6-dinitrophenol,[55] 1,3-dinitrobenzene,[51,146] and 3-nitrobenzoic acid (3-NBA).[142] The dihydroxy nitrocyclohexadienes formed by the intial dioxygenase attack at the nitro group are unstable and spontaneously rearomatize with release of nitrite to form catechols.

Aerobic bacteria attack 2,4-dinitrophenol[121,123] and 2,4,6-trinitrophenol[15,122,175] through formation of a hydride–Meisenheimer complex before elimination of the first nitro group as nitrite (Figure 2.1, pathway III). The hydride–Meisenheimer complex has also been detected in cultures provided with TNT,[61,75,208,209] but it is uncertain whether productive TNT metabolism then results.[209]

Complete reduction of the nitro group to the amine (Figure 2.1, pathway IV) does not appear to be a mechanism that is widely used by aerobic bacteria for productive metabolism. *Rhodobacter capsulatus* transforms 2,4-dinitrophenol to 2-amino-4-nitrophenol in a light-dependent reaction that takes place under anaerobic and microaerobic conditions in a reaction catalyzed by an inducible nitroreductase.[19–21] Further degradation of the 2-amino-4-nitrophenol appears to be a constitutive activity that requires light, oxygen, and additional carbon and nitrogen sources.[212] An early report that a pseudomonad degrades 3,5-dinitro-*o*-cresol via 3-amino-5-nitro-*o*-cresol remains unconfirmed.[204]

Partial reduction of the nitro group via the nitroso intermediate to the hydroxylamino derivative is a well-known chemical reaction.[63] In the last decade it has become clear that the enzyme-catalyzed partial reduction to the hydroxylamine level is an important biological mechanism as well. The oxygen-insensitive nitroreductases purified from bacteria that use partially reductive pathways specifically reduce nitroaromatic compounds to hydroxylaminoaromatic compounds[72,186] with no further conversion. Hydroxylamino compounds are subject to two separate mechanisms for elimination of the nitro group as ammonia. In one mechanism, the hydroxylamino compound is attacked by an enzyme that the authors have described as a hydroxylaminolyase,[135] resulting in the production of the corresponding catechol and elimination of ammonia (Figure 2.1, pathway V). Details of the reaction mechanism are still unclear. Bacteria degrade 4-nitrotoluene (4-NT),[79,173] 4-nitrobenzoic acid (4-NBA),[72,73,214] and 3-nitrophenol[136] via the hydroxylaminolyase mediated pathway.

The second partially reductive pathway involves a mutase-catalyzed intramolecular rearrangement of the hydroxylamino compound to an *ortho*-aminophenol (Figure 2.1, pathway VI). NB,[147] 2-chloro-5-nitrophenol,[179] 4-chloronitrobenzene,[111] 4-NT,[194] as well as 3-nitrophenol[178,180] are degraded by the mutase mediated pathway via *ortho*-aminophenols. The *ortho*-aminophenols serve as *meta*-ring cleavage substrates for highly specific dioxygenases, and the reactions result in the formation of aminomuconic semialdehydes. The amino group is released as ammonia in subsequent reactions.

Because most of the compounds mentioned above are discussed in depth else-where in this book, the remainder of this chapter will be limited to a discussion of mono- and dinitrotoluenes, NB, and mononitrobenzoates.

2.2 MONO- AND DINITROTOLUENES

2.2.1 Pathways

Dinitrotoluenes are the precursors of 2,4,6-TNT, once the most widely manu-factured military explosive in the world. The most common isomers, 2,4-DNT and 2,6-DNT, were produced in a 4:1 ratio.[165] The conversion of toluene to TNT (Figure 2.2) was carried out in three nitration steps, and safety considerations dictated that the reactors for each nitration step be housed in separate buildings. For similar reasons the production lines were separated geographically and manufacturing facil-ities were sited in remote areas. DNT contamination at military sites dates back to World War I, and due to the distributed method of manufacture of TNT, many areas at TNT production plants are contaminated with mono- and dinitrotoluenes. TNT is no longer produced in the U.S., but DNT remains an important industrial chemical as the precursor to toluene diisocyanate used in the production of polyurethane foams. Environmental release of DNT is rare in current industrial processes; how-ever, removal of DNT from industrial waste streams often requires absorption onto activated carbon.[206] In the U.S., 1.4 million pounds of DNT was released from all sources and entered into the National Toxics Release Inventory for 1988.[57] By 1997, the latest year for which the figures are available, the total release to the environment was reduced to 69,250 pounds.[206]

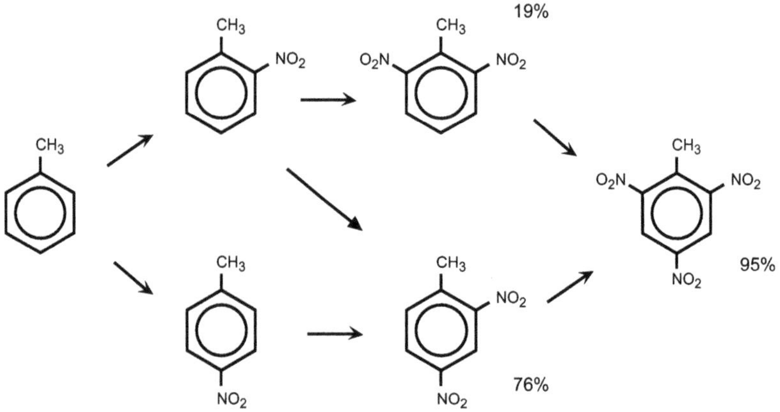

Figure 2.2 DNT is produced by sequential nitration of toluene. The first nitration produces 2- and 4-nitrotoluene in almost equal amounts, due to the *ortho-* and *para*-directing methyl group. The second nitration is also *ortho* or *para* to the methyl group, but the presence of the first nitro group dictates which carbon is free to be nitrated. Typical production batches yield 76% 2,4-DNT, 19% 2,6-DNT, and 5% other iso-mers. A third nitration produces TNT from both 2,4- and 2,6-DNT.

2.2.1.1 Aerobic Pathways

Degradation of DNT was initially observed 20 years ago (W. R. Mitchell, personal communication, 1990)[14] in mixed populations enriched from DNT-contaminated surface water, but it has only been in the last decade that the degradation mechanisms have been elucidated. Strains that grow on single DNT isomers as sole carbon, nitrogen, and energy sources have been isolated from contaminated systems worldwide.[143,149,188] More recently, strains that can grow on 2,4- and 2,6-DNT[149] have been isolated from bioreactors receiving mixtures of DNT isomers. The 2,4-DNT degradative pathway was determined in *Burkholderia* sp. strain DNT[76,192] and in strains carrying the cloned genes of the 2,4-DNT pathway.[77,198] All strains that grow on 2,4-DNT but not on 2,6-DNT appear to use the same pathway as strain DNT, although many grow considerably better than strain DNT.

An initial dioxygenase attack at the 4-nitro position converts 2,4-DNT to 4-methyl-5-nitrocatechol (4M5NC) with the release of nitrite[192] (Figure 2.3A). Studies with partially purified enzymes demonstrated that monooxygenase attack at the remaining nitro-substituted position converts 4M5NC to 2-hydroxy-5-methyl-quinone (HMQ) with the release of the second nitro group as nitrite.[77] Biochemical properties of the purified methylnitrocatechol (MNC) monooxygenase and sequence analysis of the cloned gene encoding the enzyme suggest that MNC monooxygenase is a flavoprotein with properties similar to nitrophenol oxygenases.[81] A quinone reductase converts HMQ to 2,4,5-trihydroxytoluene (THT) in a reaction requiring NADH. The product of THT ring cleavage is unstable, and attempts at purification result in decomposition to a compound that is not metabolized by the bacteria.[77] Properties of the purifed ring cleavage enzyme and genetic evidence strongly indicate

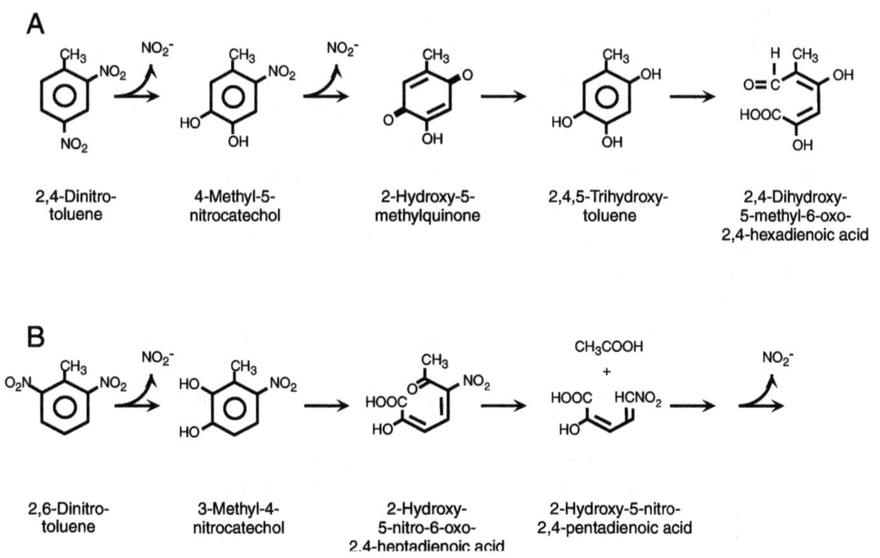

Figure 2.3 Degradation pathways for 2,4-dinitrotoluene (A) and 2,6-dinitrotoluene (B). (Modified from Nishino, S.F. et al. 2000. *Appl. Environ. Microbiol.* Vol. 66.)

that THT undergoes *meta*-ring fission catalyzed by an extradiol dioxygenase.[76] All four enzymes are induced in DNT-grown cells compared with activities in acetate-grown cells, and yeast extract appears to contain some component that acts as a nonspecific inducer for the DNT dioxygenase.[77]

In the first strain isolated and studied, *Burkholderia* sp. strain DNT, the plasmid-borne genes that encode the pathway are organized in at least three separate operons[199] that are induced sequentially. Induction of the entire pathway can take up to 3 days.[192] Although the fully induced cells will degrade concentrations of 2,4-DNT up to the solubility limit in aqueous solutions (approximately 1 mM at room temperature), the accumulation of 4M5NC above 2 μM will inhibit the 4M5NC monooxygenase in cell extracts[81,192] and slightly higher concentrations are inhibitory for whole cells.[192] The sequential induction of the 2,4-DNT pathway in this strain coupled to the sensitivity to the first pathway intermediate makes induction of the complete pathway unpredictable at 2,4-DNT concentrations above 100 μM. In a more recently isolated strain, *B. cepacia* R34,[143] the pathway gene organization differs from that of strain DNT[78] so that the pathway is induced in a shorter time and 4M5NC accumulates only briefly during induction. All of the 2,4-DNT-degrading strains that were isolated after strain DNT are much less sensitive to the presence of 4M5NC. They tolerate 100-fold higher concentrations of 4M5NC in whole cell experiments. The enzymes of a more recent isolate, *B. cepacia* JS872,[143] become induced within 4 to 8 h, and no 4M5NC accumulates during induction. The results strongly suggest that the 2,4-DNT pathway and its regulation are still evolving.

To date DNT-degrading strains have only been isolated from industrial wastes which receive regular and continuous input of DNT or from DNT-contaminated sites. DNT-degrading strains have not been isolated from activated sludges that do not normally receive DNT, nor have they been isolated from uncontaminated sites. How the DNT-degrading phenotype has evolved or been distributed at the contaminated sites is a story that is still unfolding. Genetic analysis has revealed that the initial dioxygenase in strain DNT is related to naphthalene dioxygenase (NDO).[198] Four of the five open reading frames (ORF) initially identified within the initial dioxygenase operon are highly similar to the genes that encode the NDO multicomponent enzyme system. Based upon transformation reactions catalyzed by the cloned DNT dioxygenase genes and by analogy to the similar NDO components, four of the five ORFs were assigned gene and protein designations (Table 2.2). The remaining ORF2 and recently identified ORFX (Genbank accession no. U62430) have no known function in the DNT pathway; however, recent work has shown that they are

Table 2.2 Gene and Protein Designations of DNT Dioxygenase Components

ORF	Gene Designation	% Similarity to NDO Component	Enzyme Component
ORF1	*dntAa*	59	Reductase$_{DNT}$
ORF3	*dntAb*	66	Ferredoxin$_{DNT}$
ORF4	*dntAc*	67	ISP$_{DNT}$$^\alpha$
ORF5	*dntAd*	78	ISP$_{DNT}$$^\beta$

homologous to *nagG* and *nagH* which encode salicylate 5-hydroxylase[62] in the pathway that degrades naphthalene through gentisate. Analysis of the small ISPα subunits from a number of nitroarene dioxygenases and naphthalene dioxygenases revealed that the nitroarene dioxygenases and the naphthalene dioxygenases clustered separately.[102] The tree derived from partial 16S rDNA sequences of the same strains is incongruent with the tree for the ISPα subunits. Related nitroarene dioxygenases ISPα subunits are distributed among a variety of both Gram-negative and Gram-positive strains. The results suggest that the initial dioxygenase evolved once from naphthalene dioxygenase or a common ancestor and then spread horizontally through the bacterial communities at DNT-contaminated sites. How the genes became distributed at sites as geographically separated as the U.S., Canada, Scotland, and Germany is unknown.

Recently, extended selective enrichment strategies have resulted in the isolation of 2,6-DNT-degrading bacteria from a number of 2,6-DNT-contaminated sites.[143] The delay in isolation of 2,6-DNT-degrading strains might be attributed to their sensitivity to high concentrations of both 2,4- and 2,6-DNT, but the reason for the sensitivity is not clear. Induction of the degradative pathway is slow at all 2,6-DNT concentrations and is inhibited above 100 μM; in contrast, fully induced cells will grow on 2,6-DNT at concentrations up to 500 μM. Concentrations as low as 10 μM will induce the pathway. Slow growth of bacteria on 2,6-DNT can be attributed to the inefficiency of the catabolic pathway rather than to inherent properties of the bacteria carrying the pathway. *B. cepacia* strains that degrade different DNT isomers all grow well on simple carbon sources such as succinate or acetate, but those that degrade 2,6-DNT grow slowly on 2,6-DNT, and those that grow on 2,4-DNT vary widely in how well they grow on 2,4-DNT. The regulation of DNT pathways is poorly understood and is the topic of current investigations in several laboratories.

When both DNT isomers are present, which is normally the case for environmental contamination, 2,6-DNT degradation is inhibited by high relative 2,4-DNT concentrations. Conversely, 2,6-DNT concentrations as low as 100 μM inhibit degradation of 2,4-DNT in some strains.[145] Nitrite is toxic to strain DNT at concentrations above 10 mM, and other 2,4-DNT-degrading strains are inhibited at nitrite concentrations $\geq$20 mM. 2,6-DNT-degrading strains are unaffected by nitrite concentrations up to 100 mM. The inhibitory effects of DNT and nitrite on DNT degradation are important considerations in the design of remediation strategies.

2,6-DNT is initially transformed by a dioxygenase attack to produce 3-methyl-4-nitrocatechol (3M4NC) with the loss of one of the nitro groups (Figure 2.3B). The 2,4-DNT dioxygenase will also catalyze the same reaction very slowly and only at low 2,6-DNT concentrations. The similarity of the initial dioxygenase attack in the two pathways initially suggested that the remaining steps in the 2,6-DNT degradation pathway might be analogous to those of the 2,4-DNT pathway. However, no evidence for an MNC monooxygenase has been found in extracts of 2,6-DNT-grown cells. Instead, 3M4NC undergoes direct *meta*-ring cleavage catalyzed by a catechol-2,3-dioxygenase to produce 2-hydroxy-5-nitro-6-oxo-hepta-2,4-dienoic acid.[143] Two lines of evidence support the conclusion that an extradiol dioxygenase is involved. First, the enzyme catalyzes similar reactions to those catalyzed by 3-methylcatechol-2,3-dioxygenase from the toluene dioxygenase pathway. Second, the physical properties (heat stability, peroxide sensitivity, pH range) of the MNC-2,3-dioxygenase are characteristic of catechol-2,3-dioxygenases.

A partially purified hydrolase converts 2-hydroxy-5-nitro-6-oxohepta-2,4-dienoic acid to 2-hydroxy-5-nitropenta-2,4-dienoic acid.[143] The 2-hydroxy-5-nitropenta-2,4-dienoic acid has been previously identified as a dead end metabolite in the 2,6-dinitrophenol pathway as the result of spontaneous decarboxylation of the ring-fission product;[55] however, in the 2,6-DNT pathway the reaction is clearly enzymatic. The remaining nitro group is released as nitrite in subsequent reactions that have not been characterized.[143] The 2-hydroxy-5-nitro-6-oxohepta-2,4-dienoic acid and 2-hydroxy-5-nitropenta-2,4-dienoic acid join a growing number of compounds that retain a nitro group after ring cleavage.[22,55,123] In the earliest described 2,4- and 2,6-dinitrophenol degradation pathways, the nitroaliphatic metabolites are believed to accumulate after misrouting,[55,123] whereas in the 2,6-DNT and an alternate 2,4-dinitrophenol pathway[22,143] the nitroaliphatic compounds seem to be degraded subsequently. Specific evidence for growth with the nitroaliphatic compounds has not yet been provided, nor has any mechanism for removal of the nitro group been demonstrated. Genetic studies of the 2,6-DNT-degrading strains are just beginning.

Bacteria that can grow simultaneously on both 2,4- and 2,6-DNT[149] have been isolated following prolonged selection in bioreactors that were provided mixtures of 2,4- and 2,6-DNT as the sole growth substrates. The initial inocula consisted of strains that could degrade only 2,4-DNT or 2,6-DNT. Studies (S. F. Nishino, Radford, VA, unpublished) suggest that the 2,6-DNT pathway in strains that degrade both isomers is the same as that in strains that degrade only 2,6-DNT. Preliminary results suggest that the 2,4-DNT pathway differs from that elucidated in strain DNT; 4M5NC monooxygenase activity has not been detected in extracts of 2,4-DNT-grown cells.

2-NT and 4-NT are precursors of DNT and are significant contaminants at TNT manufacturing facilities. 3-Nitrotoluene is found only in trace amounts as a consequence of the *ortho-* and *para-*directing activity of the methyl group during the nitration of toluene. A *Pseudomonas* sp. able to grow on 2-NT was isolated from groundwater contaminated with a mixture of nitroaromatic compounds.[82] The strain uses an initial dioxygenase attack similar to that of toluene dioxygenase to convert 2-NT to an unstable nitro-substituted cyclohexadiene intermediate that spontaneously rearomatizes with the release of nitrite to form 3-methylcatechol. The 3-methylcatechol is then degraded via the *meta-*ring fission pathway.[7,82] The initial enzyme system in the pathway has been purified and identified as a three component dioxygenase related to other multicomponent dioxygenases.[7] The genes encoding the initial dioxygenase have been intensively studied[154–156] and a thorough discussion of the molecular genetics of the initial dioxygenase is presented in Chapter 3.

Pseudomonas spp. strains able to grow on 4-NT were independently isolated almost simultaneously in soils from the U.S.[79] and the U.K.[173] Both strains of bacteria use an initial attack on the methyl group to convert 4-NT to 4-nitrobenzyl alcohol, then to 4-nitrobenzaldehyde, and finally to 4-nitrobenzoate (4-NBA) in a series of reactions similar to those encoded by the TOL plasmid. 4-NBA is then converted to protocatechuate with the release of ammonia via the partially reductive hydroxyl-aminolyase-mediated pathway (see below) before ring cleavage. The mode of ring cleavage differs in the two strains. One uses *ortho*-cleavage via protocatechuate-3,4-dioxygenase,[173] whereas the other uses *meta*-cleavage via protocatechuate-4,5-dioxygenase.[79] The same pathway was constructed in a *P. fluorescens* by cloning the TOL plasmid encoded upper pathway genes for toluene degradation into a 4-NBA-degrading strain.[137]

An alternative pathway for degradation of 4-NT has recently been discovered in *Mycobacterium* sp. HL 4-NT-1.[194] The *Mycobacterium* also uses a partially reductive pathway, but the initial attack is at the nitro group to form 4-hydroxyl-aminotoluene, which then undergoes a mutase-catalyzed rearrangement to 6-amino-*m*-cresol. The mode of ring cleavage has not been definitively proven, but indirect evidence indicates extradiol cleavage to form 2-amino-5-methylmuconic semialdehyde. The nitrogen is released as ammonia further down the pathway in a hydrolase-catalyzed reaction.

2.2.1.2   Anaerobic Reactions

Productive anaerobic pathways for degradation of mono- and dinitrotoluenes have not been reported to date. However, cometabolic reduction of the nitro group under both anaerobic and aerobic conditions has been reported a number of times. The reductive transformations have been attributed to the activity of non-specific nitroreductases of both the oxygen sensitive and insensitive types.[34] The cultures have generally been grown with simple sugars or alcohols to provide growth substrates and electron donors. In the earliest studies in which the products of anaerobic bacterial reduction of 2,4-DNT were carefully analyzed,[125,134] nitroso-, aminonitro-, and diamino-compounds predominated. When grown on glucose under aerobic conditions, the fungus *Mucrosporium* transformed 2,4-DNT to the same intermediates and then to dinitroazoxytoluenes and 4-acetamido-2-nitrotoluene[133] (Figure 2.4). Cometabolic reduction and acetylation of 2,4-DNT has recently been demonstrated under aerobic conditions as well as anaerobic conditions in *P. aeruginosa* cultures.[150] In general, nonspecific reduction does not lead to ring cleavage and further transformation of the metabolites.

2.2.2   Practical Applications for Degradation of DNT

DNT is a major ingredient in single-base gun propellants, generally constituting about 10% of the total components. Currently, Radford Army Ammunition Plant is a major production site, and Badger Army Ammunition Plant near Baraboo, WI was formerly the major reprocessing site that handled the single-base gun propellant formulations. At Radford the production waste stream, which is high in organic solvents and 2,4-DNT, goes into an industrial treatment system before discharge. At Badger reprocessing wastes were deposited into large in-ground pits, so that today some soils in the pits contain up to 28% 2,4-DNT by weight. Sites contaminated with a single DNT isomer seem ideal for treatment with selected strains of 2,4-DNT-degrading bacteria.

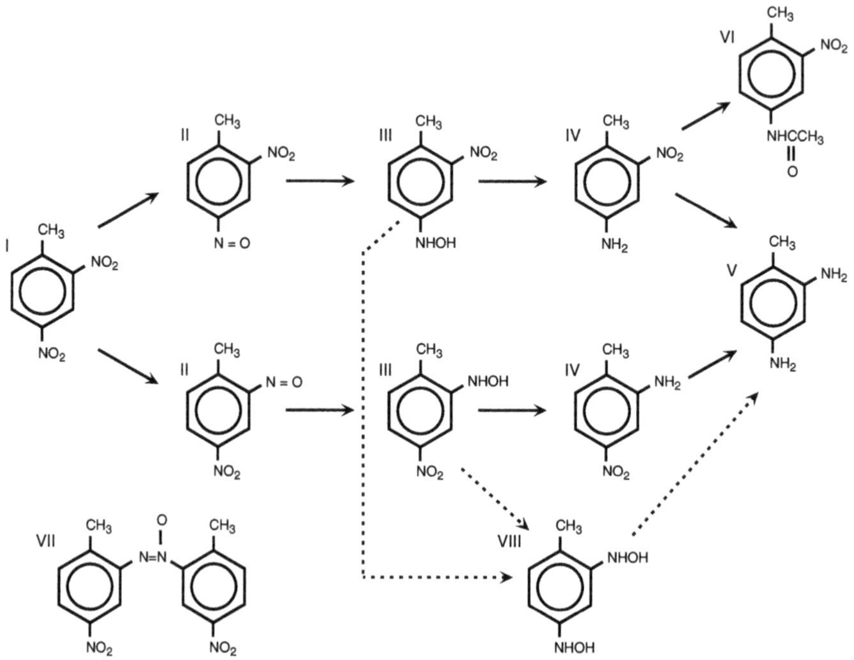

Figure 2.4 Reductive reactions for 2,4-DNT. Similar reactions also take place with 2,6-DNT. Any combination of hydroxylaminonitrotoluenes (III) can condense to form azoxynitrotoluenes (VII, only 1 shown). Acetylated compounds (VI) and azoxynitrotoluenes (VII) are not metabolized further. Dashed arrows indicate reactions that only occur under strictly anaerobic conditions.

2.2.2.1 Early Reactor Studies

Prior to the discovery of 2,6-DNT-degrading bacteria, studies with pure cultures of 2,4-DNT-degrading strains were carried out to determine their effectiveness for degradation of 2,4-DNT in aqueous systems. Strategies were devised to retain induced biomass of slow-growing, DNT-degrading bacteria within bioreactor systems. *Burkholderia* sp. DNT was immobilized in a fixed-bed reactor with 2,4-DNT supplied as the sole source of carbon, nitrogen, and energy. At a loading rate of 95 mg/L/h, removal of 92 to 97% of 2,4-DNT during a 3400-h run was achieved.[93] Similar rates (86 mg/L/h) were achieved with *B. cepacia* PR7[172] when cells were immobilized on diatomaceous earth pellets. Both studies revealed that immobilized reactor designs that retained active biomass were critical to accomplish DNT degradation within reasonable periods. The studies also highlighted the need to limit nitrite accumulation to concentrations below inhibitory levels.

The early reactor studies did not address the problem of the more commonly found situation at contaminated sites where 2,4- and 2,6-DNT are often found in combination with substantial concentrations of 2- and 4-NT as well as TNT and traces of other nitrotoluenes. At such sites the contamination may be in the groundwater, soil, or both. A number of recent studies have been conducted to address the problem of multiple nitroaromatic contaminants.

2.2.2.2 Groundwater Studies

The Volunteer Army Ammunition Plant (VAAP) near Chattanooga, TN was a TNT manufacturing plant from 1941 until 1977. At maximum capacity, 16 TNT production lines were in operation, producing over 2.8 billion pounds of TNT during the lifetime of the plant.[56] A typical production line was set up as two parallel lines of three sequential nitration reactors which converged on a single facility. DNT contamination of soil and groundwater is quite heterogeneous, and concentrations of nitrotoluenes in groundwater vary with rainfall. A series of studies has been conducted with contaminated groundwater from VAAP to determine whether and how mixtures of mono- and dinitrotoluenes are degraded.

A mixed culture of 2,4-DNT- and 2,6-DNT-degrading strains was used to inoculate a 1.5-L fluidized bed reactor (FBR).[120,185] The FBR was operated at a variety of hydraulic retention times (HRT) and fed 2,4- and 2,6-DNT. Removal efficiencies were greater than 98% (Figure 2.5) at all HRTs. The study established the minimum dissolved O_2 levels required for DNT degradation and revealed that a nitrite-oxidizing population could effectively maintain nitrite levels in the effluent below the regulatory limit for discharge. Mineralization was indicated by the fact that the total nitrite and nitrate released, other than nitrogen used for growth, was stoichiometric with the amount of DNT degraded. In addition, no aminonitrotoluenes or diaminotoluenes accumulated in the effluent.

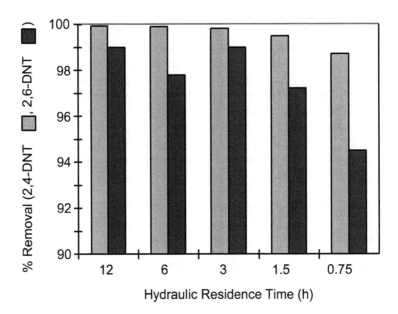

Figure 2.5 Removal of DNT in a bench-scale FBR at various hydraulic residence times. Effluent concentrations of DNT exceeded U.S. EPA standards only at the shortest HRT. (Modified from Lendenmann, U. et al. 1998. *Environ. Sci. Technol.* 32:87.)

The retention of the induced biomass was critical to the success of the system. With a large induced biomass and complete mixing of the biomass and feed, the concentration of DNT within the reactor vessel was always at a very low level. Therefore, the two DNT isomers were degraded simultaneously with no apparent inhibition by either isomer. The system was resilient to changes in flow rate and changes in influent DNT concentrations.

The FBR feed was changed to groundwater from a monitoring well at VAAP after 5 months of operation. The groundwater contained 2- and 4-NT in addition to 2,4- and 2,6-DNT. A 10-day acclimation period was required for removal rates of nitrotoluenes, with the exception of 2,6-DNT, to stabilize at ≥98%. Although the 2,6-DNT in the groundwater was at a lower concentration than in the previous tap water feed, the removal rate from the groundwater only reached 62%. The groundwater phase of the bench-scale study was too short to determine the cause of the slower 2,6-DNT degradation rate, and the question was readdressed in the subsequent field study. Nevertheless, the bench-scale studies demonstrated the feasibility of simultaneously degrading mixtures of mono- and dinitrotoluenes.

The strains provided in the inoculum all degraded either 2,4- or 2,6-DNT, but none could grow on both isomers. The FBR was sampled periodically for microbial analysis during the course of the studies. After 2 months of operation, the first strains able to grow on both 2,4- and 2,6-DNT were isolated from the FBR, where they constituted less than 1% of the total DNT-degrading isolates. After 5 months of operation, the majority of the strains isolated could degrade both 2,4- and 2,6-DNT.[148] The origin of the strains able to grow on both DNT isomers is unknown. The relative proportions of the DNT isomers did not change and the concentration of DNT was not dramatically different once acclimation occurred. Work is in progress to determine the origin of strains able to simultaneously degrade both DNT isomers.

A field study was designed based upon the parameters established by the laboratory-scale FBR. An FBR was operated at VAAP near the former site of the acid recovery house of TNT production line 4. The reactor contained a granular-activated carbon biocarrier which was inoculated with a mixed culture of 2,4-DNT-, 2,6-DNT-, 2-NT-, and 4-NT-degrading strains.[190] The removal efficiencies for 2,4-DNT, 2-NT, and 4-NT were always high (Table 2.3), and removal of TNT fluctuated in a narrow range around 50% depending on the HRT. 2,6-DNT removal started slowly, but there was a dramatic improvement at identical HRTs after

Table 2.3 Removal of Nitrotoluene by Field-Scale FBR

	2,4-DNT	2,6-DNT	TNT	2-NT	3-NT	4-NT
Influent (mg/L)	7.3–13.2	7.4–12.5	0.9–1.3	2.2–3.7	0.2–0.5	2.2–4.2
HRT in min (duration in d)			Percent Removal of Nitrotoluenes			
30 (24)	98	58	73	100	100	100
60 (20)	100	56	67	100	100	100
15 (14)	96	34	40	95	100	100
10 (9)	86	18	33	100	100	100
20 (52)	99	39	46	100	100	100
30 (9)	100	88	54	100	100	100
60 (11)	100	94	58	100	100	100

4 months of operation. Specific reasons for the acclimation are not known, nor is it known how much the removal rate might have improved if the operation continued for a longer period. It seems unlikely that new degradative strains colonized the reactor from the groundwater given the low groundwater pH. A number of studies during the course of the operation did not determine the cause of the low 2,6-DNT degradation rate. They did, however, determine that the presence of multiple nitrotoluenes at the concentrations present in VAAP groundwater was not sufficient to inhibit 2,6-DNT degradation. Therefore, some other site-specific factor was responsible for the differences in behavior of the system when groundwater was substituted for artificially contaminated water.

A major difference between the field study and the laboratory study was that the biocarrier was changed from sand- (used in the bench-scale study) to granular-activated carbon in order to increase the surface area available for bacterial colonization. An undesired consequence of the change in biocarrier was that the granular-activated carbon also required a long saturation period before nitrotoluene breakthrough in the fluidized bed permitted bacterial growth on the contaminating nitrotoluenes.

Cost analysis showed that the FBR technology is a more cost-effective treatment method for DNT than either UV/ozone treatment or liquid phase granular activated carbon adsorption when the total nitrotoluene removal rate exceeds 120 lb/day.

2.2.2.3 Degradation of DNT in Soil Slurries

Bioremediation of contaminated soil requires consideration of several factors that are not important in water. Kinetics of adsorption and desorption of contaminants to soil and organic matter[109,129] are major factors in determining whether cleanup goals can be achieved in soils. Slow desorption can be beneficial by providing gradual release of contaminants to the bacterial population to smooth out shock loads of substrate. On the other hand, extremely slow desorption can mean that a residual amount of contaminant is biologically unavailable, and the residual might remain on the soil at levels that do not meet regulatory requirements. Soils contaminated with solids such as DNT or most explosives are extremely heterogeneous with respect to contaminant distribution. Without some type of homogenization process prior to biotreatment, uneven distribution of contaminants can result in toxicity to the biocatalyst or substrate levels too low to sustain degradative capability. The additional handling of materials beyond that required for aqueous systems increases both the time and the cost of remediation process. Contaminated soil can be a source of bacteria that are well adapted to the physicochemical parameters of that particular soil which is an advantage if the bacteria use the contaminant as a growth substrate or transform it to a permanently immobilized or nontoxic form. The presence of indigenous bacteria might be a disadvantage if they catalyze nonspecific transformations to compounds that are more reactive or more toxic than the original contaminant. Soil can also present much greater extremes of pH, metals, and organic materials than are found in groundwater.

Biodegradation experiments in small slurry reactors with soils artificially contaminated with mixtures of DNT demonstrated that added DNT-degrading strains effectively degrade DNT mixtures in nonsterile soil.[149] The experiments demonstrated that organic matter in the soil does not inhibit DNT degradation and that mineralization of DNT is considerably faster than nonspecific reduction under aerobic conditions. The result led to a series of studies with field-contaminated soils. A bench-scale soil slurry system was operated with soil from the former TNT manufacturing plant at Hessisch-Lichtenau, Germany.[149] A reactor inoculated with specific DNT-degrading strains was operated in the fill and draw mode (Figure 2.6). During each cycle, the 2,4-DNT was degraded more rapidly than the 2,6-DNT despite starting from higher initial concentrations. Therefore, the slower 2,6-DNT degradation rate dictated the endpoints of the cycles. In an uninoculated reactor, 2,4-DNT degradation began after a 5-day lag, while 2,6-DNT did not degrade until 500 h. Both isomers of radiolabeled DNT were mineralized to $^{14}CO_2$ (55%), and the degradation of DNT was accompanied by stoichiometric release of nitrite. Approximately 1% of the nonradiolabeled DNT remained associated with the soil particles when all radiolabeled DNT was gone from the aqueous phase. The fraction most likely represents a slowly desorbed nonbioavailable fraction of the contamination. Only trace amounts of aminonitrotoluenes were detected, which is another indication that nonspecific reduction was only a minor factor in the aerobic soil slurry reactor.

The study showed that DNT contamination over 50 years old can be readily removed from soil by degradative bacteria. It also demonstrated that indigenous bacteria that are able to degrade specific compounds can be present at contaminated sites. The parameters that need to be controlled to provide favorable growth

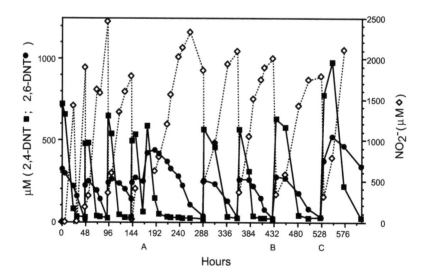

Figure 2.6   Degradation of 2,4- and 2,6-DNT in a bench-scale soil slurry reactor using field-contaminated soil. Soil was 10% (w/v) at a temperature of 30°C. (A) 10% soil added without removal of the previous slurry, (B) temperature reduced to 23°C, (C) 20% soil slurry. (Modified from Nishino, S.F. et al. 1999. *Environ. Sci. Technol.* 33:1060.)

conditions might be as simple as addition of moisture and control of pH. The consistent delay in 2,6-DNT degradation is probably attributable to the high concentrations of 2,6-DNT in the aqueous phase (up to 200 μM). 2,6-DNT-grown strains exhibit a 24-h lag period when inoculated into media containing similar concentrations of 2,6-DNT.[149]

The bench-scale experiments demonstrated that bioremediation of DNT in aged field-contaminated soils was rapid and extensive. The finding led to pilot-scale studies to establish the reactor configuration and operational conditions for scale-up of slurry reactor systems for the treatment of DNT-contaminated soils.[216] Eimco airlift bioreactors (70 L) were used to treat DNT-contaminated soils from VAAP and the Badger Army Ammunition Plant (BAAP). The airlift design of the reactors made them unsuitable for treatment of large soil particles. Therefore, the soil was washed with hot water prior to treatment of the wash water along with the silt and clay particles. The BAAP soil contained high concentrations of 2,4-DNT (11 g/kg) with much lower concentrations of 2,6-DNT (0.2 g/kg). The BAAP soil was alkaline, sandy, nutrient poor, and contained a number of petroleum hydrocarbons. The VAAP soil had similar levels of 2,4-DNT, but higher 2,6-DNT concentrations (1 g/kg) and significant amounts of TNT (0.4 g/kg). VAAP soil was acidic and clayey. Despite the differences between the soils, degradation of 2,4-DNT was rapid, predictable, and easily established for both soils (Figure 2.7A). However, nitrite toxicity became a problem at very high soil loading levels and limited the soil loading rate.

Initially, 2,6-DNT was not degraded in the airlift bioreactors (Figure 2.7A). Shake flask studies with and without soil showed that high ratios of 2,4-DNT to 2,6-DNT inhibited 2,6-DNT degradation. The problem was overcome by conducting the 2,6-DNT degradation phase in a separate reactor placed in series (Figure 2.7B). Even after separation of the two degradation processes, an acclimation period was required for 2,6-DNT degradation. A much longer acclimation period was required for the VAAP soil than for the BAAP soil. The 2,6-DNT-degrading bacteria were much more tolerant of nitrite accumulation than the 2,4-DNT-degrading cultures, and the high nitrite levels that were carried over from the 2,4-DNT reactors into the 2,6-DNT reactors had no effect on 2,6-DNT degradation. The conclusion can be drawn that when both 2,4- and 2,6-DNT are present, 2,6-DNT degradation is the limiting step, and separation of the two operations can enhance the overall biore-mediation system. The precise scale of each reactor will depend on the typical ratios of the DNT isomers. However, the results of this and other studies[149] indicated that 2,6-DNT degradation is intrinsically slower than 2,4-DNT degradation and would therefore require a larger reactor to accommodate the difference between the rates at which the two isomers can be degraded.

Bench-scale and subsequent pilot-scale experiments revealed that inoculation with specific DNT-degrading bacteria will greatly hasten the development of a stable DNT-degrading population even in the presence of an indigenous population. Inoculation will be particularly valuable for 2,6-DNT degradation where long acclimation periods appear to be the norm. In some cases, as became evident in studies with VAAP groundwater and soil, some site-specific factor(s) might also inhibit 2,6-DNT degradation and thus contribute to the overall acclimation period.

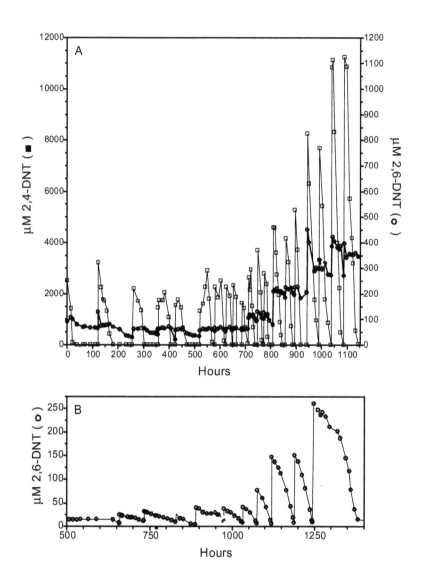

Figure 2.7 Degradation of DNT in soil from BAAP. 2,4-DNT diappeared immediately from the first reactor (A), but 2,6-DNT remained. 2,6-DNT remaining in the effluent from the first reactor was subsequently removed in a second reactor after a long acclimation period (B).

Preliminary results suggest that bioreactors eventually overcome inhibition of 2,6-DNT degradation after extended incubation periods. The mechanism for the adaptation is unknown; however, spread plates inoculated with material from the 2,6-DNT reactor after adaptation to VAAP soil are dominated by colonies that grow to a much greater size on 2,6-DNT than ever observed previously. The change in colony growth characteristics suggests that a genetic change has taken place that allows better growth on 2,6-DNT. A major question remaining is how DNT degradation is regulated. The new isolates might provide insights in that area.

The nitrite and DNT mass balances in soil slurry studies showed that rapid mineralization under aerobic conditions precludes nonspecific reduction of DNT. It is not clear what conditions encourage nonspecific, oxygen-insensitive nitroreductase activity that could convert DNT to aminoaromatic compounds. The studies with BAAP soil were begun with a high degree of uncertainty over how the high levels of phthalates and diphenylamine that also contaminated the soil would impact DNT degradation. The phthalates and diphenylamine proved not to be important factors, but microcosm studies revealed that some component of the BAAP soil supports a large indigenous bacterial population that does not attack DNT. The remaining question is whether there are common soil components that will make a site unsuitable for DNT mineralization.

A biologically unavailable soil-bound fraction of the DNT can persist in the treated solids when DNT is no longer detectable in the aqueous phase. The three field-contaminated soils examined to date (Hessisch-Lichtenau, VAAP, BAAP), but not the laboratory-contaminated soil or artificially contaminated clay aggregates,[153] all retained a low level of acetonitrile-extractable DNT after DNT was no longer detectable in the aqueous phase. Treatment of DNT-contaminated soil in bench-scale bioreactors greatly reduced the toxicity.[149] However, similar tests must to be done to determine whether the residual DNT in soils from scale-up studies requires further treatment. About half of the initial TNT in the VAAP soil disappeared during the bioreactor treatment. Some of the missing TNT was detected as aminodinitrotoluene. Strategies to remove or immobilize residual TNT and TNT metabolites need to be incorporated into the overall remediation plan when soils contain both DNT and TNT.

2.2.3 *In Situ* Remediation and Natural Attenuation of DNT

2.2.3.1 *Evidence for Natural Attenuation*

The discovery of bacteria in the environment that are able to degrade nitroaromatic compounds has raised the question of whether natural attenuation is a suitable remediation strategy for nitroaromatic contaminants. Natural attenuation refers to the stabilization or destruction of contaminants *in situ* by physical, chemical, or biological means without human intervention.[207] Biological processes that result in the actual destruction of contaminants often play the predominant role in natural attenuation,[211] particularly in the "monitored natural attenuation" (MNA) processes favored by regulatory agencies.[207]

The U.S. Environmental Protection Agency[207] recommends development of three lines of site-specific evidence during evaluation of MNA as a site remediation alternative. The lines, which provide successively more detailed information, are (1) historical groundwater or soil chemistry data that demonstrate a meaningful trend over time at appropriate monitoring points, (2) indirect hydrogeologic and geochemical data to demonstrate active, site specific natural attenuation processes, and (3) field or microcosm studies conducted with contaminated site media to directly demonstrate biological degradation processes. Whereas overwhelming evidence at level 1 or 2 may preclude the necessity for level 3 evidence, biological evidence by itself can be sufficient to suggest MNA as an effective remediation alternative.[207]

Three questions need to be addressed to determine whether biological activity will affect natural attenuation.[187] The first is whether biodegradation of the contaminant has ever been reported. Information gleaned from laboratory or field studies can help to determine whether it will be useful to look for biodegradative activity at the site. The answer is affirmative for a growing number of nitroaromatic compounds. The second question is whether suitable bacteria active against the contaminants of concern are present at the site. Again, the answer has been positive at sites with long-term contamination by nitroaromatic compounds. The third question is whether the conditions at the site are appropriate and sufficient to sustain biodegradative activity through completion of MNA.

If the contaminant is biodegradable, but degradative activity is not detected at the site or conditions are not favorable at the site for sustained biodegradative activity, then addition of degradative bacteria or enhancement of biological activity through application of nutrients or electron acceptors/donors should be considered. By definition, such engineered solutions would not be MNA, but still fall within the realm of *in situ* remediation systems.

2.2.3.2   In Situ *Processes that Affect Bioremediation*

Transformation, sequestration, and bioavailability are processes that are negligible or easily controlled in bioreactors, but might have significant impacts on *in situ* bioremediation systems.[9] Transformation, both biotic and abiotic, might be the single, most important environmental process affecting TNT.[159] As a result a number of remediation systems for explosives seek to exploit transformation (see Chapters 4, 8, 12, and 13) in combination with techniques to enhance sequestration of the transformed products. The goal is to render the nitroaromatic compound biologically unavailable so as to minimize detrimental environmental impacts. On the other hand, incomplete transformation or sequestration can limit bioavailability to degradative organisms, thereby slowing the remediation process without significant diminution of the environmental hazard.

2.2.3.3   Natural Attenuation of DNT

Groundwater at Weldon Spring Ordnance Works in Missouri is contaminated with a DNT plume[23–25] as the result of surface spills of DNT. DNT concentrations in the plume decrease in the downgradient direction. In microcosms constructed with soil from the site and incubated for 70 days, almost equal fractions of added radiolabeled 2,4-DNT were mineralized, transformed to aminonitrotoluenes, transformed to unidentified products, or remained unchanged. Only 8% of the 2,6-DNT was mineralized, 67% was unchanged, and the rest was transformed to aminonitrotoluenes and unidentified products. Although it was proposed that MNA of DNT may be suitable for the Weldon Spring site because of the mineralization of DNT,[24] the high proportion of nonspecific transformations is a negative consideration. Clearly, a cometabolic population, possibly sustained by a large reservoir of more readily degradable carbon sources (often associated with TNT production plants), was as active as the DNT-degrading population. When a population that reduces nitroaromatic compounds is active at a contaminated site, immobilization of reduced metabolites could be an alternative strategy to mineralization if absence of toxicity of the reduced products can be established. Reduced products of DNT are still toxic and do not bind as well to soil as the more thoroughly studied TNT reduction products.[117]

The situation at BAAP is considerably different from that at Weldon Spring. Waste materials from the reprocessing of single-base propellants were deposited into large in-ground waste pits. Six waste pits, each roughly 40 ft in diameter and extending 100 ft down to the water table, contain soil heavily contaminated with 2,4-DNT.[195] One of the waste pits is the source of a DNT-contaminated groundwater plume. As at Weldon Spring, the first line of evidence for MNA is that DNT concentrations in the groundwater plume decrease in downgradient wells. 2,3-Dinitrotoluene (2,3-DNT) does not decrease at the same rate as 2,4- and 2,6-DNT. 2,3-DNT has not been demonstrated to be biodegradable and can thus be considered a conservative tracer that reflects the effects of abiotic processes. The much greater decrease in 2,4- and 2,6-DNT concentrations can therefore be attributed to biological activity. 2,4-DNT-degrading bacteria have been isolated from monitoring well water from the site, and 2,4-DNT disappears with stoichiometric release of nitrite from microcosms constructed with DNT-contaminated soil from the site.[148] The understanding of the 2,4-DNT catabolic pathway taken with laboratory studies with soil from the site, disappearance of DNT from the monitoring wells, and the isolation of bacteria able to degrade DNT from the same wells provides evidence that MNA might be applicable at the site. Stoichiometric release of nitrite demonstrates complete mineralization of DNT which precludes formation of significant amounts of amino compounds. MNA might not be suitable for bulk DNT in the soil at BAAP where 2,4-DNT occurs at concentrations up to 28% by weight.[195]

2.2.4 Sequential Anaerobic/Aerobic Strategies

Wastewater discharged from the Radford Army Ammunition Plant (RAAP) contains high levels of 2,4-DNT, ethanol, and ether.[150] The alternate carbon sources can diminish DNT degradation by blocking induction of the DNT pathway, by stimulating growth of a microbial population capable of cometabolic transformation of DNT,[60] and by depleting the available oxygen. Sequential anaerobic/aerobic strategies seek to avoid the difficulty of maintaining a DNT-degrading population in the presence of ethanol by maximizing cometabolic transformation of DNT to the corresponding diaminotoluenes in an anaerobic phase, followed by treatment of the diaminotoluene in the aerobic phase.

The anaerobic/aerobic strategy was applied to a system designed for removal of DNT from the RAAP waste stream. Anaerobic FBRs were tested with artificial waste streams that approximated the RAAP waste stream.[16] 2,4-DNT was completely converted to 2,4-diaminotoluene (2,4-DAT) when the ethanol concentration was twice that of 2,4-DNT or greater. The effluent containing 2,4-DAT, which still contained significant amounts of ethanol, was subsequently treated aerobically in a batch reactor. 2,4-DAT (16 mg/L) disappeared in the aerobic reactor in 9 h and the total nitrogen balance was stoichiometric. The authors suggested that 2,4-DAT was mineralized during the aerobic treatment.

Details about the fate of 2,4-DAT in remediation systems remain to be elucidated. Acclimated activated sludges are reported to biodegrade 2,4-DAT,[11,163] but to date rigorous proof of mineralization is lacking. A *Pseudomonas aeruginosa* strain has been reported to use DAT (isomer unreported) as a sole carbon, but not a sole nitrogen source.[163] The fact that nitrogen must be supplied to the culture in the form of ammonium sulfate argues that a significant portion of the DAT ring remains unavailable to the bacterium. Nothing is known about potential toxicity of the 2,4-DAT metabolites. When wastewater from the treatment system at RAAP was fed repeated additions of 2,4-DAT, each addition disappeared within 2 to 4 days.[60] The presence of ethanol did not affect DAT disappearance. The increase in volatile suspended solids as DAT disappeared was taken as presumptive evidence that DAT was used as a growth substrate, but confirmation with a more direct measure was not done.

An alternative fate for 2,4-DAT is further transformation. Recently, a *P. aeruginosa* strain that rapidly converts 2,4-DNT and 2,4-DAT to acetylated compounds in the presence of oxygen and ethanol has been isolated from a rotating biological contactor at RAAP.[150] The fate as well as toxicity of the acetylated compounds remains to be determined.

Anaerobic/aerobic strategies have also been applied to TNT in soil systems via cometabolic reduction of TNT to triaminotoluene under anaerobic conditions either in slurries[42,124] (see Chapter 4) or composting systems[26,29] (see Chapter 13), followed by exposure of triaminotoluene to oxygen to give rise to polymerization and covalent binding to soil.[174] Soils often contain mixtures of DNT and TNT. No effective biological system has been developed to mineralize TNT, and high concentrations of TNT adversely affect mineralization of DNT. A single system to simultaneously reduce and detoxify the di- and trinitrotoluenes through immobilization on soil could greatly simplify and reduce the cost of explosives remediation. However, little information is available on DAT interactions with soil. One study[41] has shown that although DAT preferentially partitions to loam soil from the water phase, the binding is not completely irreversible. During year-long incubations with radiolabeled DAT, the percentage of radiolabel recovered from soil declined from 90 to 60% without commensurate increases in radiolabeled carbon dioxide or the water associated pool. The authors attribute the discrepancy to leakage of carbon dioxide, but the long-term fate of bound DAT remains unknown. Another study revealed that DAT was removed biologically under nitrate-reducing conditions, but not methanogenic- or sulfate-reducing conditions.[117] Disappearance began following a 22-week lag period and was not complete after 30 more weeks. The authors could not determine whether nitrate consumption was linked to DAT disappearance. The study also indicated that 2,4- and 2,6-DAT are much less strongly sorbed than aminodinitrotoluene or triaminotoluene to a municipal landfill soil and equilibrium concentrations in water remain high. In a third study[124] an anaerobic/aerobic regime was applied to a soil slurry containing TNT, 2,4-DNT, and hexahydro-1,3,5-trinitro-1,3,5-triazine (RDX). The nitroaromatic compounds were completely reduced to aminoaromatic compounds during the anaerobic phase, although 2,4-DNT reduction took five to six times as long as TNT reduction. The resulting 2,4-DAT slowly but irreversibly bound to the soil components during the aerobic phase. Clearly, much remains to be learned about the biological fate of DAT as well as the interactions of DAT with different soil types. Remediation systems based on binding and immobilization of reduced DNT products remain questionable.

A basic assumption of sequential anaerobic/aerobic strategies is that DNT is completely reduced to DAT. Recent work indicated that *Clostridium acetybutylicum* cells reduced DNT only to the dihydroxylaminotoluene level.[99] DNT was transformed via a variety of hydroxylaminonitrotoluenes to the final dihydroxylamino-toluene. Further reduction to DAT was only possible with cell extracts. The authors suggest that the inability of intact cells to further reduce dihydroxylaminotoluenes is related to the shift of *C. acetybutylicum* from acetogenic to solventogenic growth because a similar change in the ability to reduce TNT was correlated with growth phase.[113] The accumulation of dihydroxylaminotoluenes and possible decomposition and interaction products complicates the assessment of the efficacy of anaerobic/aerobic systems. The fate of the partially reduced products in aqueous-, soil-, and slurry-based remediation systems is unknown (Figure 2.4).

2.3 NITROBENZENE

NB is the simplest of the nitroaromatic compounds, yet organisms able to degrade it were not isolated until recently. First synthesized in 1834,[53] by 1995 NB ranked 49th among the 50 top industrial chemicals produced worldwide at 1.65 billion pounds,[35] with production expected to rise to 2 billion pounds by the year 2000.[85] Lethal toxic doses, estimated to be 1 to 5 g for the average adult, are readily absorbed via inhalation, cutaneous contact, and ingestion. Although NB is only slightly volatile at room temperature, absorption through the lungs is twice as fast as through the skin and increases with humidity levels.[53] The toxicity of airborne NB requires remediation strategies that decrease or minimize NB transfer to the atmosphere. Such minimization of volatilization is a recurrent theme among the NB biodegradation strategies discussed below.

2.3.1 Pathways

Early investigations sought NB disappearance in soil and sewage sludge with results that hinted at but did not prove degradation. Some studies led the authors to conclude that NB was recalcitrant,[5] while others revealed disappearance after long acclimation periods.[68,114,164] One report indicated that NB was converted to aniline under anaerobic conditions.[83] One author reported that pyridine-degrading bacteria could utilize NB as the sole source of carbon, nitrogen, and energy, but provided no further details.[140] Recently, however, three pathways have been described for NB degradation. The earliest known and by far the most widespread pathway is a partially reductive pathway; the second pathway is oxidative; and the third is an engineered hybrid pathway, which is also oxidative.

2.3.1.1 Partial Reduction of Nitrobenzene

Bacteria that degrade NB through the partially reductive pathway[144] have been isolated from a number of geographically distinct sites.[147] NB-degrading bacteria can grow on NB in the presence of a secondary carbon source such as succinate. The pathway has been the same in all strains examined to date that release ammonia during growth on NB (Figure 2.9A). However, simultaneous induction studies and experiments with cell extracts suggest that the enzymes involved might differ among the strains.[147] Enzymes in *P. pseudoalcaligenes* JS45 initially reduce NB through nitrosobenzene to hydroxylaminobenzene. A mutase catalyzes the conversion of hydroxylaminobenzene to 2-aminophenol which undergoes *meta*-ring cleavage to 2-aminomuconic semialdehyde.[144] The semialdehyde is oxidized to 2-aminomuconate and subsequently deaminated to 4-oxalocrotonic acid which is degraded to pyruvate and acetaldehyde.[90]

Several of the genes involved in NB degradation have been cloned and sequenced, and a number of the enzymes have been purified. The oxygen-insensitive NB nitroreductase is a flavoprotein with a tightly bound FMN cofactor, but with no detectable requirement for a metal cofactor.[186] NB and nitrosobenzene are reduced by the purified enzyme in the presence of NADPH, but hydroxylaminobenzene is not; the enzyme is unable to reduce the nitro group all the way to the amine. Preliminary results indicate that the enzyme can catalyze the reduction of a wide range of

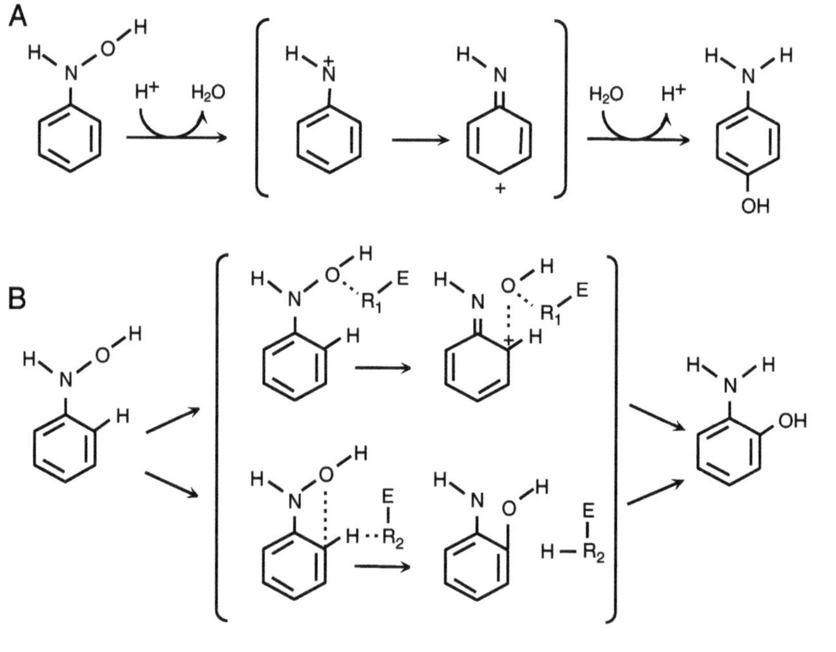

Figure 2.8 Proposed mechanisms for (A) acid-catalyzed and (B) enzyme-catalyzed Bamberger rearrangements of hydroxylaminobenzene. E-R_n: mutase and functional groups.

Table 2.4 Microorganisms that Convert Nitroaromatic Compounds to Aminophenols

Microorganism	Substrate	Metabolite	Ref.
Rhodosporidium sp.	4-Chloronitrobenzene	2-Amino-5-cholorophenol (o, $-$)[a]	Corbett and Corbett 1981
Pseudomonas pseudoalcaligenes JS45	Nitrobenzene	2-Aminophenol (o, +)	Nishino and Spain 1993
Coryneform sp.	Nitrobenzene	2-Aminophenol (o, +)	Lu 1997
Ralstonia eutropha JMP134	3-Nitrophenol 2-Chloro-5-nitrophenol	Aminohydroquinone (o, $-$) 2-Amino-5-chlorohydroquinone (o, $-$)	Schenzle et al. 1997 Schenzle et al. 1999
Mycobacterium sp.	4-Nitrotoluene	6-Amino-m-cresol (o, +)	Spiess et al. 1998
Clostridium acetobutylicum	2,4,6-Trinitrotoluene	4-Amino-6-hydroxylamino-3-methyl-2-nitrophenol (p, $-$) 6-Amino-4-hydroxylamino-3-methyl-2-nitrophenol (o, $-$)	Hughes et al. 1998
Bacterium strain LW1	1-Chloro-4-nitrobenzene	2-Amino-5-chlorophenol (o, +)	Katsivela et al. 1999

[a] o, *ortho*-aminophenol; p, *para*-aminophenol; +, the aminophenol directly subject to ring cleavage as shown in Figure 2.5; $-$, the aminophenol not directly subject to ring cleavage or not reported.

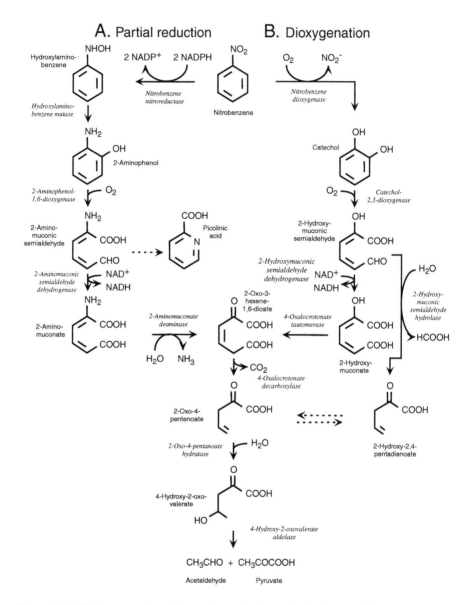

Figure 2.9   Nitrobenzene degradation pathways in *P. pseudoalcalingenes* JS45 and *Comamonas* sp. JS765.

nitroaromatic compounds to hydroxylamino compounds.[186] Experiments with *P. pseudoalcalignes* JS52, a strain derived from JS45, demonstrated that both NB-induced cells and purified NB nitroreductase convert TNT to a variety of novel hydroxylamino compounds.[58]

The second step in the partially reductive pathway is the key reaction that leads to productive metabolism. Mutases catalyze an intramolecular rearrangement of hydroxylamino compounds to form *ortho*-aminophenols. Reactions that appear to be mutase mediated have been reported for a variety of organisms and compounds (Table 2.4). Recent experiments with $H_2^{18}O$ clearly demonstrate that the hydroxyl moiety in the *ortho*-aminophenol is from the original hydroxylamino group[87] (Figure 2.8). The enzyme-mediated rearrangement has been termed a Bamberger or Bamberger-like rearrangement[39,40] because of the similarity of the reaction to the acid-catalyzed chemical Bamberger rearrangement. However, the similarity might be superficial because the chemical reaction is an intermolecular rearrangement that derives the hydroxyl group from water and produces primarily *para*-aminophenols.[184] Elucidation of the details of the mutase reaction mechanism will require complete purification of the enzymes.

The mutase in the 3-nitrophenol degradation pathway found in *Ralstonia eutropha* JMP134 has been purified.[180] The subunit molecular mass is 62 kDa, but the native molecular mass was not determined due to the high hydrophobicity of the enzyme. The N-terminal amino acid sequence is similar to those of glutamine synthetases. The substrates and reactions catalyzed by the two types of enzymes are different, and it is not clear whether there is any similarity in the catalytic mechanisms of the two types of enzymes. In addition to aminohydroquinone, the mutase also attacks hydroxylaminobenzene, 4-hydroxylaminobenzoate, and 2-chloro-5-hydroxylaminophenol. The authors proposed that the reaction mechanism is analogous to the acid-catalyzed Bamberger rearrangement based on the lack of a requirement for oxygen or added cofactors.

Genetic analysis has revealed two hydroxylaminobenzene mutase genes (*habA, habB*) in JS45.[45] The deduced amino acid sequences are 44% identical to each other and show no similarity to the partial amino acid sequence of the mutase from JMP134 or to those of glutamine synthetases. There is only one similar sequence in the database. Gene Rv3078 of cosmid MTCY22D7 from *Mycobacterium tuberculosis* H37Rv encodes a protein that is 52% identical to HabB, but the function of the gene in *Mycobacterium* is unknown. When gene Rv3078 was cloned into *Escherichia coli*, the resultant strains expressed low levels of hydroxylaminobenzene mutase activity (John Davis, 1997, unpublished results). The results suggest that the protein might catalyze a similar activity in *Mycobacterium*. It is not clear whether both *habA* and *habB* are expressed in strain JS45. The fact that they are divergently transcribed suggests that they are not on the same operon. The mutases have been only partially purified due to the difficulty of preventing aggregation of the proteins in cell-free preparations. The evolutionary origins of the mutase enzymes are currently under investigation.

2-Aminophenol-1,6-dioxygenase converts 2-aminophenol to 2-aminomuconic semialdehyde.[144] The purified enzyme[119] is a multicomponent nonheme iron dioxygenase, closely related to the 2-aminophenol-1,6-dioxygenase from *Pseudomonas* sp. AP-3, a strain which grows on 2-aminophenol but not on NB.[202] The 2-aminophenol-1,6-dioxygenases have only limited amino acid sequence similarity to other extradiol ring cleavage dioxygenases. The enzyme from JS45 has a narrow substrate range with a marked preference for *ortho*-aminophenols over catechols.

In vivo, 2-aminomuconic semialdehyde is oxidized to 2-aminomuconate by 2-aminomuconic semialdehyde dehydrogenase in a reaction that requires NAD^+.[90] *In vitro*, the aminomuconic semialdehyde and the substituted aminomuconic semial-dehydes that result from cleavage of substituted *ortho*-aminophenols spontaneously rearrange to form picolinic acids with the loss of water.[119,144] The genes that encode 2-aminophenol-1,6-dioxygenase and 2-aminomuconic semialdehyde dehydrogenase have been cloned and sequenced.[44] The genes appear to be coordinately transcribed with ORF1, which encodes a protein of unknown function. Biochemical and genetic analyses indicate that the 2-aminomuconic semialdehyde dehydrogenase and 2-hydroxymuconic semialdehyde dehydrogenase of the catechol degradation pathway are homologous.[86]

Hydrolytic deamination of 2-aminomuconate to 4-oxalocrotonate is catalyzed by 2-aminomuconate deaminase (Figure 2.9A) in a reaction that requires no co-factors.[90] The protein was purified and characterized to determine whether 2-hydroxymuconate or 2-oxohex-3-ene-1,6-dioate is the direct product of the deam-ination during degradation of NB in JS45.[89] The 2-aminomuconate deaminase has a molecular mass of 100 kDa with six identical subunits. The N-terminal amino acid sequence is not significantly similar to any sequence in the databases.[89] Com-parative spectrophotometric investigation indicated that 2-aminomuconate is not deaminated to 2-hydroxymuconate; instead, it is converted directly to 2-oxohex-3-ene-1,6-dioate. This observation suggests that the deamination mechanism involves an imine intermediate and that the deamination reaction is analogous to the tau-tomerization of 2-hydroxymuconate in the catechol degradative pathway. To date, the deaminase is the only purified protein that catalyzes deamination of a linear unsaturated amino acid.

The foregoing reveals that the pathway comprises a number of unusual steps that seem to be conserved. The evolutionary origin of the pathway is not known, but it was likely assembled recently in response to the presence of NB in the biosphere. The pathway has only been examined in detail in one strain. Comparative analysis of other isolates will provide insight about the origin of the pathway. To date nothing is known about the regulation of the pathway or the mechanisms involved in induction.

2.3.1.2 Oxidation of Nitrobenzene

Comamonas sp. JS765 was isolated from powdered activated carbon from an industrial waste treatment system.[147] The strain degrades NB via an initial dioxy-genation which leads to the production of catechol and the liberation of nitrite (Figure 2.9B). *meta*-Cleavage of catechol yields 2-hydroxymuconic semialdehyde, which is oxidized to 2-hydroxymuconic acid or hydrolyzed to 2-oxo-4-pentanoic acid. The initial dioxygenase has a very broad substrate specificity, which means that induced cells transform a wide range of nitroaromatic compounds.[147]

The NB dioxygenase is inducible, although it is present in succinate-grown cells at low constitutive levels.[147] The postulated product of the reaction, nitrobenzenedi-hydrodiol, would spontaneously rearomatize to catechol with the release of nitrite. Preliminary studies indicate that the initial dioxygenase is closely related to the nitrotoluene dioxygenases and more distantly related to the naphthalene dioxygen-ases. The genes that encode NB dioxygenase have recently been cloned, and efforts to understand the molecular biology of the unusual dioxygenase are just beginning (see Chapter 3).

Catechol-2,3-dioxygenase activity is present at low levels in succinate-grown cells, but is induced to 17- to 40-fold higher levels in NB-grown cells.[147] The genes that encode a catechol-2,3-dioxygenase have been cloned from JS765,[157] but it is not yet clear whether they are the genes for the ring cleavage dioxygenase of the NB pathway.

The ring cleavage product, 2-hydroxymuconic semialdehyde, is converted by both an NAD$^+$-dependent dehydrogenase and a hydrolase to 2-hydroxymuconic acid and 2-oxo-4-pentanoic acid, respectively. Both enzymes are inducible.[147] The hydrolase, but not the dehydrogenase, was partially purified from cell extracts of JS765. 4-Oxalocrotonate tautomerase, 2-oxo-4-pentenoate tautomerase, and 2-oxo-4-pentenoate hydralase activities were present in the same extract. All are enzymes typically involved in the conversion of 2-hydroxymuconic semialdehyde to pyruvate and acetaldehyde in the *meta*-ring fission pathways for catechol in other organisms.[88]

2.3.1.3 Hybrid Nitrobenzene Pathway

A hybrid pathway for the oxidation of NB was created by introducing the genes for toluene dioxygenase (*todC1C2BA*) into *P. putida* mt-2 which carries the TOL plasmid pWW0.[118] Toluene dioxygenase converts NB to NB dihydrodiol,[80] which the authors postulate,[103] after loss of nitrite, is redirected down the TOL pathway as catechol for ring cleavage and subsequent degradation. The mechanism of how the XylL protein of the TOL pathway could catalyze the removal of the nitro group from NB dihydrodiol is not clear and the authors provide no quantitative results.

2.3.1.4 Convergence of the Lower Pathways

The two NB degradative pathways and the pathway for degradation of 2-aminophenol by *Pseudomonas* sp. AP-3 are similar from the ring fission reaction to the end of the pathway. The similarities exist both at the functional and genetic levels and might provide insight into the evolution of the NB degradation pathways (Figure 2.9).

Pseudomonas sp. AP-3 grows on 2-aminophenol as the sole source of carbon, nitrogen, and energy. The pathway for the degradation of 2-aminophenol in AP-3[201] is similar to the one in JS45, except that the decarboxylation step is proposed to occur prior to the deamination reaction (Figure 2.10). If the sequence of the two reactions were reversible, the two pathways would be virtually identical. There are examples in bacterial metabolic pathways where sequential enzyme-catalyzed reactions can occur in either order.[97] The sequence of deamination and decarboxylation is not reversible in JS45,[91] and the question has not been addressed for AP-3.

In strain AP-3 the formation and consumption of the key intermediates, 2-aminomuconate and 2-amino-2,4-pentadienoate, were not directly observed in either resting cells or cell extracts. The proposed pathway was primarily based on the GC-MS identification of a methyl-derivative of 2-amino-2,4-pentadienoate and on the observation that 2-hydroxymuconic semialdehyde is converted to pyruvate and acetaldehyde by cell extracts of *Pseudomonas* sp. AP-3.[201] More detailed examination

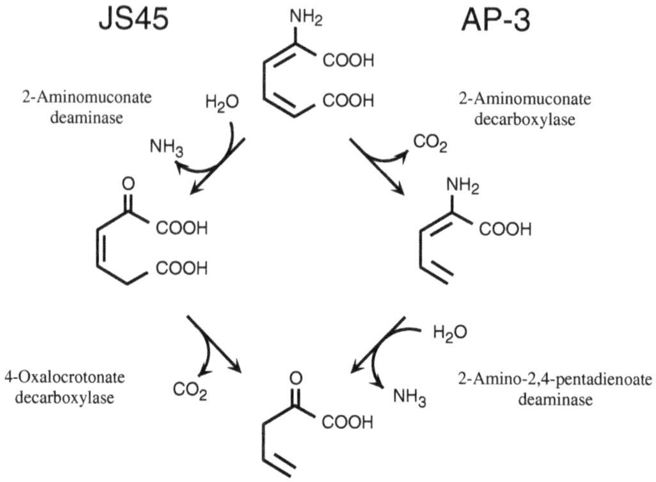

Figure 2.10 The deamination and decarboxylation reactions following dehydrogenation of 2-aminomuconic semialdehyde are reported to occur in the opposite order in *P. pseudoalcaligenes* JS45 and *Pseudomonas* sp. AP-3.

of the three steps of dehydrogenation, decarboxylation, and deamination in AP-3 will clarify the alternative pathway and provide a biochemical basis for further comparative investigation of the two pathways.

2-Aminophenol-1,6-dioxygenase catalyzes the ring cleavage of 2-aminophenol and several substituted *ortho*-aminophenols.[119,202] *ortho*-Aminophenols are analogs of catechols in which one of the two hydroxyl groups is replaced by an amino group. The amino and hydroxyl groups have comparable sizes and similar effects on electrophilic aromatic substitution. Therefore, comparison of the ring cleavage dioxygenases from the reductive and oxidative NB degradation pathways is of interest both for understanding the biochemistry of the dioxygenation reactions and the evolution of genes that encode the dioxygenases.

Biochemical characterization and sequence analysis of 2-aminophenol-1,6-dioxygenases from JS45 and AP-3 show that the enzymes share some general characteristics with extradiol dioxygenases, but differ in detail.[44,119,202] 2-Aminophenol-1,6-dioxygenase catalyzes the extradiol ring cleavage of catechol at a low rate. Catechol, however, is an efficient suicide substrate of the enzyme,[44] indicating that there is a strong interaction between catechol and the active site of 2-aminophenol-1,6-dioxygenase. Further investigation of the kinetics of the reactions using substrate analogs will increase the understanding of the catalytic mechanism of 2-aminophenol-1,6-dioxygenase. The molecular mass of 2-aminophenol-1,6-dioxygenase is the same as that of catechol 2,3-dioxygenase from a *P. putida*,[115] but the aminophenol dioxygenase is a heterotetramer with an $\alpha_2\beta_2$ structure, whereas the catechol dioxygenase is a homotetramer. Ferrous iron is present in both enzymes as a cofactor, but there is one molecule of ferrous iron for each subunit of the catechol dioxygenases and only one for every two subunits of the 2-aminophenol dioxygenases. The β subunit of 2-aminophenol dioxygenase, but not the α subunit, shares regions of limited amino acid sequence similarity with homoprotocatechuate 2,3-dioxygenases of *E. coli* W and C.[166,176] The β subunit shares lower similarity with 2,3-dihydroxyphenylpropionate-1,2-dioxygenase from *E. coli*[193] and catechol 2,3-dioxygenase from *Alcaligenes eutrophus*.[105] The α and β subunits of the 2-aminophenol-1,6-dioxygenases from both JS45 and AP-3 share a significant degree of identity with each other. Four highly conserved active site relevant histidines are present in the β subunit, but are not present in the α subunit. The lack of conserved histidines indicates the absence of an Fe^{2+}-binding site in the α subunit, which explains the observation of only one Fe^{2+} per two subunits in the 2-aminophenol-1,6-dioxygenases. The biochemical and substrate differences between the catechol and aminophenol dioxygenases might make it appropriate to distinguish *ortho*-aminophenol dioxygenases as extradiol-like ring cleavage dioxygenases.

2-Aminomuconic semialdehyde dehydrogenase in JS45 carries out a reaction analogous to that of hydroxymuconic semialdehyde dehydrogenase in the catechol pathway. Purified 2-aminomuconic semialdehyde dehydrogenase catalyzes dehydrogenation of both 2-hydroxymuonic semialdehyde and 2-aminomuconic semialdehyde.[86] *Pseudomonas* sp. AP-3 is also reported to harbor the same enzyme, but no details about the enzyme from *Pseudomonas* sp. AP-3 have been reported. In contrast to the substantial difference between 2-aminophenol-1,6-dioxygenase and catechol 2,3-dioxygenase, the two types of semialdehyde dehydrogenases show close relationships in both biochemical properties and sequence similarity.[86] The dehydrogenases share 43 to 55% amino acid sequence identity. Functionally important amino acid residues for aldehyde dehydrogenase, including the NAD^+-binding site and the catalytic glutamate and cysteine residues,[1,95,210] are conserved in both types of enzymes. Thus, it is clear that the two types of enzymes are homologous.

P. pseudoalcaligenes JS45 genes *habA* and *habB,* encoding hydroxylaminobenzene mutase, *amnBA* encoding 2-aminophenol-1,6-dioxygenase, and *amnC* encoding 2-aminomuconic semialdehyde dehydrogenase have been cloned and sequenced.[44,45,86] The *habA* and *habB* genes are on a separate operon from *amnBAC*.[43] *amnBA* and *amnC* are contiguous, but unusually far apart.[44] Downstream of *amnC* are two putative ORFs with similarity to genes encoding ATP-dependent RNA helicases. Although it is not known if either of the ORFs produces a functional protein, the fact that no pathway-relevant genes are encoded in the sequence flanking *amnBAC* distinguishes the organization of the genes in JS45 from the gene organization of the extradiol cleavage pathway in other microorganisms. The results clearly indicate that the genes for the NB degradative pathway in *P. pseudoalcaligenes* JS45 are located in at least three operons.

Genetic analyses suggests that the genes for NB degradation in JS765 are also organized in more than one operon. The properties of a catechol 2,3-dioxygenase and 2-hydroxymuconic semialdehyde dehydrogenase that are expressed by genes cloned from *Comamonas* JS765 into *E. coli* are similar to those expressed by NB-grown JS765.[157] The results suggest but do not prove that the cloned genes are involved in NB degradation. Between the cloned dioxygenase and dehydrogenase there are unusual ORFs that are not found in sequences encoding enzymes of the catechol pathway in other microorganisms. The cloned gene encoding a 2-hydroxymuconic semialdehyde hydrolase has been altered by a single-base deletion, leading to two genes whose functions, if any, are unknown.[91] The wild-type JS765 synthesizes 2-hydroxymuconic semialdehyde hydrolase; therefore, a functional hydrolase gene must be present in another operon somewhere in the JS765 genome. Additional investigations about the regulation and organization of the genes involved in NB degradation will yield insight about their functions and evolutionary origins.

2.3.1.5 Biocatalytic Applications of NB Degradation Enzymes

The JS45 NB degradation pathway involves several novel enzymatic steps. Hydroxylaminobenzene mutase, 2-aminophenol-1,6-dioxygenase, 2-aminomuconic semialdehyde dehydrogenase, and 2-aminomuconate deaminase have not been previously reported in any prokaryote. Therefore, insight into the mechanisms and specificities of these enzymatic reactions might facilitate the application of the enzymes as biocatalysts for use in synthetic reactions. For example, 2-aminomuconate, a chemical which has a potential use in biomedical research, can be prepared from 2-aminophenol by the coupled action of 2-aminophenol-1,6-dioxygenase and 2-aminomuconic semialdehyde dehydrogenase from JS45 (Figure 2.11A).[91] It seems likely that substituted 2-aminomuconates could be prepared by a similar strategy. By itself, 2-aminophenol-1,6-dioxygenase provides a one-step method for biosynthesis of picolinic acids from corresponding *ortho*-aminophenols (Figure 2.11B).[119] Finally, the first two enzymes in the partially reductive NB pathway provide a biosynthetic route to substituted *ortho*-aminophenols (Figure 2.11C). 2-Amino-5-phenol was synthesized from 4-hydroxylaminobiphenyl ether with such a strategy.[141]

2.3.2 Aerobic Reactors

2.3.2.1 Remediation of Nitrobenzene-Contaminated Groundwater in a Fluid Bed Reactor

A field demonstration to remediate NB and aniline in groundwater at the site of a former dye manufacturing plant resulted in 99% removal of the primary contaminants.[70] Removal of all contaminants was to levels low enough to allow reinjection of the treated groundwater. The basis of the treatment system was an FBR with activated carbon for the biocarrier. The reactor was inoculated with organisms that had been isolated based on the ability to degrade NB and aniline. The authors did not report how the distinction was made between biodegradation of the contaminants and carbon sorption. Cost analysis and treatment success of the pilot study led to installation of a full-scale remediation system at the site. For both the pilot- and full-scale systems, stripping of volatile contaminants in the groundwater was minimized by dissolving pure oxygen into the aqueous phase without bubble formation. It is likely that the inoculation was responsible for the initial performance of the reactor. In the long-term, however, recruitment of degradative bacteria from the groundwater probably played a major role in the performance of the system. NB-degrading bacteria[147] and aniline-degrading bacteria[2,213] are readily isolated from contaminated sites. Recruitment of

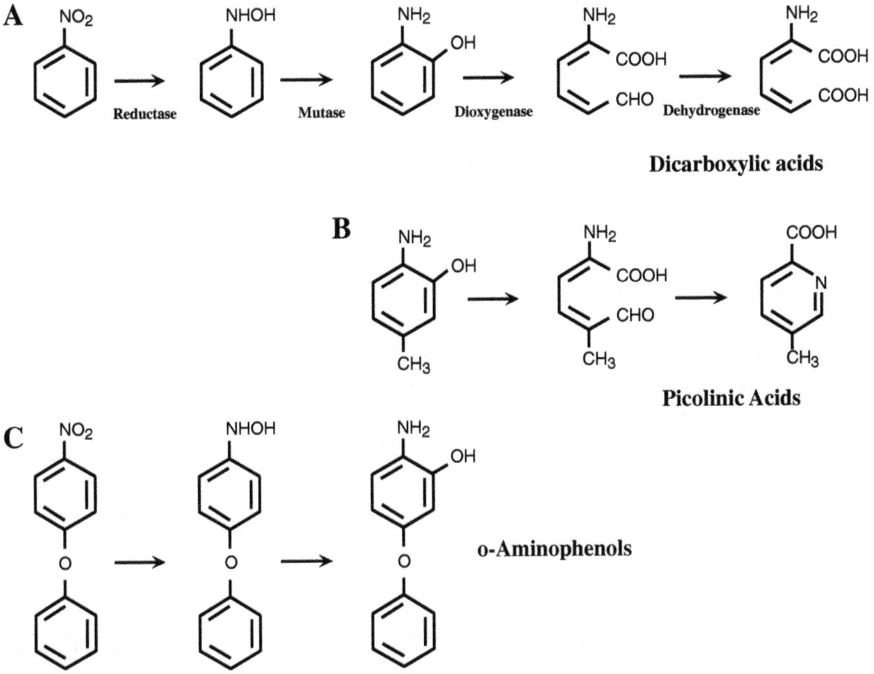

Figure 2.11 Potential biocatalyic applications for enzymes expressed in the partially reductive NB degradation pathway. Examples are 2-aminomuconate (via A), picolinic acid or 5-methylpicolinic acid (via B), or 2-amino-5-phenoxyphenol (via C).

indigenous bacteria is a likely mechanism for removal of the BTEX compounds and chlorobenzenes that also contaminated the groundwater.

2.3.2.2 Membrane Reactor Technology for Treatment of an Industrial Effluent

The composition of industrial effluents can preclude biological approaches to wastewater cleanup without prior dilution or other pretreatment.[126] Dilution is undesirable because it increases the volume of the hazardous material, and sufficient dilution to reduce the concentration of an inorganic component of a waste stream to biologically acceptable levels can be prohibitively expensive. The NB-bearing waste stream generated in the production of 3-chloronitrobenzene (3-CNB) has a low pH and high salt content.[126] A membrane bioreactor was designed to take advantage of the solvent properties of NB and 3-CNB. When the waste stream was pumped through silicone tubing, the aromatic compounds diffused into the culture fluid, while the acid and salt remained behind. NB and 3-CNB removal rates were close to 100%. Stoichiometric chloride release and recovery of 80% of the carbon entering the system as carbon dioxide suggest complete mineralization of NB and 3-CNB. The organisms responsible for removal of the nitroaromatic compounds were not reported, and the mechanism of 3-CNB degradation is not known.

The discovery of a 4-chloronitrobenzene (4-CNB)-degrading bacterium supports the possibility of the existence of 3-CNB-degrading strains. Bacterial strain LW1 uses 4-CNB as a sole source of carbon, nitrogen, and energy.[111] Strain LW1 uses a partially reductive pathway to transform 4-CNB to the Bamberger rearrangement product, 2-amino-5-chlorophenol, which is cleaved in a reaction analogous to the 2-aminophenol-1,6-dioxygenase-catalyzed cleavage reaction in the partial reductive pathway for NB degradation.[119,144] Ammonium and chloride ions are released after ring cleavage. No information was provided on whether strain LW1 could also grow on or transform 3-CNB, although it did grow on NB.[111]

An alternative proposal for degradation of mixtures of NB and CNB[158] involves cometabolic reduction and acetylation of CNB by an NB-degrading strain, followed by degradation of the acetylated intermediates by a second strain. However, the authors' conclusion that the second organism degrades acetylated compounds was based on very preliminary evidence. Acetylated metabolites have been reported for a variety of nitroaromatic compounds, including 4-CNB, 3-nitrophenol, 4-chloro-2-nitrophenol, 2,4-DNT, 2,6-DNT, and TNT.[18,39,58–60,66,111,178] To date, however, further metabolism of the acetylated products has not been demonstrated, and, where the fate of the acetylated products has been carefully investigated, the acetylated intermediates have been deemed dead end metabolites.

2.3.2.3 Removal of Nitrobenzene Vapors with a Trickling Air Biofilter

The use of a trickling air biofilter has been proposed for situations where NB has already volatilized into the air. In one such system a mixed culture of NB-degrading strains was inoculated onto a perlite-packed glass column.[151] NB-contaminated air was circulated downward through the column where 80 to 90% of the NB was removed by the column with >98% of the nitrogen trapped as ammonia. The system described is attractive for its simplicity. Recirculation of the aqueous medium minimized fluid handling and waste, pH control was unnecessary, and addition of oxygen was automatic with the air stream. The critical element with this system is to ensure that the NB-degrading population uses the partially reductive pathway. Organisms that release nitrite via the oxidative NB pathway would quickly acidify the system, which would then require pH control and lead to potential problems with salt buildup. Nitrite is regulated in aqueous waste streams, and removal from the aqueous phase would require additional reactors for either anaerobic denitrification or aerobic oxidation to nitrate.

2.3.3 Anaerobic Treatments for Nitrobenzene

Anaerobic strategies for treatment of NB effectively result in treatment of aniline. Strategies for anaerobic treatment of NB have been proposed for two major reasons: (1) treatment or handling of aniline is considered easier than treatment of NB or (2) the contaminated site or waste stream is already anaerobic. Aniline is somewhat less toxic than NB, and the lethal dose for human adults is estimated to be 15 to 30 g.[84] Although pure aniline has a vapor pressure approximately twice that of NB, aniline is 20 times more soluble than NB in water. Therefore, conversion of NB to

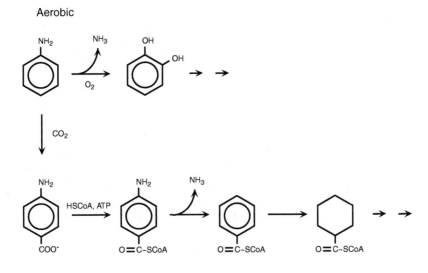

Figure 2.12 Aniline degradation pathways. Aerobic degradation produces catechol via oxidative deamination. The anaerobic pathway used by a sulfate-reducing bacterium activates the molecule with sulfhydryl-CoA prior to reductive deamination. (Modified from Schnell, S. and B. Schink. 1991. *Arch. Microbiol.* 155:183.)

aniline will reduce the overall transfer of toxic chemicals to the vapor phase. Under aerobic conditions in pond water inoculated with activated sewage sludge, 95% of aniline loss was attributed to biodegradation, with only 0.4% accounted for by volatilization.[130] Aniline-degrading bacteria were isolated decades earlier than NB-degrading bacteria, and both aerobic[10] and anaerobic[182] aniline degradation pathways have been reported (Figure 2.12).

Aerobic degradation of aniline proceeds by oxidative deamination via dioxygenase attack at the 1,2-position with the simultaneous release of ammonia.[10,12] The catechol produced then undergoes *ortho*-ring fission.[65] In sewage sludge, aniline degradation begins with little or no lag period.[130]

Desulfobacterium anilini will grow on aniline under sulfate-reducing conditions.[182] The initial step is a carboxylation to 4-aminobenzoate, which is activated to 4-aminobenzoyl-CoA. Reductive deamination produces benzoyl-CoA with the release of ammonia. The benzoyl-CoA enters the normal anaerobic benzoate degradation pathway. The aniline degradation pathway might not be widespread in sulfate-reducing bacteria. When anaerobic sewage sludge and anaerobic estuarine sediments were tested for aniline degradation, evidence was found for carboxylation of aniline, but only under nitrate-reducing conditions.[46] Another study reported similar results for microcosms constructed with soil from an anaerobic wetland.[200] The failure to detect widespread distribution of anaerobic aniline-degrading bacteria has led to a focus on aerobic aniline degradation once the initial reduction of NB has been accomplished.

2.3.3.1 Sequential Reductive/Oxidative Reactors

A two-stage system for degradation of NB was proposed to take advantage of the properties of aniline. An anaerobic reactor inoculated with anaerobic sewage digester sludge was used to nonspecifically reduce NB in an aqueous waste stream to aniline. The waste stream was then transferred to a stirred tank reactor where aerobic bacteria from activated sludge mineralized the accumulated aniline.[50] The cometabolic anaerobic reduction was supported by the addition of glucose or a mixture of simple alcohols. Although NB-degrading bacteria could have been applied to mineralize NB directly, the authors suggested that transformation of NB to the much more water-soluble amine could greatly reduce volatilization of NB in the specific wastewater treatment plant application. Another potential advantage of the anaerobic/aerobic system is that specially adapted inocula are not required for NB reduction. Aerobic aniline-degrading bacteria are widespread,[10,12,108] and the activity was easily selected in the sewage treatment plant activated sludge used in the demonstration.[50]

A variation of the anaerobic/aerobic strategy involved the use of three aerobic reactors to degrade NB via aniline.[160] One reactor contained a cometabolic NB-reducing mixed culture with glucose, succinate, and NB; a second contained an aniline-mineralizing strain; and the third received the outflow from the first two reactors. NB was reduced to aniline in the first reactor, then mineralized in the third reactor by the aniline-grown culture. The strength of the system was that precise control was maintained over the biological elements in the first two reactors. The weaknesses of the system were twofold: (a) significant amounts of NB were lost to air stripping in both the first and third reactors, and therefore the system had to be operated at less than optimal growth conditions; and (2) a separate aniline feed was required to maintain the activity of the second reactor.

2.3.3.2 Natural Attenuation

The widespread distribution of NB-degrading bacteria at NB-contaminated sites[147] makes the compound a good candidate for remediation by MNA. The potential for MNA was studied in a wetland contaminated with NB and aniline from a chemical manufacturing plant.[200] The wetland is predominantly anoxic, but experiences fluctuating water levels and periods of dryness. Monitoring data that showed a decrease in contaminant level over time led to microbiological studies and physicochemical characterization of the site. Aerobic enrichment cultures could grow on NB or aniline, and both compounds were mineralized in aerobic microcosms. Under anaerobic conditions, soil microcosms stoichiometrically reduced NB to aniline under all redox conditions, but the aniline was degraded only under nitrate-reducing conditions. The studies demonstrated that the biology for aerobic degradation of the two contaminants is present at the site, but the authors believed that the prevailing anoxic conditions would preclude significant aerobic degradation. The reported subsurface redox potentials would not support a nitrate-reducing population, nor were there significant levels of nitrate in the soil. Given the anaerobic conditions and the lack of anaerobic aniline-degrading organisms, natural attenuation is not likely to be an effective remediation strategy at the site during flood periods. However, the fact that aerobic bacteria at the site have evolved to degrade both NB and aniline strongly suggests that significant degradation of the two compounds occurs during less anoxic periods. The possibility is supported by the finding of moderate levels of ammonia in the soil which would be released by degradation of either compound.

2.3.3.3 Zero Valent Iron

The initial reduction of NB to aniline can also be driven by abiotic reactions (see Reference 74 and Chapter 12 for reviews). Use of zero valent iron as the electron donor in the reductive stage of nitroaromatic treatment systems is initially appealing for its simplicity. Rapid, first-order reduction can occur under a variety of conditions that do not require optimization to support biocatalysis.[3] Metallic iron is a readily available commodity that requires no special preparation for use in columns[47] or *in situ* as sheet piles or curtains. The major drawback is the loss of reactivity of the metallic iron after the initial exposure to reducible compounds. The iron surface can be passivated by oxides or carbonates or by sorbed aniline.[3,47] The sorption of aniline is a function of pH.[3] If sorption of aniline is not the desired endpoint, then the system must be designed to desorb and transport the aniline to an aerobic reactor for mineralization. The role of iron-reducing bacteria in the regeneration of active sites[92] on metallic iron remains to be fully developed.

2.4 NITROBENZOATES

2.4.1 Pathways

Bacteria able to grow on nitrobenzoic acids were some of the earliest nitroaromatic-degrading organisms isolated. Strains able to grow on 2- and 4-nitrobenzoic acid (2-NBA, 4-NBA) were reported in 1958,[31] and 3-nitrobenzoic acid (3-NBA)-degrading strains were found the following year.[32] 4-Nitrocatechol or protocatechuic acid were proposed intermediates in degradative pathways for all three isomers,[32] but many questions remained about what role reductive pathways played in the degradation of NBA. Follow-up studies[30,33] concluded that in the strains studied, reduction was due to nonspecific reactions and aminobenzoic acids were not direct intermediates in 2- or 4-NBA degradation. Release of the nitro group as nitrite rather than ammonia was the major evidence for the oxidative pathway. More recently, $^{18}O_2$ incorporation experiments[142] provided rigorous proof of an oxidative pathway for 3-NBA.

A partially reductive pathway for degradation of 4-NBA was discovered in *C. acidovorans* NBA-10.[72,73] 4-NBA is partially reduced to 4-hydroxylaminobenzoate before conversion to protocatechuate by a hydroxylaminolyase. The genes for 4-NBA degradation have been cloned from *P. pickettii* YH105, which employs the same pathway.[214] Further discussion of the molecular biology can be found in Chapter 6. Some bacteria that grow on 4-nitrotoluene oxidize the methyl group and then degrade the resultant 4-NBA by the above pathway.[79,173]

A *P. pseudoalcaligenes* strain grows on 2-NBA with stoichiometric release of ammonia.[138] 2-Aminobenzoic acid was not detected in cultures. There was no anthranilate oxidase activity in whole cells or cell extracts, nor was there any catechol-1,2- or 2,3-dioxygenase activity. The authors assumed that reduction of the nitro group occurred after hydroxylation of the aromatic ring. The authors did not examine the possibility of partial reduction of 2-NBA to 2-hydroxylaminobenzoate followed by a Bamberger rearrangement to 3-hydroxyanthranilate. 3-Hydroxyanthranilate might be cleaved or routed into the niacin synthesis pathway.

2.4.1.1 Immobilized Cell Reactor for Mixed NBA Degradation

A process developed for cleaning of metal parts prior to plating generates large quantities of 3-NBA in the waste stream. Other industrial waste streams are characterized by alternating and fluctuating levels of the 3- and 4-NBA isomers. A study was undertaken to develop a system tolerant of frequent changes in substrate mix and concentration.[69] *Comamonas* strain JS47, which grows on 4-NBA via the partially reductive pathway, and *Comamonas* strain JS46, which grows on 3-NBA via the oxidative pathway, were jointly and separately immobilized in alginate beads. The beads were loaded in airlift reactors and fed different regimes of alternating NBA isomers or mixed NBA isomers. When the two strains were coimmobilized in the same alginate beads, recovery of the system from changes in influent composition was faster than recovery when beads carrying separate strains were mixed together in the reactor. The authors suggest that cross-feeding of NBA metabolites within beads kept the biomass healthy during periods of substrate deprivation and that such metabolites were not available to bacteria immobilized in different beads.

In another system, *B. cepacia* strain PB4 was immobilized on diatomaceous earth in a packed bed column and fed equimolar mixtures of 4-NBA and 4-aminobenzoate.[161] The strain possesses separately inducible pathways for both aromatic compounds, but each compound apparently serves as a gratuitous inducer for the other. Cultures induced on either substrate degraded 4-NBA and 4-aminobenzoate simultaneously and without any lag. The advantage of the system is that selection to maintain the degradative strain will occur whether 4-NBA or 4-aminobenzoate is present.

2.5 FUTURE WORK

During the past decade bacteria able to degrade a range of nitroaromatic compounds by aerobic pathways once thought unlikely have been discovered. Degradative bacteria are found only in contaminated areas and not in nearby uncontaminated sites. A number of new reactions have been discovered in the new pathways. In many instances, the enzymes and genes responsible for the reactions and pathways have been isolated and studied. Understanding of the pathways by which specific nitroaromatic compounds are degraded has led to the development of effective, self-sustaining bioremediation systems for a number of nitroaromatic compounds.

A major question that remains to be answered is how degradative nitroaromatic pathways are regulated and the related problem of identification of the factors that control the biodegradation of mixtures of nitroaromatic compounds. Additionally, the study of the evolutionary origin and distribution of organisms and degradative pathways is still in the preliminary stages. Understanding the above aspects of nitroaromatic compound biodegradation will provide valuable insight into the occurrence and use of natural attenuation as an environmental remediation strategy.

ACKNOWLEDGMENTS

This work was supported by grants from the Air Force Office of Scientific Research and from the Strategic Environmental Research Defense Program.

REFERENCES

1. Abriola, D. P., R. Fields, S. Stein, A. D. MacKerell, Jr., and R. Pietruszko. 1987. Active site of human liver aldehyde dehydrogenase. *Biochemistry.* 26:5679–5684.
2. Aelion, C. M., C. M. Swindoll, and F. K. Pfaender. 1987. Adaptation to and biodegradation of xenobiotic compounds by microbial communities from a pristine aquifer. *Appl. Environ. Microbiol.* 53:2212–2217.
3. Agrawal, A. and P. G. Tratnyek. 1995. Reduction of nitro aromatic compounds by zero-valent iron metal. *Environ. Sci. Technol.* 30:153–160.
4. Alexander, M. 1981. Biodegradation of chemicals of environmental concern. *Science.* 211:132–138.
5. Alexander, M. and B. K. Lustigman. 1966. Effect of chemical structure on microbial degradation of substituted benzenes. *J. Agric. Food Chem.* 14:410–413.
6. Ali-Sadat, S., K. S. Mohan, and S. K. Walia. 1995. A novel pathway for the biodegradation of 3-nitrotoluene in *Pseudomonas putida*. *FEMS Microbiol. Ecol.* 17:169–176.
7. An, D., D. T. Gibson, and J. C. Spain. 1994. Oxidative release of nitrite from 2-nitrotoluene by a three-component enzyme system from *Pseudomonas* sp. strain JS42. *J. Bacteriol.* 176:7462–7467.
8. Anderson, K. E., G. J. Hammons, F. F. Kadlubar, J. D. Potter, K. R. Kaderlik, K. F. Ilett, R. R. Minchin, C. H. Teitel, H.-C. Chou, M. V. Martin, F. P. Guengerich, G. W. Barone, N. P. Lang, and L. A. Peterson. 1997. Metabolic activation of aromatic amines by human pancreas. *Carcinogenesis.* 18:1085–1092.
9. Anderson, W. C., R. C. Loehr, and B. P. Smith (ed.). 1999. *Environmental Availability of Chlorinated Organics, Explosives, and Metals in Soil*. American Academy of Environmental Engineers, Annapolis, MD.
10. Aoki, K., R. Shinke, and H. Nishira. 1983. Metabolism of aniline by *Rhodococcus erythropolis* AN-13. *Agric. Biol. Chem.* 47:1611–1616.
11. Asakura, S. and S. Okazaki. 1995. Biodegradation of toluene diamine (TDA) in activated sludge acclimated with aniline and TDA. *Chemosphere.* 30:2209–2217.
12. Bachofer, R., F. Lingens, and W. Schäfer. 1975. Conversion of aniline into pyrocatechol by a *Nocardia* sp.: incorporation of oxygen-18. *FEBS Lett.* 50:288–290.

13. Barkay, T. and H. Pritchard. 1988. Adaptation of aquatic microbial communities to pollutant stress. *Microbiol. Sci.* 5:165–169.

14. Bausum, H. T., W. R. Mitchell, and M. A. Major. 1992. Biodegradation of 2,4- and 2,6-dinitrotoluene by freshwater microorganisms. *J. Environ. Sci. Health A.* 27:663–695.

15. Behrend, C. and K. Heesche-Wagner. 1999. Formation of hydride-Meisenheimer complexes of picric acid (2,4,6-trinitrophenol) and 2,4-dinitrophenol during mineralization of picric acid by *Nocardioides* sp. strain CB 22–2. *Appl. Environ. Microbiol.* 65:1372–1377.

16. Berchtold, S. R., S. L. VanderLoop, M. Suidan, and S. W. Maloney. 1995. Treatment of 2,4-dinitrotoluene using a two-stage system: fluidized-bed anaerobic granular activated carbon reactors and aerobic activated sludge reactors. *Water Environ. Res.* 67:1081–1091.

17. Bester, K., R. Gatermann, H. Hühnerfuss, W. Lange, and N. Theobald. 1998. Results of nontarget screening of lipophilic organic pollutants in the German Bight. IV: Identification and quantification of chloronitrobenzenes and dichloronitrobenzenes. *Environ. Pollut.* 102:163–169.

18. Beunink, J. and H.-J. Rehm. 1990. Coupled reductive and oxidative degradation of 4-chloro-2-nitrophenol by a co-immobilized mixed culture system. *Appl. Microbiol. Biotechnol.* 34:108–115.

19. Blasco, R. and F. Castillo. 1997. Characterization of 2,4-dinitrophenol uptake by *Rhodobacter capsulatus*. *Pestic. Biochem. Physiol.* 58:1–6.

20. Blasco, R. and F. Castillo. 1993. Characterization of a nitrophenol reductase from the phototrophic bacterium *Rhodobacter capsulatus* E1F1. *Appl. Environ. Microbiol.* 59:1774–1778.

21. Blasco, R. and F. Castillo. 1992. Light-dependent degradation of nitrophenols by the phototrophic bacterium *Rhodobacter capsulatus* E1F1. *Appl. Environ. Microbiol.* 58:690–695.

22. Blasco, R., E. Moore, V. Wray, D. Pieper, K. Timmis, and F. Castillo. 1999. 3-Nitroadipate, a metabolic intermediate for mineralization of 2,4-dinitrophenol by a new strain of a *Rhodococcus* species. *J. Bacteriol.* 181:149–152.

23. Bradley, P. M., F. H. Chapelle, and J. E. Landmeyer. 1995. Degradation of 2,4-DNT, 2,6-DNT and 2,4,6-TNT by indigenous aquifer microorganisms, p. 267–271. In R. E. Hinchee, D. B. Anderson, and R. E. Hoeppel (eds.), *Bioremediation of Recalcitrant Organics: International In Situ On Site Bioreclamation Symposium.* Battelle Press, Columbus, OH.

24. Bradley, P. M., F. H. Chapelle, J. E. Landmeyer, and J. G. Schumacher. 1997. Potential for intrinsic bioremediation of a DNT-contaminated aquifer. *Ground Water.* 35:12–17.

25. Bradley, P. M., R. H. Chapelle, J. E. Landmeyer, and J. G. Schumacher. 1994. Microbial transformation of nitroaromatics in surface soils and aquifer materials. *Appl. Environ. Microbiol.* 60:2170–2175.

26. Breitung, J., D. Bruns-Nagel, K. Steinbach, L. Kaminski, D. Gemsa, and E. von Löw. 1996. Bioremediation of 2,4,6-trinitrotoluene-contaminated soils by two different aerated compost systems. *Appl. Microbiol. Biotechnol.* 44:795–800.

27. Bringmann, G. and R. Kühn. 1971. Biologischer Abbau von Nitrotoluolen und Nitrobenzolen mittels *Azotobacter agilis*. *Gesund. Ing.* 92:273–276.

28. Bruhn, C., R. C. Bayly, and H.-J. Knackmuss. 1988. The *in vivo* construction of 4-chloro-2-nitrophenol assimilatory bacteria. *Arch. Microbiol.* 150:171–177.

29. Bruns-Nagel, D., O. Drzyzga, K. Steinbach, T. C. Schmidt, E. von Löw, T. Gorontzy, K.-H. Blotevogel, and D. Gemsa. 1998. Anaerobic/aerobic composting of 2,4,6-trinitrotoluene-contaminated soil in a reactor system. *Environ. Sci. Technol.* 32:1676–1679.

30. Cain, R. B. 1966. Induction of anthranilate oxidation system during the metabolism of *ortho*-nitrobenzoate by certain bacteria. *J. Gen. Microbiol.* 42:197–217.

31. Cain, R. B. 1958. The microbial metabolism of nitroaromatic compounds. *J. Gen. Microbiol.* 19:1–14.

32. Cartwright, N. J. and R. B. Cain. 1959. Bacterial degradation of nitrobenzoic acids. *Biochem. J.* 71:248–261.

33. Cartwright, N. J. and R. B. Cain. 1959. Bacterial degradation of the nitrobenzoic acids. 2. Reduction of the nitro group. *Biochem. J.* 73:305–314.

34. Cerniglia, C. E. and C. C. Somerville. 1995. Reductive metabolism of nitroaromatic and nitropolycyclic aromatic hydrocarbons, p. 99–115. In J. C. Spain (ed.), *Biodegradation of Nitroaromatic Compounds*. Plenum Publishing Corp., New York.

35. Chemical & Engineering News. 1995. Facts and figures for the chemical industry. *Chem. Eng. News.* June 24, 1995:41.

36. Chemical & Engineering News. 1999. Facts and figures for the chemical industry. *Chem. Eng. News.* June 20, 1999:32–73.

37. Chemical Online. 1999. Air products DNT production facility will produce 500 million pounds annually. http://news.chemicalonline.com/industry-news/19990311-9566.html. Access date 9/21/99.

38. Cheng, J., Y. Kanjo, M. T. Suidan, and A. D. Venosa. 1996. Anaerobic biotransformation of 2,4-dinitrotoluene with ethanol as primary substrate: mutual effect of the substrates on their biotransformation. *Water Res.* 30:307–314.

39. Corbett, M. D. and B. R. Corbett. 1981. Metabolism of 4-chloronitrobenzene by the yeast *Rhodosporidium* sp. *Appl. Environ. Microbiol.* 41:942–949.

40. Corbett, M. D., B. R. Corbett, and D. R. Doerge. 1982. Hydroxamic acid production and active-site induced Bamberger rearrangement from the action of α-ketoglutarate dehydrogenase on 4-chloronitrosobenzene. *J. Chem. Soc. Perkin Trans.* 1:345–350.

41. Cowen, W. F., A. M. Gastinger, C. E. Spanier, J. R. Buckel, and R. E. Bailey. 1998. Sorption and microbial degradation of toluenediamines and methylenedianiline in soil under aerobic and anaerobic conditions. *Environ. Sci. Technol.* 32:598–603.

42. Daun, G., H. Lenke, M. Reuss, and H.-J. Knackmuss. 1998. Biological treatment of TNT-contaminated soil. 1. Anaerobic cometabolic reduction and interaction of TNT and metabolites with soil components. *Environ. Sci. Technol.* 32:1956–1963.

43. Davis, J. unpublished results.

44. Davis, J. K., Z. He, C. C. Somerville, and J. C. Spain. 1999. Genetic and biochemical comparison of 2-aminophenol-1,6-dioxygenase of *Pseudomonas pseudoalcaligenes* JS45 to *meta*-cleavage dioxygenases: divergent evolution of 2-aminophenol *meta*-cleavage pathway. *Arch. Microbiol.* 172:330–339.

45. Davis, J. K., C. C. Somerville, and J. C. Spain. 1997. *Pseudomonas pseudoalcaligenes* JS45 possesses two genes encoding hydroxylaminobenzene mutase, abstr. Q-342, p. 512, Abstracts, 97th General Meeting of the American Society for Microbiology. American Society for Microbiology, Washington, D.C.

46. De, M. A., O. A. O'Connor, and D. S. Kosson. 1994. Metabolism of aniline under different anaerobic electron-accepting and nutritional conditions. *Environ. Toxicol. Chem.* 13:233–239.

47. Devlin, J. F., J. Klausen, and R. P. Schwarzenbach. 1998. Kinetics of nitroaromatic reduction on granular iron in recirculating batch experiments. *Environ. Sci. Technol.* 32:1941–1947.
48. Dey, S. and S. H. Godbole. 1986. Biotransformation of *m*-dinitrobenzene by *Candida pulcherrima*. *Indian J. Exp. Biol.* 24:29–33.
49. Dey, S., P. Kanekar, and S. H. Godbole. 1986. Aerobic microbial degradation of *m*-dinitrobenzene. *Indian J. Environ. Health.* 29(2):118–128.
50. Dickel, O., W. Haug, and H.-J. Knackmuss. 1993. Biodegradation of nitrobenzene by a sequential anaerobic-aerobic process. *Biodegradation.* 4:187–194.
51. Dickel, O., and H.-J. Knackmuss. 1991. Catabolism of 1,3-dinitrobenzene by *Rhodococcus* sp. QT-1. *Arch. Microbiol.* 157:76–79.
52. Donlon, B. A., E. Razo-Flores, G. Lettinga, and J. A. Field. 1996. Continuous detoxification, transformation, and degradation of nitrophenols in upflow anaerobic sludge blanket (USAB) reactors. *Biotechnol. Bioeng.* 51:439–449.
53. Dorigan, J., and J. Hushon. 1976. Air pollution assessment of nitrobenzene. Final MTR-7228. U.S. Environmental Protection Agency, Washington, D.C.
54. Ebert, S., P.-G. Rieger, and H.-J. Knackmuss. 1999. Function of coenzyme F_{420} in aerobic catabolism of 2,4,6-trinitrophenol and 2,4-dinitrophenol by *Nocardioides simplex* FJ2–1A. *J. Bacteriol.* 181:2669–2674.
55. Ecker, S., T. Widmann, H. Lenke, O. Dickel, P. Fischer, C. Bruhn, and H.-J. Knackmuss. 1992. Catabolism of 2,6-dinitrophenol by *Alcaligenes eutrophus* JMP134 and JMP222. *Arch. Microbiol.* 158:149–154.
56. Envirogen Inc. 1997. Technology demonstration work plan for National Environmental Technology Test Sites Volunteer Army Ammunition Plant: Fluid bed reactor study for the mineralization of DNT. U.S. Army Corps of Engineers, Baltimore District, Baltimore, MD.
57. Environmental Health Center. 1997. 2,4-Dinitrotoluene and 2,6-dinitrotoluene ($Ch_3C_6H_3(NO_2)_2$) chemical backgrounder, July 1, 1997 ed. National Safety Council. http://safety.webfirst.com/ehc/ew/chems/dintrot.html. Access date 9/21/99.
58. Fiorella, P. D. and J. C. Spain. 1997. Transformation of 2,4,6-trinitrotoluene by *Pseudomonas pseudoalcaligenes* JS52. *Appl. Environ. Microbiol.* 63:2007–2015.
59. Freedman, D. L. and D. R. Noguera. 1995. An evaluation of biotransformation products from 2,4-dinitrotoluene under nitrate-reducing conditions. Final Report. U.S. Army Research Office, Research Triangle Park, NC.
60. Freedman, D. L., R. S. Shanley, and R. J. Scholze. 1996. Aerobic biodegradation of 2,4-dinitrotoluene, aminonitrotoluene isomers, and 2,4-diaminotoluene. *J. Hazard. Mater.* 49:1–14.
61. French, C. E., S. Nicklin, and N. C. Bruce. 1998. Aerobic degradation of 2,4,6-trinitrotoluene by *Enterobacter cloacae* PB2 and by pentaerythritol tetranitrate reductase. *Appl. Environ. Microbiol.* 64:2864–2868.
62. Fuenmayer, S. L., M. Wild, A. L. Boyes, and P. A. Williams. 1998. A gene cluster encoding steps in conversion of naphthalene to gentisate in *Pseudomonas* sp. strain U2. *J. Bacteriol.* 180:2522–2530.
63. Furniss, B. S., A. J. Hannaford, P. W. G. Smith, and A. R. Tatchell. 1989. *Vogel's Textbook of Practical Organic Chemistry*, p. 953–956, 5th ed. Longman Scientific & Technical, New York.
64. Gibson, D. T. 1976. Initial reactions in the bacterial degradation of aromatic hydrocarbons. *Zentrabul. Bakteriol. Hyg. Abt. 1 Orig. Reihe B.* 162:157–168.

65. Gibson, D. T. and V. Subramanian. 1984. Microbial degradation of aromatic hydrocarbons, p. 181–252. In D. T. Gibson (ed.), Microbial Degradation of Organic Compounds. Marcel Dekker, New York.

66. Gilcrease, P. C. and V. G. Murphy. 1995. Bioconversion of 2,4-diamino-6-nitrotoluene to a novel metabolite under anoxic and aerobic conditions. *Appl. Environ. Microbiol.* 61:4209–4214.

67. Gisi, D., G. Stucki, and W. Hanselmann. 1997. Biodegradation of the pesticide 4,6-dinitro-ortho-cresol by microorganisms in batch cultures and in fixed-bed column reactors. *Appl. Microbiol. Biotechnol.* 48:441–448.

68. Gomólka, E. and B. Gomólka. 1979. Ability of activated sludge to degrade nitrobenzene in municipal wastewater. *Acta Hydrochim. Hydrobiol.* 7:605–622.

69. Goodall, J. L., S. M. Thomas, J. C. Spain, and S. W. Peretti. 1998. Operation of mixed-culture immobilized cell reactors for the metabolism of *meta-* and *para-*nitrobenzoate by *Comamonas* sp. JS46 and *Comamonas* sp. JS47. *Biotechnol. Bioeng.* 59:21–27.

70. Greene, M., M. Del Vecchio, S. Frisch, T. Beauregard, and R. Meyer. 1999. Presented at the Water Environment Federation 72nd Annual Conference and Exposition, New Orleans.

71. Griest, W. H., A. J. Stewart, A. A. Vass, and C.-H. Ho. 1998. Chemical and toxicological characterization of slurry reactor biotreatment of explosives-contaminated soils. Final SFIM-AEC-ET-CR-96186. U.S. Army Environmental Center (USAEC), Aberdeen Proving Ground, MD.

72. Groenewegen, P. E. J., P. Breeuwer, J. M. L. M. van Helvoort, A. A. M. Langenhoff, F. P. de Vries, and J. A. M. de Bont. 1992. Novel degradative pathway of 4-nitrobenzoate in *Comamonas acidovorans* NBA-10. *J. Gen. Microbiol.* 138:1599–1605.

73. Groenewegen, P. E. J. and J. A. M. de Bont. 1992. Degradation of 4-nitrobenzoate via 4-hydroxylaminobenzoate and 3,4-dihydroxybenzoate in *Comamonas acidovorans* NBA-10. *Arch. Microbiol.* 158:381–386.

74. Haderlein, S. B. and R. P. Schwarzenbach. 1995. Environmental processes influencing the rate of abiotic reduction of nitroaromatic compounds in the subsurface, p. 199–225. In J. C. Spain (ed.), *Biodegradation of Nitroaromatic Compounds*. Plenum Publishing Corp., New York.

75. Haïdour, A. and J. L. Ramos. 1996. Identification of products resulting from the biological reduction of 2,4,6-trinitrotoluene, 2,4-dinitrotoluene, and 2,6-dinitrotoluene by *Pseudomonas* sp. *Environ. Sci. Technol.* 30:2365–2370.

76. Haigler, B. E., G. R. Johnson, W.-C. Suen, and J. C. Spain. 1999. Biochemical and genetic evidence for *meta*-ring cleavage of 2,4,5-trihydroxytoluene in *Burkholderia* sp. strain DNT. *J. Bacteriol.* 181:3965–3972.

77. Haigler, B. E., S. F. Nishino, and J. C. Spain. 1994. Biodegradation of 4-methyl-5-nitrocatechol by *Pseudomonas* sp. strain DNT. *J. Bacteriol.* 176:3433–3437.

78. Haigler, B. E., C. C. Somerville, J. C. Spain, and R. K. Jain. 1997. Sequence analysis differences between the 2,4-dinitrotoluene degrading genes of *Burkholderia* sp. strain R34 and strain DNT, abstr. Q-343, p. 512, Abstracts, 97th General Meeting of the American Society for Microbiology. American Society for Microbiology, Washington, D.C.

79. Haigler, B. E. and J. C. Spain. 1993. Biodegradation of 4-nitrotoluene by *Pseudomonas* sp. strain 4NT. *Appl. Environ. Microbiol.* 59:2239–2243.

80. Haigler, B. E. and J. C. Spain. 1991. Biotransformation of nitrobenzene by bacteria containing toluene degradative pathways. *Appl. Environ. Microbiol.* 57:3156–3162.

81. Haigler, B. E., W.-C. Suen, and J. C. Spain. 1996. Purification and sequence analysis of 4-methyl-5-nitrocatechol oxygenase from *Burkholderia* sp. strain DNT. *J. Bacteriol.* 178:6019–6024.
82. Haigler, B. E., W. H. Wallace, and J. C. Spain. 1994. Biodegradation of 2-nitrotoluene by *Pseudomonas* sp. strain JS42. *Appl. Environ. Microbiol.* 60:3466–3469.
83. Hallas, L. E. and M. Alexander. 1983. Microbial transformation of nitroaromatic compounds in sewage effluent. *Appl. Environ. Microbiol.* 45:1234–1241.
84. Hazardous Substances Data Bank. 1999. Aniline. N.L.M. Toxicology Data Network. http://sis.nlm.nih.gov/cgi-bin/sis/search/f?/temp/NAAAn5aamW:1.
85. Hazardous Substances Data Bank. 1999. Nitrobenzene. N. L. M. Toxicology Data Network. http://sis.nlm.nih.gov/cgi-bin/sis/search/f?/temp/NAAAYIa4+w:1.
86. He, Z., J. K. Davis, and J. C. Spain. 1998. Purification, characterization, and sequence analysis of 2-aminomuconic 6-semialdehyde dehydrogenase from *Pseudomonas pseudoalcaligenes* JS45. *J. Bacteriol.* 180:4591–4595.
87. He, Z., L. J. Nadeau, and J. C. Spain. 1999. Characterization of hydroxylaminobenzene mutase from *Pseudomonas pseudoalcaligenes* JS45, abstr. Q-251, p. 581, Abstracts, 99th General Meeting of the American Society for Microbiology. American Society for Microbiology, Washington, D.C.
88. He, Z. and J. C. Spain. 1999. Comparison of the downstream pathways for degradation of nitrobenzene by *Pseudomonas pseudoalcaligenes* JS45 (2-aminophenol pathway) and by *Comamonas* sp. JS765 (catechol pathway). *Arch. Microbiol.* 171:309–316.
89. He, Z. and J. C. Spain. 1998. A novel 2-aminomuconate deaminase in the nitrobenzene degradation pathway of *Pseudomonas pseudoalcaligenes* JS45. *J. Bacteriol.* 180:2502–2506.
90. He, Z. and J. C. Spain. 1997. Studies of the catabolic pathway of degradation of nitrobenzene by *Pseudomonas pseudoalcaligenes* JS45: removal of the amino group from 2-aminomuconic semialdehyde. *Appl. Environ. Microbiol.* 63:4839–4843.
91. He, Z. and J. C. Spain. unpublished results.
92. Heijman, C. G., E. Grieder, C. Holliger, and R. P. Schwarzenbach. 1995. Reduction of nitroaromatic compounds coupled to microbial iron reduction in laboratory aquifer columns. *Environ. Sci. Technol.* 29:775–783.
93. Heinze, L., M. Brosius, and U. Wiesmann. 1995. Biologischer Abbau von 2,4-Dinitrotoluol in einer kontinuierlich betriebenen Versuchsanlage und kinetische Untersuchungen. *Acta Hydrochim. Hydrobiol.* 23:254–263.
94. Heitkamp, M. A. and R. G. Orth. 1998. Aerobic biodegradation potential of a constructed biozone for Lasagna™ soil remediation technology. *Environ. Eng. Sci.* 14:153–161.
95. Hempel, J., H. Nicholas, and R. Lindahl. 1993. Aldehyde dehydrogenases: widespread structural and functional diversity within a shared framework. *Protein Sci.* 2:1890–1900.
96. Hess, T. F., S. K. Schmidt, J. Silverstein, and B. Howe. 1990. Supplemental substrate enhancement of 2,4-dinitrophenol mineralization by a bacterial consortium. *Appl. Environ. Microbiol.* 56:1551–1558.
97. Hsu, T., M. F. Lux, and H. L. Drake. 1990. Expression of an aromatic-dependent decarboxylase which provides growth-essential CO_2 equivlents for the acetogenic (wood) pathway of *Clostridium thermoaceticum*. *J. Bacteriol.* 172:5901–5907.
98. Hughes, J. B., C. Wang, K. Yesland, A. Richardson, R. Bhadra, G. Bennett, and F. Rudolph. 1998. Bamberger rearrangement during TNT metabolism by *Clostridium acetobutylicum*. *Environ. Sci. Technol.* 32:494–500.

99. Hughes, J. B., C. Y. Wang, and C. Zhang. 1999. Anaerobic biotransformation of 2,4-dinitrotoluene and 2,6-dinitrotoluene by *Clostridium acetobutylicum:* a pathway through dihydroxylamino intermediates. *Environ. Sci. Technol.* 33:1065–1070.

100. Jain, R. K., J. H. Dreisbach, and J. C. Spain. 1994. Biodegradation of *p*-nitrophenol via 1,2,4-benzenetriol by an *Arthrobacter* sp. *Appl. Environ. Microbiol.* 60:3030–3032.

101. Jensen, H. L. and G. Lautrup-Larsen. 1967. Microorganisms that decompose nitroaromatic compounds, with special reference to dinitro-ortho-cresol. *Acta Agric. Scand.* 17:115–126.

102. Johnson, G. R. and J. C. Spain. 1997. Phylogenetic analysis of nitrotoluene dioxygenases, abstr. Q-344, p. 512, Abstr. Annu. Meet. Am. Soc. Microbiol. 1997. American Society for Microbiology, Washington, D.C.

103. Jung, K.-H., J.-Y. Lee, and H.-S. Kim. 1995. Biodegradation of nitrobenzene through a hybrid pathway in *Pseudomonas putida. Biotechnol. Bioeng.* 48:625–630.

104. Kaake, R. H., D. J. Roberts, T. O. Stevens, R. L. Crawford, and D. L. Crawford. 1992. Bioremediation of soils contaminated with the herbicide 2-*sec*-butyl-4,6-dinitrophenol (Dinoseb). *Appl. Environ. Microbiol.* 58:1683–1689.

105. Kabisch, M. and P. Fortnagel. 1990. Nucleotide sequence of the metapyrocatechase II (catechol 2,3-oxygenase II) gene *mpc*II from *Alcaligenes eutrophus* JMP 222. *Nucleic Acids Res.* 18:5543.

106. Kadiyala, V., B. F. Smets, K. Chandran, and J. C. Spain. 1998. High affinity *p*-nitrophenol oxidation by *Bacillus sphaericus* JS905. *FEMS Microbiol. Lett.* 166:115–120.

107. Kadiyala, V. and J. C. Spain. 1998. A two-component monooxygenase catalyzes both the hydroxylation of *p*-nitrophenol and the oxidative release of nitrite from 4-nitrocatechol in *Bacillus sphaericus* JS905. *Appl. Environ. Microbiol.* 64:2479–2484.

108. Kaminski, U., D. Janke, H. Prauser, and W. Fritsche. 1983. Degradation of aniline and monochloroanilines by *Rhodococcus* sp. An 117 and a pseudomonad: a comparative study. *Z. Allg. Mikrobiol.* 23:235–246.

109. Kan, A. T., G. Fu, M. Hunter, W. Chen, C. H. Ward, and M. B. Tomson. 1998. Irreversible sorption of neutral hydrocarbons to sediments: experimental observations and model predictions. *Environ. Sci. Technol.* 32:892–902.

110. Kaplan, D. L. 1992. Biological degradation of explosives and chemical agents. *Curr. Opin. Biotechnol.* 3:253–260.

111. Katsivela, E., V. Wray, D. H. Pieper, and R.-M. Wittich. 1999. Initial reactions in the biodegradation of 1-chloro-4-nitrobenzene by a newly isolated bacterium, strain LW1. *Appl. Environ. Microbiol.* 65:1405–1412.

112. Keith, L. H. and W. A. Telliard. 1979. Priority pollutants I. A perspective view. *Environ. Sci. Technol.* 13:416–423.

113. Khan, T. A., R. Bhadra, and J. Hughes. 1997. Anaerobic transformation of 2,4,6-TNT and related nitroaromatic compounds by *Clostridium acetobutylicum. J. Ind. Micrbiol. Biotechnol.* 18:198–203.

114. Kincannon, D. F., E. L. Stover, V. Nichols, and D. Medley. 1983. Removal mechanisms for toxic priority pollutants. *J. Water Pollut. Control Fed.* 55:157–163.

115. Kobayashi, T., T. Ishida, K. Horiike, Y. Takahara, N. Numao, A. Nakazawa, T. Nakazawa, and M. Nozaki. 1995. Overexpression of *Pseudomonas putida* catechol 2,3-dioxygenase with high specific activity by genetically engineered *Escherichia coli. J. Biochem.* 117:614–622.

116. Kruchoski, M. P. 1997. New lower estimate for soils contaminated with secondary explosives and the associated implications. Final SFIM-AEC-ET-CR97026. U.S. Army Environmental Center, Aberdeen Proving Ground, MD.

117. Krumholz, L. R., W. W. Clarkson, G. G. Wilber, and J. M. Suflita. 1997. Transformations of TNT and related aminotoluenes in groundwater aquifer slurries under different electron-accepting conditions. *J. Ind. Microbiol. Biotechnol.* 18:161–169.

118. Lee, J.-Y., K.-H. Jung, S. H. Choi, and H.-S. Kim. 1995. Combination of the *tod* and the *tol* pathways in redesigning a metabolic route of *Pseudomonas putida* for the mineralization of a benzene, toluene, and p-xylene mixture. *Appl. Environ. Microbiol.* 61:2211–2217.

119. Lendenmann, U. and J. C. Spain. 1996. 2-Aminophenol 1,6-dioxygenase: a novel aromatic ring cleavage enzyme purified from *Pseudomonas pseudoalcaligenes* JS45. *J. Bacteriol.* 178:6227–6232.

120. Lendenmann, U., J. C. Spain, and B. F. Smets. 1998. Simultaneous biodegradation of 2,4-dinitrotoluene and 2,6-dinitrotoluene in an aerobic fluidized-bed biofilm reactor. *Environ. Sci. Technol.* 32:82–87.

121. Lenke, H. and H.-J. Knackmuss. 1996. Initial hydrogenation and extensive reduction of substituted 2,4-dinitrophenols. *Appl. Environ. Microbiol.* 62:784–790.

122. Lenke, H. and H.-J. Knackmuss. 1992. Initial hydrogenation during catabolism of picric acid by *Rhodococcus erythropolis* HL 24–2. *Appl. Environ. Microbiol.* 58:2933–2937.

123. Lenke, H., D. H. Pieper, C. Bruhn, and H.-J. Knackmuss. 1992. Degradation of 2,4-dinitrophenol by two *Rhodococcus erythropolis* strains, HL 24–1 and HL 24–2. *Appl. Environ. Microbiol.* 58:2928–2932.

124. Lenke, H., J. Warrelmann, G. Daun, K. Hund, U. Sieglen, U. Walter, and H.-J. Knackmuss. 1998. Biological treatment of TNT-contaminated soil II: Biologically induced immobilization of the contaminants and full-scale application. *Environ. Sci. Technol.* 32:1964–1971.

125. Liu, D., K. Thomson, and A. C. Anderson. 1984. Identification of nitroso compounds from biotransformation of 2,4-dinitrotoluene. *Appl. Environ. Microbiol.* 47:1295–1298.

126. Livingston, A. G. 1993. A novel membrane bioreactor for detoxifying industrial wastewater: II. Biodegradation of 3-chloronitrobenzene in an industrially produced wastewater. *Biotechnol. Bioeng.* 41:927–936.

127. Livingston, A. G. and A. Willacy. 1991. Degradation of 3,4-dichloroaniline in synthetic and industrially produced wastewaters by mixed cultures freely suspended and immobilized in a packed-bed reactor. *Appl. Microbiol. Biotechnol.* 35:551–557.

128. Lu, M.-F. 1997. Use of nitrobenzene by a coryneform bacteria as a nitrogen source, abstr. Q-349, p. 513, Abstracts, 97th General Meeting of the American Society for Microbiology. American Society for Microbiology, Washington, D.C.

129. Luthy, R. G., G. R. Aiken, M. L. Brusseau, S. D. Cunningham, P. M. Gschwend, J. J. Pignatello, M. Reinhard, S. J. Traina, W. J. Weber, Jr., and J. C. Westall. 1997. Sequestration of hydrophobic organic contaminants by geosorbents. *Environ. Sci. Technol.* 31:3341–3347.

130. Lyons, C. D., S. Katz, and R. Bartha. 1984. Mechanisms and pathways of aniline elimination from aquatic environments. *Appl. Environ. Microbiol.* 48:491–496.

131. Manning, J. F., R. Boopathy, and E. R. Breyfogle. 1996. Field demonstration of slurry reactor biotreatment of explosives-contaminated soils SFIM-AEC-ET-CR-96178. U.S. Army Environmental Center, Aberdeen Proving Ground, MD.

132. Mason, R. P. and P. D. Josephy. 1985. Free radical mechanism of nitroreductase, p. 121–140. In D. E. Rickert (ed.), *Toxicity of Nitroaromatic Compounds*. Hemisphere Publishing Corporation, Washington, D.C.

133. McCormick, N. G., J. H. Cornell, and A. M. Kaplan. 1978. Identification of biotransformation products from 2,4-dinitrotoluene. *Appl. Environ. Microbiol.* 35:945–948.

134. McCormick, N. G., F. F. Feeherry, and H. S. Levinson. 1976. Microbial transformation of 2,4,6-trinitrotoluene and other nitroaromatic compounds. *Appl. Environ. Microbiol.* 31:949–958.

135. Meulenberg, R. and J. A. M. de Bont. 1995. Microbial production of catechols from nitroaromatic compounds, p. 37–52. In J. C. Spain (ed.), *Biodegradation of Nitroaromatic Compounds*. Plenum Publishing Corp., New York.

136. Meulenberg, R., M. Pepi, and J. A. M. de Bont. 1996. Degradation of 3-nitrophenol by *Pseudomonas putida* B2 occurs via 1,2,4-benzenetriol. *Biodegradation.* 7:303–311.

137. Michán, C., A. Delgado, A. Haïdour, G. Lucchesi, and J. L. Ramos. 1997. In vivo construction of a hybrid pathway for metabolism of 4-nitrotoluene in *Pseudomonas fluorescens. J. Bacteriol.* 63:3036–3038.

138. Mironov, A. D., V. Y. Krest'yaninov, V. I. Korzhenevich, I. Y. Evtushenko, and A. L. Barkovskii. 1992. Degradation of 2-nitrobenzoic acid and other aromatic compounds by a *Pseudomonas pseudoalcaligenes* strain. *Appl. Biochem. Microbiol.* 27:433–438.

139. Mitchell, W. R. and W. H. Dennis, Jr. 1982. Biodegradation of 1,3-dinitrobenzene. *J. Environ. Sci. Health A.* A17:837–853.

140. Moore, F. W. 1949. The utilization of pyridine by micro-organisms. *J. Gen. Microbiol.* 3:143–147.

141. Nadeau, L., Z. He, and J. C. Spain. Unpublished results.

142. Nadeau, L. J. and J. C. Spain. 1995. Bacterial degradation of *m*-nitrobenzoic acid. *Appl. Environ. Microbiol.* 61:840–843.

143. Nishino, S. F., G. Paoli, and J. C. Spain. 2000. Aerobic biodegradation of dinitrotoluenes and pathway for bacterial degradation of 2,6-dinitrotoluene. *Appl. Environ. Microbiol.* 66:2139–2147.

144. Nishino, S. F. and J. C. Spain. 1993. Degradation of nitrobenzene by a *Pseudomonas pseudoalcaligenes. Appl. Environ. Microbiol.* 59:2520–2525.

145. Nishino, S. F. and J. C. Spain. 1995. The effect of 2,6-dinitrotoluene on 2,4-dinitrotoluene degradation, abstr. PW253, p. 277, Abstracts, Second SETAC World Congress. Society of Environmental Toxicology and Chemistry, Pensacola, FL.

146. Nishino, S. F. and J. C. Spain. 1992. Initial steps in the bacterial degradation of 1,3-dinitrobenzene, abstr. Q-135, p. 358, Abstr. Annu. Meet. Am. Soc. Microbiol. 1992. American Society for Microbiology, Washington, D.C.

147. Nishino, S. F. and J. C. Spain. 1995. Oxidative pathway for the biodegradation of nitrobenzene by *Comamonas* sp. strain JS765. *Appl. Environ. Microbiol.* 61:2308–2313.

148. Nishino, S. F. and J. C. Spain. unpublished results.

149. Nishino, S. F., J. C. Spain, H. Lenke, and H.-J. Knackmuss. 1999. Mineralization of 2,4- and 2,6-dinitrotoluene in soil slurries. *Environ. Sci. Technol.* 33:1060–1064.

150. Noguera, D. R. and D. L. Freedman. 1996. Reduction and acetylation of 2,4-dinitrotoluene by a *Pseudomonas aeruginosa* strain. *Appl. Environ. Microbiol.* 62:2257–2263.

151. Oh, Y.-S. and R. Bartha. 1997. Removal of nitrobenzene vapors by a trickling air biofilter. *J. Ind. Microbiol. Biotechnol.* 18:293–296.

152. Organization for Economic Co-Operation and Development. 1999. OECD high production volume chemicals programme — Phase 2. SIDS initial assessment report. 2,4-dinitrotoluene. Organization for Economic Co-Operation and Development. http://www.chem.unep.ch/irptc/sids/volume4/partI/dinitrotoluene/oecd_rep.html.

153. Ortega-Calvo, J.-J. 1999. Biodegradation of sorbed 2,4-dinitrotoluene in a clay-rich, aggregated porous medium. *Environ. Sci. Technol.* 33:3737–3742.

154. Parales, J. V., A. Kumar, R. E. Parales, and D. T. Gibson. 1996. Cloning and sequencing of the genes encoding 2-nitrotoluene dioxygenase from *Pseudomonas* sp. JS42. *Gene.* 181:57–61.

155. Parales, J. V., R. E. Parales, S. M. Resnick, and D. T. Gibson. 1998. Enzyme specificity of 2-nitrotoluene 2,3-dioxygenase from *Pseudomonas* sp. strain JS42 is determined by the C-terminal region of the α subunit of the oxygenase component. *J. Bacteriol.* 180:1194–1199.

156. Parales, R. E., M. D. Emig, N. A. Lynch, and D. T. Gibson. 1998. Substrate specificities of hybrid naphthalene and 2,4-dinitrotoluene dioxygenase enzyme systems. *J. Bacteriol.* 180:2337–2344.

157. Parales, R. E., T. A. Ontl, and D. T. Gibson. 1997. Cloning and sequence analysis of a catechol 2,3-dioxygenase gene from the nitrobenzene-degrading strain *Comamonas* sp. JS765. *J. Ind. Microbiol. Biotechnol.* 19:385–391.

158. Park, H.-S., S.-J. Lim, Y. K. Chang, A. G. Livingston, and H.-S. Kim. 1999. Degradation of chloronitrobenzenes by a coculture of *Pseudomonas putida* and a *Rhodococcus* sp. *Appl. Environ. Microbiol.* 65:1083–1091.

159. Pennington, J. C. 1999. Explosives, p. 85–109. In W. C. Anderson, R. C. Loehr, and B. P. Smith (ed.), *Environmental Availability of Chlorinated Organics, Explosives, and Metals in Soils.* American Academy of Environmental Engineers, Annapolis, MD.

160. Peres, C. M., H. Naveau, and S. N. Agathos. 1998. Biodegradation of nitrobenzene by its simultaneous reduction into aniline and mineralization of the aniline formed. *Appl. Microbiol. Biotechnol.* 49:343–349.

161. Peres, C. M., B. van Aken, H. Naveau, and S. N. Agathos. 1999. Continuous degradation of mixtures of 4-nitrobenzoate and 4-aminobenzoate by immobilized cells of *Burkholderia cepacia* strain PB4. *Appl. Microbiol. Biotechnol.* 52:440–445.

162. Perkins, R. E., J. S. Rajan, and F. S. Sariaslani. December 26, 1995. U.S. Patent 5,478,743.

163. Pesce, S. F. and D. A. Wunderlin. 1997. Biodegradation of 2,4- and 2,6-diaminotoluene by acclimated bacteria. *Water Res.* 31:1601–1608.

164. Pitter, P. 1976. Determination of biological degradability of organic substances. *Water Res.* 10:231–235.

165. Popp, J. A. and T. B. Leonard. 1985. The hepatocarcinogenicity of dinitrotoluenes, p. 53–60. In D. E. Rickert (ed.), *Toxicity of Nitroaromatic Compounds.* Hemisphere Publishing Corp., Washington, D.C.

166. Prieto, M. A., E. Díaz, and J. L. Garciá. 1996. Molecular characterization of the 4-hydroxyphenylacetate catabolic pathway of *Escherichia coli* W: engineering a mobile aromatic degradative cluster. *J. Bacteriol.* 178:111–120.

167. Rajan, J., K. Valli, R. E. Perkins, F. S. Sariaslani, S. M. Barns, A.-L. Reysenbach, S. Rehm, M. Ehringer, and N. R. Pace. 1996. Mineralization of 2,4,6-trinitrophenol (picric acid): characterization and phylogenetic identification of microbial strains. *J. Ind. Microbiol.* 16:319–324.

168. Rajan, J. S. and F. S. Sariaslani. August 6, 1996 1996. U.S. Patent 5,543,324.

169. Rani, L. and D. Lalithakumari. 1994. Degradation of methyl parathion by *Pseudomonas putida. Can. J. Microbiol.* 40:1000–1006.

170. Ray, P., M. Ait Oubelli, and C. Löser. 1999. Aerobic 4-nitrophenol degradation by microorganisms fixed in a continuously working aerated solid-bed reactor. *Appl. Microbiol. Biotechnol.* 51:284–290.
171. Raymond, D. G. M. and M. Alexander. 1971. Microbial metabolism and cometabolism of nitrophenols. *Pestic. Biochem. Physiol.* 1:123–130.
172. Reardon, K. F. and J. C. Spain. 1993. Immobilized cell bioreactor for 2,4-dinitrotoluene degradation, Abstr. Biot86. Abstracts, 205th ACS National Meeting. American Chemical Society, Washinton, D.C.
173. Rhys-Williams, W., S. C. Taylor, and P. A. Williams. 1993. A novel pathway for the catabolism of 4-nitrotoluene by *Pseudomonas. J. Gen. Microbiol.* 139:1967–1972.
174. Rieger, P.-G. and H.-J. Knackmuss. 1995. Basic knowledge and perspectives on biodegradation of 2,4,6-trinitrotoluene and related nitroaromatic compounds in contaminated soil, p. 1–18. In J. C. Spain (ed.), *Biodegradation of Nitroaromatic Compounds*. Plenum Publishing Corp., New York.
175. Rieger, P.-G., V. Sinnwell, A. Preuß, W. Francke, and H.-J. Knackmuss. 1999. Hydride-Meisenheimer complex formation and protonation as key reactions of 2,4,6-trinitrophenol biodegradation by *Rhodococcus erythropolis. J. Bacteriol.* 181:1189–1195.
176. Roper, D. I., T. Fawcett, and R. A. Cooper. 1993. The *Escherichia coli* C homoprotocatechuate degradative operon: *hpc* gene order, direction of transcription and control of expression. *Mol. Gen. Genet.* 237:241–250.
177. Schäfer, A., H. Harms, and A. J. B. Zehnder. 1996. Biodegradation of 4-nitroanisole by two *Rhodococcus* spp. *Biodegradation.* 7:249–255.
178. Schenzle, A., H. Lenke, P. Fischer, P. A. Williams, and H.-J. Knackmuss. 1997. Catabolism of 3-nitrophenol by *Ralstonia eutropha* JMP 134. *Appl. Environ. Microbiol.* 63:1421–1427.
179. Schenzle, A., H. Lenke, J. C. Spain, and H.-J. Knackmuss. 1999. Chemoselective nitro group reduction and reductive dechlorination initiate degradation of 2-chloro-5-nitrophenol by *Ralstonia eutropha* JMP134. *Appl. Environ. Microbiol.* 65:2317–2323.
180. Schenzle, A., H. Lenke, J. C. Spain, and H.-J. Knackmuss. 1999. 3-Hydroxylaminophenol mutase from *Ralstonia eutropha* JMP134 catalyzes a Bamberger rearrangement. *J. Bacteriol.* 181:1444–1450.
181. Schmidt, S. K. and M. J. Gier. 1989. Dynamics of microbial populations in soil: indigenous microorganisms degrading 2,4-dinitrophenol. *Microb. Ecol.* 18:285–296.
182. Schnell, S. and B. Schink. 1991. Anaerobic aniline degradation via reductive deamination of 4-aminobenzoyl-CoA in *Desulfobacterium anilini. Arch. Microbiol.* 155:183–190.
183. Serdar, C. M., D. T. Gibson, D. M. Munnecke, and J. H. Lancaster. 1982. Plasmid involvement in parathion hydrolysis by *Pseudomonas diminuta. Appl. Environ. Microbiol.* 44:246–249.
184. Shine, H. J. 1967. The rearrangement of phenylhydroxylamines, p. 182–190, *Aromatic Rearrangements*. Elsevier, Amsterdam.
185. Smets, B. F., R. G. Riefler, U. Lendenmann, and J. C. Spain. 1999. Kinetic analysis of simultaneous 2,4-dinitrotoluene (DNT) and 2,6-DNT biodegradation in an aerobic fluidized-bed biofilm reactor. *Biotechnol. Bioeng.* 63:642–653.
186. Somerville, C. C., S. F. Nishino, and J. C. Spain. 1995. Purification and characterization of nitrobenzene nitroreductase from *Pseudomonas pseudoalcaligenes* JS45. *J. Bacteriol.* 177:3837–3842.

187. Spain, J. 1997. Synthetic chemicals with potential for natural attenuation. *Bioremed. J.* 1:1–9.
188. Spain, J. C. (ed.). 1995. *Biodegradation of Nitroaromatic Compounds*. Plenum Publishing Corp., New York.
189. Spain, J. C. and D. T. Gibson. 1991. Pathway for biodegradation of *p*-nitrophenol in a *Moraxella* sp. *Appl. Environ. Microbiol.* 57:812–819.
190. Spain, J. C., S. F. Nishino, M. R. Green, J. E. Forbert, N. A. Nogalski, R. Unterman, W. M. Riznychok, S. E. Thompson, P. M. Sleeper, and M. A. Boxwell. 1999. Field demonstration of FBR for treatment of nitrotoluenes in groundwater, p. 7–14. In B. C. Alleman and A. Leeson (eds.), *Bioremediation of Nitroaromatic and Haloaromatic Compounds*. Battelle Press, Columbus, OH.
191. Spain, J. C., O. Wyss, and D. T. Gibson. 1979. Enzymatic oxidation of *p*-nitrophenol. *Biochem. Biophys. Res. Commun.* 88:634–641.
192. Spanggord, R. J., J. C. Spain, S. F. Nishino, and K. E. Mortelmans. 1991. Biodegradation of 2,4-dinitrotoluene by a *Pseudomonas* sp. *Appl. Environ. Microbiol.* 57:3200–3205.
193. Spence, E. L., M. Kawamukai, J. Sanvoisin, H. Braven, and T. D. H. Bugg. 1996. Catechol dioxygenases from *Escherichia coli* (MhpB) and *Alcaligenes eutrophus* (MpcI): sequence analysis and biochemical properties of a third family of extradiol dioxygenases. *J. Bacteriol.* 178:5249–5256.
194. Spiess, T., F. Desiere, P. Fischer, J. C. Spain, H.-J. Knackmuss, and H. Lenke. 1998. A new 4-nitrotoluene degradation pathway in a *Mycobacterium* strain. *Appl. Environ. Microbiol.* 64:446–452.
195. Stone & Webster Environmental Technology & Services. 1998. Draft alternative feasibility study Propellant Burning Ground and Deterrent Burning Ground, Waste Pits, Subsurface Soil, Badger Army Ammunition Plant, Baraboo, Wisconsin. Draft. U.S. Army Corps of Engineers, Omaha District, Omaha, NE.
196. Struijs, J. and J. Stoltenkamp. 1986. Ultimate biodegradation of 2-, 3- and 4-nitrotoluene. *Sci. Total. Environ.* 57:161–170.
197. Sudhakar-Barik, R. Siddaramappa, P. A. Wahid, and N. Sethunathan. 1978. Conversion of *p*-nitrophenol to 4-nitrocatechol by a *Pseudomonas* sp. *Antonie van Leeuwenhoek.* 44:171–176.
198. Suen, W.-C., B. E. Haigler, and J. C. Spain. 1996. 2,4-Dinitrotoluene dioxygenase from *Burkholderia* sp. strain DNT: similarity to naphthalene dioxygenase. *J. Bacteriol.* 178:4926–4934.
199. Suen, W.-C. and J. C. Spain. 1993. Cloning and characterization of *Pseudomonas* sp. strain DNT genes for 2,4-dinitrotoluene degradation. *J. Bacteriol.* 175:1831–1837.
200. Swindoll, C. M., R. E. Perkins, J. T. Gannon, M. Holmes, and G. A. Fisher. 1993. Assessment of bioremediation of a contaminated wetland, p. 163–169. In R. E. Hinchee, J. T. Wilson, and D. C. Downey (eds.), *Intrinsic Bioremediation*. Battelle Press, Columbus, OH.
201. Takenaka, S., S. Murakami, R. Shinke, and K. Aoki. 1998. Metabolism of 2-aminophenol by *Pseudomonas* sp. AP-3: modified *meta*-cleavage pathway. *Arch. Microbiol.* 170:132–137.
202. Takenaka, S., S. Murakami, R. Shinke, K. Hatakeyama, H. Yukawa, and K. Aoki. 1997. Novel genes encoding 2-aminophenol 1,6-dioxygenase from *Pseudomonas* species AP-3 growing on 2-aminophenol and catalytic properties of the purified enzyme. *J. Biol. Chem.* 272:14727–14732.

203. Takeo, M., T. Nagayama, K. Takatani, Y. Maeda, and M. Nakaoka. 1997. Mineralization and desulfonation of 3-nitrobenzenesulfonic acid by *Alcaligenes* sp. GA-1. *J. Ferm. Bioeng.* 83:505–509.

204. Tewfik, M. S. and W. C. Evans. 1966. The metabolism of 3,5-dinitro-o-cresol (DNOC) by soil microorganisms. *Biochem. J.* 99:31–32.

205. Tokiwa, H. and Y. Ohnishi. 1986. Mutagenicity and carcinogenicity of nitroarenes and their sources in the environment. *CRC Crit. Rev. Toxicol.* 17:23–60.

206. Toxics Release Inventory Reports. 1999. TOXNET Toxics Release Inventory. Toxicology Data Network. http://sis.nlm.nih.gov/sis1.

207. U.S. Environmental Protection Agency. 1999. Use of monitored natural attenuation at Superfund, RCRA corrective action, and underground storage tank sites. Final Directive OSWER Directive 9200.4–17P. U.S. Environmental Protection Agency Office of Solid Waste and Emergency Response, Washington, D.C.

208. Vorbeck, C., H. Lenke, P. Fischer, and H.-J. Knackmuss. 1994. Identification of a hydride-Meisenheimer complex as a metabolite of 2,4,6-trinitrotoluene by a *Mycobacterium* strain. *J. Bacteriol.* 176:932–934.

209. Vorbeck, C., H. Lenke, P. Fischer, J. C. Spain, and H.-J. Knackmuss. 1998. Initial reductive reactions in aerobic microbial metabolism of 2,4,6-trinitrotoluene. *Appl. Environ. Microbiol.* 64:246–252.

210. Wierenga, R. K., P. Terpstra, and W. G. Hol. 1986. Prediction of the occurrence of the ADP-binding beta alpha beta-fold in proteins, using an amino acid sequence fingerprint. *J. Mol. Biol.* 187:101–107.

211. Williams, R., K. Shuttle, J. Kunkler, E. Madsen, and S. Hooper. 1997. Intrinsic bioremediation in a solvent-contaminated alluvial groundwater. *J. Ind. Microbiol. Biotechnol.* 18:177–188.

212. Witte, C.-P., R. Blasco, and F. Castillo. 1998. Microbial photodegradation of aminoarenes. Metabolism of 2-amino-4-nitrophenol by *Rhodobacter capsulatus*. *Appl. Biochem. Biotechnol.* 69:191–202.

213. Wyndham, R. C. 1986. Evolved aniline catabolism in *Acinetobacter calcoaceticus* during continuous culture of river water. *Appl. Environ. Microbiol.* 51:781–789.

214. Yabannavar, A. V. and G. J. Zylstra. 1995. Cloning and characterization of the genes for *p*-nitrobenzoate degradation from *Pseudomonas pickettii* YH105. *Appl. Environ. Microbiol.* 61:4284–4290.

215. Zeyer, J. and P. C. Kearney. 1984. Degradation of *o*-nitrophenol and *m*-nitrophenol by a *Pseudomonas putida*. *J. Agric. Food Chem.* 32:238–242.

216. Zhang, C., S. F. Nishino, J. C. Spain, and J. B. Hughes. Sequential aerobic biodegradation of 2,4- and 2,6-dinitrotoluene in a pilot-scale slurry reactor system. *Environ. Sci. Technol.* Submitted.

CHAPTER 3

Molecular Biology of Nitroarene Degradation

Rebecca E. Parales

CONTENTS

3.1 INTRODUCTION

Nitroaromatic compounds are rarely produced by biological processes. Consequently, the vast majority of nitroaromatic compounds present in the environment are solely the result of industrial activities over the last century. Nitroaromatic compounds are the starting materials for the synthesis of all aromatic compounds with nitrogen attached to the aromatic ring.[18] Nitroarenes such as nitrobenzene and mono- and dinitrotoluenes are produced on a large scale for use as solvents and in the production of dyes, pigments, explosives, pesticides, herbicides, and polymers.[18] Total annual production of nitrobenzene was >1 billion pounds in 1980,[18] and in 1997 alone, total reported environmental releases of nitrobenzene amounted to over 700,000 pounds. The majority of this waste was injected into underground wells, but significant amounts were released into the air and surface waters.[39] Trinitrotoluene (TNT) is produced at government facilities, and waste from these sites have been found to contain dinitrotoluene isomers.[50] The mass production and use of TNT in World Wars I and II caused extensive soil and water contamination by TNT and dinitrotoluenes.[56,62] The release of nitroaromatic compounds into the environment either intentionally (pesticides and herbicides) or through accidental spills has caused significant environmental damage. As a result, nitrobenzene and 2,4- and 2,6-dinitrotoluene have been included on the U.S. Environmental Protection Agency's list of priority pollutants.[26]

Nitroarenes are toxic and, in some cases, mutagenic and/or carcinogenic.[4,27,32,63,76] Toxic effects can result from exposure by inhalation, ingestion, or absorption through the skin. Dinitrotoluenes are suspected human carcinogens and confirmed animal carcinogens;[72] TNT is considered to be a possible human carcinogen. In addition, these compounds are volatile, flammable, and explosive.

The combination of the electron-withdrawing character of nitro groups and the stability of the benzene ring make nitroaromatic compounds extremely stable and resistant to oxidative attack.[56,62] Since these xenobiotic compounds have been present in the environment for a relatively short time, bacteria have had little time to evolve pathways for their degradation. Although numerous reports of reductive transformations of nitroaromatic compounds have appeared in the literature,[3,51] it has only been within the last decade that bacteria capable of mineralizing compounds such as nitrobenzene and nitrotoluenes have been isolated.[61,62] Such strains have been isolated from sites with a history of exposure to nitroaromatic compounds and do not seem to be widely distributed in the environment.[61] Most isolates grow on a very narrow range of nitroaromatic compounds, typically degrading only the compound on which they were isolated. This chapter will focus primarily on the genetics and molecular biology of the degradation of mono- and dinitroarene compounds. Further details on the physiology of nitroarene-degrading organisms are presented in Chapter 2.

3.2 DIOXYGENASE-CATALYZED NITROARENE DEGRADATION PATHWAYS

Bacterial strains capable of degrading nitrobenzene, 2-nitrotoluene (2NT), and 2,4-dinitrotoluene (2,4-DNT) have been isolated and characterized. In each case,

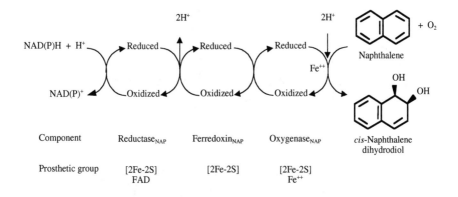

Figure 3.1 Reaction catalyzed by the three-component naphthalene dioxygenase system. Prosthetic groups present in each protein component are indicated.

degradation is initiated by multicomponent dioxygenase systems that are very similar to the well-studied naphthalene dioxygenase system (Figure 3.1). Each enzyme consists of an iron-sulfur flavoprotein reductase and an iron-sulfur ferredoxin that transfer electrons from NAD(P)H to the catalytic oxygenase component. Each oxygenase consists of large (α) and small (β) subunits. The crystal structure of the oxygenase component of naphthalene dioxygenase was recently solved,[25] and provides an excellent model for predicting the structures of the related nitroarene dioxygenases. Based on the structure,[25] it is expected that the native conformation of each nitroarene dioxygenase is $\alpha_3\beta_3$.

3.2.1 Nitrobenzene Degradation

Comamonas sp. JS765 was isolated from an industrial waste-treatment plant in New Jersey after selective enrichment with nitrobenzene as the sole source of carbon and nitrogen.[37] Nitrobenzene is initially converted to catechol and nitrite by nitrobenzene 1,2-dioxygenase. Catechol is degraded by a *meta*-cleavage pathway (Figure 3.2). The genes encoding nitrobenzene dioxygenase have recently been cloned and sequenced.[29] An unlinked catechol 2,3-dioxygenase gene from JS765 was also cloned and sequenced.[47] Although nitrobenzene was the only nitroarene growth substrate identified for JS765, preliminary studies indicate that the nitrobenzene dioxygenase has a broad substrate range and catalyzes the oxidation of several compounds including nitrophenols and nitrotoluenes.[37]

3.2.2 2-Nitrotoluene Degradation

Pseudomonas sp. JS42 was isolated by selective enrichment of a soil-groundwater sample from a nitrobenzene production facility in Mississippi with 2NT as the carbon and nitrogen source.[17] 2NT is converted by 2NT 2,3-dioxygenase to 3-methylcatechol and nitrite. 3-Methylcatechol is a substrate for *meta*-cleavage (Figure 3.2). 2NT was originally reported to be the only nitroaromatic compound

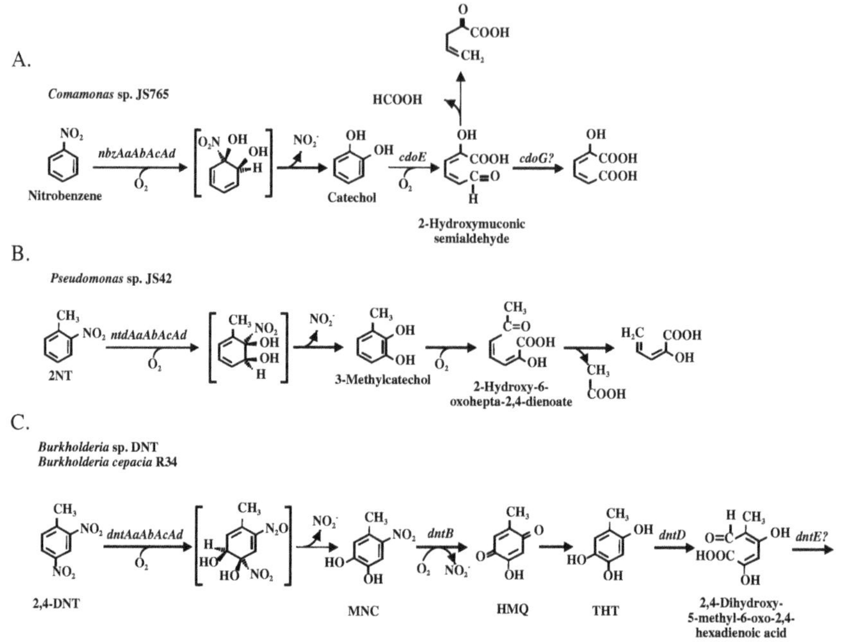

Figure 3.2 Pathways for the degradation of nitroarene compounds. Enzymatic steps for which genes have been cloned are indicated by gene designations above the arrows. (A) Nitrobenzene pathway in *Comamonas* sp. strain JS765.[37] *nbzAaAbAcAd* encode polypeptides of the nitrobenzene 1,2-dioxygenase system.[29] *cdoE* and *cdoG* encode catechol 2,3-dioxygenase[47] and 2-hydroxymuconic semialdehyde dehydrogenase,[19] respectively, which may be involved in nitrobenzene degradation. 2-Hydroxymuconic semialdehyde hydrolase (HMSH) activity has been demonstrated in JS765,[37] but no gene with this function has been cloned. (B) 2NT degradation in *Pseudomonas* sp. strain JS42.[17] The 2NT dioxygenase system is encoded by *ntdAaAbAcAd*.[41] (C) 2,4-DNT degradation in *Burkholderia* sp. strain DNT and *B. cepacia* R34.[64] The 2,4-DNT dioxygenase system is encoded by *dntAaAbAcAd*.[66] 4-Methyl-5-nitrocatechol (MNC) oxygenase is encoded by *dntB*.[16] This enzyme converts MNC to 2-hydroxy-5-methylquinone (HMQ). An NADP-dependent quinone reductase (whose gene has not yet been cloned) converts HMQ to 2,4,5-trihydroxytoluene (THT) (13). THT oxygenase (encoded by *dntD*) cleaves the aromatic ring at the *meta*-position.[12,21] *dntE* encodes a semialdehyde dehydrogenase that may be responsible for catalyzing the next step in the pathway.[22]

that is a growth substrate for the strain. However, we have observed weak growth with nitrobenzene.[46] Genes encoding 2NT dioxygenase were cloned and sequenced,[41] and the proteins have been purified and characterized.[1,48]

3.2.3 2,4-Dinitrotoluene Degradation

The 2,4-DNT-degrading strains, *Burkholderia* sp. strain DNT and *B. cepacia* R34, were isolated from water samples obtained near the Volunteer Army Ammunition Plant in Tennessee and the Radford Army Ammunition Plant in West Virginia, respectively.[34,35,64] 2,4-DNT is oxidized to 4-methyl-5-nitrocatechol (MNC) by 2,4-DNT dioxygenase[64] (Figure 3.2). A monooxygenase and a quinone reductase convert MNC sequentially to 2-hydroxy-5-methylquinone and 2,4,5-trihydroxytoluene (THT).[13] Ring cleavage catalyzed by THT oxygenase occurs at the *meta*-position.[12] Genes encoding the 2,4-DNT dioxygenase, MNC monooxygenase, and THT oxygenase have been cloned from both strains.[14,67] The MNC monooxygenase and THT oxygenase from strain DNT have been purified and characterized.[12,16] 2,4-DNT-degrading strains are not capable of growth with 2,6-DNT.[35]

3.2.4   2,6-Dinitrotoluene Degradation

Bacterial strains that grow on 2,6-DNT were isolated from sites contaminated with DNT. These strains use a dioxygenation reaction to carry out the initial step in the degradation of 2,6-DNT.[35,36,38] The isolates grow slowly following long acclimation periods, and none of the strains grow with 2,4-DNT. All of the strains convert 2,6-DNT to 3-methyl-4-nitrocatechol with the release of nitrite. 3-Methyl-4-nitrocatechol undergoes *meta*-cleavage to form 2-hydroxy-5-nitro-6-oxohepta-2,4-dienoic acid.[35] Little information is available about the genes encoding 2,6-DNT degradation. However, genomic DNA from several 2,6-DNT-degrading strains was subjected to polymerase chain reaction (PCR) amplification with primers designed from conserved regions of dioxygenase α subunit genes. Sequence analysis of amplified fragments suggests that dioxygenase genes present in these strains are very closely related to dioxygenase α subunit genes from 2NT- and 2,4-DNT-degrading isolates.[23] Due to the current lack of information regarding the molecular biology of 2,6-DNT degradation, the pathway will not be discussed further.

3.3 MOLECULAR BIOLOGY OF DIOXYGENASE PATHWAYS

3.3.1   Dioxygenase Genes

As mentioned in the previous section, genes encoding nitrobenzene, 2NT, and 2,4-DNT dioxygenases have been cloned and sequenced.[14,29,41,66] Genes for 2,4-DNT degradation (*dnt* genes) in *Burkholderia* sp. strain DNT are located on a large (180-kb) plasmid, pJS1.[67] Enzymes of the DNT degradation pathway in *B. cepacia* R34 are encoded on a 216-kb plasmid, pJS311.[21] The location of the 2NT (*ntd*) and nitrobenzene (*nbz*) pathway genes in *Pseudomonas* sp. strain JS42 and *Comamonas* sp. strain JS765 have not been determined. However, evidence for the presence of one or more plasmids together with the observation that JS765 can easily lose the ability to degrade nitrobenzene suggest that the nitrobenzene pathway may be plasmid encoded.[46]

The gene orders of the nitrobenzene, 2NT, and DNT dioxygenase gene clusters from all four nitroarene-degrading strains show a striking similarity to each other and to the naphthalene dioxygenase (*nag; nah*) genes from *Pseudomonas* sp. U2 and *C. testosteroni* GZ42 (Figure 3.3).[8,10] In each gene cluster, the reductase gene is separated from the ferredoxin gene by either one or two open reading frames

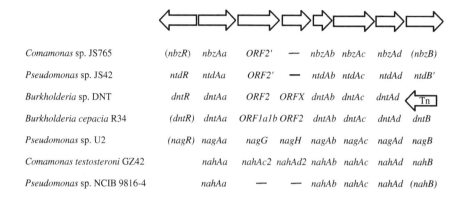

Comamonas sp. JS765	*(nbzR)*	*nbzAa*	*ORF2'*	—	*nbzAb*	*nbzAc*	*nbzAd*	*(nbzB)*
Pseudomonas sp. JS42	*ntdR*	*ntdAa*	*ORF2'*	—	*ntdAb*	*ntdAc*	*ntdAd*	*ntdB'*
Burkholderia sp. DNT	*dntR*	*dntAa*	*ORF2*	*ORFX*	*dntAb*	*dntAc*	*dntAd*	Tn
Burkholderia cepacia R34	*(dntR)*	*dntAa*	*ORF1a1b* *ORF2*		*dntAb*	*dntAc*	*dntAd*	*dntB*
Pseudomonas sp. U2	*(nagR)*	*nagAa*	*nagG*	*nagH*	*nagAb*	*nagAc*	*nagAd*	*nagB*
Comamonas testosteroni GZ42		*nahAa*	*nahAc2 nahAd2*		*nahAb*	*nahAc*	*nahAd*	*nahB*
Pseudomonas sp. NCIB 9816-4		*nahAa*	—	—	*nahAb*	*nahAc*	*nahAd*	*(nahB)*

Figure 3.3   Gene order in the dioxygenase gene clusters from nitroarene- and naphthalene-degrading bacteria. Gene order in the nitrobenzene, 2NT, and 2,4-DNT dioxygenase gene clusters (*nbz*, *ntd*, and *dnt* genes, respectively) are compared to the order of naphthalene dioxygenase (*nah* and *nag*) genes from three naphthalene-degrading strains. Parentheses indicate genes that have not been completely sequenced. (') denotes a truncated gene; Tn is a transposase gene; and (–) indicates that the gene is not present.

(ORFs) that are not required for dioxygenase activity. This gene order differs from that reported for all other naphthalene dioxygenase gene clusters, including those from *Pseudomonas* sp. NCIB 9816-4 and *P. putida* G7.[60] However, in all cases, the genes encoding the α and β subunits of the oxygenase are located downstream of the ferredoxin gene (Figure 3.3). Amino acid sequence identities of the reductase, ferredoxin, and α and β subunits from the nitroarene dioxygenases, two naphthalene dioxygenases, and the distantly related toluene dioxygenase from *P. putida* F1[81] are compared in Tables 3.1 to 3.4. It is especially clear in comparisons of reductases and β subunits (Tables 3.2 and 3.4) that the proteins from the nitroarene-degrading enzyme systems are more closely related to the Nag components than the Nah components. All Nbz, Ntd, Dnt, Nag, and Nah proteins are distantly related to those from the toluene dioxygenase (Tod) enzyme system (Tables 3.1 to 3.4).

3.3.2   The Function of Open Reading Frames

The ORFs present in nitroarene dioxygenase gene clusters (Figure 3.3) have homology with dioxygenase α and β subunit genes. The deduced amino acid sequences of products of ORF2 from JS765, ORF2 from JS42, ORF2 from strain DNT, and ORFs 1a-1b from strain R34 share 93% identity with NagG from *Pseudomonas* sp. strain U2. Each of these amino acid sequences shares approximately 30% identity with naphthalene dioxygenase α subunits. NagG also shares approximately 30% amino acid identity with the nitroarene and naphthalene dioxygenase α subunits (Table 3.1). ORF2 from JS765 and JS42 appear to be truncated forms of *nagG* and only encode approximately the first half of the protein. The products of ORFX from strain DNT and ORF2 from strain R34 are 98% identical

Table 3.1 Amino Acid Identities between Oxygenase α Subunits

	NagAc	NahAc	NbzAc	NtdAc	DntAc$_{DNT}$	DntAc$_{R34}$	TodC1	NagG
NagAc	100							
NahAc	90	100						
NbzAc	88	82	100					
NtdAc	90	84	95	100				
DntAc$_{DNT}$	88	80	87	88	100			
DntAc$_{R34}$	91	84	89	91	94	100		
TodC1	35	35	36	34	34	35	100	
NagG	29	28	29	28	29	27	31	100

Table 3.2 Amino Acid Identities between Oxygenase β Subunits

	NagAd	NahAd	NbzAd	NtdAd	DntAd$_{DNT}$	DntAd$_{R34}$	TodC2	NagH
NagAd	100							
NahAd	78	100						
NbzAd	92	78	100					
NtdAd	89	76	95	100				
DntAd$_{DNT}$	94	78	96	92	100			
DntAd$_{R34}$	93	78	95	93	99	100		
TodC2	24	22	26	25	24	24	100	
NagH	20	<16	18	17	17	18	25	100

Table 3.3 Amino Acid Identities between Ferredoxin Components

	NagAb	NahAb	NbzAb	NtdAb	DntAb$_{DNT}$	DntAb$_{R34}$	TodB
NagAb	100						
NahAb	79	100					
NbzAb	83	72	100				
NtdAb	83	72	100	100			
DntAb$_{DNT}$	76	66	79	79	100		
DntAb$_{R34}$	85	74	87	87	87	100	
TodB	36	35	36	36	36	35	100

Table 3.4 Amino Acid Identities between Reductase Components

	NagAa	NahAa	NbzAa	NtdAa	DntAa$_{DNT}$	DntAa$_{R34}$	TodA
NagAa	100						
NahAa	66	100					
NbzAa	99	67	100				
NtdAa	99	67	100	100			
DntAa$_{DNT}$	91	61	91	91	100		
DntAa$_{R34}$	99	67	99	99	91	100	
TodA	22	20	21	21	21	22	100

to NagH, which has low sequence similarity to dioxygenase β subunits (Table 3.2). A function for the ORFs present in the nitroarene dioxygenase gene clusters has not been identified. However, the genes (*nagG* and *nagH*) in *Pseudomonas* sp. U2 encode the hydroxylase component of a multicomponent salicylate 5-hydroxylase. The hydroxylase shares the naphthalene dioxygenase ferredoxin and reductase components (NagAa and NagAb) and converts salicylate to gentisate,[8] a different route than that in other naphthalene-degrading organisms for which genes have been cloned (Figure 3.4). Genes *nahAc2* and *nahAd2* from *C. testosteroni* GZ42 are nearly identical in sequence to *nagG* and *nagH*, suggesting that GZ42 may also use a gentisate pathway for the degradation of naphthalene.[8,80]

None of the *nagG*- or *nagH*-like ORFs present in the nitroarene-degrading strains appears to encode a functional oxygenase. No salicylate hydroxylase activity was detected in clones from strain DNT or R34.[22] In strain DNT, ORF2 appears to have two frameshift mutations that result in a short region of nonidentity with NagG and probably causes the loss of enzyme activity.[22] In the R34 gene cluster, there is a frameshift that results in a truncated ORF1a. A second partial ORF (ORF1b) starts at a nearby methionine codon, but it is unlikely to be translated since no obvious ribosome binding site is present.[22]

3.3.3 Other Pathway Genes

Genes (*nbzB* and *ntdB*) located downstream of the *nbzAd* and *ntdAd* genes are similar to the *cis*-naphthalene diol dehydrogenase-encoding genes *nahB* and *nagB*.[29,42] The role of these genes is unclear, since *cis*-diol dehydrogenases are not necessary for the degradation of either nitrobenzene or 2NT. NtdB has no dehydrogenase activity with *cis*-naphthalene dihydrodiol as a substrate, suggesting that *ntdB* probably does not encode a functional dehydrogenase. Also, the *ntdB* sequence goes out of frame

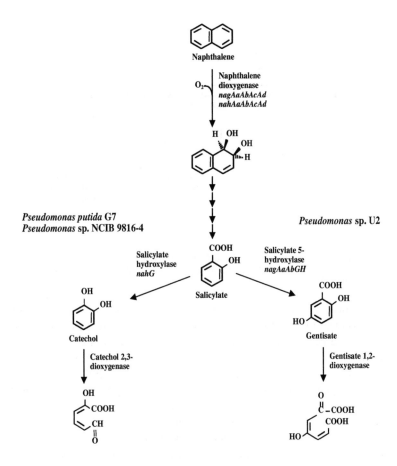

Figure 3.4 Degradation of naphthalene through the catechol *meta*-cleavage pathway and the gentisate pathway by different *Pseudomonas* strains.

and the C-terminus of the protein that would be formed has no similarity to a *cis*-diol dehydrogenase.[42] These data suggest that the nitroarene gene clusters evolved from a gene cluster for a pathway that requires a diol dehydrogenase reaction.

Genes encoding the MNC oxygenase (*dntB*) and THT oxygenase (*dntD*) from strain DNT are located on the same plasmid as the dioxygenase genes, but are not closely linked.[67] In contrast, several of the DNT pathway genes in strain R34 are located in a large gene cluster (Figure 3.5).[14] In strain R34, the MNC oxygenase gene, *dntB*, is located just downstream of the dioxygenase genes, and *dntD*, the gene encoding THT oxygenase, is located 15 kb downstream of *dntB* on the opposite strand of the DNA. In the region between *dntB* and *dntD* there is an ORF that appears to encode an intradiol dioxygenase (ORF3) with no known role in DNT degradation, an ORF (ORF4) that may encode a transposase, and several genes of unknown function (*dntFaFbG*). Just downstream of *dntD* is *dntE*, whose product is similar to hydroxymuconic semialdehyde dehydrogenases and may catalyze the oxidation of the THT ring cleavage product (Figure 3.2).[22] Interestingly, MNC oxygenases from the two strains are

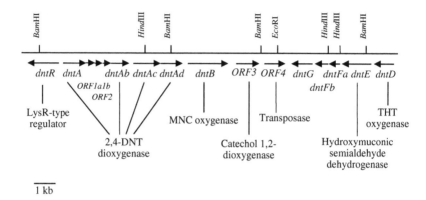

Figure 3.5 Map of the *B. cepacia* R34 2,4-DNT dioxygenase gene cluster. Known or putative functions of gene products are indicated and are described in more detail in the text.

only 53% identical in amino acid sequence.[22] Similarly, THT oxygenases are 60% identical.[21] It is clear that the two sets of genes encoding the proteins in strains DNT and R34 have diverged significantly, while the *dntA* genes have not.

A catechol 2,3-dioxygenase gene (*cdoE*) that may be required for the degradation of nitrobenzene has been cloned from JS765.[47] Associated with the catechol dioxygenase gene are genes encoding a LysR-type regulatory protein and a plant-type ferredoxin (*cdoR* and *cdoT*, respectively) (Figure 3.6). Further upstream are genes that appear to encode a multicomponent phenol hydroxylase (*orfLMNOP*).[45] Wild-type JS765 does not grow on phenol, but spontaneous phenol-degrading mutants of JS765 are easily isolated.[45] The presence of the phenol hydroxylase genes may partially explain this observation, but the mutation(s) that causes this phenotype has not been identified. Downstream of the catechol dioxygenase gene are the remaining *meta*-cleavage pathway genes, which are organized in an order that is unlike that of other *meta*-cleavage pathways.[19] The region between the catechol dioxygenase gene and the hydroxymuconic semialdehyde dehydrogenase (HMSD) gene (*cdoG*) contains several ORFs of unknown function (*cdoX1X2X3*), a possible *lysR*-type regulatory gene (*cdoR2*), and what appears to be the remnant of a hydroxymuconic semialdehyde hydrolase (HMSH) gene (*cdoFaFb*) which contains a frameshift that results in truncation of the gene (Figure 3.6). Since JS765 has HMSH activity[37] and the clone does not, there is a functional HMSH gene (and possibly another *meta*-cleavage pathway) encoded elsewhere on the JS765 genome. There is no firm evidence indicating that this gene cluster is required for nitrobenzene degradation. It may be required for phenol degradation, nitrobenzene degradation, or both.

3.3.4 Regulation of Dioxygenase Gene Clusters

An apparent regulatory gene that is transcribed divergently is located upstream of the reductase genes in the *nbz*, *ntd*, *dnt*, and *nag* gene clusters (Figure 3.3).

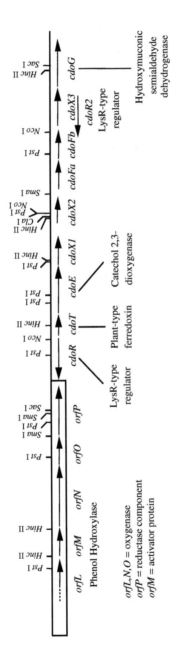

Figure 3.6 Map of the catechol dioxygenase gene cluster from *Comamonas* sp. strain JS765. Known or putative functions of gene products are indicated and are described in more detail in the text.

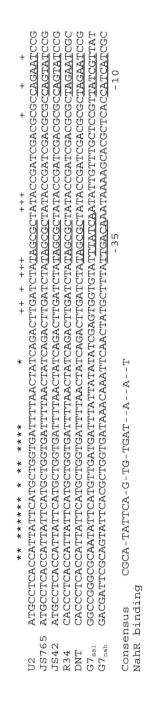

Figure 3.7 Sequence alignment at the naphthalene and nitroarene dioxygenase gene promoters. A consensus NahR binding sequence is shown below the alignment. Nucleotides in the NahR binding region that are conserved in all promoters are indicated by asterisks. Other conserved nucleotides in the region are indicated by plus signs. The −35 and −10 sites of the *sal* and *nah* promoters from *P. putida* G7 are underlined.[59] Possible nitroarene dioxygenase −35 and −10 sites that correspond to the G7 sites are also underlined. These differ from those suggested previously.[8,66]

The putative regulatory genes were identified by sequence similarity to the deduced amino acid sequence of NahR, the activator of the naphthalene pathway in *P. putida* G7.[58] The sequences of *ntdR* from JS42 and *dntR* from strain DNT were completed,[33] and the deduced amino acid sequences of the two proteins are 97% identical. NtdR and DntR are 61 and 62% identical to NahR from *P. putida* G7, respectively. The consensus binding sequence for NahR[59] was identified upstream of the *nbzAa*, *ntdAa*, *dntAa*, and *nagAa* genes.[8] A comparison of the sequences upstream of the reductase gene in each system is shown in Figure 3.7. No significant identity to the NahR binding consensus sequence is seen upstream of *dntB* in *Burkholderia* sp. strain DNT,[16] suggesting that this gene may not be regulated by DntR. To date there is no biochemical evidence available to support a role for these regulators in nitroarene degradation.

The synthesis of nitrobenzene dioxygenase in JS765 is induced when cultures are grown with nitrobenzene.[37] Similarly, enzymes in the 2,4-DNT degradation pathway in strains DNT and R34 are inducible during growth with 2,4-DNT.[13,22,64] In contrast, the *ntd* genes in JS42 are expressed constitutively.[17] At present, regulation in these systems is not understood. The *nbz* and *ntd* promoter regions are identical, and there are very few differences in the sequence of the *nbz*/*ntd* promoter region from that of the *dnt* gene cluster (Figure 3.7). In addition, the deduced amino acid sequences of *ntdR* and *dntR* are nearly identical. The inducer of the nitrobenzene genes is not known, and the only inducer identified for the *dnt* genes thus far is salicylate.[22] This result seems to indicate that the *dnt* genes are still controlled like those in the progenitor gene cluster (see Section 3.6.1) such as that in *Pseudomonas* sp. U2, which oxidizes naphthalene through gentisate, with salicylate serving as the inducing molecule.[8] However, this does not explain the induction of the *dnt* genes, since salicylate is not an intermediate in 2,4-DNT degradation. Lower pathway genes for the degradation of 2,4-DNT do not seem to be coordinately regulated, since pathway intermediates transiently accumulate during growth on 2,4-DNT.[64] Clearly, many questions regarding the regulation of the nitroarene degradation genes remain to be addressed.

3.4 SUBSTRATE SPECIFICITIES OF 2NT AND 2,4-DNT DIOXYGENASES

3.4.1 Substrate Specificity Is Controlled by the C-Terminal Portion of the Oxygenase α Subunit

The substrate specificities of 2NT and 2,4-DNT dioxygenases have been compared to that of the related enzyme naphthalene dioxygenase[43,66] and significant differences have been identified (Table 3.5). Construction of hybrid enzymes demonstrated that the β subunit does not play a significant role in determining the substrate specificity of naphthalene dioxygenase, 2NT dioxygenase, or DNT dioxygenase (Figure 3.8).[43,44] Hybrid enzymes containing the N-terminal portion of the 2NT dioxygenase α subunit and the C-terminal portion of the DNT dioxygenase α subunit have substrate specificities similar to DNT dioxygenase (Figure 3.8), indicating that the C-terminal half of the α subunit controls substrate specificity.[43]

Table 3.5 Substrate Specificity of Naphthalene, 2NT, and 2,4-DNT Dioxygenases

Substrate	NDO	2NTDO	DNTDO
Naphthalene	(dihydrodiol structure) >99% (+)-cis-(1R,2S)	(dihydrodiol structure) 70% (+)-cis-(1R,2S)	(dihydrodiol structure) 96% (+)-cis-(1R,2S)
Nitrobenzene	None Detected	(catechol structure, OH OH)	None Detected
2-Nitrotoluene	CH₂OH, NO₂	CH₃, OH OH — 90%	CH₂OH, NO₂ — 10%	CH₂OH, NO₂
3-Nitrotoluene	CH₂OH, NO₂	CH₃, OH OH — 59%	CH₃, OH OH — 27%	CH₂OH, NO₂ — 14%	CH₂OH, NO₂
4-Nitrotoluene	CH₂OH, NO₂	CH₂OH, NO₂ — 76%	CH₃, OH — 24%	CH₃, OH — 89%	CH₂OH, NO₂ — 11%
2,4-Dinitrotoluene	None Detected	None Detected	CH₃, NO₂, HO OH

3.4.2 Site-Directed Mutagenesis Studies

Genetic and biochemical analyses described in previous sections indicate that 2NT dioxygenase is closely related to naphthalene dioxygenase. Therefore, the crystal structure of naphthalene dioxygenase provides a useful model for the structure of 2NT dioxygenase. We have initiated site-directed mutagenesis studies to identify amino acids at the active site of 2NT dioxygenase that are critical in determining substrate specificity. Amino acids near the active site iron in the α subunit of naphthalene dioxygenase were identified,[25] and amino acid substitutions were made at the corresponding positions identified in amino acid sequence alignments in 2NT dioxygenase. These studies are at an early stage, but a few interesting results are

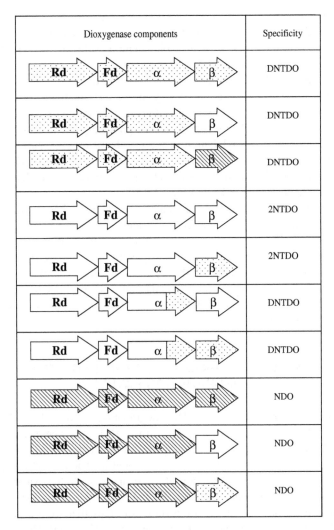

Dioxygenase components	Specificity
Rd Fd α β	DNTDO
Rd Fd α β	DNTDO
Rd Fd α β	DNTDO
Rd Fd α β	2NTDO
Rd Fd α β	2NTDO
Rd Fd α β	DNTDO
Rd Fd α β	DNTDO
Rd Fd α β	NDO
Rd Fd α β	NDO
Rd Fd α β	NDO

Figure 3.8 Schematic diagram of constructed hybrid dioxygenases and summary of their substrate specificities.[43,44] Open arrows represent the *ntd* genes from *Pseudomonas* sp. strain JS42, which encode the 2NT dioxygenase components. Shaded arrows represent *dnt* genes from *Burkholderia* sp. strain DNT, which encode 2,4-DNT dioxygenase components. Hatched arrows represent the *nah* genes from *Pseudomonas* sp. strain NCIB 9816–4, which encode the naphthalene dioxygenase components. Labels within the arrows indicate the gene product encoded: Rd, reductase; Fd, ferredoxin; α, oxygenase α subunit; β, oxygenase β subunit. The substrate specificity of each hybrid enzyme is indicated: 2NTDO, 2NT dioxygenase; DNTDO, 2,4-DNT dioxygenase; NDO, naphthalene dioxygenase.

worth mentioning. A valine present at position 260 in naphthalene dioxygenase corresponds to an asparagine in 2NT dioxygenase. When a valine was substituted at position 260 in 2NT dioxygenase, a significant change in the specificity of the enzyme resulted. The enzyme lost the ability to oxidize nitrobenzene, oxidized 2NT only at the methyl group, and formed 60% (−)-*cis*-(1*S*,2*R*)-naphthalene dihydrodiol from naphthalene.[40] This change made the enzyme specificity more like that of naphthalene dioxygenase with 2NT and nitrobenzene as substrates, but very different from naphthalene dioxygenase with naphthalene as substrate (see Table 3.5). 2,4-DNT dioxygenase also has a valine at this position. However, 2NT dioxygenase with the valine substitution did not convert 2,4-DNT to MNC, suggesting that other amino acids at the active site are necessary for the enzyme to catalyze this reaction. The position corresponding to 316 in naphthalene dioxygenase has a tryptophan in both naphthalene and 2NT dioxygenases and a phenylalanine in 2,4-DNT dioxygenase. Substitution of a phenylalanine in the α subunit of 2NT dioxygenase at this position resulted in no change in specificity with nitrobenzene, 2NT, or 2,4-DNT, and only a minor change in the stereochemistry of *cis*-naphthalene dihydrodiol formed from naphthalene [80% (+)-(1R,2S)].[40]

While the crystal structure of naphthalene dioxygenase has been useful in identifying which amino acids may affect substrate specificity, we have not been able to predict the specific effects of particular amino acid substitutions. Currently, efforts are being made to characterize the substrate specificity of nitrobenzene dioxygenase. With this information, together with the crystal structure of naphthalene dioxygenase and sequence alignments of the four related α subunits, we hope to identify residues that control the regio- and enantioselectivities of these enzymes. Future goals are to design enzymes with the desired specificities.

3.5 MOLECULAR BIOLOGY OF THE 4-NITROTOLUENE DEGRADATION PATHWAY

The pathway for 4-nitrotoluene (4NT) degradation in two isolates, *Pseudomonas* sp. strain TW3 and *Pseudomonas* sp. strain 4NT, has been characterized[15,55] (Figure 3.9). Unlike other oxidative pathways for the degradation of nitroarenes, the degradation of 4NT is not initiated by a dioxygenase. The methyl group of 4NT is initially oxidized in a series of reactions that are analogous to the oxidation of toluene by the TOL plasmid-encoded pathway.[15,55] Following the formation of 4-nitrobenzoate, the nitro group is reduced by a series of steps that were described for the degradation of 4-nitrobenzoate in *C. acidovorans* NBA-10.[11] The nitro group is reduced to form the hydroxylamino-substituted compound which is then converted to protocatechuate (Figure 3.9). However, pathways present in strains TW3 and 4NT

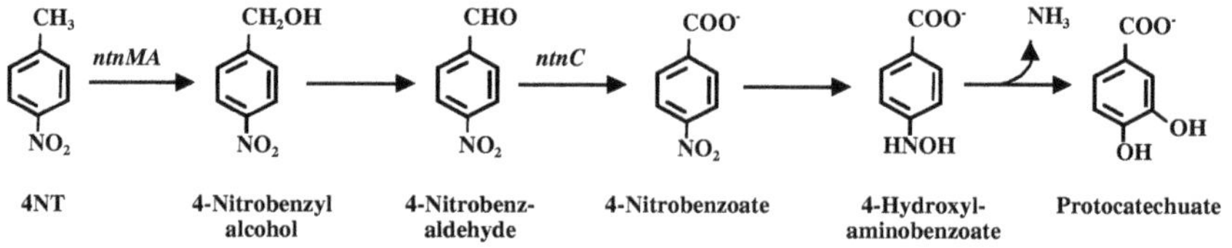

Figure 3.9 Pathway for the degradation of 4NT in *Pseudomonas* sp. strain TW3. A similar pathway was identified in *Pseudomonas* sp. strain 4NT.

differ at the ring cleavage step. Strain TW3 carries out an *ortho*-cleavage of proto-catechuate, whereas strain 4NT uses a *meta*-cleavage pathway.[15,55]

Genes for the conversion of 4NT to 4-nitrobenzoate have been cloned from strain TW3,[20] and the gene order is comparable to the upper pathway genes from the TOL plasmid (Figure 3.10). *ntnMA* encodes a two-component enzyme, 4NT monooxygenase, that converts 4NT to 4-nitrobenzyl alcohol, and *ntnC* encodes an aldehyde dehydrogenase that converts 4-nitrobenzaldehyde to 4-nitrobenzoate.[20] Neither *xylW* in the TOL pathway operon, nor its counterpart, *ntnW*, appears to play an essential role in their respective pathways, and *ntnB** (Figure 3.10) encodes a nonfunctional truncated alcohol dehydrogenase.[20,79] The *ntn* pathway was shown to be required for degradation of toluene as well as 4NT in *Pseudomonas* sp. strain TW3.[20] The molecular biology of 4-nitrobenzoate degradation has not been reported for strains TW3 and 4NT. However, Chapter 6 reports molecular studies of nitroben-zoate-degrading isolates.

3.6 EVOLUTION OF NITROARENE DEGRADATION PATHWAYS

3.6.1 Evolution of Dioxygenase-Catalyzed Pathways from a *nag* Gene Cluster?

Based on the gene order and sequence identities (Figure 3.3, Tables 3.1 to 3.4), it appears that nitroarene dioxygenase genes evolved from a parental gene cluster like that present in *Pseudomonas* sp. strain U2 or *C. testosteroni* GZ42. Thus far, all of the sequenced nitroarene dioxygenase genes contain homologs of at least one of the salicylate monooxygenase genes. It was proposed that during the evolution of *Pseudomonas* sp. U2, a strain carrying an archetype naphthalene dioxygenase (*nah*) gene cluster such as that present in *P. putida* G7 acquired *nagGH* genes, allowing the utilization of naphthalene through a gentisate pathway rather than a catechol *meta*-cleavage pathway.[8] A phylogenetic tree with α subunits from naphthalene and nitroarene dioxygenases also supports this proposal (Figure 3.11). The α subunits of naphthalene dioxygenases from strains using a catechol dioxygenase pathway are separate from the branch carrying NagAc, and the three nitroarene dioxygenase α subunits arise from the NagAc branch. *C. testosteroni* GZ42 and *Pseudomonas* sp. U2 were isolated in New Jersey and Venezuela, respectively, suggesting that the pathway is widespread. It is not clear what selective advantage the alternate pathway may have provided or why strains that degrade nitroarenes appear to have evolved from *nag* pathways rather than from the archetype *nah* pathways.

Evolution of the nitroarene-oxidizing phenotype does not appear to require the presence of a functional salicylate monooxygenase, since the *nagGH* gene homologs in all of the nitroarene-degrading strains appear to be defective. It is possible that fewer mutations in the *nag*-type naphthalene dioxygenase genes are needed to generate a dioxygenase capable of oxidizing the aromatic ring of nitroarene compounds.[8] This possibility is supported by the report that the naph-thalene dioxygenase from *C. testosteroni* GZ42 can convert 2,4-DNT to a

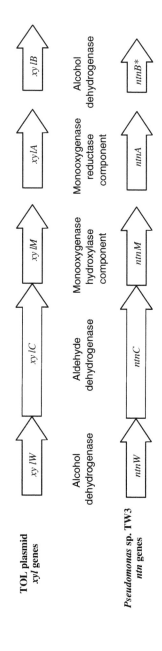

Figure 3.10 Comparison of the gene order for the *xyl* genes for toluene degradation from the TOL plasmid and the *ntn* genes for 4NT degradation from *Pseudomonas* sp. strain TW3. Functions of gene products are indicated between the maps. *xylW* does not appear to be necessary for the degradation of toluene. A role for *ntnW* in 4NT degradation has not been established. The *ntnB** gene is truncated and nonfunctional.

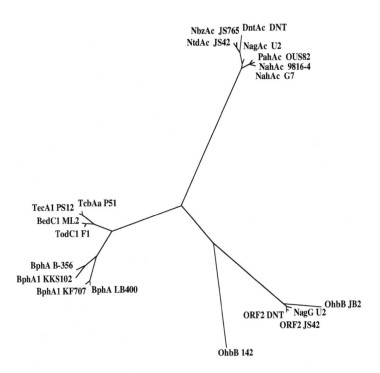

Figure 3.11. Phylogenetic tree of dioxygenase α subunits, salicylate 5-hydroxylase α subunits, and related ORFs. The tree was generated using CreatePhyloTree (cbrg@inf.ethz.ch). The proteins listed are from the following enzyme systems: NbzAc, nitrobenzene dioxygenase from *Comamonas* sp. strain JS765;[29] NtdAc, 2NT dioxygenase from *Pseudomonas* sp. strain JS42;[41] DntAc, 2,4-DNT dioxygenase from *Burkholderia* sp. strain DNT;[66] NagAc, naphthalene dioxygenase from *Pseudomonas* sp. strain U2;[8] PahAc, polycyclic aromatic hydrocarbon dioxygenase from *P. putida* OUS82;[70] NahAc, naphthalene dioxygenase from *Pseudomonas* sp. strain NCIB 9816-4[28,41] and *P. putida* G7;[60] TcbAa, chlorobenzene dioxygenase from *Pseudomonas* sp. strain P51;[78] TecA1, tetrachlorobenzene dioxygenase from *Burkholderia* sp. strain PS12;[2] BedC1, benzene dioxygenase from *P. putida* ML2;[71] TodC1, toluene dioxygenase from *P. putida* F1;[81] BphA, biphenyl dioxygenase from *C. testosteroni* B-356[68] and *Pseudomonas* sp. strain LB400;[7] BphA1, biphenyl dioxygenase from *Pseudomonas* sp. strain KKS102[9] and *P. pseudoalcaligenes* KF707;[69] OhbB, o-halobenzoate dioxygenase from *P. aeruginosa* 142[77] and *P. aeruginosa* JB2 (Genbank accession number AF087482); ORF2 from *Pseudomonas* sp. strain JS42[41] and *Burkholderia* sp. strain DNT.[66]

catechol.[10] NagG is most closely related to the defective ORFs from the 2NT and 2,4-DNT dioxygenase gene clusters (Figure 3.11). The only other proteins of known function that cluster with this group are α subunits (OhbB) from o-haloben-zoate dioxygenases (Figure 3.11) as reported previously.[77]

3.6.2 Evolution of the 4NT Pathways

The pathways for the degradation of 4NT appear to have evolved by recruitment of a TOL-like upper pathway, a 4-nitrobenzoate pathway such as that found in *C. acidovorans* NBA-10,[11] and a protocatechuate ring cleavage pathway. The two strains that have been characterized have similar hybrid pathways except for the final ring cleavage portion.[15,55] The TOL pathway enzymes are active with substrates carrying nitro groups in either the *meta-* or *para-*position,[5] so no change in substrate range would have been necessary. However, XylR, the regulator of the TOL pathway genes, does not recognize nitroaromatic effectors.[5] The *ntn* genes are induced during growth with either 4NT or toluene in strain TW3.[20] Therefore, a modified set of regulatory controls has evolved to allow for 4NT degradation in TW3.

3.6.3 Possible Horizontal Transfer of Nitroarene Genes

Three of the four sequenced nitroarene dioxygenase gene clusters have associated transposase or insertion sequence (IS) elements. A sequence with low sequence similarity to IS elements from *Escherichia coli* and *Bacillus subtilis* is located upstream of the *nbz* genes,[29] and a transposase gene transcribed in the opposite direction is present just downstream of *dntAd* in strain DNT (Figure 3.3). A transposase gene is also located within the R34 gene cluster (Figure 3.5).[22] In addition, an insertion probably left after a transposition event is present in the *ntn* gene cluster in *Pseudomonas* sp. strain TW3.[20] These results suggest that many of the genes for nitroarene degradation may have been acquired by horizontal transfer.

The fact that there has only been recent selective pressure for the development of the nitroarene-degrading capacity suggests that the pathways evolved in the not too distant past. Vestigial genes are present in all nitroarene-degrading strains (*nbzB* and the defective *ntdB* in JS765 and JS42, respectively; defective *nagG* or *nagGH* homologs in JS765, JS42, DNT, and R34; *ntnB** in TW3). The presence of vestigial genes, the lack of coordinate regulation, and the long acclimation times indicate that the pathways have not been optimized.

3.7 ENGINEERED PATHWAYS

Several recent reviews have focused on designing bacteria for the degradation of toxic and xenobiotic compounds.[49,54,73–75] This pathway engineering approach has met with success in the construction of pathways for the simultaneous degradation of methyl- and chloroaromatics and to expand the substrate range of the TOL plasmid to include alkylbenzoates.[53,54,57] A few efforts at constructing pathways for the degradation of nitroaromatic compounds have been reported. The introduction of the upper TOL pathway into a *P. fluorescens* strain with a 4-nitrobenzoate pathway resulted in an engineered pathway for 4-nitrotoluene degradation.[31] This constructed pathway is analogous to the naturally occurring pathways found in *Pseudomonas* sp. strains TW3 and 4NT.[15,55]

In another study, the combination of the TOL plasmid-encoded pathway for toluene degradation and an initial dioxygenation reaction was reported to result in a complete pathway for the degradation of nitrobenzene.[24] In this pathway, *cis*-toluate diol dehydrogenase from the TOL pathway was reported to catalyze the elimination of nitrite from the nitrodihydrodiol formed by the dioxygenation reaction. The mechanism for this unusual reaction was not presented. Induction studies indicated that the cells only degraded nitrobenzene when they were induced with toluene and this is probably why sustained growth on nitrobenzene was not reported.

A *Pseudomonas* strain that apparently utilized TNT as the sole nitrogen source was isolated and the introduction of the TOL plasmid into this strain resulted in a hybrid organism that was reported to grow on TNT.[6,52] At high initial cell densities, slow growth and nitrite release were observed. However, evidence for sustained growth and degradation were not presented. In addition, amino compounds were formed, indicating that degradation was inefficient and reduction of nitro groups was occurring.

Overall, these early attempts to engineer nitroarene pathways did not result in robust strains that would be useful for bioremediation purposes. Clearly, more work needs to be done. It is important that appropriate regulatory systems be in place to allow efficient degradation in the absence of added (and potentially toxic) inducers. In addition, the constructed strains should completely metabolize the compounds to low levels without producing toxic side products.

3.8 FUTURE DIRECTIONS

Since nitroarene compounds have been present in the environment for a relatively short period of time, it is likely that genes have been recruited from other pathways and these newly composed pathways are not optimized. It is probable that efficient regulatory systems have not yet evolved, especially in the more complex pathways where two or more gene modules are required for complete degradation of the compound. The long lag periods and slow degradation rates typical of *Burkholderia* sp. DNT and most of the 2,6-DNT degrading organisms[35,36,64] may be due to the presence of several gene clusters that are not coordinately regulated. Further evidence for this is seen in the transient accumulation of pathway intermediates during growth on nitroaromatic compounds.[35,64]

Most contaminated environments have combinations of nitroaromatic compounds present, and this will complicate bioremediation efforts, especially in light of the fact that most isolates are only capable of degrading a single nitroaromatic compound. Since most dioxygenases have very broad substrate specificities, another problem that may be encountered during attempts to degrade mixtures of nitroaromatic compounds is the possibility of forming toxic dead end metabolites.

Some of the above problems might be overcome by designing efficient pathways for the degradation of nitroaromatic compounds. Using a genetic engineering approach, it should be possible to combine pathways for the degradation of different nitroaromatic compounds into a single organism to increase the range of nitroaromatic compounds degraded. Other possible ways to improve pathways include the introduction of an appropriate system of regulation, modification of enzyme specificity to allow degradation of a wider range of nitroaromatics by site-directed or random mutagenesis, and the use of gene shuffling[65] to combine useful aspects of related pathways and evolve new functions.

Besides the obvious gaps in our knowledge of nitroarene metabolism that are apparent from discussions in this and other chapters in this book, three aspects have received little or no attention. At this time, very little is known about the regulation of pathways for the degradation of nitroaromatic compounds, nothing is known about whether these compounds are transported into the cell or if they enter the cell solely by diffusion, and nothing is known about whether these compounds can elicit a chemotactic response in bacteria. These are areas that deserve to be studied.

A final potential area of study is the possibility of producing valuable products from nitroaromatic wastes. It will always be difficult to prevent the release of toxic wastes into the environment. However, the development of biocatalytic processes that exploit relatively inexpensive nitroaromatic feedstocks for the production of commercially useful products could potentially reduce levels of waste at the source. Such a strategy for the production of catechols from nitroaromatic compounds was previously described.[30]

ACKNOWLEDGMENTS

The author thanks Jim Spain, Glenn Johnson, Shirley Nishino, Zhongqi He, Juan Parales, and Dan Lessner for providing data prior to publication; Juan Parales for making figures; and Gerben Zylstra, Peter Williams, Juan Parales, and David Gibson for suggestions and helpful discussions. This work was supported by grants F49620-96-1-0115 and FA49620-97-1-0402 from the U.S. Air Force Office of Scientific Research and DAAD19–99-1-0285 from the U.S. Army Research Office.

REFERENCES

1. An, D., D. T. Gibson, and J. C. Spain. 1994. Oxidative release of nitrite from 2-nitrotoluene by a three-component enzyme system from *Pseudomonas* sp. strain JS42. *J. Bacteriol.* 176:7462–7467.
2. Beil, S., B. Happe, K. N. Timmis, and D. H. Pieper. 1997. Genetic and biochemical characterization of the broad spectrum chlorobenzene dioxygenase from *Burkholderia* sp. strain PS12 — dechlorination of 1,2,4,5-tetrachlorobenzene. *Eur. J. Biochem.* 247:190–199.
3. Cerniglia, C. E. and C. C. Somerville. 1995. Reductive metabolism of nitroaromatic and nitropolycyclic aromatic hydrocarbons, p. 99–115. In J. C. Spain (ed.), *Biodegradation of Nitroaromatic Compounds*, Vol. 49. Plenum Press, New York.
4. Couch, D. B., P. F. Allen, and D. J. Abernethy. 1981. The mutagenicity of dinitrotoluenes in *Salmonella typhimurium*. *Mutat. Res.* 90:373–383.
5. Delgado, A., M. G. Wubbolts, M.-A. Abril, and J. L. Ramos. 1992. Nitroaromatics are substrates for the TOL plasmid upper pathway enzymes. *Appl. Environ. Microbiol.* 58:415–417.

6. Duque, E., A. Haidour, F. Godoy, and J. L. Ramos. 1993. Construction of a *Pseudomonas* hybrid strain that mineralizes 2,4,6-trinitrotoluene. *J. Bacteriol.* 175:2278–2283.

7. Erickson, B. D. and F. J. Mondello. 1992. Nucleotide sequencing and transcriptional mapping of the genes encoding biphenyl dioxygenase, a multicomponent polychlorinated biphenyl-degrading enzyme in *Pseudomonas* strain LB400. *J. Bacteriol.* 174:2903–2912.

8. Fuenmayor, S. L., M. Wild, A. L. Boyles, and P. A. Williams. 1998. A gene cluster encoding steps in the conversion of naphthalene to gentisate in *Pseudomonas* sp. strain U2. *J. Bacteriol.* 180:2522–2530.

9. Fukuda, M., Y. Yasukochi, Y. Kikuchi, Y. Nagata, K. Kimbara, H. Horiuchi, M. Takagi, and K. Yano. 1994. Identification of the *bphA* and *bphB* genes of *Pseudomonas* sp. strain KKS102 involved in degradation of biphenyl and polychlorinated biphenyls. *Biochem. Biophys. Res. Commun.* 202:850–856.

10. Goyal, A. K. and G. J. Zylstra. 1997. Genetics of naphthalene and phenanthrene degradation by *Comamonas testosteroni*. *J. Ind. Microbiol. Biotechnol.* 19:401–407.

11. Groenewegen, P. E. J., P. Breeuwer, J. M. L. M. van Helvoort, A. A. M. Langenhoff, F. P. de Vries, and J. A. M. de Bont. 1992. Novel degradative pathway of 4-nitrobenzoate in *Comamonas acidovorans* NBA-10. *J. Gen. Microbiol.* 138:1599–1605.

12. Haigler, B. E., G. R. Johnson, W.-C. Suen, and J. C. Spain. 1999. Biochemical and genetic evidence for *meta*-ring cleavage of 2,4,5-trihydroxytoluene in *Burkholderia* sp. strain DNT. *J. Bacteriol.* 181:965–972.

13. Haigler, B. E., S. F. Nishino, and J. C. Spain. 1994. Biodegradation of 4-methyl-5-nitrocatechol by *Pseudomonas* sp. strain DNT. *J. Bacteriol.* 176:3433–3437.

14. Haigler, B. E., C. C. Sommerville, J. C. Spain, and R. K. Jain. 1997. Sequence analysis differences between the 2,4-dinitrotoluene degrading genes of *Burkholderia* sp. strain R34 and strain DNT. Q-343, p. 512. In Abstracts of the 97th General Meeting of the American Society for Microbiology 1997. American Society for Microbiology, Washington, D.C.

15. Haigler, B. E. and J. C. Spain. 1993. Biodegradation of 4-nitrotoluene by *Pseudomonas* sp. strain 4NT. *Appl. Environ. Microbiol.* 59:2239–2243.

16. Haigler, B. E., W.-C. Suen, and J. C. Spain. 1996. Purification and sequence analysis of 4-methyl-5-nitrocatechol oxygenase from *Burkholderia* sp. strain DNT. *J. Bacteriol.* 178:6019–6024.

17. Haigler, B. E., W. H. Wallace, and J. C. Spain. 1994. Biodegradation of 2-nitrotoluene by *Pseudomonas* sp. strain JS42. *Appl. Environ. Microbiol.* 60:3466–3469.

18. Hartter, D. R. 1985. The use and importance of nitroaromatic chemicals in the chemical industry, p. 1–14. In D. E. Rickert (ed.), *Toxicity of Nitroaromatic Compounds*. Chemical Industry Institute of Toxicology Series. Hemisphere Publishing Corporation, Washington, D.C.

19. He, Z., G. R. Johnson, R. E. Parales, D. T. Gibson, and J. C. Spain. Unpublished data.

20. James, K. D. and P. A. Williams. 1998. *ntn* genes determine the early steps in the divergent catabolism of 4-nitrotoluene and toluene in *Pseudomonas* sp. strain TW3. *J. Bacteriol.* 180:2043–2049.

21. Johnson, G. R., R. K. Jain, and J. C. Spain. 2000. Properties of the trihydroxytoluene oxygenase from *Burkholderia cepacia* R34: an extradiol dioxygenase from the 2,4-dinitrotoluene pathway. *Arch. Microbiol.* 173:86–90.

22. Johnson, G. R. and J. C. Spain. Personal communication.

23. Johnson, G. R. and J. C. Spain. 1997. Phylogenetic analysis of nitrotoluene dioxygenases. Q-344, p. 512. In Abstracts of the 97th General Meeting of the American Society for Microbiology 1997. American Society for Microbiology, Washington, D.C.

24. Jung, K.-H., J.-Y. Lee, and H.-S. Kim. 1995. Biodegradation of nitrobenzene through a hybrid pathway in *Pseudomonas putida. Biotechnol. Bioeng.* 48:625–630.
25. Kauppi, B., K. Lee, E. Carredano, R. E. Parales, D. T. Gibson, H. Eklund, and S. Ramaswamy. 1998. Structure of an aromatic ring-hydroxylating dioxygenase-naphthalene 1,2-dioxygenase. *Structure.* 6:571–586.
26. Keith, L. H. and W. A. Telliard. 1979. Priority pollutants I-a perspective view. *Environ. Sci. Technol.* 13:416–423.
27. Klausmeier, R. E., J. L. Osman, and D. R. Walls. 1974. The effect of trinitrotoluene on microorganisms. *Dev. Ind. Microbiol.* 15:309–317.
28. Kurkela, S., H. Lehvaslaiho, E. T. Palva, and T. H. Teeri. 1988. Cloning, nucleotide sequence and characterization of genes encoding naphthalene dioxygenase of *Pseudomonas putida* strain NCIB 9816. *Gene.* 73:355–362.
29. Lessner, D. J., R. E. Parales, G. J. Johnson, J. C. Spain, and D. T. Gibson. Unpublished data.
30. Meulenberg, R. and J. A. M. de Bont. 1995. Microbial production of catechols from nitroaromatic compounds, p. 37–53. In J. C. Spain (ed.), *Biodegradation of Nitroaromatic Compounds.* Vol. 49. Plenum Press, New York.
31. Michan, C., A. Delgado, A. Haïdour, G. Lucchesi, and J. L. Ramos. 1997. In vivo construction of a hybrid pathway for metabolism of 4-nitrotoluene in *Pseudomonas fluorescens. J. Bacteriol.* 179:3036–3038.
32. Mirsalis, J. C. and B. E. Butterworth. 1982. Induction of unscheduled DNA synthesis in rat hepatocytes following *in vivo* treatment with dinitrotoluene. *Carcinogenesis.* 3:241–245.
33. Narayan, S., R. E. Parales, and D. T. Gibson. Unpublished data.
34. Nishino, S. and J. C. Spain. Unpublished data.
35. Nishino, S. F., G. C. Paoli, and J. C. Spain. 2000. Aerobic degradation of dinitrotoluenes and the pathway for bacterial degradation of 2,6-dinitrotoluene. *Appl. Environ. Microbiol.* 66:2139–2147.
36. Nishino, S. F. and J. C. Spain. 1996. Degradation of 2,6-dinitrotoluene by bacteria. Q-380, p. 452. In Abstracts of the 96th General Meeting of the American Society for Microbiology 1996. American Society for Microbiology, Washington, D.C.
37. Nishino, S. F. and J. C. Spain. 1995. Oxidative pathway for the biodegradation of nitrobenzene by *Comamonas* sp. strain JS765. *Appl. Environ. Microbiol.* 61:2308–2313.
38. Nishino, S. F. and J. C. Spain. 1997. Ring-cleavage of 3-methyl-4-nitrocatechol by 2,6-dinitrotoluene-degrading bacteria. Q-348, p. 513. In Abstracts of the 97th General Meeting of the American Society for Microbiology 1997. American Society for Microbiology, Washington, D.C.
39. Office of Pollution Prevention and Toxics. 1997. Toxic Release Inventory, U.S. Environmental Protection Agency. http://www.epa.gov/tri/.
40. Parales, J. V. and D. T. Gibson. 1999. Identification of single amino acid changes that alter the substrate specificity and enantioselectivity of 2-nitrotoluene 2,3-dioxygenase. Q-249, p. 579. In Abstracts of the 99th General Meeting of the American Society for Microbiology 1999. American Society for Microbiology, Washington, D.C.
41. Parales, J. V., A. Kumar, R. E. Parales, and D. T. Gibson. 1996. Cloning and sequencing of the genes encoding 2-nitrotoluene dioxygenase from *Pseudomonas* sp. JS42. *Gene.* 181:57–61.
42. Parales, J. V., R. E. Parales, and D. T. Gibson. Unpublished data.

43. Parales, J. V., R. E. Parales, S. M. Resnick, and D. T. Gibson. 1998. Enzyme specificity of 2-nitrotoluene 2,3-dioxygenase from *Pseudomonas* sp. JS42 is determined by the C-terminal region of the α subunit of the oxygenase component. *J. Bacteriol.* 180:1194–1199.

44. Parales, R. E., M. D. Emig, N. A. Lynch, and D. T. Gibson. 1998. Substrate specificities of hybrid naphthalene and 2,4-dinitrotoluene dioxygenase enzyme systems. *J. Bacteriol.* 180:2337–2344.

45. Parales, R. E. and D. T. Gibson. 1998. Selection of spontaneous mutants of the nitrobenzene-degrading strain *Comamonas* sp. JS765 that grow with phenol and cresols. K-146, p. 350. In Abstracts of the 98th General Meeting of the American Society for Microbiology 1998. American Society for Microbiology, Washington, D.C.

46. Parales, R. E. and D. T. Gibson. Unpublished data.

47. Parales, R. E., T. A. Ontl, and D. T. Gibson. 1997. Cloning and sequence analysis of a catechol 2,3-dioxygenase gene from the nitrobenzene-degrading strain *Comamonas* sp. JS765. *J. Ind. Microbiol. Biotechnol.* 19:385–391.

48. Parales, R. E., J. V. Parales, R. An, A. Kumar, S. M. Resnick, J. C. Spain, and D. T. Gibson. 1995. 2-Nitrotoluene dioxygenase from *Pseudomonas* sp. strain JS42: protein purification, molecular biology and substrate specificity. D-12, p. 95 In Abstracts, *Pseudomonas* 1995, Fifth International Symposium on *Pseudomonas*: Biotechnology and Molecular Biology. Tsukuba, Japan.

49. Peiper, D. H., K. N. Timmis, and J. L. Ramos. 1996. Designing bacteria for the degradation of nitro- and chloroaromatic pollutants. *Naturwissenschaften.* 83:201–213.

50. Pereira, W. E., D. L. Short, D. B. Manigold, and P. K. Roscio. 1979. Isolation and characterization of TNT and its metabolites in groundwater by gas chromatograph-mass spectrometer-computer techniques. *Bull. Environ. Contam. Toxicol.* 21:554–562.

51. Preuß, A. and P.-G. Rieger. 1995. Anaerobic transformation of 2,4,6-trinitrotoluene and other nitroaromatic compounds, p. 69–85. In J. C. Spain (ed.), *Biodegradation of Nitroaromatic Compounds.* Vol. 49. Plenum Press, New York.

52. Ramos, J. L., A. Haïdour, A. Delgado, M.-D. Fandila, M. Gil, and G. Piñar. 1995. Potential of toluene-degrading systems for the construction of hybrid pathways for nitrotoluene metabolism. In J. C. Spain (ed.), *Biodegradation of Nitroaromatic Compounds.* Vol. 49. Plenum Press, New York.

53. Ramos, J. L., A. Wasserfallen, K. Rose, and K. N. Timmis. 1987. Redesigning metabolic routes: manipulation of TOL plasmid pathway for catabolism of alkylbenzoates. *Science.* 235:593–596.

54. Reineke, W. 1998. Development of hybrid strains for the mineralization of chloroaromatics by patchwork assembly. *Annu. Rev. Microbiol.* 52:287–331.

55. Rhys-Williams, W., S. C. Taylor, and P. A. Williams. 1993. A novel pathway for the catabolism of 4-nitrotoluene by *Pseudomonas*. *J. Gen. Microbiol.* 139:1967–1972.

56. Rieger, P.-G. and H.-J. Knackmuss. 1995. Basic knowledge and perspectives on biodegradation of 2,4,6-trinitrotoluene and related nitroaromatic compounds in contaminated soil, p. 1–18. In J. C. Spain (ed.), *Biodegradation of Nitroaromatic Compounds.* Vol. 49. Plenum Press, New York.

57. Rojo, F., D. H. Pieper, K.-H. Engesser, H.-J. Knackmuss, and K. N. Timmis. 1987. Assemblage of ortho cleavage route for simultaneous degradation of chloro- and methylaromatics. *Science.* 238:1395–1398.

58. Schell, M. A. 1990. Regulation of naphthalene degradation genes of plasmid NAH7: example of a generalized positive control system in *Pseudomonas* and related bacteria, p. 165–177. In A. M. Chakrabarty, S. Kaplan, B. Iglewski, and S. Silver (eds.), *Pseudomonas: Biotransformations, Pathogenesis, and Evolving Biotechnology.* American Society for Microbiology, Washington, D.C.

59. Schell, M. A. and E. F. Poser. 1989. Demonstration, characterization, and mutational analysis of NahR protein binding to *nah* and *sal* promoters. *J. Bacteriol.* 171:837–846.

60. Simon, M. J., T. D. Osslund, R. Saunders, B. D. Ensley, S. Suggs, A. Harcourt, W.-C. Suen, D. L. Cruden, D. T. Gibson, and G. J. Zylstra. 1993. Sequences of genes encoding naphthalene dioxygenase in *Pseudomonas putida* strains G7 and NCIB 9816-4. *Gene.* 127:31–37.

61. Spain, J. C. 1995. Bacterial biodegradation of nitroaromatic compounds under aerobic conditions, p. 19–35. In J. C. Spain (ed.), *Biodegradation of Nitroaromatic Compounds.* Vol. 49. Plenum Press, New York.

62. Spain, J. C. 1995. Biodegradation of nitroaromatic compounds. *Annu. Rev. Microbiol.* 49:523–555.

63. Spanggord, R. J., K. E. Mortelmans, A. F. Griffin, and V. F. Simmon. 1982. Mutagenicity in *Salmonella typhimurium* and structure-activity relationships of wastewater components emanating from the manufacture of trinitrotoluene. *Environ. Mutagenesis.* 4:163–179.

64. Spanggord, R. J., J. C. Spain, S. F. Nishino, and K. E. Mortelmans. 1991. Biodegradation of 2,4-dinitrotoluene by a *Pseudomonas* sp. *Appl. Environ. Microbiol.* 57:3200–3205.

65. Stemmer, W. P. C. 1994. DNA shuffling by random mutagenesis and reassembly: *in vitro* recombination for molecular evolution. *Proc. Natl. Acad. Sci. U.S.A.* 91:10747–10751.

66. Suen, W.-C., B. E. Haigler, and J. C. Spain. 1996. 2,4-Dinitrotoluene dioxygenase from *Burkholderia* sp. strain DNT: similarity to naphthalene dioxygenase. *J. Bacteriol.* 178:4926–4934.

67. Suen, W.-C. and J. C. Spain. 1993. Cloning and characterization of *Pseudomonas* sp. strain DNT genes for 2,4-dinitrotoluene degradation. *J. Bacteriol.* 175:1831–1837.

68. Sylvestre, M., M. Sirois, Y. Hurtubise, J. Bergeron, D. Ahmad, F. Shareck, D. Barriault, I. Guillemette, and J. M. Juteau. 1996. Sequencing of *Comamonas testosteroni* strain B-356-biphenyl/chlorobiphenyl dioxygenase genes: evolutionary relationships among Gram-negative bacterial biphenyl dioxygenases. *Gene.* 174:195–202.

69. Taira, K., J. Hirose, S. Hayashida, and K. Furukawa. 1992. Analysis of *bph* operon from the polychlorinated biphenyl-degrading strain of *Pseudomonas pseudoalcaligenes* KF707. *J. Biol. Chem.* 267:4844–4853.

70. Takizawa, N., N. Kaida, S. Torigoe, T. Moritani, T. Sawada, S. Satoh, and H. Kiyohara. 1994. Identification and characterization of genes encoding polycyclic aromatic hydrocarbon dioxygenase and polycyclic aromatic hydrocarbon dihydrodiol dehydrogenase in *Pseudomonas putida* OUS82. *J. Bacteriol.* 176:2444–2449.

71. Tan, H.-M., H.-Y. Tang, C. L. Joannou, N. H. Abdel-Wahab, and J. R. Mason. 1993. The *Pseudomonas putida* ML2 plasmid-encoded genes for benzene dioxygenase are unusual in codon usage and low in G+C content. *Gene.* 130:33–39.

72. Threshold limit values (TLVs) for chemical substances and physical agents. Biological exposure indices for 1998. American Conference of Governmental Industrial Hygenists, Cincinnati, OH.

73. Timmis, K. N. and D. H. Pieper. 1999. Bacteria designed for bioremediation. *Trends Biotechnol.* 17:201–204.

74. Timmis, K. N., F. Rojo, and J. L. Ramos. 1989. Design of new pathways for the catabolism of environmental pollutants. *Adv. Appl. Biotechnol.* 4:61–82.
75. Timmis, K. N., R. J. Steffan, and R. Unterman. 1994. Designing microorganisms for the treatment of toxic wastes. *Annu. Rev. Microbiol.* 48:525–557.
76. Tokiwa, H. and Y. Ohnishi. 1986. Mutagenicity and carcinogenicity of nitroarenes and their sources in the environment. *CRC Crit. Rev. Toxicol.* 17:23–60.
77. Tsoi, T. V., E. G. Plotnikova, J. R. Cole, W. F. Guerin, M. Bagdasarian, and J. M. Tiedje. 1999. Cloning, expression, and nucleotide sequence of the *Pseudomonas aeruginosa* 142 *ohb* genes coding for oxygenolytic *ortho* dehalogenation of halobenzoates. *Appl. Environ. Microbiol.* 65:2151–2162.
78. Van Der Meer, J. R., A. R. W. Van Neerven, E. J. De Vries, W. M. De Vos, and A. J. B. Zehnder. 1991. Cloning and characterization of plasmid-encoded genes for the degradation of 1,2-dichloro-, 1,4-dichloro-, and 1,2,4-trichlorobenzene of *Pseudomonas* sp. strain P51. *J. Bacteriol.* 173:6–15.
79. Williams, P. A., L. M. Shaw, C. W. Pitt, and M. Vrecl. 1996. *xylUW*, two genes at the start of the upper pathway operon of TOL plasmid pWW0, appear to play no essential part in determining its catabolic phenotype. *Microbiology.* 143:101–107.
80. Zylstra, G. J. Personal communication.
81. Zylstra, G. J. and D. T. Gibson. 1989. Toluene degradation by *Pseudomonas putida* F1: nucleotide sequence of the *todC1C2BADE* genes and their expression in *E. coli. J. Biol. Chem.* 264:14940–14946.

Perspectives of Bioelimination of Polynitroaromatic Compounds

Hiltrud Lenke, Christof Achtnich, and Hans-Joachim Knackmuss

CONTENTS

4.1 CHEMICAL STRUCTURE AND BIODEGRADABILITY OF POLYNITROAROMATIC COMPOUNDS

A fundamental strategy of microbial catabolism of aromatic hydrocarbons is the initial activation of the molecules by oxygenation. This generates catechols (Figure 4.1) or 1,4-dihydroxyarenes (not shown) which are subject to oxygenolytic ring cleavage. Subsequently, isomerization and hydrolysis allow assimilation and complete oxidation of the carbon skeleton. Alkenes are subject to monooxygenation (Figure 4.1b), which requires subsequent detoxification of the highly reactive alkene-oxides by enzymatic hydrolysis[93] or dehydrogenation.[10]

Many synthetic aromatic compounds carry substituents or structural elements that are rare or even absent among natural compounds. Such xenobiotic structures (xenophors) include the trifluoromethyl group, certain substitution patterns with halogen, azo and sulfonic acid groups, and nitro substituents. The electron with-drawing character of the substituents generates rather high redox potentials (Figure 4.2) and renders the π-electron system of the unsaturated hydrocarbon electron deficient, particularly when several such xenophors are present as substituents. As a consequence, the xenobiotic compounds can resist initial electrophilic attack by oxygen species associated with oxygenases of aerobic bacteria.

Theoretically, an initial reductive reaction rather than an oxidative one is more likely to occur because electron deficient xenobiotics undergo such initial reductive

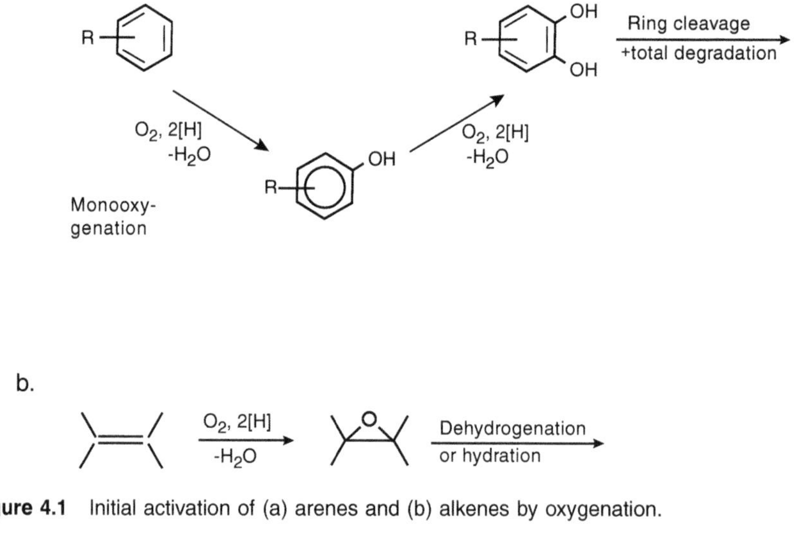

Figure 4.1   Initial activation of (a) arenes and (b) alkenes by oxygenation.

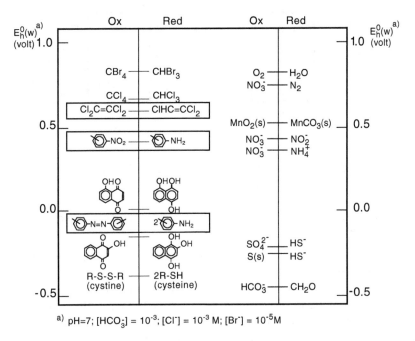

Figure 4.2 Half-reaction reduction potential of selected organic (left side) and some bio-geochemically important redox couples (right side). (Modified from Haderlein and Schwarzenbach, 1995.[38])

transformations, particularly in anoxic microbial systems which harbor an impressive array of nucleophiles and reducing species. Reductive dechlorination, hydrogenolysis of the azo bond, hydrogenolytic denitrations, and reduction of nitro groups are all transformations that reduce the electron deficiency of the xenobiotic carbon skeleton. Increased reactivity of the reduction products toward oxygen can facilitate a subsequent oxidative treatment process.

Thus, a general concept for biological elimination of electron deficient xenobiotic compounds is a two-stage reductive and subsequent oxidative treatment. For example, *cis*-dichloroethene and vinyl chloride generated from reductive dechlorination of tetra- or trichloroethene can be cooxidized by bacterial cultures that use the structural analogue, ethene, as the auxiliary substrate (Figure 4.3a).[56] Anaerobic/aerobic treatment can also be used for bioelimination of azo dyes (Figure 4.3b). The aromatic amines from the reductive cleavage of the azo bond can be subject to productive aerobic catabolism, as demonstrated for the sulfonated azo dye Mordant Yellow.[43] In principle, a two-stage reductive/oxidative process can also be used for complete degradation of nitroaromatic compounds such as 4-chloro-2-nitrophenol or nitrobenzene.[7,25,78] The latter compound could be reduced to aniline by anaerobic bacteria. Subsequently, aniline can readily be mineralized by aerobic microorganisms that are ubiquitous in activated sludge. Even partial reduction of a nitroaromatic compound to hydroxylamino derivatives can initiate catabolic sequences that finally allow complete productive oxygen-dependent mineralization (see Chapter 2). An unusual hybrid reductive/oxidative pathway has recently been

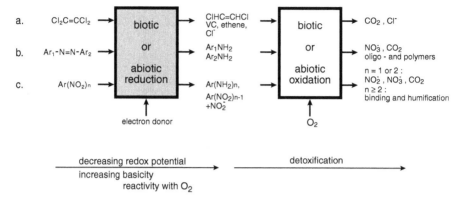

Figure 4.3 Degradation of xenobiotic compounds with high redox potential by integrated anaerobic/aerobic processes.

identified in certain aerobic Gram-positive bacterial strains that utilize picrate and 2,4-dinitrophenol (see below).

In principle, polynitroaromatic compounds can also be treated by two-stage anaerobic/aerobic processes. During successive reduction of the nitro groups, the redox potential decreases drastically and the basicity increases so that eventually the polyamino aromatic compounds become amenable to oxidative treatment (Figure 4.3c). It is meaningless to differentiate between biotic and abiotic reactions in the two-step treatment process, particularly if nondefined mixed cultures from anaerobic or aerobic sludge are used. In both anaerobic and aerobic systems, low molecular weight reductants or oxidants can mediate the transfer of reduction equivalents between bacterial cells and the contaminating chemicals (see Chapter 12).

Biodegradability of xenobiotic compounds with high redox potential and electron deficiency on the unsaturated carbon skeleton is underestimated by the standardized test conditions when either aerobic or anaerobic mixed cultures are used as inocula. Completely different findings on the biodegradability would emerge from a two-step anaerobic/aerobic treatment. According to the scheme in Figure 4.3, reductive pretreatment by use of anaerobic sludge would generate reduction products that can readily be oxidized in a subsequent aerobic treatment with activated sludge. The two-step anaerobic/aerobic treatment allows for the fact that redox conditions in microhabitats of soils and sediments can change tremendously so that distinctly different physiological groups of microorganisms, e.g., anaerobes and aerobes, can be active within a small sample of soil or mud. For example, redox conditions that allow reductive and oxidative catabolic processes within a microenvironment were responsible for the complete degradation of the azo dye Mordant Yellow by a microbial consortium immobilized in calcium-alginate beads.[58]

The above, more general considerations make clear that the use of dual stage reductive and oxidative processes provides not only a more realistic assessment of biodegradability of xenobiotic compounds in the environment, but also opens a hitherto largely unexplored potential to eliminate electron deficient compounds from the environment. The following sections will elaborate on the use of anaerobic/aerobic treatment strategies for the destruction of polynitroaromatic compounds.

4.2 REDUCTION OF TNT UNDER AEROBIC CONDITIONS

As outlined above, persistence of trinitrotoluene (TNT) is not due to a general lack of reactivity in microbial systems. Although the electron deficiency of the aromatic ring prevents electrophilic attack by oxygenases of aerobic organisms, numerous publications that describe initial reductive transformations have been cited in a recent comprehensive review and evaluated with respect to their relevance for remediation technologies.[65]

The majority of investigations carried out with aerated microbial mixed cultures in activated sludge, compost, or soil have revealed little or insignificant mineralization of TNT. All transformations appear to be gratuitous reactions involving sequential reduction of nitro groups to the isomeric aminodinitrotoluenes and diaminonitrotoluenes (Figure 4.4a). Studies with pure or mixed cultures indicate a preferential reduction of the nitro group in the 4-position yielding 4-amino-2,6-dinitrotoluene (4-ADNT).[19,21,65] The susceptibility of the 4-nitro group to reductive transformation increases with increasing substitution by nitro groups on the aromatic ring. Thus, reaction rates decrease in the order 2,4,6-TNT > 2,4-DNT > 2,6-DNT > 4-NT. The reduction rates of nitrotoluenes can be correlated with the molecular electrostatic potential, which is a measure of the magnitude of partial charges on the molecules.[69] The correlation may help to predict nitro group reduction rates for other nitrotoluenes. The work was carried out with a single bacterial isolate and, for generalization, must be verified for other organisms and bacterial consortia. Nevertheless, the observations seem to be representative because the propensity of nitro groups to accept electrons from different kinds of reductants has been substantiated by very recent publications[46] and appears to be a nonspecific capability harbored by many bacterial strains and consortia[15,21] (see Chapter 8).

Furthermore, in the majority of the reported pathways, regio- and chemoselectivity were the same as for the chemical reduction of aromatic nitro groups.

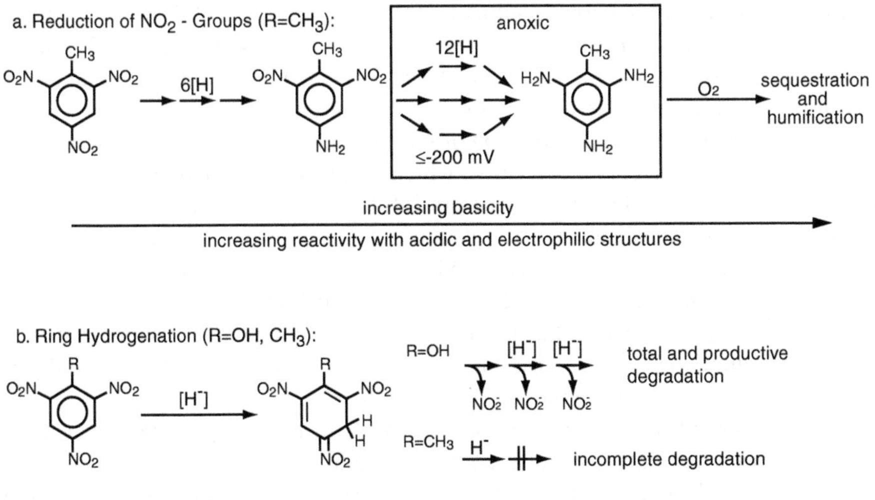

Figure 4.4   Perspectives of biodegradation of trinitroaromatic compounds.

Therefore, both transformation by nonspecific reductases and abiotic reduction through low molecular weight redox mediators[46] can account for the generation of the partially reduced nitrotoluenes.

The formation of azoxy-tetranitrotoluenes and covalently bound TNT-derived residues in cell material such as lipids and proteins[17,19,64,68,71] or in the organic fraction of soil (see below) indicates that highly reactive early reduction products such as hydroxylamino or nitroso derivatives of TNT are responsible for further chemical transformations (see also Figure 4.7). Both electrophilic and nucleophilic reactions of the TNT reduction products may be considered for covalent bond formation.

A novel pathway for aerobic TNT cometabolism has been observed in *Pseudomonas pseudoalcaligenes* JS52.[33] The enzymes involved in productive degradation of nitrobenzene can also transform TNT. Strain JS52 can grow on nitrobenzene via chemoselective reduction to hydroxylaminobenzene and subsequent rearrangement to 2-aminophenol (see Chapter 2). The nitrobenzene grown cells of strain JS52 cometabolize TNT, yielding 2,4-dihydroxylamino-6-nitrotoluene and 2-hydroxylamino-4-amino-6-nitrotoluene. The authors have shown further oxygen-dependent metabolism and the release of nitrite with concomitant formation of an additional, highly polar, yellow metabolite and other unidentified transformation products. *P. aeruginosa* MAO1 catalyzes a similar oxygen dependent formation of polar metabolites from TNT and aminodinitrotoluene, which suggests that ring cleavage accounts for the metabolites.[5]

Since the amino groups generated during reduction of nitro substituents weaken the electron deficiency originally present in TNT, nitro group transformation rates decrease with progressive reduction of nitro groups. Because of its reduced electron deficiency, further oxidative metabolism of aminonitrotoluenes should be possible. A few recent observations support this hypothesis.[1]

Under drastic conditions ($Na_2Cr_2O_7/H_2SO_4$) the methyl group of TNT can be oxidized to produce 2,4,6-trinitrobenzoic acid. A corresponding biological reaction could explain why 2,4,6-trinitrobenzoate can be detected as a minor contaminant in groundwater from aged TNT contamination of soil.[90] Occasionally, 4-amino-2,6-dinitrobenzoic acid was observed as a metabolite of 4-ADNT.[99] Also, 2,4,6-trinitrobenzaldehyde, 2-nitro-4-hydroxybenzoic acid, 4-hydroxybenzaldhyde, and 4-hydroxybenzoic acid as metabolites of TNT[32] indicate that the methyl group of TNT can be oxidized before or after removal of nitro groups.

The products of partial reduction of nitro groups and of methyl group oxidation of TNT can be subject to methylation, which explains the formation of 2,4-diamino-6-nitromethoxybenzyl methyl ether from 2,4-diamino-6-nitrotoluene (2,4-DANT).[99] The function and mechanism of acetylation and formation of TNT reduction products are still unknown both in bacteria[37] and fungi.[45] These reactions were discussed as detoxification mechanisms of microbes during a dual-stage anaerobic/aerobic composting process. Although the metabolites almost completely disappeared under aerobic conditions,[15] it is rather unlikely that acetylation is part of a productive catabolic pathway by bacteria.

It is noteworthy that fungi, which are capable of generating highly efficient oxidants, prior to oxidation initially carry out partial reduction of TNT to hydroxylaminodinitrotoluenes and aminodinitrotoluenes (see Chapter 9). On the one hand this reductive transformation reduces electron deficiency of the dinitroaromatic ring and, thus, facilitates oxidative attack. On the other hand ligninolytic fungi are sensitive to TNT[95] and particularly to 4-hydroxylamino-2,6-dinitrotoluene (4-HADNT) which inhibits lignin peroxidase.[73] Reduction to the aminodinitrotoluenes relieves the toxicity.

4.3 TRANSFORMATION OF TNT UNDER ANAEROBIC CONDITIONS

High electron deficiency of the π-electron system explains why even under aerobic conditions some of the nitro groups of TNT can be reduced and why part of the TNT originally present at contamination sites exists today as aminonitrotoluenes. As long as readily degradable organic matter is limiting and more or less oxidative conditions exist, reductive transformation of TNT stops at the level of aminodinitrotoluenes or diaminonitrotoluenes. Anoxic and strict anaerobic conditions can be generated (see below) only in the presence of an excess of usable organic matter, such as sugars or readily degradable biopolymers. Such conditions favor progressive reduction of all nitro groups and intermediary products of TNT (Figures 4.4 and 4.5). The diversity of partially reduced intermediates formed during TNT transformation under aerobic or semianaerobic conditions has been reviewed comprehensively in a recent article.[65] It is clear that under anaerobic conditions the abundance of products generated under aerobic conditions decreases and a few recent studies have shown that the reductive sequences finally converge to 2,4,6-triaminotoluene (TAT) (see Figure 10 in Reference 86). Even the binuclear azoxy compounds, which are generated chemically from spontaneous condensation of hydroxylamino and nitroso intermediates, can be further reduced and finally bound to soil organic matter in the anaerobic stage (see below).

Recent work using refined analytical techniques such as solid phase microextraction, GC-MS and LC-MS[44] verified older work on the appearance and disappearance of intermediates of TNT reduction (Figures 4.5 and 4.6). As in aerobic cultures the reduction under anaerobic conditions is regioselective with a preference for the nitro group in the 4-position. The selectivity can vary, however, with the relative contributions of biotic and abiotic reactions to the overall process.[103]

The final product of complete reduction of TNT is TAT. There is a lot of controversy about TAT as a potential key metabolite of productive mineralization of TNT. It is clear that TAT disappears after prolonged incubation under anaerobic culture conditions; however, less than 0.1% is mineralized.[17] The disappearance of TAT under strict anaerobic conditions was recently ascribed to the formation of tetraaminoazobenzenes and even polynuclear azo compounds.[44] The authors discuss the possibility that metal ions in the strictly anaerobic sludge might have triggered the biological oxidation of TAT and speculate on an "anonymous electron acceptor" in the anaerobic bioprocess. In contrast to previous work, hydrolysis products of TAT such as 2,4,6-trihydroxytoluene or 4-hydroxytoluene[35] were not detected while the microcosms remained strictly anaerobic at pH $\geq$ 7.2.[44] Hydroxydiaminotoluene and dihydroxyaminotoluene and trace amounts of trihydroxytoluene were generated only upon exposure to air, particularly, under slightly acidic conditions (pH < 6) which enhance chemical hydrolysis of TAT. The function of oxygen in the conversion of TAT to hydrolysis products is not clear because [18]O labeling showed that oxygen was not the source of the OH-groups generated. Although toluene has been reported

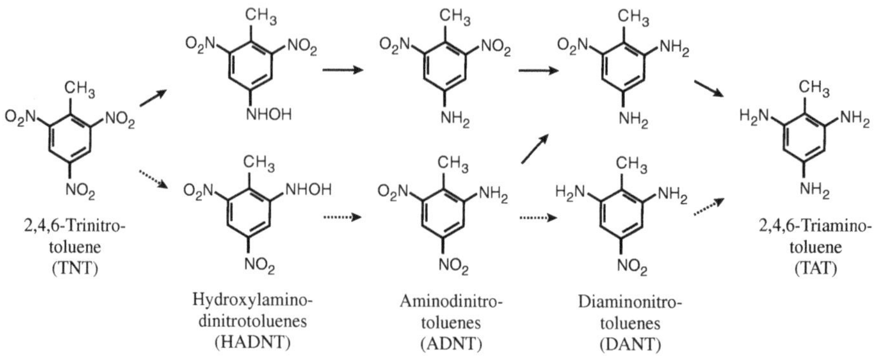

Figure 4.5   Complete reduction of TNT and intermediates (main degradative route: solid arrows).

as a metabolite during TNT reduction under anaerobic conditions,[11] Hawari et al.[44] could not detect any products that indicate reductive denitration reactions of [^{13}CH$_3$]-TNT in anaerobic sludge. From the data currently available, it is clear that TAT exhibits chemical reactivity with respect to hydrolysis to phenolic compounds and, thus, could be a key metabolite for complete degradation of TNT under anaerobic conditions. However, evidence for its productive breakdown remains elusive.

Less attractive is the degradation of TAT by aerobic bacteria or activated sludge. TAT is readily oxidized by oxygen particularly in the presence of metal ions such as Mn^{2+}.[83] The generation of dark polymers as autooxidation products obviously precludes productive breakdown by aerobic organisms. Recent observations indicate that TAT can be misrouted from mineralization by other side reactions such as oxidative dimerization and polymerization to azo and polyazo compounds even in the absence of oxygen.[44] Furthermore, 1,2- or 1,4-additions and condensation reactions with quinoid structures become important if humic substances are present during soil remediation (see below).

Despite its high reactivity and propensity to being routed into unproductive catabolic sequences, TAT holds a key position as a metabolite. Complete reduction to TAT under anaerobic conditions could be an appropriate and simple option for biological remediation, particularly in aged contaminants containing a complex mix of partially reduced TNT derivatives. A cost effective remediation solution would become available if productive metabolism of TAT or just irreversible chemical reactions like fixation to an inert matrix lead to detoxification (see below).

4.4 INTERACTION OF TNT AND METABOLITES WITH SOIL AND SOIL COMPONENTS

4.4.1 Sorption to Clay Minerals

TNT and its metabolites interact with soil components by different mechanisms. The chemical structure of the sorbate, the nature of the sorbing agents, and the

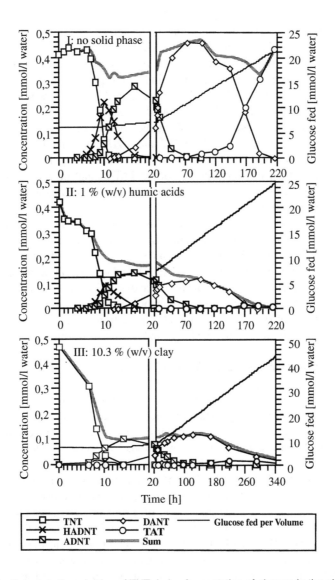

Figure 4.6 Cometabolic reduction of TNT during fermentation of glucose in the absence and the presence of humic acids or clay.

environmental or incubation conditions determine whether physical or van der Waals interactions, hydrogen bondings, or ionic or covalent bonds are the predominant sorption mechanisms. Clay minerals and the soil organic matter are known as the major sorbents in soil. Formation of coplanar electron donor-acceptor complexes between oxygen ligands at the external siloxane surfaces of clay minerals and the electron deficient π-system of nitroaromatic compounds causes specific physisorption of nitroaromatic compounds.[39,104] With TNT the reduction of an electron withdrawing nitro group to an electron donating amino group leads to a decrease in the

k_d values.[40] The higher adsorption of 2-ADNT as compared to 4-ADNT can be explained by the *ortho*-effect of the methyl group.

Although the adsorption of TNT, 2-ADNT, and 4-ADNT to montmorillonite is fully reversible, the adsorption of 2,4-DANT is only partially reversible, which indicates that the basic amino groups bind to the polyanion surfaces of the clay minerals.[21] TNT and 2-ADNT sorbed more than 2,4-DANT, which might be due to a higher polarity of the latter compound. Considerable attention has been paid to 2-ADNT, 4-ADNT, and 2,4-DANT as metabolites of TNT, although the interaction of clay minerals with 2-HADNT and 4-HADNT and TAT is more important. Thus, sorption of TAT to montmorillonite is very fast and irreversible especially under aerobic conditions. Sorbed TAT cannot be extracted with different organic solvents nor can it be released by alkaline or acidic hydrolysis or by methanol/HCl mixtures.[21]

When cometabolic reduction of TNT during fed batch fermentation of glucose was followed in the presence of montmorillonite, a strong interaction of 2-HADNT and 4-HADNT became obvious (Figure 4.6).[21] In contrast to the control experiment without clay, only low amounts of 2-HADNT and 4-HADNT were detectable. The cometabolic reduction of TNT in the presence of clay led to nearly complete removal of free and extractable TNT and its metabolites. Results were similar in abiotic experiments where Fe^{2+} was used for TNT reduction.[14] Mass balance studies indicated the formation of nonextractable or unknown transformation products.

4.4.2   Interaction with Soil Organic Matter

The interaction of TNT and its metabolites with the soil organic matter due to covalent binding of amino groups to humic substances seems to be more important than sorption to clay minerals. The amino group reacts as a nucleophile with α,β-unsaturated carbonyl compounds. 1,4-Additions lead to the formation of nonhydrolyzable C–N bonds,[76,98,100] and Schiff's base formation leads to C=N bonds.[76] In addition, a variety of radical reactions of aromatic amines correspond to the known autooxidation reactions of phenols in natural humification processes.[9,67,106]

Little is known about the interaction of TNT and its metabolites with humic substances. Li et al.[66] investigated the interaction of TNT, 2-ADNT, and 2,6-DANT with humic acids. Both TNT and 2,6-DANT bound slowly to the humic acid, but the functional relationship between bound and free TNT was different from that of 2,6-DANT. Therefore, the authors assumed that the two compounds bind by different mechanisms. The experimental design did not allow determination of whether part of the contaminants could be subject to an irreversible binding to humic acids. Such chemisorption could become important when two or more nitro groups of TNT become reduced. Other studies revealed that 2,4-DANT became coupled to guaiacol dimers, a model for humic substances, whereas 4-ADNT did not react to a significant extent.[22] TAT became irreversibly bound to humic acids much faster than TNT, 4-ADNT, and 2,4-DANT.[21] The isomeric hydroxylaminodinitrotoluenes interacted rapidly with humic acids. The evidence is compelling that 2,4-DANT and TAT sorption occurs through binding of the amino groups, but the mechanism of interaction of hydroxylaminodinitrotoluenes with humic acids is unclear.

Cometabolic reduction of TNT in the presence of humic acids leads to a complete disappearance of free TNT and its metabolites (Figure 4.6). The bound metabolites of TNT could not be desorbed by alkaline or acidic hydrolysis or by methanolic saponification.[21] The observation indicates an irreversible binding of the reduction products of TNT to humic substances, as was observed for clay minerals. Deviation from the 100% mass balance during initial TNT reduction indicates that early metabolites such as hydroxylaminodinitrotoluenes interact quickly with the solid phase (Figure 4.6). Because hydroxylaminodinitrotoluenes are often initial TNT metabolites in various microbial systems under anaerobic or aerobic conditions, they appear to play a crucial role in the initial binding process (see below).

4.4.3   Sorption Studies for Bioremediation of TNT-Contaminated Soils

Binding studies with humic acids and clay minerals indicate that an irreversible binding of certain reduction products of TNT, such as the isomeric hydroxylamino-dinitrotoluenes and TAT, with soil components is inevitable during reductive transformation of TNT (Figures 4.6 and 4.7). On one hand, the binding is a major hurdle because the metabolites become unavailable for further microbial degradation and mineralization. On the other hand, cometabolic reduction of TNT and related compounds opens a new possibility of soil remediation through cometabolically induced

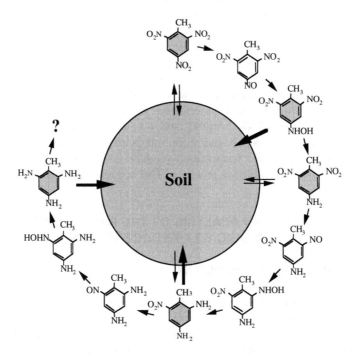

Figure 4.7   Interaction of TNT and its metabolites with soil (reversible sorption: thin black arrows; irreversible sorption: bold black arrows; metabolites that accumulate: gray structures).

immobilization. Because biological processes that lead to a complete mineralization of TNT are not yet available, an increasing number of decontamination techniques follow our previous suggestion of a biologically induced immobilization of the contaminants in soil.[20,86] The binding of the contaminants is achieved by activation of TNT through partial or complete reduction of the nitro groups. The different biological treatment systems mostly differ in the duration and intensity of the anaerobic or aerobic incubation, in the amount and type of external carbon sources added, and in the degree of water saturation.

It is well known that TNT is partially bound to soil organic matter when TNT-contaminated soil is treated in compost systems.[18,47,48,77] Combined "anaerobic/aerobic composting" was proposed recently in order to achieve a more complete incorporation of metabolized TNT in soil.[15] A few reports deal with an anaerobic incubation of TNT-contaminated soil in laboratory-scale slurry systems. In one of the remediation procedures tested with highly TNT-contaminated soil slurries (1% w/w), the addition of an external carbon source allows the creation of anaerobic conditions. Once such conditions were established, the transformation took place. Due to the fact that no $^{14}CO_2$ was produced in experiments with radiolabeled TNT, the fate of TNT is unclear. TAT, 2,4,6-trihydroxytoluene, and o-cresol were postulated as metabolites of TNT, but only in soil-free experiments.[35]

In order to optimize the soil slurry process, Roberts et al.[88] proposed combining the anaerobic treatment with an aerobic treatment to lower the amount of total organic carbon in the aqueous phase. The authors only described the disappearance of TNT and its reduction products 4-ADNT and 2,4-DANT from the aqueous phase, whereas the fate of TNT in the anaerobic or anaerobic/aerobic soil slurry remained unclear. Lenke et al.[63] carried out a similar anaerobic/aerobic treatment process which led to nitro group reduction and immobilization of TNT in soil. When aged contaminated soil from a former ammunition plant was treated in a laboratory soil slurry reactor (50% soil w/v), anaerobic bacteria fermented glucose to ethanol, acetate, and propionate. Simultaneously, TNT, 2-ADNT, and 4-ADNT present as contaminants were completely transformed and bound to the soil. A subsequent aerobic treatment was required to mineralize the fermentation products of glucose and to complete the immobilization and humification process of TNT and related compounds (see below).

4.5 STABILITY AND ANALYSIS OF THE BINDING STRUCTURES OF BIOLOGICALLY REDUCED TNT IN SOIL

4.5.1 Investigations with ^{14}C-TNT

If immobilization of the contaminants in soil is to be accepted as an option of remediation, the binding of the contaminants has to be analyzed carefully with regard to the stability and safety of the product. Biologically induced immobilization of TNT must be based on stable and covalent bonds between the contaminants and soil organic matter. The binding of the reduced TNT must also be stable under rigorous extraction conditions as well as under conditions of natural turnover of humic material. When aged TNT-contaminated soil (350 mg of TNT per kilogram soil) spiked with ^{14}C-TNT was treated by the anaerobic/aerobic process[63] in a soil slurry experiment, the cometabolic reduction of TNT led to a complete disappearance of extractable contaminants originated from TNT.[4] Whereas extractable 2-ADNT, 4-ADNT, and 2,4-DANT transiently appeared, only insignificant amounts of 2-HADNT/4-HADNT or TAT accumulated. This is in contrast to what has been observed in soil-free experiments.[21] As described above, 2-HADNT and 4-HADNT interact rapidly with soil components so that they appear only transiently when highly contaminated soil is treated anaerobically. In lightly contaminated soil the sorption process is so fast that free hydroxylaminodinitrotoluenes cannot be detected.

Formation of highly polar radiolabeled products in the early stage of the anaerobic/aerobic treatment (see above and Achtnich et al.[4]) indicated that in addition to the well known reduction of the nitro groups other secondary reactions may take place. Vanderberg et al.[99] reported that the oxidation of the methyl group of metabolites of TNT yielded 4-amino-2,6-dinitrobenzoic acid. The observation corresponds to the identification of trinitrobenzoic acids as well as aminonitrobenzoic acids in the drain water of TNT-contaminated areas or in TNT-contaminated soil.[90,91] Achtnich et al.[4] observed that in the early stage of the treatment, where the conditions were still aerobic or semi-aerobic, oxidative reactions can generate highly polar metabolites that react with humic acids. Characteristically, at the end of the entire anaerobic/aerobic treatment, only insignificant amounts of highly polar metabolites are still present in the supernatant.

TNT is not mineralized during the anaerobic/aerobic treatment process.[4] Similarly, mineralization rates of TNT (<0.3%) were insignificant in microcosms simulating bioslurry reactors with 40% soil content under anaerobic conditions.[92] Subsequent aerobic treatment did not improve the mineralization rate of TNT. In contrast to the soil slurry studies carried out under anaerobic[92] or combined anaerobic/aerobic conditions,[4] a laboratory study of a soil slurry bioremediation process indicated that 20% of the added ^{14}C-TNT was transformed to ^{14}CO$_2$.[105] The degradative mechanism for mineralization of TNT is not clear, although 2,3-butanediol, which was postulated as a metabolite of TNT without experimental evidence, would suggest substantial mineralization.[105] Some of the radiolabeled TNT was also bound to the soil fraction or in the biomass, indicating that immobilization of metabolized TNT cannot be avoided even if some of the TNT is mineralized.

Mass balances of the anaerobic/aerobic treatment described above indicate that 87% of the total radioactivity was immobilized.[4] After subsequent aerobic treatment, up to 98% of the radioactivity was associated with the soil and only ≤2% of the total radioactivity was still extractable from the soil. The extractable radioactivity was not TNT or any of its reduction products and seemed to be associated with solubilized soil organic matter. Thus, the biological anaerobic/aerobic treatment led to an almost complete binding of reduced TNT derivatives to soil. To our knowledge, this is the first study in which the total amount of radioactivity added to the soil slurry experiment was quantitatively bound to the soil. Other studies under anaerobic or anaerobic/aerobic conditions revealed a maximum immobilization of radiolabeled TNT of 72[26,92] or 84%,[28] indicating that extractable radioactivity was still present in the soil. Incomplete binding may be due to less extensive reduction of the nitro groups, which renders the metabolites less reactive for chemisorption.

During the early stage of the anaerobic treatment, physically bound TNT meta-bolites or residues can still be extracted by water or methanol.[4] In contrast, a strong and irreversible binding characterized the interaction of TNT derivatives during the late anaerobic and particularly during the subsequent aerobic incubation. Attempts to liberate TNT metabolites by breaking organic-inorganic aggregates of the treated soil failed, indicating that the binding of reduced metabolites of TNT cannot be explained by physical entrapment (sequestration). Rigorous solubilizing conditions and even strong hydrolytic conditions could mobilize only low amounts of the bound residues. According to Thorn et al.,[98] 1,4-addition reactions between the amino groups and the soil organic matter could explain the formation of bonds that resist hydrolysis.

Fractionation of the soil after the anaerobic/aerobic treatment showed that the majority of the TNT residues were bound to the humin fraction (85%) and only smaller amounts were found in the humic acids (8%) or in the fulvic acid fraction (7%). When the humic acid fraction was analyzed by size exclusion chromatography, the bulk of the radioactivity (95%) was bound to molecules larger than 5000 Da.[4] The results provide compelling evidence that the reduced metabolites of TNT are an integral part of humic substances and that they bind nonspecifically to the entire spectrum of molecular masses of humic acids.

Long-term leaching of the treated soil for 21 months simulating 2000 mm of rain per year,[4] again confirmed the irreversibility of the bound metabolites. Only low amounts of radioactivity could be eluted at the end of the treatment. The radioactivity eluted during the first month of the leaching study and did not corre-spond to TNT or its known metabolites. During that time, the eluate showed a slightly brownish color indicating that the released radioactivity arose from small molecules of the fulvic or humic acids that had incorporated some radiolabeled metabolites of TNT.

4.5.2 Characterization of Chemisorption of TNT Metabolites

Bioremediation strategies based on immobilization of reduced TNT in soil require a careful analysis of the stability of the bound residues, ideally by direct proof of irreversible chemisorption. Three different kinds of interactions between TNT and its metabolites are possible: (1) physisorption, (2) sequestration, and (3) chemisorption, i.e., covalent binding to soil organic matter. Only when TNT and its metabolites are bound through covalent linkages can they be considered an integral part of the humus and, thus, not present a potential hazard to the environment. In contrast, physisorption or sequestration is highly unlikely to explain strong immo-bilization of reduced TNT in soil.

A clear distinction between chemisorption (covalent binding) and physisorption, or sequestration, was required to demonstrate the covalent and irreversible character of the bound metabolites to soil components. A gentle derivatization technique (silylation) developed by Haider et al.[41,42] substitutes active hydrogens of functional groups in soil organic matter with trimethylsilyl residues. This causes a disintegration of the micelle structure of the organic matter and thus leads to a release of physically entrapped molecules into the organic solvent. The method was used to examine the binding character and particularly the sequestration of pesticides.[23,50,51]

The method described above can also be used for the analysis of soil bound residues formed during the biological treatment of TNT-contaminated soil. Aged TNT-contaminated soil from a former ammunition plant was spiked with ^{14}C-TNT. High doses of ^{15}N-TNT (4 g/kg) were added before the anaerobic/aerobic treatment of the soil slurry to allow analysis of the bound residues by ^{15}N NMR spectroscopy (see below)[1]. The anaerobic (51 days) and aerobic (32 days) phases led to a complete removal of TNT and the transiently formed metabolites 2-ADNT, 4-ADNT, and 2,4-DANT (Figure 4.8). In contrast to results in soil slurry experiments with lower TNT concentrations, significant amounts of the isomeric hydroxylaminodinitrotoluenes accumulated after 1 day of anaerobic incubation. The disappearance of the latter compounds was accompanied by the formation of azoxy compounds which disappeared during the ongoing anaerobic treatment (Figure 4.8).

All of the radioactivity was bound to the soil after the anaerobic/aerobic process which correlated well with the results the ^{14}C-TNT experiment described above. The amount of radioactivity detected in the fulvic and humic acid fraction was rather low (11 to 16%) and did not change significantly, whereas the radioactivity bound to the humin increased to 71%, indicating that the majority of the radioactivity was bound to the humin fraction. Since humin is extremely recalcitrant to further decomposition in soil, exhibiting a turnover time of hundreds of years,[96] it must be expected that the reduced metabolites of TNT covalently bound to soil organic matter also persist for a prolonged time span.

When a silylation procedure with trimethylchlorosilane was used to solubilize soil organic matter from the biologically treated ^{14}C-TNT-spiked soil,[2] 54% of the total radioactivity was found in the organic solvent (DMSO). Analysis of the extract by thin layer chromatography combined with radiocounting demonstrated that neither free TNT nor its metabolites were liberated by this method. The results clearly indicate that TNT or its metabolites were covalently bound to soil organic substances

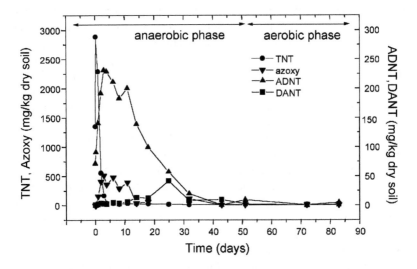

Figure 4.8 Reduction of TNT and its metabolites during biological anaerobic/aerobic treatment.

rather than sequestered. In contrast, when highly TNT-contaminated soil was only briefly treated anaerobically (8 days instead of 51 days) followed by an aerobic incubation, free TNT or its metabolites liberated by the silylation procedure were present at the end of the treatment.[2] The results indicate that metabolites of TNT might be physically entrapped (sequestered) during insufficient anaerobic treatment. In addition, the results also confirmed that the silylation technique allows a clear distinction between covalent binding and sequestration of the contaminants.

Unequivocal evidence for covalent binding required the application of solid-state ^{15}N NMR spectroscopy, which allows the analysis of the bulk sample without major pretreatment. Previous studies on soils and sediments indicated that it can be used to elucidate the chemical structure of immobilized soil organic nitrogen formed during the humification of biogenic precursors.[52,53] Application of the technique to a humic acid isolated from a soil incubated after addition of ^{15}N-TNT allows the determination of TNT-bound residues.[55] Soil bound residues of [13]C- or ^{15}N-enriched pesticides can also be analyzed by the above described silylation procedure which allows subsequent liquid-state NMR analysis.[23,50] Liquid-state NMR analysis was also used to demonstrate covalent bindings of anilines via 1,4 addition reactions to quinoid structures of humic acids.[98] Whereas the solid-state ^{15}N NMR spectroscopy allows the analysis of whole soil samples, ^{15}N NMR measurements of soil extracts have the advantage of a higher resolution and more precise assignment of the signals.

When TNT-contaminated soil spiked with ^{14}C-TNT and ^{15}N-TNT was treated by the above described anaerobic/aerobic process, all the radioactivity was immobilized in the soil.[1,4] In addition, all extractable TNT and its metabolites which transiently accumulated were totally removed after the complete treatment (Figure 4.8).

The solid-state ^{15}N NMR data confirmed that a considerable amount of TNT was already reduced to aminodinitrotoluenes and diaminonitrotoluenes during the first days of the anaerobic incubation.[54] Most of the products were extractable from the soil by methanol (Figure 4.8). Some of the nitrogen functions of the TNT metabolites were involved in condensation reactions as indicated by signals of azoxy-N; N-heterocyclic aromatic compounds such as indoles, pyrroles, quinolines, and anilidines; enamines; or amides. Although a considerable fraction of the azoxy compounds such as 2,2′,6,6′-tetranitro-4,4′-azoxytoluene and 4,4′,6,6′-tetranitro-2,2′-azoxytoluene were extractable by methanol (Figure 4.8), the major fraction of the immobilized residues was also due to azoxy species.

At the end of the anaerobic phase, azoxy-N and nitro groups were no longer detectable in the solid-state ^{15}N NMR spectrum, which indicated that they were almost completely reduced and/or transformed. In contrast, recent analysis of an anaerobic/aerobic compost treatment by solid-state ^{15}N NMR spectroscopy demonstrated that nitro groups were still present at the end of the process (see Chapter 13), indicating that extensive anaerobic treatment of the soil slurry is required for complete reduction of the nitro groups.

Condensed TNT metabolites and aromatic amines make up the majority of the bound residues in the soil sample at the end of the anaerobic phase. The relative decrease of the aromatic amines during the aerobic phase may be caused by oxidative transformations of free amino groups and subsequent condensation reactions.[54] Such condensation reactions of free amino groups are known as addition reactions of aniline to quinones or ketones of the native soil organic matter.[97,98] Our results indicate that the nitro groups of TNT are completely reduced and the resulting products are covalently incorporated into the soil organic matter.

The findings described above were confirmed by liquid-state ^{15}N NMR spectra of silylated whole soil samples taken after 1 day of incubation and after the complete treatment process.[1] Because the chemical entities revealed (*ortho* and *para* nitro groups of TNT, azoxynitrotoluenes, and aminonitrotoluenes) (Figure 4.9, spectrum A) are extractable by methanol and exhibited narrow line width of the ^{15}N NMR signals, covalent binding of early transformation products can be excluded. In contrast, the broad resonance signals of spectrum B after 83 days of incubation (Figure 4.9) clearly indicate that the reduced metabolites of TNT from extensive anaerobic treatment are covalently bound to soil organic matter. The spectrum of

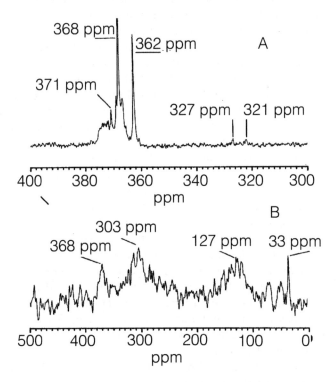

Figure 4.9 Inverse-gated decoupled ^{15}N NMR spectra of silylated whole soil: (A) sample after 1 day of the anaerobic/aerobic, (B) sample after 83 days of the anaerobic/aerobic incubation.
Signal assignments of spectrum A: 368 ppm, *ortho* nitro group of TNT; 362 ppm, *para* nitro group of TNT; 371, 327, and 321 ppm, azoxynitrotoluenes and aminonitrotoluenes.
Signal assignments of spectrum B: 365–375 ppm, nitro groups; 290–320 ppm, azoxy and imine; 110–140 ppm, amides, tertiary amines; 33 ppm, aliphatic amines or ammonia.

the silylated soil sample after the complete treatment shown in Figure 4.9B is representative of the NMR data measured in different soil fractions such as humic acids as well as in the silylated humin.

The liquid-state NMR spectra of the humic acid fraction indicate that presumably ionic bound azoxy species were formed during the early stage of the anaerobic treatment. Nitro and azoxy compounds disappeared completely during the ongoing anaerobic incubation, and all nitrogen was covalently bound to humic acids as substituted amines or amides at the end of the treatment process. Correspondingly, amides or tertiary amines were found in the silylated whole soil after 83 days of incubation (Figure 4.9B).

The NMR spectrum of silylated humin obtained from soil samples after the above described anaerobic/aerobic treatment suggests the formation of azoxy compounds and imine linkages. In contrast to the situation in the humic acid fraction, bound metabolites possessing nitro groups were still present after the anaerobic as well as aerobic treatment. Solid-state ^{15}N NMR analysis revealed that primary amines produced during the anaerobic incubation disappeared during aerobic treatment. Simultaneously, the amount of amide and tertiary amines increased. Nitro and azoxy residues of bound molecules were still present in humin at the end of the complete treatment. Aromatic amines of covalently bound molecules along with azoxy, imine, and nitro groups were also observed in the NMR spectra of the silylated whole soil sample measured at the end of the treatment. Nitro and azoxy groups were completely reduced and/or transformed when the whole soil sample was analyzed by solid-state ^{15}N NMR (Figure 4.9B). However, with the silylation procedure the extraction of soil organic matter is incomplete and, therefore, the amount of nitro and azoxy groups may be overestimated by the liquid-state ^{15}N NMR measurements.

Despite the above assumption, even extensive anaerobic incubation did not allow a complete transformation of all nitro groups present in the humin fraction. The transient formation of hydroxylaminodinitrotoluenes and the presence of free and bound azoxy compounds[1,54] indicate that condensation of the reactive hydroxylamines followed by subsequent integration of the azoxy species in soil organic matter is the first important binding mechanism (Figure 4.10). Aromatic amines produced from TNT were further transformed and incorporated to soil organic matter, leading to tertiary amines and amides. This appears to be the second important binding mechanism. Also, nitro groups of metabolites bound to the soil seemed to be further reduced and became available for further binding to soil organic matter. This explains why the process of complete reduction of nitro groups is rather slow during the later stage of the anaerobic treatment.[3]

Based on the results achieved by ^{14}C investigations and ^{15}N NMR, it can be concluded that the anaerobic/aerobic treatment leads to increasing sorption and finally to covalent binding of TNT in soil (Figure 4.10). The results indicate a rather efficient and safe process to remediate soil contaminated with TNT. Recent investigations revealed a rapid disappearance of TNT in the presence of Fe^{2+} sorbed to clay surfaces at reduced redox potentials.[84] Clearly, iron-reducing conditions led to a fast and complete reduction of TNT to TAT without significant accumulation of partially reduced TNT metabolites.[46] These observations suggest that the anaerobic/aerobic process could be improved by creating iron-reducing conditions which

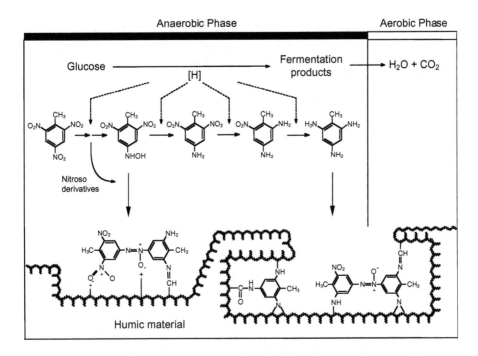

Figure 4.10 Proposed scheme for the reduction and covalent incorporation of TNT and its metabolites to soil organic matter during and after the anaerobic/aerobic treatment according to NMR studies.[1]

may also reduce nitro groups of metabolites that were already bound in the early stage of the anaerobic phase.

4.5.3 Immunological Investigations

The observation made by NMR measurements that nitro groups of TNT metabolites covalently bound to soil organic matter can be further reduced under anaerobic conditions was confirmed by an immunoassay. Pfortner et al.[81,82] developed an enzyme immunoassay system able to detect TNT covalently bound to humic acids. For this purpose, a conjugate was synthesized from N-(2,4,6-trinitrophenyl)-4-aminobutyric acid and then coupled to humic acids by amide bonds. A highly specific monoclonal antibody was generated against the TNT conjugate. The structure recognized by the antibody is a trinitrophenyl residue which is bound to soil organic matter via the methyl group and can serve as a model of a bound residue. Chemical changes of the TNT conjugate lead to structures that are not detected by the monoclonal antibody.[81,82] In addition, the experiment was designed to ignore free TNT or its metabolites completely. The TNT conjugate was treated biologically under anaerobic conditions in the presence of low concentrations of free TNT (0.03 mM) in order to test whether trinitrophenyl moieties can be converted during the cometabolic reduction process.[3] In contrast to the direct and fast reduction of free

TNT, the immunoassay signal due to a trinitrophenyl residue decreased significantly only after the major part of free TNT was reduced to 2,4-DANT. Although the reduction of the nitro groups of the trinitrophenyl conjugate proceeded much slower than the reduction of free TNT, the results clearly indicate that covalently bound trinitrophenyl residues were subject to further reductive transformations. Whereas the reduction of the free TNT may occur in the bacterial cells, the trinitrophenyl conjugate can only be transformed by reducing species outside the cells. This explains the slower reduction rate of the bound trinitrophenyl residues, especially because the humic acids bound to TNT might lower the reduction potential of the nitro groups.

Interestingly, aged TNT-contaminated soil from a former ammunition plant in Germany showed a strong signal in the immunoassay, indicating the presence of bound trinitrophenyl residues.[82] The residues could originate from reactions of soil organic matter with 2,4,6-trinitrobenzoic acid or the highly reactive 2,4,6-trinitrobenzaldehyde. Products with an oxidized methyl group were observed in aged TNT-contaminated soils.[16] When the aged TNT-contaminated soil was treated by the biological anaerobic/aerobic slurry process, complete transformation of TNT was accompanied by a slow decrease of the immunoassay signal. As observed with the TNT conjugate, the transformation of the covalently bound trinitrophenyl residues was much slower than the reduction of free TNT or its metabolites, which were preferentially transformed in the aqueous phase. The results clearly indicate that even TNT metabolites bound to humic acids can be further reduced during the anaerobic/aerobic process.

4.6 PERSPECTIVES OF BIOREMEDIATION OF TNT-CONTAMINATED SOIL

4.6.1 Technical Application

Most of the recent investigations on biodegradation of nitroaromatic compounds are based on the interest in remediation of soils contaminated with explosives such as TNT. Biological treatments are attractive alternatives to incineration because they are ecologically harmless due to low energy consumption, low emissions, and preservation of the biological activity of the soil.

Based on the experimental results described above, a two-stage anaerobic/aerobic process for the bioremediation of TNT-contaminated soil was tested on a technical scale at a former TNT production site in Germany.[63] A sludge reactor was filled with 18 m^3 of contaminated soil (TNT, 2,4-dinitrotoluene, hexahydro-1,3,5-trinitro-1,3,5-triazine [RDX] as major contaminants) and 10 m^3 of water. Anaerobic conditions were maintained in the soil slurry by periodic feeding of sucrose. As a result of the microbial reduction of the nitro group, TNT was completely and irreversibly bound to the soil. Subsequently, the sludge was dewatered and treated aerobically to aid the mineralization of the fermentation products formed from sucrose. It was clear that the aerobic step completed the bioremediation process through intensified immobilization and humification of the reduced contaminants as demonstrated by the analysis of the binding structures of reduced TNT via NMR spectroscopy.[1,54]

Chemical analysis of the contaminants revealed an overall decrease of more than 98% for TNT and the isomeric aminodinitrotoluenes that are normally found as side contaminants of TNT within 30 weeks. In addition, 2,4-DNT and RDX disappeared during the anaerobic/aerobic treatment process. Formation of 2,4-diaminotoluene indicated complete reduction of the nitro groups of 2,4-DNT via 2-amino-4-nitro-toluene and 4-amino-2-nitrotoluene as described previously.[11,75] The resulting 2,4-diaminotoluene bound irreversibly to soil components. Compared with TAT the interaction was slower and clearly dependent on the presence of oxygen for complete and irreversible binding. The results are consistent with those of Krumholz et al.[57] who reported a low sorption capacity of the diaminotoluenes to sediments under anaerobic conditions.

Unknown metabolites of RDX transiently accumulated in the soil slurry experiment, but disappeared completely during the anaerobic treatment. This corresponds to observations reported previously.[35] According to previous investigations,[72] it can be deduced that after successive reduction of the nitro groups, noncyclic degradation products are formed. Additionally, ^{14}C-RDX is partially mineralized under anaerobic conditions,[49,92] indicating that RDX is also partially degraded during the anaerobic/aerobic treatment described above.

An anaerobic full-scale bioremediation of TNT-contaminated soil by the SABRE™ (SIMPLOT) process was described by Funk et al.[36] Instead of sucrose a starchy carbon source was added in order to achieve anaerobic conditions by the oxidative utilization of starch by aerobic heterotrophs. Subsequently, TNT was degraded under anaerobic conditions. Although the conditions of anaerobic transformation of TNT were very similar to those described above,[63] different conclusions were drawn. Without presenting conclusive data, Funk et al. postulated[36] that the SABRE process led to a complete degradation of TNT under anaerobic conditions.

4.6.2   Ecotoxicological Analysis of the Treated Soil

Chemical analyses of the biologically treated soil demonstrated that the anaerobic/aerobic process led to complete disappearance of the contaminants from the soil. Since the analyses of the soil are limited to known and defined contaminants, ecotoxicological tests which identify the sum of all bioavailable toxic substances are recommended for the final assessment of residual toxins in the treated soil.[24] Thus, aquatic ecotoxicolagal tests with luminescence bacteria (*Vibrio fischeri*), daphnids (*Daphnia magna*), and green algae (*Scenedesmus subspicatus*) were carried out with soil eluates.[63] The tests were used for the determination of mobile contaminants or metabolites that can leach out of the soil. In contrast to the originally contaminated soil, no residual toxicity was detectable in the treated soil by aquatic tests systems. The observation is in accordance with chemical and spectroscopic analyses which indicate that the biologically reduced products of TNT are irreversibly bound in the soil and cannot become bioavailable as toxins. Similar results were obtained when the habitat function of the biologically treated soil was assessed using terrestrial ecotoxicological tests with plants (*Lepidivum sativum*) and earthworms. In addition, respiration and nitrification activity of the soil bacteria were undisturbed in the treated soil.[63] Therefore, according to a German guideline,[24] the soil can be recommended for reuse in green spaces, parks, and recreation areas.

4.7 PERSPECTIVES OF MINERALIZATION OF POLYNITROAROMATIC COMPOUNDS

Whereas mono- and occasionally dinitroaromatic compounds can be attacked by oxygenases and mineralized via oxidative pathways (see Chapter 2) only initial reductive mechanisms have been described so far for productive catabolism of trinitroaromatic compounds. High redox potential (Figure 4.2) and strong electron deficiency of the π-electron system favor initial reductive attack not only on the nitro groups (Figure 4.4a), but also on the aromatic ring (Figure 4.4b).

A case in point is picric acid, which has been used widely as a high explosive in shells. Because of its high solubility in water it has been identified as a contaminant in groundwater at former military sites and production facilities. Picric acid and 2,4-dinitrophenol (2,4-DNP) are also byproducts during large scale synthesis of compounds for a variety of industrial applications, including the manufacture of aniline and color-fast dyes. As documented by recent patents,[79,80] biodegradation of high loads of picric acid and 2,4-DNP in waste streams is of major industrial concern.

Under standardized conditions of biodegradation, such as the DOC Die-Away test with activated sludge, picrate is classified as a xenobiotic compound not readily biodegradable.[31] Nevertheless, under highly selective conditions using low concentrations of 2,4-DNP or picrate as the sole nitrogen source and a readily degradable carbon and energy source, Gram-positive bacteria such as *Rhodococcus erythropolis*[60] can be isolated. These organisms, which are only marginally present in unadapted activated sludge, harbor a highly efficient and unique catabolic system for 2,4-DNP and picrate. This unusual catabolic system became obvious when 2,4-DNP grown cells of HL PM-1 were exposed to picrate and a temporary orange-red color appeared in the culture medium.[60] The red color was not due to the reduction of a nitro group and formation of orange-red picramate. The red metabolite is also generated during growth of *R. erythropolis* strain HL PM-1 with picrate as the sole nitrogen source. Originally, the strain was isolated as a mutant of a 2,4-DNP-degrading microorganism, but it can also utilize picrate as a carbon and energy source. The red metabolite is seldom observed during aerobic degradation of picrate by other isolates. However, the catabolic mechanism of picrate seems to be the same or very similar in all of the isolates studied so far.[89]

Due to their negative mesomeric and also negative inductive effect, the three nitro groups of picrate cause a highly electron deficient aromatic system. This favors reductive attack on the π-electron systems with formation of an orange-red metabolite, the hydride–Meisenheimer complex of picrate (H⁻-picrate).[60] The formation of the latter complex has also been proven by *in vitro* studies with NADPH as the source of hydride.[87] For unequivocal evidence of the hydride–Meisenheimer complex as an initial key metabolite of picrate it was necessary to synthesize larger amounts of the pure complex chemically by kinetically controlled hydride transfer from

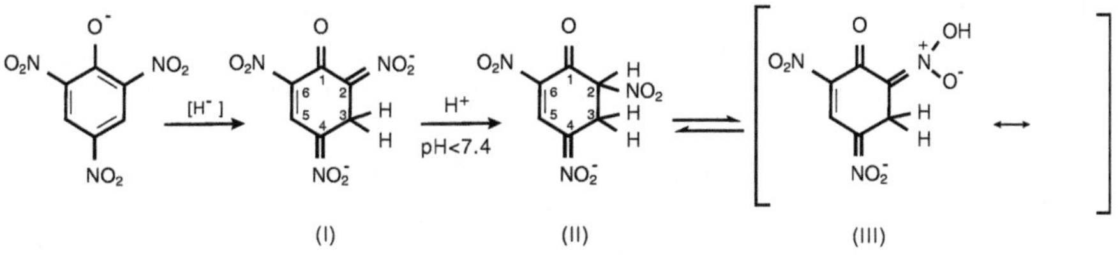

Figure 4.11 Formation of the hydride–Meisenheimer complex of picric acid (I), its protonated form (II), and the possible tautomeric aci-nitro form (III).

borohydrides.[87] Final structural evidence came from ^{1}H and ^{13}C NMR studies, particularly by analyzing the protonated hydride complex (compound II in Figure 4.11), which is sufficiently expressive for structural analysis. By addition of a proton at C2 the protonated form loses its planarity due to a different hydridization at C2 and displays different resonances in the region of the methylene proton. The existence and stability of H$^-$ picrate and its pH-dependent protonation could also be proven by ^{13}C NMR and UV-Vis spectroscopy. Biological hydride transfer to C3 of picrate is not followed by spontaneous rearomatization and nitrite elimination. Even protonation at C2 of the hydride complex does not promote spontaneous rearomatization because liberation of nitrite and 2,4-DNP formation did not occur under physiological or acidic conditions (pH $\geq$ 2).

Nitrite elimination from C2 of complex I is unfavorable because the complex is stabilized by resonance. The protonated species II, which is present under physiological conditions (pH $\leq$ 7.4), seems to be the reactive intermediate for enzymatic nitrite elimination.[87] Obviously, the initial key reaction for picrate degradation is the enzymatic hydride transfer to C3 of the aromatic nucleus, which is followed by protonation at C2 and subsequent enzymatic nitrite elimination. This chemoselective hydrogenation of the aromatic ring circumvents metabolic misrouting through nitro group reduction as observed with anaerobic sludge[89] and particularly with TNT as an acceptor of reduction equivalents (see below).

4.7.1   Hydride Transfer to Picrate and 2,4-Dinitrophenol Via Coenzyme F$_{420}$

A deeper insight into the biochemical mechanism of the hydride transfer to picrate has been gained from an enzyme system which has recently been purified from *Nocardioides simplex* FJ2–1A. It catalyzes the hydride transfer from NADPH to picrate or 2,4-DNP and consists of two protein components A and B.[31] Catalytic activity of hydride transfer is clearly dependent on NADPH as hydride donor and coenzyme F$_{420}$. The latter and component A, a F$_{420}$ reductase, are typical for methanogenic archaea. Protein A, factor F$_{420}$, plus a novel hydride transferase (component B) are part of an enzyme system that catalyzes hydride transfer from NADPH to the electron deficient trinitroaromatic ring. Interestingly, the N-terminal amino acid sequence of the F$_{420}$ reductase (component A) shows high similarity to the corresponding enzymes in methanogenic archaea in spite of the large evolutionary distance between the organisms.

Recently, the genes encoding protein component A and B in *R. erythropolis* HL PM-1 were cloned and sequenced. The amino acid sequence of the enzyme corresponding to component B shows homology to the N-terminal sequence of component B analyzed in *N. simplex* FJ2–1A as well as to F_{420}-dependent enzymes from archaea (see Chapter 5).

Despite its unusual mechanism the catabolic reaction sequence appears to be widespread in picrate and 2,4-DNP degrading bacteria.[89] All strains isolated so far from different places have been Gram-positive and exhibit the same characteristic blue-green fluorescence as *N. simplex* FJ2–1A, probably due to cofactor F_{420} (Ebert et al., unpublished).

The fate of 2,4-DNP is still unclear. The purified hydride transferase adds a hydride ion not only to picrate, but also to 2,4-DNP. 4,6-Dinitrohexanoate, which has been identified by earlier *in vivo* experiments with cells from *R. erythropolis* HL 24–2,[62] is also formed *in vitro* with the partially–purified enzymes from *N. simplex* FJ2–1A.[31] The stoichiometry calls for two hydride ions to be transferred. Hydride ion transfer to 2,4-DNP is supported by recent work from Behrend and Heesche-Wagner,[6] who described a hydride–2,4-DNP complex on the basis of NMR spectroscopic data. Thereafter, additional hydride transfer could generate a dihydride complex of 2,4-DNP, which after protonation would yield 2,4-dinitrocyclohexanone. According to the chemical reactivity described for similar structures the α-nitroketo group of the protonated dihydride complex of 2,4-DNP would be easily hydrolyzed to 4,6-dinitrohexanoate. This mechanism also explains the acid catalyzed formation of 1,3,5-trinitropentane from the $2H^-$–picrate complex, which is a metabolite of an unproductive side route of picrate transformation by various *R. erythropolis* strains.[60]

Additional support for the postulated catabolic sequence of picric acid and 2,4-DNP (Figure 4.12) comes from transformation of substituted 2,4-DNP with –Cl, –CH₃, or –NH₂ in the free *ortho*-position.[61] 2-Chloro-4,6-dinitrophenol (2-Cl-4,6-dinitrophenols) is metabolized with liberation of stoichiometric amounts of chloride and nitrite. In the absence of oxygen, resting cells transiently accumulated 2,4-dinitrophenols in the culture fluid, indicating reductive dechlorination of 2-Cl-4,6-DNP. The final product of this transformation is 4,6-dinitrohexanoate, which only accumulated in the absence of oxygen. Obviously, 2-Cl-4,6-DNP, like picrate, is subject to ring hydrogenation which leads to a hydride–Meisenheimer complex. That chloride instead of nitrite is eliminated is consistent with the fact that chlorine is the better anionic leaving group. Reductive chloride elimination is the crucial catabolic step of 2-Cl-4,6-DNP and allows its utilization as the sole source of nitrogen, carbon, and energy by *R. erythropolis* HL 24-2.[61]

In contrast, transformation of 2-amino-4,6-dinitrophenol (picramate) by resting cells of *R. erythropolis* HL 24-1 leads to reductive removal of one nitro group and 2-amino-6-nitrophenol as a dead end product. Apparently, the reaction generates a hydride complex which during subsequent rearomatization eliminates nitrite rather than ammonia. In this case preferential elimination of nitrite precludes utilization of picramate as a carbon and energy source.

2-Methyl-4,6-dinitrophenol (DNOC) clearly follows the initial catabolic sequence of 2,4-DNP (Figure 4.12). The methyl group in 2-position, however, interrupts the metabolic pathway at the level of 4,6-dinitrohexanoate. Consequently,

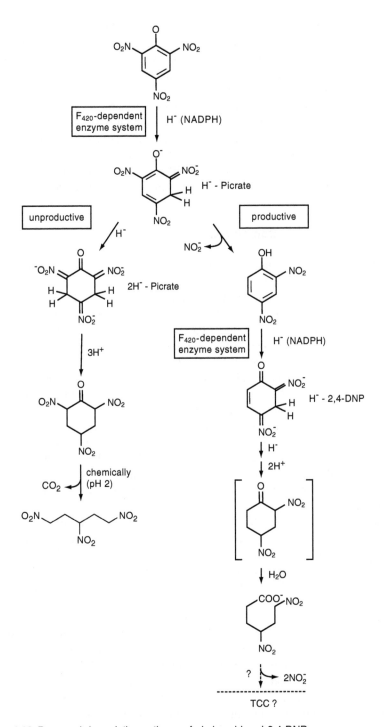

Figure 4.12 Proposed degradative pathway of picric acid and 2,4-DNP.

the diastereomeric 4,6-dinitro-2-methylhexanoates accumulated as deadend meta-bolites.[61] DNOC, by being funneled into the initial catabolic sequence of 2,4-DNP, serves as a chemical probe indicating that two hydride ions must be transferred. Subsequent protonation then gives rise to the α-nitroketo grouping which can easily hydrolyze to 4,6-dinitro-2-methylhexanoates as discussed above. That a mixture of diastereomers rather than a pure optical isomer is generated can easily be explained by the acidity of the hydrogen in α-position to the nitro group at carbon atom 4.

Further investigations and quantitative data with respect to the stoichiometry of hydride transfer to the dinitroaromatic ring system must clarify whether 4,6-dinitro-hexanoate is a real metabolite of 2,4-DNP or whether alternative routes with 3-nitroadipate[8] or nitrohydroquinone[6] as intermediates must be considered for the lower pathway of 2,4-DNP.

4.7.2   TNT-Hydride Complexes

Ring hydrogenation and subsequent rearomatization with elimination of nitrite, as described for picric acid (see above), could also represent an effective pathway for TNT degradation. Reductive denitration would yield 2,4- or 2,6-dinitrotoluenes which are biodegradable.[59,74] Such a reductive denitration mechanism has been proposed for TNT;[30] however, in our hands all attempts to reproduce the formation of the H$^-$–TNT complex and to identify nitrite or denitration products with TNT-utilizing strains were unsuccessful. Neither *Pseudomonas* sp. clone A (2NT$^-$)[30] nor any isolates enriched with TNT as the sole nitrogen source generated the H$^-$–TNT complex.[102] Also TNT-grown resting cells were unable to carry out ring hydrogenation or to generate major amounts of nitrite from TNT. Instead of the proposed dinitrotoluenes as products of reductive denitration, only products of gratuitous nitro group reduction were identified as dead end products. Therefore, with the bacterial strains exhibiting some growth with TNT as the sole nitrogen source, initial hydride transfer and subsequent nitrite elimination could not be identified as the major mechanism used by the organisms to provide themselves with nitrogen.

In our hands the ability to grow on highly pure TNT as the sole nitrogen source is attenuated with each transfer to a fresh mineral salts medium containing TNT. The observation indicates that the increasing inhibitory effect is not due to the toxicity of TNT itself. However, an increasing red-brown coloration of the medium and of the cells, which has also been observed in TNT-metabolizing cultures by other authors,[13,35] indicates diverse metabolic and chemical misrouting of TNT. These dead end metabolites and reactive intermediates such as hydroxylamino or nitroso intermediates may be subject to (auto)oxidation, condensation, and interaction with lipids, proteins, sugars, and even carboxylic acids of living cells.[17,19,64,68,71]

Thus, a major barrier to productive metabolism of TNT seems to be the toxic effects of metabolites and secondary products that are generated by gratuitous partial reduction of nitro groups. The fact that TNT-enriched strains can utilize part of the nitrogen of nitro groups as the sole source of nitrogen indicates that besides reductive elimination of nitrite other mechanisms of making nitrogen available to the bacterial cells must exist. Hydroxylaminodinitrotoluenes and even further reduced products may be key intermediates for the elimination of ammonia by some kind of Bamberger rearrangement or nitrite release by oxygenases (see Chapters 2 and 8).

Another possibility to accomplish productive breakdown of TNT is to use microorganisms that do not grow with TNT, but can generate the hydride-Meisenheimer complex of TNT.[102] *Mycobacterium* sp. HL 4-NT-1, a 4-nitrotoluene utilizing bacterium,[94] can reduce TNT to the H⁻–TNT complex.[101] The H⁻–TNT complex is further reduced to a yellow metabolite (Figure 4.13),[102] which by its spectroscopic properties has been identified as a protonated 3,5-dihydride complex of TNT (2H⁻–TNT). Conversions of TNT and of synthetic H⁻–TNT by picrate-induced cells of *R. erythropolis* HL PM-1 were so fast that H⁻–TNT did not accumulate in the culture fluid.

Although the H⁻–picrate complex undergoes enzymatic denitration and rearomatization, the corresponding reaction with the TNT complex yielding nitrite and dinitrotoluene does not occur. Instead, a second hydride ion is added to the H⁻–TNT complex.[102] The route of dihydride complex formation, however, is unproductive both for the catabolism of picric acid and TNT in *R. erythropolis*. Productive degradation of picric acid (Figure 4.12, pathway sequence to the right) requires that gratuitous reduction of nitro groups be prevented by hydride transfer to the aromatic nucleus. In the case of TNT the latter reaction is also the predominant initial reaction so that only ≤15% of the TNT is misrouted to hydroxylamino or amino derivatives. Obviously, denitration of H⁻–TNT as observed with H⁻–picrate is of minor importance. As a consequence, the misrouting to the TNT dihydro complex prevents productive breakdown of TNT. In unbuffered systems, however, the yellow 2H⁻–TNT complex is chemically unstable and gives rise to a spontaneous hydrolysis product. Although it could not be characterized completely, it lacks at least one nitro group originally present in TNT. Therefore, ring hydrogenation of TNT might provide the opportunity for developing a catabolic pathway that allows partial utilization of the xenobiotic, at least, as a nitrogen source. Unusual but attractive substrates for selecting such a productive pathway could be H⁻–TNT or even 2H⁻–TNT.

At the present time it is unknown whether a similar hydride transfer system to that described in *N. simplex* FJ2–1A[31] or *R. erythropolis* HL PM-1 (see Chapter 5) is present in bacteria such as *Enterobacter cloacae* PB2[34] or *Mycobacterium* sp. strain HL 4-NT-1[102] that gratuitously generates hydride complexes of TNT. The strain PB2 was isolated on the basis of its ability to utilize pentaerythritol tetranitrate (PETN) or glycerol trinitrate. The purified PETN reductase catalyzes the hydrogenolytic cleavage of the nitric acid ester bond and liberates nitrite plus the corresponding alcohol. It also reduces TNT to its hydride–Meisenheimer complex and even further to the dihydro complex. Reduction of TNT was also catalyzed by the 4-nitrotoluene utilizing *Mycobacterium* sp. strain HL 4-NT-1.[101-102] A reductive activity induced in 4-nitrotoluene grown cells catalyzes ring hydrogenation, although its function in 4-nitrotoluene catabolism is not obvious. In contrast to the system for productive catabolism of 2,4-DNP or picrate by *N. simplex* FJ2–1A or the *Rhodococcus* strains described above, the reductase systems of *E. cloacae* PB2 or *Mycobacterium* sp. strain HL 4-NT-1 were not identified as part of a complete metabolic sequence. Therefore, it will be interesting to see whether an F_{420} dependent system can be detected in other nonmethanogenic bacteria (see Chapter 7).

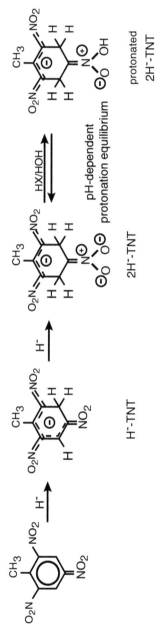

Figure 4.13 Hydride complexes of TNT.

4.8 CONCLUSION AND PROSPECTS FOR FURTHER DEVELOPMENT

From the current status of knowledge, mineralization and productive breakdown of polynitroaromatic compounds is still problematic. Chemical and biochemical misrouting is a major hurdle to the use of microbial systems to destroy polynitroaromatic compounds such as TNT. The high electron deficiency of the π-electron system favors initial reductive rather than oxidative transformations. As a consequence, nonspecific nitro group reduction gives rise to various modes of metabolic misrouting. First, products of nitro group reduction from TNT, particularly hydroxylaminodinitrotoluenes and triaminotoluene, are highly reactive toward oxidizing agents and electrophiles. Thus, the metabolites can generate di- and polynuclear compounds or form stable covalent bonds with the organic matrix of soil. Such transformations are irreversible and the products are biologically rather inert.

Second, electron deficiency of the π-electron system of the trinitroaromatic system also allows the reduction of the aromatic ring, i.e., the formation of hydride complexes. In the case of TNT the reaction seems to be another dead end pathway. An exception is hydride complex formation from picrate and 2,4-DNP. Certain Gram-positive bacteria such as *Rhodococcus* or *Nocardioides* strains have adopted an F_{420} dependent hydride transfer system from methanogenic bacteria as part of a complete catabolic pathway. This integrated reductive/oxidative sequence allows mineralization and utilization of the trinitroaromatic ring as the sole source of carbon, nitrogen, and energy.

In principle, TNT should be biodegradable by a two-stage reductive/oxidative process. Candidates for key metabolites in such a pathway are the products of reduction of the nitro groups or the aromatic ring system. For instance, the hydroxylamino toluenes could be substrates for mutases or hydroxyaminolyases which generate aminophenols or catechols. Such compounds and the triaminotoluene from complete reduction of TNT or its hydride complexes harbor structural features and sufficient chemical reactivities to be potential parent compounds of complete, mainly oxidative pathways. The potential function of TAT as a key metabolite in a two-stage reductive/oxidative process is limited, however, by its high reactivity toward oxygen, which gives rise to extensive misrouting and finally the generation of dark high molecular weight products.

In soil, however, the prospects for complete and productive catabolism of TNT are rather poor because aged contaminations always contain partly reduced derivatives, particularly the rather unreactive aminodinitrotoluenes. A way out of the dilemma is microbially induced immobilization of the contaminants in soil. The basis of such a novel approach is the reductive treatment which generates highly reactive species, particularly in the subsequent oxidative stage of the process that allows active binding even to a rather inert matrix like humin. An alternative, cost effective process could be composting where a highly reactive matrix (e.g., quinoid or semiquinoid humic acid precursors) may even bind less reactive species like aminonitrotoluenes. Such an approach, however, requires a better understanding of the mechanisms and kinetics of chemisorption. Therefore, engineering of the matrix by selecting the most appropriate additives for irreversible covalent binding of pollutants must be a major objective of future research.

REFERENCES

1. Achtnich, C., E. Fernandes, J.-M. Bollag, H.-J. Knackmuss, and H. Lenke. 1999. Covalent bindings of reduced metabolites of [^{15}N$_3$]TNT to soil organic matter during a bioremediation process analyzed by ^{15}N NMR spectroscopy. *Environ. Sci. Technol.* 33:4448–4456.

2. Achtnich, C., H. Lenke, U. Klaus, M. Spiteller, and H.-J. Knackmuss. 2000. Immobilization of TNT in soil as a function of nitro group reduction. *Environ. Sci. Technol.* submitted.

3. Achtnich, C., P. Pfortner, M. G. Weller, R. Niessner, H. Lenke, and H.-J. Knackmuss. 1999. Immunological detection and reductive transformation of bound trinitrophenyl residues during a biological treatment process. *Environ. Sci. Technol.* 33:3421–3426.

4. Achtnich, C., U. Sieglen, H.-J. Knackmuss, and H. Lenke. 1999. Irreversible binding of biologically reduced metabolites of TNT to soil. *Environ. Toxicol. Chem.* 18:2416–2423.

5. Alvarez, M. A., L. K. Christopher, J. L. Botsford, and P. J. Unkefer. 1995. *Pseudomonas aeruginosa* strain MAO1 aerobically metabolizes the aminodinitrotoluene produced by 2,4,6-trinitrotoluene nitro group reduction. *Can. J. Microbiol.* 41:984–991.

6. Behrend, C. and K. Heesche-Wagner. 1999. Formation of hydride-Meisenheimer complexes of picric acid (2,4,6-trinitrophenol) and 2,4-dinitrophenol during mineralization of picric acid by *Nocardioides* sp. strain CB 22-2. *Appl. Environ. Microbiol.* 65:1372–1377.

7. Beunik, J. and H.-J. Rehm. 1990. Coupled reductive and oxidative degradation of 4-chloro-2-nitrophenol by a co-immobilized culture system. *Appl. Microbiol. Biotechnol.* 34:108–115.

8. Blasco, R., E. Moore, V. Wray, D. Pieper, K. Timmis, and F. Castillo. 1999. 3-Nitroadipate, a metabolic intermediate for mineralization of 2,4-dinitrophenol by a new strain of a *Rhodococcus* species. *J. Bacteriol.* 181:149–152.

9. Bollag, J.-M., R. D. Minard, and S. Y. Liu. 1983. Cross-linkage between anilines and phenolic humus constituents. *Environ. Sci. Technol.* 17:72–80.

10. Bont, J. A. M. de and W. Harder. 1978. Metabolism of ethylene by *Mycobacterium* E20. *FEMS Microbiol. Lett.* 3:89–93.

11. Boopathy, R. and C. F. Kupla. 1993. Nitroaromatic compounds serve as nitrogen source for *Desulfovibrio* sp. (B. strain). *Can. J. Microbiol.* 34:430–433.

12. Boopathy, R., J. Manning, and C. F. Kulpa. 1998. A laboratory study of the bioremediation of 2,4,6-trinitrotoluene-contaminated soil using aerobic/anoxic soil slurry reactor. *Water Environ. Res.* 70:80–86.

13. Boopathy, R., J. Manning, C. Montemagno, and C. F. Kulpa. 1994. Metabolism of 2,4,6-trinitrotoluene by a *Pseudomonas* consortium under aerobic conditions. *Curr. Microbiol.* 28:131–137.

14. Brannon, J. M., C. B. Price, and C. Hayes. 1998. Abiotic transformation of TNT in montmorillonite and soil suspensions under reducing conditions. *Chemosphere.* 36:1453–1462.

15. Bruns-Nagel, D., O. Drzyzga, K. Steinbach, T. C. Schmidt, E. v. Löw, T. Gorontzy, K.-H. Blotevogel, and D. Gemsa. 1998. Anaerobic/aerobic composting of 2,4,6-trinitrotoluene-contaminated soil in a reactor system. *Environ. Sci. Technol.* 32:1676–1679.

16. Bruns-Nagel, D., T. C. Schmidt, O. Drzyzga, E. v. Löw, and K. Steinbach. 1999. Identification of oxidized TNT metabolites in soil samples of a former ammunition plant. *Environ. Sci. Pollut. Res.* 6:7–10.

17. Carpenter, D. F., N. G. McCormick, J. H. Cornell, and A. M. Kaplan. 1978. Microbial transformation of ^{14}C-labeled 2,4,6-trinitrotoluene in an activated-sludge system. *Appl. Environ. Microbiol.* 35:949–954.

18. Caton, J. E., C. Ho, R. T. Williams, and W. H. Griest. 1994. Characterization of insoluble fractions of TNT transformed by composting. *J. Environ. Sci. Health A.* 29:659–670.

19. Channon, H. J., G. T. Mills, and R. T. Williams. 1944. The metabolism of 2,4,6-trinitrotoluene (α-TNT). *Biochem. J.* 38:70–85.

20. Daun, G., H. Lenke, F. Desiere, H. Stolpmann, J. Warrelmann, M. Reuss, and H.-J. Knackmuss. 1995. Biological treatment of TNT-contaminated soil by a two-stage anaerobic/aerobic process, p. 337–346. In W. J. van den Brink, R. Bosman, and F. Arendt (eds.), *Contaminated Soil '95*, Kluwer Academic Publishers, The Netherlands.

21. Daun, G., H. Lenke, M. Reuss, and H.-J. Knackmuss. 1998. Biological treatment of TNT-contaminated soil I: Anaerobic cometabolic reduction and interaction of TNT and metabolites with soil components. *Environ. Sci. Technol.* 32:1956–1963.

22. Dawel, G., M. Kästner, J. Michels, W. Poppitz, W. Günther, and W. Fritsche. 1997. Structure of a laccase-mediated product of coupling of 2,4-diamino-6-nitrotoluene to guaiacol, a model for coupling of 2,4,6-trinitrotoluene metabolites to a humic organic matrix. *Appl. Environ. Microbiol.* 63:2560–2665.

23. Dec, J., K. Haider, A. Schaffer, E. Fernandes, and J.-M. Bollag. 1997. Use of silylation procedure and ^{13}C-NMR spectroscopy to characterize bound and sequestered residues of cyprodinil in soil. *Environ. Sci. Technol.* 31:2991–2997.

24. DECHEMA. 1995. Bioassays for Soils/Ad-Hoc-Committee "Methods for Toxicological/Ecotoxicological Assessment of Soils." DECHEMA, Deutsche Gesellschaft für Chemisches Apparatewesen, Chemische Technik und Biotechnologie e.V.

25. Dickel, O., W. Haug, and H.-J. Knackmuss. 1993. Biodegradation of nitrobenzene by a sequential anaerobic-aerobic process. *Biodegradation.* 4:187–194.

26. Drzyzga, O., D. Bruns-Nagel, T. Gorontzy, K.-H. Blotevogel, and E. v. Löw. 1999. Anaerobic incorporation of the radiolabeled explosive TNT and metabolites into the organic soil matrix of contaminated soil after different treatment procedures. *Chemosphere.* 38:2081–2095.

27. Drzyzga, O., D. Bruns-Nagel, T. Gorontzy, K.-H. Blotevogel, D. Gemsa, and E. v. Löw. 1998. Mass balance studies with ^{14}C-labeled 2,4,6-trinitrotoluene (TNT) mediated by an anaerobic *Desulfovibrio* species and an aerobic *Serratia* species. *Curr. Microb.* 37:380–386.

28. Drzyzga, O., D. Bruns-Nagel, T. Gorontzy, K.-H. Blotevogel, D. Gemsa, and E. v. Löw. 1998. Incorporation of ^{14}C-labeled 2,4,6-trinitrotoluene metabolites into different soil fractions after anaerobic and anaerobic-aerobic treatment of soil/molasses mixtures. *Environ. Sci. Technol.* 32:3529–3535.

29. Dunnivant, F. M., R. P. Schwarzenbach, and D. L. Macalady. 1992. Reduction of substituted nitrobenzenes in aqueous solutions containing natural organic matter. *Environ. Sci. Technol.* 26:2133–2141.

30. Duque, E., A. Haidour, F. Godoy, and J. L. Ramos. 1993. Construction of a *Pseudomonas* hybrid strain that mineralizes 2,4,6-trinitrotoluene. *J. Bacteriol.* 175:2278–2283.

31. Ebert, S., P.-G. Rieger, and H.-J. Knackmuss. 1999. Function of coenzyme F_{420} in aerobic catabolism of 2,4,6-trinitrophenol and 2,4-dinitrophenol by *Nocardioides simplex* FJ2–1A. *J. Bacteriol.* 181:2669–2674.

32. Esteve-Núñez, A. and J. L. Ramos. 1998. Metabolism of 2,4,6-trinitrotoluene by *Pseudomonas* sp. JLR11. *Environ. Sci. Technol.* 32:3802–3808.

33. Fiorella, P. D. and J. C. Spain. 1997. Transformation of 2,4,6-trinitrotoluene by *Pseudomonas pseudoalcaligenes* JS52. *Appl. Environ. Microbiol.* 63:2007–2015.

34. French, C. E., S. Nicklin, and N. C. Bruce. 1998. Aerobic degradation of 2,4,6-trinitrotoluene by *Enterobacter cloacae* PB2 and by pentaerythritol tetranitrate reductase. *Appl. Environ. Microbiol.* 64:2864–2868.

35. Funk, S. B., D. J. Roberts, D. L. Crawford, and R. L. Crawford. 1993. Initial-phase optimization for bioremediation of munition compound-contaminated soils. *Appl. Environ. Microbiol.* 59:2171–2177.

36. Funk, S. B., D. L. Crawford, R. L. Crawford, G. Mead, and W. Davis-Hoover. 1995. Full-scale anaerobic bioremediation of trinitrotoluene contaminated soil. *Appl. Biochem. Biotech.* 51/52:625–633.

37. Gilcrease, P. C. and V. G. Murphy. 1995. Bioconversion of 2,4-diamino-6-nitrotoluene to a novel metabolite under anoxic and aerobic conditions. *Appl. Environ. Microbiol.* 61:4209–4214.

38. Haderlein, S. B. and R. P. Schwarzenbach. 1995. Environmental processes influencing the rate of abiotic reduction of nitroaromatic compounds in the subsurface, p. 199–226. In J. C. Spain (ed.), *Biodegradation of Nitroaromatic Compounds*. Plenum Publishing Co., New York.

39. Haderlein, S. B., K. W. Weissmahr, and R. P. Schwarzenbach. 1993. Adsorption of substituted nitrobenzenes and nitrophenols to mineral surfaces. *Environ. Sci. Technol.* 27:316–326.

40. Haderlein, S. B., K. W. Weissmahr, and R. P. Schwarzenbach. 1996. Specific adsorption of nitroaromatic explosives and pesticides to clay minerals. *Environ. Sci. Technol.* 30:612–622.

41. Haider, K., M. Spiteller, A. Wais, and M. Fild. 1993. Evaluation of the binding mechanism of anilazine and its metabolites in soil organic matter. *Int. J. Environ. Anal. Chem.* 51:125–137.

42. Haider, K., M. Spiteller, K. Reichert, and M. Fild. 1992. Derivatization of humic compounds: an analytical approach for bound organic residues. *Int. J. Environ. Anal. Chem.* 46:201–212.

43. Haug, W., A. Schmidt, B. Nörtemann, D. C. Hempel, A. Stolz, and H.-J. Knackmuss. 1991. Mineralization of the sulfonated azo dye Mordant Yellow 3 by a 6-aminonaphthalene-2-sulfonate-degrading bacterial consortium. *Appl. Environ. Microbiol.* 57:3144–3149.

44. Hawari, J., A. Halasz, L. Paquet, E. Zhou, B. Spencer, G. Ampleman, and S. Thiboutot. 1998. Characterization of metabolites in the biotransformation of 2,4,6-trinitrotoluene with anaerobic sludge: role of triaminotoluene. *Appl. Environ. Microbiol.* 64:2200–2206.

45. Hawari, J., A. Halasz, S. Beaudet, L. Paquet, G. Ampleman, and S. Thiboutot. 1999. Biotransformation of 2,4,6-trinitrotoluene with *Phanerochaete chrysosporium* in agitated cultures at pH 4.5. *Appl. Environ. Microbiol.* 65:2977–2986.

46. Hofstetter, T. B., C. G. Heijman, S. B. Haderlein, C. Holliger, and R. P. Schwarzenbach. 1999. Complete reduction of TNT and other (poly)nitroaromatic compounds under iron-reducing subsurface conditions. *Environ. Sci. Technol.* 33:1479–1487.

47. Isbister, J. D., G. L. Anspach, J. F. Kitchens, and R. C. Doyle. 1984. Composting for decontamination of soils containing explosives. *Microbiologica.* 7:47–73.

48. Kaplan, D. L. and A. M. Kaplan. 1982. Thermophilic biotransformation of 2,4,6-trinitrotoluene under stimulated composting conditions. *Appl. Environ. Microbiol.* 44:757–760.

49. Kitts, C. L., D. P. Cunningham, and P. J. Unkefer. 1994. Isolation of three hexahydro-1,3,5-trinitro-1,3,5-triazine-degrading species of the family Enterobacteriaceae from nitramine explosive-contaminated soil. *Appl. Environ. Microbiol.* 60:4608–4711.

50. Klaus, U., S. Mohamed, M. Volk, and M. Spiteller. 1998. Interactions of aquatic humic substances with anilazine and its derivatives: the nature of the bound residues. *Chemosphere.* 37:341–361.

51. Klaus, U., T. Oesterreich, M. Volk, and M. Spiteller. 1998. Interactions of aquatic dissolved organic matter with amitrole and its derivatives: the nature of the bound residues. *Acta Hydrochim. Hydrobiol.* 26:311–317.

52. Knicker, H. and H.-D. Lüdemann. 1995. N-15 and C-13 CPMAS and solution NMR studies of N-15 enriched plant material during 600 days of microbial degradation. *Org. Geochem.* 23:329–341.

53. Knicker, H. and I. Kögel-Knabner. 1998. Soil organic nitrogen formation examined by means of NMR spectroscopy. p. 339–357 In B.A. Stankiewicz and P.F. van Bergen (eds.), *Fate of Nitrogen Containing Macromolecules in the Biosphere and Geosphere.* ACS Symposium Series. American Chemical Society, Washington, D.C.

54. Knicker, H., C. Achtnich, and H. Lenke. 1999. Solid-state ^{15}N NMR spectroscopic investigations on chemical alteration of TNT in soil induced by an anaerobic/aerobic bioremediation process. *J. Environ. Qual.*, submitted.

55. Knicker, H., D. Bruns-Nagel, O. Drzyga, E. v. Löw, and K. Steinbach. 1999. Characterization of ^{15}N-TNT residues after an anaerobic/aerobic treatment of soil/molasses mixtures by solid-state ^{15}N NMR spectroscopy. I. Determination and optimization of relevant NMR spectroscopic parameters. *Environ. Sci. Technol.* 33:343–349.

56. Koziollek, P., D. Bryniok, and H.-J. Knackmuss. 1999. Ethene as an auxiliary substrate for the cooxidation of *cis*-1,2-dichloroethene and vinyl chloride. *Arch. Microbiol.* L172:240–246.

57. Krumholz, L. R., J. Li, W. W. Clarkson, G. G. Wilson, and J. M. Suflita. 1997. Transformations of TNT and related aminotoluenes in groundwater aquifer slurries under different electron-accepting conditions. *J. Ind. Microbiol. Biotechnol.* 18:161–169.

58. Kudlich, M., P. L. Bishop, H.-J. Knackmuss, and A. Stolz. 1996. Simultaneous anaerobic and aerobic degradation of the sulfonated azo dye Mordant Yellow 3 by immobilized cells from a naphthalenesulfonate-degrading mixed culture. *Appl. Microbiol. Biotechnol.* 46:597–603.

59. Lendenmann, U., J. C. Spain, and B. F. Smets. 1998. Simultaneous biodegradation of 2,4-dinitrotoluene and 2,6-dinitrotoluene in an aerobic fluidized-bed biofilm reactor. *Environ. Sci. Technol.* 32:82–87.

60. Lenke, H. and H.-J. Knackmuss. 1992. Initial hydrogenation during catabolism of picric acid by *Rhodococcus erythropolis* HL 24-2. *Appl. Environ. Microbiol.* 58:2933–2937.

61. Lenke, H. and H.-J. Knackmuss. 1996. Initial hydrogenation and extensive reduction of substituted 2,4-dinitrophenols. *Appl. Environ. Microbiol.* 62:748–790.

62. Lenke, H., D. H. Pieper, C. Bruhn, and H.-J. Knackmuss. 1992. Degradation of 2,4-dinitrophenol by two *Rhodococcus erythropolis* strains, HL 24–1 and HL 24–2. *Appl. Environ. Microbiol.* 58:2928–2932.

63. Lenke, H., J. Warrelmann, G. Daun, K. Hund, U. Sieglen, U. Walter, and H.-J. Knackmuss. 1998. Biological treatment of TNT-contaminated soil. II. Biological induced immobilization of the contaminants and full-scale application. *Environ. Sci. Technol.* 32:1964–1971.

64. Leung, K. H., M. Yao, R. Stearns, and S.-H. Lee Chiu. 1995. Mechanism of bio-activation and covalent binding of 2,4,6-trinitroluene. *Chem. Biol. Interact.* 97:37–51.
65. Lewis, T. A., M. M. Ederer, R. L. Crawford, and D. L. Crawford. 1997. Microbial transformation of 2,4,6-trinitrotoluene. *J. Ind. Microbiol. Biotechnol.* 18:89–96.
66. Li, A. Z., K. A. Marx, J. Walker, and D. L. Kaplan. 1997. Trinitrotoluene and metabolites binding to humic acid. *Environ. Sci. Technol.* 31:584–589.
67. Liu, S. Y., R. D. Minard, and J.-M. Bollag. 1987. Soil-catalyzed complexation of the pollutant 2,6-diethylaniline with syringic acid. *J. Environ. Qual.* 16:48–53.
68. Liu, Y.-Y., A. Y. H. Lu, R. A. Staerns, and S.-H. Lee Chiu. 1992. In vivo covalent binding of [¹⁴C]trinitrotoluene to proteins in the rat. *Chem. Biol. Interact.* 82:1–19.
69. Lotrario, J. B. and S. L. Woods. 1997. Comparison of the correlation of MEP and the free energies of reaction with nitro group reduction of nitrotoluene congeners. *Bioremediation J.* 1:115–122.
70. McCormick, N. G., F. E. Feeherry, and H. S. Levinson. 1976. Microbial transformation of 2,4,6-trinitrotoluene and other nitroaromatic compounds. *Appl. Environ. Microbiol.* 31:949–958.
71. McCormick, N. G., J. H. Cornell, and A. M. Kaplan. 1978. Identification of biotransformation products from 2,4-dinitrotoluene. *Appl. Environ. Microbiol.* 35:945–948.
72. McCormick, N. G., J. H. Cornell, and A. M. Kaplan. 1981. Biodegradation of hexahydro-1,3,5-trinitro-1,3,5-triazine. *Appl. Environ. Microbiol.* 42:817–823.
73. Michels, J. and G. Gottschalk. 1994. Inhibition of the lignin peroxidase of *Phanerochaete chrysosporium* by hydroxylamino-dinitrotoluene, an early intermediate in the degradation of 2,4,6-trinitrotoluene. *Appl. Environ. Microbiol.* 60:187–194.
74. Nishino, S. F., J. C. Spain, H. Lenke, and H.-J. Knackmuss. 1999. Mineralization of 2,4- and 2,6-dinitrotoluene in soil slurries. *Environ. Sci. Technol.* 33:1060–1064.
75. Noguera, D. R. and D. L. Freedman. 1996. Reduction and acetylation of 2,4-dinitrotoluene by a *Pseudomonas aeruginosa* strain. *Appl. Environ. Microbiol.* 62:2257–2263.
76. Parris, G. E. 1980. Covalent binding of aromatic amines to humates. I. Reaction with carbonyl and quinones. *Environ. Sci. Technol.* 14:1099–1106.
77. Pennington, J. C., C. A. Hayes, K. F. Myers, M. Ochman, D. Gunnison, D. R. Felt, and E. F. McCormick. 1994. Fate of 2,4,6-trinitrotoluene in a simulated compost system. *Chemosphere.* 30:429–438.
78. Peres, C. M., H. Naveau, and S. N. Agathos. 1998. Biodegradation of nitrobenzene by its simultaneous reduction into aniline and mineralization of the aniline formed. *Appl. Microbiol. Biotechnol.* 49:343–349.
79. Perkins, R. E., J. Rajan, and F. S. Sariaslani. 1995. Microbially mediated degradation of nitrogen-containing phenol compounds by single bacterial isolates. U.S. Patent 5,478,743.
80. Perkins, R. E., J. Rajan, and F. S. Sariaslani. 1995. Single bacterial isolates for the degradation of nitrogen-containing phenol compounds. Patent WO 95/24465.
81. Pfortner, P., M. G. Weller, and R. Niessner. 1998. Detection of bound nitroaromatic residues in soil by immunoassay. *Fresenius Z. Anal. Chem.* 360:781–783.
82. Pfortner, P., M. G. Weller, and R. Niessner. 1998. Immunological method for detection of nitroaromatic residues bound to humic acids. Fresenius J. Anal. Chem. 360:192–198.
83. Preuss, A., J. Fimpel, and G. Diekert. 1993. Anaerobic transformation of 2,4,6-trinitrotoluene (TNT). *Arch. Microbiol.* 159:345–353.
84. Price, C. B., J. M. Brannon, and C. A. Hayes. 1997. Effect of redox potential and pH on TNT transformation in soil-water slurries. *J. Environ. Eng.* 123:988–992.

85. Rajan, J., K. Valli, R. E. Perkins, F. S. Sariaslani, S. M. Barns, A.-L. Reysenbach, S. Rehm, M. Ehringer, and N. R. Pace. 1996. Mineralization of 2,4,6-trinitrophenol (picric acid): characterization and phylogenetic identification of microbial strains. *J. Ind. Microbiol.* 16:319–324.

86. Rieger, P.-G. and H.-J. Knackmuss. 1995. Basic knowledge and perspectives on biodegradation of 2,4,6-trinitrotoluene and related nitroaromatic compounds in contaminated soil, p. 1–18. In J. C. Spain (ed.), *Biodegradation of Nitroaromatic Compounds.* Plenum Publishing Co., New York.

87. Rieger, P.-G., V. Sinnwell, A. Preuß, W. Francke, and H.-J. Knackmuss. 1999. Hydride-Meisenheimer complex formation and protonation as key reactions of 2,4,6-trinitrophenol biodegradation by *Rhodococcus erythropolis. J. Bacteriol.* 181:1189–1195.

88. Roberts, D. J., F. Ahmad, and S. Pendharkar. 1996. Optimization of an aerobic polishing stage to complete the anaerobic treatment of munitions-contaminated soil. *Environ. Sci. Technol.* 30:2021–2026.

89. Russ, R., S. Ebert, and H.-J. Knackmuss. 1998. Transformation of picrate and 2,4-dinitrophenol by microorganisms. Abstr. Q-85, p. 435. Abstract 98th Annu. Meet. Am. Soc. Microbiol., Atlanta, GA.

90. Schmidt, T. C., K. Steinbach, E. v. Löw, and G. Stork. 1998. Highly polar metabolites of nitroaromatic compounds in ammunition wastewater. *Chemosphere.* 37:1079–1091.

91. Schmidt, T. C., M. Petersmann, L. Kaminski, E. v. Löw, and G. Stork. 1997. Analysis of aminobenzoic acids in waste water from a former ammunition plant with HPLC and combined diode array and fluorescence detection. *Fresenius Z. Anal. Chem.* 357:121–126.

92. Shen, C. F., S. G. Giot, S. Thiboutot, G. Ampleman, and J. Hawari. 1998. Fate of explosives and their metabolites in bioslurry treatment processes. *Biodegradation.* 8:339–347.

93. Small, F. J. and S. A. Ensign. 1995. Carbon dioxide fixation in the metabolism of propylene and propylene oxide by *Arthrobacter* strain Py2. *J. Bacteriol.* 177:6170–6175.

94. Spiess, T., F. Desiere, P. Fischer, J. C. Spain, H.-J. Knackmuss, and H. Lenke. 1998. A novel degradative pathway of 4-nitrotoluene by a *Mycobacterium* sp. strain. *Appl. Environ. Microbiol.* 64:446–452.

95. Spiker, J. K., D. L. Crawford, and R. L. Crawford. 1992. Influence of 2,4,6-trinitrotoluene (TNT) concentration on the degradation of TNT in explosive-contaminated soils by the white-rot fungus *Phanerochaete chrysosporium. Appl. Environ. Microbiol.* 58:3199–3202.

96. Stevenson, F. J. 1989. *Humus Chemistry: Genesis, Composition, Reaction.* Wiley-Interscience Publ., John Wiley & Sons, New York.

97. Thorn, K. A. 1997. Covalent binding of the reductive degradation products of TNT to humic substances examined by N-15 NMR, p. 305–306. In Abstracts, The 213th ACS National Meeting. American Chemical Society, San Francisco, CA.

98. Thorn, K. A., P. J. Pettigrew, W. S. Goldenberg, and E. J. Weber. 1996. Covalent binding to humic substances. 2. ^{15}N NMR studies of nucleophilic addition reactions. *Environ. Sci. Technol.* 30:2764–2775.

99. Vanderberg, L. A., J. J. Perry, and P. J. Unkefer. 1995. Catabolism of 2,4,6-trinitrotoluene by *Mycobacterium vaccae. Appl. Microbiol. Biotechnol.* 42:937–945.

100. Volkel, W., A. Lienrack, and F. Andreux. 1995. Additions-reactions of herbicides and amino-carrying metabolites on methylcatechol, a precursor of humic substances. *C. R. Acad. Sci.* 320:103–108.

101. Vorbeck, C., H. Lenke, P. Fischer, and H.-J. Knackmuss. 1994. Identification of a hydride-Meisenheimer complex as a metabolite of 2,4,6-trinitrotoluene by a *Mycobacterium* strain. *J. Bacteriol.* 176:932–934.

102. Vorbeck, C., H. Lenke, P. Fischer, J. C. Spain, and H.-J. Knackmuss. 1998. Initial reductive reactions in aerobic microbial metabolism of 2,4,6-trinitrotoluene. *Appl. Environ. Microbiol.* 64:246–252.

103. Weber, E. J., M. S. Elcovitz, D. G. Truhlar, J. C. Cramer, and S. E. Barrows. 1996. Factors controlling regioselectivity in the reduction of polynitroaromatics in aqueous solution. *Environ. Sci. Technol.* 30:3028–3038.

104. Weissmahr, K. W., S. B. Haderlein, and R. P. Schwarzenbach. 1998. Complex formation of soil minerals with nitroaromatic explosives and other π-acceptors. *Soil Sci. Soc. Am. J.* 62:369–378.

105. Widrig, D. L., R. Boopathy, and J. F. Manning. 1997. Bioremediation of TNT-contaminated soil: a laboratory study. *Environ. Toxicol. Chem.* 16:1141–1148.

106. Ziechmann, W. 1994. *Huminstoffe*. Verlag Chemie, Weinheim.

CHAPTER 5

Identification of Genes Involved in Picric Acid and 2,4-Dinitrophenol Degradation by mRNA Differential Display

Rainer Russ, Dana M. Walters, Hans-Joachim Knackmuss,
and Pierre E. Rouvière

CONTENTS

5.1 INTRODUCTION

Nitrophenols have been used extensively in the industrial production of pesticides, dyes, and explosives. The U.S. Environmental Protection Agency (EPA) lists several mononitrophenols, dinitrophenols, and 2,4,6-trinitrophenol (picric acid) on its Emergency Planning and Community Right-to-Know Act (EPCRA) list describing hazardous and toxic chemicals (EPA, 1998). The toxicity of 2,4-dinitrophenol (2,4-DNP) and picric acid is described in more detail later in this chapter.

Today, one of the main sources of dinitrophenol and picric acid in industrial effluents is the production of nitrobenzene from benzene via adiabatic nitration (727,000 t/year in 1999 in the U.S., SRI International, Menlo Park, CA). During the process, 0.1% of the benzene consumed ends up as picric acid and dinitrophenols

(SRI International, Menlo Park, CA).[35] Picric acid is used for the production of 2-amino-4,6-dinitrophenol (picramic acid), which is a feedstock for the synthesis of azo dyes.[16] In addition, 2,4-DNP is used for manufacturing of dyes.[16]

The environmental release of picric acid and 2,4-DNP, as well as other toxic chemicals in the U.S., is listed in the Toxic Release Inventory (TRI) of the EPA and is posted on their World Wide Web pages. In the U.S., the release of picric acid has decreased dramatically from 618 t in 1988 to 20 to 50 t in 1997 (TRI Explorer, http://www.epa.gov/ceisweb1/ceishome/ceisdata/xplor-tri/explorer.htm).

Multiple mechanisms contribute to the toxicity of picric acid. Through external contacts it causes irritations, burns, and allergies. Upon ingestion, acute toxicity is probably due to acidosis (LD_{50} in rats of 200 to 290 mg/kg).[63] Chronic exposure may cause damage to red blood cells, liver, and kidneys. In animals the metabolism of picric acid is limited to reduction of nitro groups to amines with subsequent acetylation of the amino function. In rats, almost 40% of the picric acid is excreted as derivatives of picramic acid (2-amino-4,6-dinitrophenol).[62,63] In contrast to picric acid, picramic acid, as well as the nitroso or hydroxylamino intermediates of picric acid and picramic acid, are mutagenic.[30,54,55,59,63] Because of its charge, picric acid does not accumulate in lipoid tissues.[9,63] Picric acid has much less of an uncoupling effect than 2,4-DNP because it has a higher acidity (pKa of 0.3) than 2,4-DNP (pKa of 4.1) (see below).[32]

While the environmental release of 2,4-DNP in the U.S. has dramatically decreased from 217 t in 1989 to 24 t in 1997 (TRI Explorer, 1999), the rising rank of 2,4-DNP on the Comprehensive Environmental Response and Compensation Liability Act (CERCLA) priority list indicates that its relative importance as a pollutant is increasing. 2,4-DNP is much more toxic than picric acid. The LD_{50} value for rats is 30 mg/kg (U.S. DHHS), only around one tenth of the value for picric acid. Long-term exposure can lead to skin, eye, bone marrow, central nervous system, and cardiovascular system damage (reference dose [RfD] is 0.002 mg/kg/d). The acute toxicity comes from the ability of 2,4-DNP to function as an "uncoupler."[3,19] As a weak acid, 2,4-DNP can cross the membrane in its protonated form, acting as an H^+ carrier (H^+ ionophore), dissipating the electrochemical gradient across cell membranes and thus uncoupling the oxidative phosphorylation pathway (the energy production of the cell as ATP) without blocking oxygen consumption (cellular respiration).[2] The uncoupling effect of 2,4-DNP is supported by the fact that its toxicity is dependent on the external pH, with a higher external pH resulting in a decreased toxicity of 2,4-DNP. This has been demonstrated for a variety of organisms including the bacterium *Escherichia coli*,[32] the ciliate *Tetrahymena*,[34] the yeast *Saccharomyces cerevisiae*,[51] and plants.[46,49,50]

Because polynitrophenols are pollutants with a high toxicity to humans, animals, plants, and bacteria, it is important to analyze the fate of these compounds in the environment and to establish remediation strategies for polluted wastewater. Nitrophenols are more water soluble than the nitrotoluenes and can be leached into the groundwater.[46] Sequestration in soil organic matter results from the reduction of nitro groups by biological (biotic) as well as abiotic reactions (see Chapter 4). Finally, the photochemical degradation of different nitrophenols, including 2,4-DNP and picric acid, was described recently.[37]

Complete removal of nitrophenols from the environment by bacterial mineralization has been characterized extensively. Mononitrophenols are degraded by
various bacteria via oxidative removal of the nitro group as nitrite or partial
reduction to the respective hydroxylamine and subsequent release of ammonium
(see Chapters 2 and 6). A dioxygenase catalyzes the initial oxidative attack on
2,6-DNP in *Ralstonia eutropha*.[15]

None of the above oxidative mechanisms could be demonstrated for the productive degradation of picric acid or 2,4-DNP by bacteria. Instead, Lenke and
Knackmuss described an initial reduction of the aromatic ring involving the transfer
of a hydride ion, resulting in the formation of a hydride–Meisenheimer complex of
picric acid.[23,24] The complex rearomatizes by enzymatic elimination of nitrite to
produce 2,4-DNP.[24,41] Very recent publications indicate that the initial reduction
mechanism is also used for the transformation of 2,4-DNP to its corresponding
Meisenheimer complex (Figure 5.1).[4,14] The downstream reaction sequence leading
to the complete mineralization of picric acid is still unknown but it is proposed to
involve 4,6-dinitrohexanoate as an intermediate (see Chapter 4).

With one exception, only high G+C Gram-positive bacteria (*Actinobacteria*)
have been reported to degrade picric acid or 2,4-DNP. Hess et al. described the only
exception, a *Janthinobacterium* which used 2,4-DNP as a nitrogen source, albeit at
a concentration lower than 0.1 mM.[20] Because of their very thick cell envelope,[53]
Actinobacteria seem *a priori* more likely to tolerate the strong uncoupling property
of 2,4-DNP and the weaker one of picric acid. In fact, the ability to tolerate picric
acid was proposed as a taxonomical feature to characterize mycobacterial species.[26,45]
The resistance to the uncoupling effect of 2,4-DNP and picric acid, nevertheless,

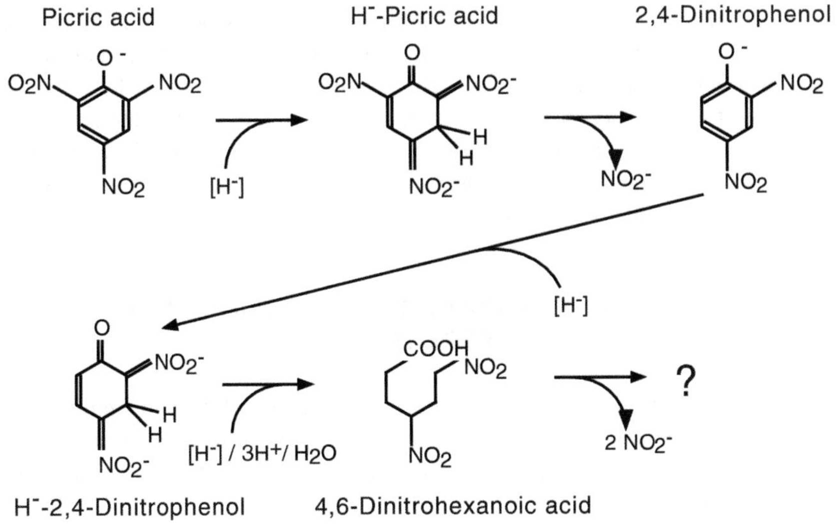

Figure 5.1   Proposed sequence for the degradation of picric acid and 2,4-DNP based on the
literature.[4,14,23,24,41,44] H⁻–Picric acid and H⁻-2,4-DNP represent the Meisenheimer
complexes of picric acid and 2,4-DNP, respectively.

may not be the reason for the "specialization" of high G+C Gram-positive bacteria to degrade the two compounds. Although 2,6-DNP is also an uncoupling agent of strength intermediate between that of 2,4-DNP and picric acid,[6,8] Gram-negative bacteria able to degrade 2,6-DNP have been isolated.[15,43,44] Similarly, a recent publication describes a mixed culture able to degrade 2-methyl-4,6-DNP and 2,4-DNP, but not 2,6-DNP.[18]

It seems, rather, that the difference in the ability of bacteria to degrade the two dinitrophenols is based on the ability of dioxygenases to attack the compounds. The productive degradation of 2,6-DNP is initiated by a dioxygenase,[15] but the described strain is not able to oxidize 2,4-DNP (Lenke, personal communication). The same is true for other Gram-negative strains (Russ, unpublished). On the other hand, nine bacterial strains have been reported to degrade 2,4-DNP by a reductive attack on the aromatic ring.[4,5,24,44] The observations provide a strong indication that dioxygenase attack on 2,4-DNP is, if not impossible, at least very difficult.

Recently, Ebert et al. reported that in *Nocardoides simplex* the reduction of picric acid or 2,4-DNP involves the unusual redox cofactor deazaflavin F_{420}[14] (see Chapter 4). The unusual ability of F_{420}-dependent enzymes to perform the hydride transfer could explain the phylogenetic distribution of picric acid and 2,4-DNP degraders. Cofactor F_{420} has so far been described only in archaea, cyanobacteria, and high G+C Gram-positive bacteria, but, to our knowledge, never in Gram-negative bacteria.[11,38,39]

Although the degradation pathway of 2,4-DNP and picric acid is not completely understood so far, the compounds can be mineralized by bacteria.[40] Thus, these organisms could be used for the remediation of industrial effluents resulting from nitrobenzene production. To develop a technical process that completely mineralizes the nitrophenols, an understanding of the underlying biochemical reactions and of the genes encoding the enzymes of the pathway is necessary. In addition to biochemical investigations which have not completely elucidated the pathway yet, molecular approaches can be used. The identified genes should deliver suitable sequences for constructing gene probes to detect polynitrophenol degrading organisms in ecological systems such as wastewater treatment plants. Thus, the rest of this chapter covers the development of a new molecular approach to identify the genes involved in the degradation of picric acid and 2,4-DNP.

5.2 IDENTIFICATION OF METABOLIC GENES
BY DIFFERENTIAL DISPLAY

The identification of metabolic genes is an arduous task and no single approach is guaranteed success. Until recently, two strategies have been utilized for the identification of these metabolic genes: a "direct" genetics approach and a "reverse" genetics approach.

The direct genetic strategy relies on the expression of genes *in vivo*, either by the isolation of specific mutants of the original strain that are defective in the genes encoding the enzymes of interest.[12] or by the expression of these genes in a heterologous host.[52,64] The strength of the direct genetic approach is that it allows the direct identification of DNA fragments carrying the genes of interest. However, this approach can only be used for organisms that have a developed genetic system or whose genes can be expressed in related heterologous hosts. With respect to bacteria encountered in microbial biodegradation applications, heterologous hosts are usually restricted to *E. coli* and a few Pseudomonads (*Pseudomonas aeruginosa, P. putida*). Successful identification of degradation genes by complementation of Gram-negative heterologous hosts is less likely when the source of the DNA is a high G+C Gram-positive organism such as *Arthrobacter, Mycobacterium, Nocardioides*, or *Rhodococcus*. Not only do the regulatory components of gene expression differ from those of Gram-negative organisms, but their high G+C content limits the recognition of promoter-like sequences in A+T rich regions. Finally, even if a gene can be successfully expressed in a Gram-negative host, it is rare that all the genes of a pathway can be successfully expressed to confer the degradative capabilities to the host. Genetic tools are being developed for environmental isolates of high G+C Gram-positive organisms,[10] but they are still primitive. To our knowledge, in the field of nitroaromatic degradation, no genes from this class of microbes have been identified by a direct genetic approach.

The second traditional approach is termed "reverse genetics." It involves the purification of the enzyme of interest followed by the identification of its gene. In principle, this approach can be used with any microorganism and is very general. The early reverse genetics approach used the purified enzyme to raise antibodies that could be used to screen libraries.[7] A major limitation to this approach is the necessity for expression of the cloned fragments in the cloning host.

An alternative method which does not rely on expression of the cloned genes consists of obtaining some amino acid sequence information, both from the N terminus and from proteolytic digests of the pure protein. Degenerate oligonucleotide primers can then be designed from the amino acid sequence[33] and used to direct the amplification of a small DNA fragment to be used as a probe to screen clones in a library.

The reverse genetics strategy may be derailed if the enzyme activity cannot be assayed easily, if the enzyme resists purification, if the enzyme activity is labile, or if the enzyme is not abundant. It can also fail if insufficient amino acid information is obtained or if it leads to the synthesis of an oligonucleotide probe which is too degenerate and fails to isolate the correct DNA fragment. The case of the picric acid and 2,4-DNP degradation pathway is an example where both direct and reverse genetic strategies have failed to rapidly identify the genes involved (see below).

Recently, a new approach to gene discovery was developed that does not rely on a genetic system or on protein purification. This technique, called mRNA differential display, is a comparative technique that looks for differences in gene expression at the level of the mRNA between a control culture and a culture induced to express the enzymes of interest, such as those of a degradation pathway for xenobiotic compounds. mRNA species present in the induced culture, but not in the control culture, are candidates for the inducible genes encoding the metabolic activity studied.

The comparison is performed by sampling the mRNA pool of each culture, using a reverse transcriptase enzyme that copies the mRNA into a complementary DNA

molecule (cDNA). In the absence of sequence information to design the oligonucleotide used to initiate the reverse transcription (RT) reaction, primers of arbitrary
sequence are used under conditions of low stringency to allow base pairing when
the overall sequence of the primer and that of the nucleic acid template match only
partially.[28,58,60] From this single-strand DNA molecule, a second complementary
strand is synthesized in a polymerase chain reaction (PCR) step using the same
arbitrary primer under low stringency conditions. The double-stranded DNA molecule can then be amplified by PCR to yield enough material for further manipulation
and analysis (Figure 5.2). The latter step is essentially identical to that of random
amplification of polymorphic DNA (RAPD).[60] The outcome of each RT-PCR reaction is the generation of a discrete (10 to 30) number of DNA fragments, each
representing an individual sampling of the RNA population. These DNA fragments,
usually between 300 and 800 bp in length, can be separated by gel electrophoresis
to yield a pattern of DNA bands (Figure 5.3). Each RT-PCR directed by a different
primer will yield a different sampling of the RNA population. Typically, a series of
reactions, each directed by a different primer of arbitrary sequence, is performed.[58]

When the RNA population of the control culture and the induced culture are
sampled in parallel by RT-PCR, most of the RNAs sampled will be the same,
corresponding to genes equally expressed in both cultures. However, a particular
primer may sample a gene only expressed in the induced culture and yield a DNA
fragment not found in the pattern obtained from the control culture mRNA
(Figure 5.3). The DNA is then eluted from the gel, reamplified by PCR, and cloned.
The sequences of the clones are then determined and analyzed.

The overwhelming majority of applications of differential display have been with
eukaryotes and make use of the poly-A tails at the end of eukaryotic mRNA

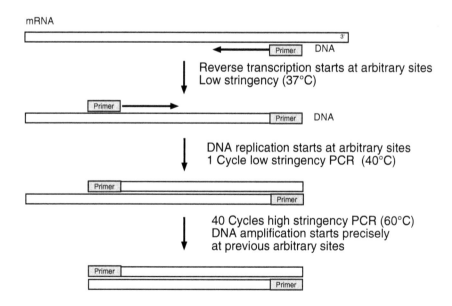

Figure 5.2   Sampling RNA by low stringency reverse transcription and PCR.

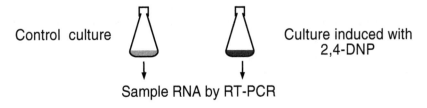

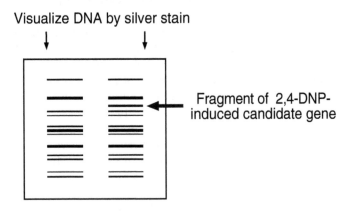

Figure 5.3   Schematic representation of a differential display experiment.

molecules.[27,28] The approach may be useful to identify the genes of fungi that degrade nitroaromatic compounds, but it cannot be applied to prokaryotes because their mRNA lacks poly-A tails. The few applications of differential display to prokaryotes involve the use of primers of arbitrary sequence for the RT step as described above.[1,17,42,47,61,65,66] Many variations of protocol have been described. The variations deal with (1) the use of separate[1,17,65,66] or identical[47,61] (Rouvière, unpublished results) primers for the RT and PCR step, (2) the use of a collection of hexamer nucleotides for the RT step to completely transcribe the mRNA pool,[65] (3) the number of arbitrary primers used in parallel reactions (from a few[61] to a series of 240) (Russ et al., unpublished results), (4) the use of distinct reactions for the RT and PCR steps vs. a single tube reaction (Russ et al., unpublished results), (5) the gel used for separation of the RAPD DNA fragments (agarose gels,[65] polyacrylamide sequencing gels,[1,17,42,61] short horizontal precast polyacrylamide gels) (Russ et al., unpublished results), and (6) the method used for visualization of the DNA (ethidium bromide,[65] 35S labeled dATP,[17,61] and silver stain[13,29,31]).

The use of differential display for prokaryotic applications is still very limited, with fewer than ten published applications. This lack of use of differential display by microbiologists explains why, in part, no protocol has yet become standard. Very few novel bacterial genes have been identified with this approach. Beyond the work we report here, only Fleming et al. have succesfully used the approach to identify genes of a degradation pathway. They describe the identification of a salicylate-inducible naphthalene dioxygenase gene in *Burkholderia cepacia* JS150.[17] We believe that differential display is particularly adapted to the identification of new biodegradation genes in prokaryotes because catabolic genes are often inducible, thus lending themselves to comparative analysis, and are often strongly expressed, which increases the chance of sampling by RT-PCR. We have used the approach to rapidly identify a gene cluster involved in the degradation of picric acid and 2,4-DNP.

5.3 IDENTIFICATION OF A GENE CLUSTER INVOLVED IN THE DEGRADATION OF PICRIC ACID AND 2,4-DINITROPHENOL

The organization of the genes that encode the enzymes involved in the degradation of picric acid and 2,4-DNP should be elucidated for the following reasons: (1) after 10 years of biochemical investigations, large parts of the metabolic pathway are still unknown; (2) the genes should deliver suitable DNA sequences to design gene probes to detect polynitrophenol degrading organisms in various ecosystems; and (3) to optimize the functioning of wastewater plants treating effluents containing nitrophenols.

However, until recently the identification of the genes involved in the degradation of picric acid and 2,4-DNP has not been successful. Because all isolates capable of degrading picric acid belong to genera of high G+C Gram-positive bacteria, it has not been possible to use the approach of genetics to identify their genes. Furthermore, the biochemistry of picric acid reduction had been rather difficult to tackle. Ring reduction of picric acid with NADPH as an electron donor proceeds readily in crude extract, but the activity disappears upon fractionation of the extracts,[4,41] suggesting a multicomponent system. Only in the last year have Ebert et al. started to dissect this enzymatic system[14] (see Chapter 4). Thus, the identification of the genes involved in the degradation of picric acid and 2,4-DNP was a good case to test the alternative approach of gene discovery by differential display because traditional approaches had not yet been successful. We chose the strain *Rhodococcus erythropolis* HL PM-1,[24] in which the degradation pathway of picric acid and 2,4-DNP is inducible in contrast to other strains such as *N. simplex* FJ2-1A, where the pathway is constitutive (Russ et al., unpublished results). As in all differential display experiments, the RNA was extracted from cultures where the induction had been thoroughly characterized.

A series of 240 RT-PCR reactions, each directed by a different primer of arbitrary sequence, was used to sample the RNA population of a control culture growing on acetate and of a culture induced to degrade picric acid and 2,4-DNP by addition of 2,4-DNP. Forty-eight bands were identified as differentially amplified from the mRNA of the 2,4-DNP-induced culture. All the bands were reamplified, cloned, and their nucleotide sequence determined. When the resulting sequences were aligned, several contiguous sequences were assembled from sequences of DNA fragments generated in separate RT-PCR reactions using different primers. They correspond to instances where a single multicistronic mRNA is sampled repeatedly, suggesting that this sequence corresponds to the mRNA of an operon induced by 2,4-DNP.

Sequence analysis of the diffentially expressed DNA fragments revealed that two of the assembled contiguous sequences coded for F_{420}-dependent enzymes with

Table 5.1   Homologies of Contiguous Sequences Assembled from More than One Band and More than One Primer

Sequence Similarity	Multiplicity of Sampling	Contiguous Sequence Size	Part of Picric Acid Gene Cluster
Dehydratase/F_{420}-dependent dehydrogenase/aldehyde dehydrogenase	6 Primers / 8 Bands	1.7 kb	Yes
Aldehyde dehydrogenase	3 Primers/4 bands	0.7 kb	Yes
F_{420}-dependent oxidoreductase	3 Primers/3 bands	1.1 kb	Yes
RNA polymerase α subunit	4 Primers/4 bands	1.1 kb	No
16S rRNA	4 Primers/4 bands	1.1 kb	No
23S rRNA	4 Primers/4 bands	1.2 kb	No
ATP synthase	3 Primers/3 bands	0.9 kb	No
Transcriptional regulator	2 Primers/3 bands	0.8 kb	Yes
Transcription factor	2 Primers/2 bands	0.7 kb	No

the putative functions of a reductase/dehydrogenase and an F_{420}:NADPH oxidoreductase (Table 5.1). Oligonucleotide probes designed against these genes led to the identification of a 12.5-kb sequence that contained four of the contiguous sequences identified by differential display (Table 5.1). The deduced amino acid sequences of all the genes encoded on the gene cluster were significantly similar to sequences in the database (Figure 5.4). All the genes are transcribed in the same direction, which suggests an operon structure. Three of the genes (ORF3, 7, and 8) were similar in sequence to oxidoreduction enzymes that use the deazaflavin F_{420} as electron donor or acceptor. ORF7 is homologous to an F_{420}:NADPH oxidoreductase of methanogenic bacteria (Table 5.2). The N-terminal sequence of ORF7 is almost identical to that of the corresponding enzyme purified by Ebert et al.[14] Cloning and expression of ORF7 in *E. coli* allowed for subsequent production of a pure enzyme capable of reducing F_{420} from *Methanobacterium thermoautotrophicum* with NADPH (Russ et al., unpublished results). Two other genes, ORF3 and ORF8, have sequences similar to those of members of the F_{420}-dependent reductase/dehydrogenase family (Table 5.2). The N-terminal sequence of ORF8 is also closely related to that of Component B isolated by Ebert et al.[14] Extracts of *E. coli* expressing ORF8 were

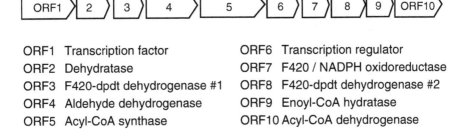

ORF1	Transcription factor	ORF6	Transcription regulator
ORF2	Dehydratase	ORF7	F420 / NADPH oxidoreductase
ORF3	F420-dpdt dehydrogenase #1	ORF8	F420-dpdt dehydrogenase #2
ORF4	Aldehyde dehydrogenase	ORF9	Enoyl-CoA hydratase
ORF5	Acyl-CoA synthase	ORF10	Acyl-CoA dehydrogenase

Figure 5.4   Gene cluster involved in picric acid and 2,4-DNP degradation.

Table 5.2 Sequence Similarities of ORFs of the Picric Acid Degradation Gene Cluster

ORF	aa	Similarity Identified	Identity	Similarity	E-Value
1	532	sp|Q10550| Putative regulatory protein (two domains) [*Mycobacterium tuberculosis*]	32% + 45%	45% + 58%	3e-25 + 1e-13
2	381	(AE001036) L-Carnitine dehydratase [*Archaeoglobus fulgidus*]	34%	52%	9e-51
3	296	pir|E64491| N5,N10-Methylene-tetrahydromethanopterin reductase [*Methanococcus jannaschii*]	24%	42%	6e-12
4	508	(U24215) *p*-Cumic aldehyde dehydrogenase [*Pseudomonas putida*]	44%	60%	2e-99
5	537	sp|P39062| Acetate-CoA ligase [*Bacillus subtilis*]	27%	42%	5e-42
6	252	(AE000277) Transcriptional regulator [*Escherichia coli*]	26%	42%	3e-11
7	227	sp|O26350| F_{420}-dependent NADP reductase [*Methanobacterium thermoautotrophicum*]	32%	44%	1e-18
8	350	gi|2649522| N5,N10-Methylene-tetrahydromethanopterin reductase [*Archaeoglobus fulgidus*]	28%	46%	7e-26
9	237	gi|97441| Enoyl-CoA hydratase [*Rhodobacter capsulatus*]	26%	38%	9e-08
10	361	gi|2649289| Acyl-CoA dehydrogenase (acd-9) [*Archaeoglobus fulgidus*]	32%	54%	5e-44

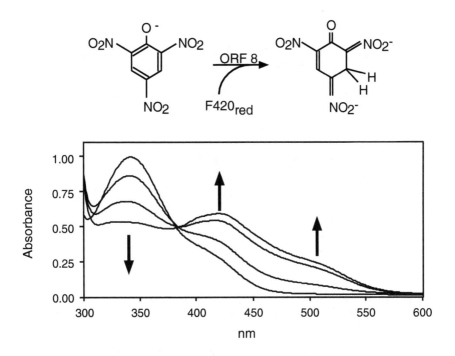

Figure 5.5   Reduction of picric acid by cell extracts of *E. coli* expressing the F_{420}-dependent picric acid reductase (ORF8). The same extracts reduced 2,4-DNP.

capable of reducing picric acid using electrons from reduced F_{420} (Figure 5.5). 2,4-DNP was also reduced by extracts of *E. coli* expressing ORF8 (Russ et al., unpublished results). The precise stoichiometry of these reactions is not yet known. However, spectral changes following addition of extracts expressing ORF8 were identical to those reported by Behrend and Heesche-Wagner, with crude extracts of *Nocardioides* sp. CB 22-2 where the spectra were due to the respective Meisenheimer complexes of picric acid and 2,4-DNP.[4]

The role of the second F_{420}-dependent reductase (ORF3) is not known at this time. Sequence similarities indicate that it is most closely related to a large number of hypothetical proteins from *Mycobacterium* for which there has yet been no biochemical validation. Interestingly, a *Mycobacterium* was shown to transform trinitrotoluene (TNT) to the Meisenheimer complex of TNT.[57] The gene encoded by ORF3 is also related to LmbY, a dehydrogenase proposed to be involved in the F_{420}-dependent synthesis of propylproline, an intermediate of lincomycin synthesis in *Streptomyces lincolnensis*.[36] The degradation pathway of picric acid needs a third reduction step to yield the proposed dinitro-cyclohexanone from the 2,4-DNP Meisenheimer complex (Figure 5.6). The reaction might be catalyzed by an enzyme encoded by ORF3.

The identification of genes of the picric acid degradation cluster and their sequences now allows us to propose functions by comparison to other genes in the

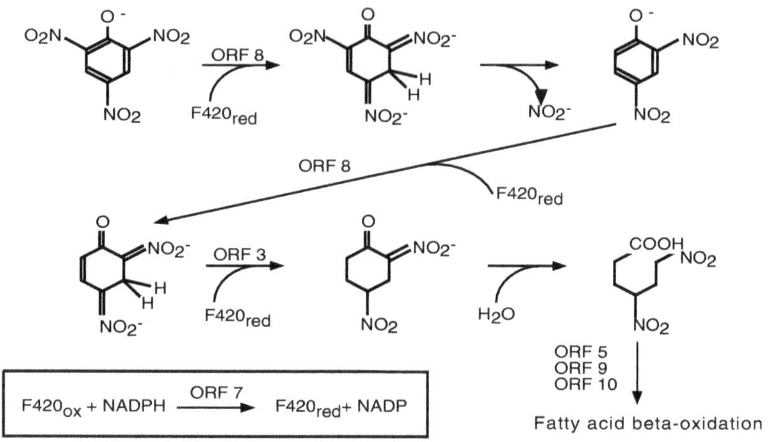

Figure 5.6   Proposed functional assignment for some of the ORFs of the picric acid gene cluster. Enzymatic activity has been validated only for the products of ORF7 and ORF8.

databases (Table 5.2) together with the established information and hypotheses regarding the degradation pathway.

Two genes, having homologies to transcription factors, are likely to have regulatory functions (ORF1 and 6). Such genes are expected since the pathway is inducible. Three other genes (ORF5, 9, and 10) code for enzymes typical of the beta oxidation pathway of a fatty acid derivative. A proposed function might be in the oxidation of 4,6-dinitrohexanoate (Figure 5.6.)

Another enzymatic function that must be fulfilled in the proposed metabolic pathway is the removal of the nitro group from the picric acid–Meisenheimer complex. A possible candidate enzyme for this reaction is ORF2, a homologue of various dehydratases and racemases. Other members of this enzyme family are involved in other group transfers such as the transfer of a Coenzyme A moiety from formyl-Coenzyme A to oxalate in *Oxalobacter*.[48] Alternatively, the protein encoded by ORF2 may be involved in the enzymatic hydrolysis of 2,4-dinitrocyclohexanone if the nonenzymatic reaction is too slow under physiological conditions.

4,6-Dinitrohexanoate is a metabolite of 2,4-DNP as well as picric acid.[24,25] Blasco et al.[5] described 3-nitroadipic acid as a metabolite of 2,4-DNP degradation, but did not propose 4,6-dinitrohexanoate as an intermediate of the productive degradation. However, the oxidation of 4,6-dinitrohexanoate by a nitroalkane oxidase, as described for the oxidation of 1-nitropropane to 1-propanone,[21] might be possible in the catabolism of 2,4-DNP. Further oxidation of the corresponding aldehyde to 3-nitroadipate might explain the finding of an aldehyde dehydrogenase gene (ORF4).

The functions proposed above can now be validated by cloning each gene in *E. coli* or another expression host and testing its enzymatic activity.

Other pathways for the bacterial degradation of picric acid and 2,4-DNP have been proposed. From their identification of the 2,4-DNP–Meisenheimer complex, Behrend and Heesche-Wagner argue that the Meisenheimer complex is possibly a substrate for a monooxygenase.[4] This is an interesting idea since the addition of a hydride to 2,4-DNP should facilitate an electrophilic attack. The monooxygenation would take place at one of the carbon atoms bearing a nitro group. One could imagine that the following elimination of nitrite with rearomatization would lead to 4-nitrocatechol or nitrohydroquinone. Since the authors have not found any 4-nitrocatechol converting activity with the *Nocardioides* strain studied, they propose nitrohydroquinone as a metabolite. There are no experimental data supporting the hypothesis. The 2,4-DNP degrading strains *Rhodococcus erythropolis* HL 24–1 and HL 24–2 (parental strain of HL PM-1) do not transform nitrohydroquinone or 4-nitrocatechol.[22]

Valli et al. detected 4-nitrophenol and 3-nitroadipic acid during the conversion of picric acid by *N. simplex* Nf. From that finding the authors proposed the removal of a nitro group from 2,4-DNP via the same reductive mechanism as for picric acid, leading to 4-nitrophenol with further transformation to 3-nitroadipic acid by an unknown mechanism.[56] This mechanism can be ruled out for the degradation of 2,4-DNP by *R. erythropolis* HL 24–1 since 4-nitrophenol is transformed by that strain to 4-nitrocatechol as a dead end metabolite.[22,25]

Our work on the application of differential display has demonstrated that differential display constitutes a powerful alternative to classic gene discovery techniques. Based on the sole knowledge that the degradation of 2,4-DNP was inducible, we were able to rapidly identify a cluster of genes involved in the degradation of both picric acid and 2,4-DNP. Because of the inherent difficulties in identifying the metabolic intermediates and purifying the enzymes, the identification of the gene cluster will greatly accelerate the elucidation of the degradation pathway of picric acid by providing hypotheses that can be tested following expression of the genes.

In view of the above results, we anticipate that differential display for the identification of metabolic genes in prokaryotes will become a widespread and generalized tool. We also anticipate that the discovery of metabolic pathways will be extended from pure cultures to more complex enrichment cultures and natural microbial communities (Rouvière, unpublished results).[17]

ACKNOWLEDGMENTS

We thank Dr. Hiltrud Lenke for suggesting the early presentation of our work, for many helpful suggestions, and for a critical reading of this manuscript. We also thank Dr. Vasantha Nagarajan for originally bringing to our attention the application of mRNA differential display to bacteria and for initiating a fruitful collaboration between our laboratories.

REFERENCES

1. Abu Kwaik, Y. and L. L. Pederson. 1996. The use of differential display-PCR to isolate and characterize a *Legionella pneumophila* locus induced during the intracellular infection of macrophages. *Mol. Microbiol.* 21:543–556.

2. Alberts, B., D. Bray, L. Lewis, M. Raff, K. Roberts, and J. D. Watson. 1989. *Molecular Biology of the Cell*, 2nd ed. Garland, New York.

3. ATSDR. 1995. Toxicological Profile for Dinitrophenols. Agency for Toxic Substances and Disease Registry (ATSDR), Atlanta, GA.

4. Behrend, C. and K. Heesche-Wagner. 1999. Formation of hydride-Meisenheimer complexes of picric acid (2,4,6-trinitrophenol) and 2,4-dinitrophenol during mineralization of picric acid by *Nocardioides* sp. strain CB 22–2. *Appl. Environ. Microbiol.* 65:1372–1377.

5. Blasco, R., E. Moore, V. Wray, D. Pieper, K. Timmis, and F. Castillo. 1999. 3-Nitroadipate, a metabolic intermediate for mineralization of 2,4-dinitrophenol by a new strain of a *Rhodococcus* species. *J. Bacteriol.* 181:149–152.

6. Bruhn, C., H. Lenke, and H.-J. Knackmuss. 1987. Nitrosubstituted aromatic compounds as nitrogen source for bacteria. *Appl. Environ. Microbiol.* 53:208–210.

7. Bryant, C., L. Hubbard, and W. D. McElroy. 1991. Cloning, nucleotide sequence, and expression of the nitroreductase gene from *Enterobacter cloacae*. *J. Biol. Chem.* 266:4126–4130.

8. Burke, J. F. and M. W. Whitehouse. 1967. Concerning the differences in uncoupling activity of isomeric dinitrophenols. *Biochem. Pharmacol.* 16:209–211.

9. Cooper, K. R., D. T. Burton, W. L. Goodfellow, and D. H. Rosenblatt. 1984. Bioconcentration and metabolism of picric acid (2,4,6-trinitrophenol) and picramic acid (2-amino-4,6-dinitrophenol) in rainbow trout *Salmo gairdneri*. *J. Toxicol. Environ. Health.* 14:731–747.

10. Dabbs, E. R. 1998. Cloning of genes that have environmental and clinical importance from rhodococci and related bacteria. *Antonie van Leeuwenhoek. J. Microbiol. Serol.* 74:155–167.

11. Daniels, L., N. Bakhiet, and K. Harmon. 1985. Widespread distribution of a 5-deazaflavin cofactor in *Actinomycetes* and related bacteria. *Syst. Appl. Microbiol.* 6:12–17.

12. Dennis, J. J. and G. J. Zylstra. 1998. Plasposons: modular self-cloning minitransposon derivatives for rapid genetic analysis of gram-negative bacterial genomes. *Appl. Environ. Microbiol.* 64:2710–2715.

13. Doss, R. P. 1996. Differential display without radioactivity — a modified procedure. *Biotechniques.* 21:408–410, 412.

14. Ebert, S., P.-G. Rieger, and H.-J. Knackmuss. 1999. Function of coenzyme F420 in aerobic catabolism of 2,4,6-trinitrophenol and 2,4-dinitrophenol by *Nocardioides simplex* FJ2–1A. *J. Bacteriol.* 181:2669–2674.

15. Ecker, S., T. Widmann, H. Lenke, O. Dickel, P. Fischer, C. Bruhn, and H.-J. Knackmuss. 1992. Catabolism of 2,6-dinitrophenol by *Alcaligenes eutrophus* JMP 134 and JMP 222. *Arch. Microbiol.* 158:149–154.

16. Elvers, B., S. Hawkins, and G. Schulz (eds.). 1991. *Ullmann's Encyclopedia of Industrial Chemistry*, 5th ed. Verlag Chemie, Weinheim.

17. Fleming, J. T., W. H. Yao, and G. S. Sayler. 1998. Optimization of differential display of prokaryotic mRNA: application to pure culture and soil microcosms. *Appl. Environ. Microbiol.* 64:3698–3706.

18. Gisi, D., G. Stucki, and K. W. Hanselmann. 1997. Biodegradation of the pesticide 4,6-dinitro-ortho-cresol by microorganisms in batch cultures and in fixed-bed column reactors. *Appl. Microbiol. Biotechnol.* 48:441–448.

19. Hanstein, W. G. 1976. Uncoupling of oxidative phosphorylation. *Biochim. Biophys. Acta.* 456:129–148.

20. Hess, T. F., S. K. Schmidt, and J. Silverstein. 1990. Supplemental substrate enhancement of 2,4-dinitrophenol mineralization by a bacterial consortium. *Appl. Environ. Microbiol.* 56:1551–1558.

21. Kido, T., K. Hashizume, and K. Soda. 1978. Purification and properties of nitroalkane oxidase from *Fusarium oxysporum. J. Bacteriol.* 133:53–58.

22. Lenke, H. 1990. Doctoral dissertation. University of Stuttgart, Stuttgart, Germany.

23. Lenke, H. and H.-J. Knackmuss. 1996. Initial hydrogenation and extensive reduction of substituted 2,4-dinitrophenols. *Appl. Environ. Microbiol.* 62:784–790.

24. Lenke, H. and H.-J. Knackmuss. 1992. Initial hydrogenation during catabolism of picric acid by *Rhodococcus erythropolis* HL 24-2. *Appl. Environ. Microbiol.* 58:2933–2937.

25. Lenke, H., D. H. Pieper, C. Bruhn, and H.-J. Knackmuss. 1992. Degradation of 2,4-dinitrophenol by two *Rhodococcus erythropolis* strains, HL 24-1 and HL 24-2. *Appl. Environ. Microbiol.* 58:2928–2932.

26. Levy-Frebault, V. V. and F. Portaels. 1992. Proposed minimal standards for the genus *Mycobacterium* and for description of new slowly growing *Mycobacterium* species. *Int. J. Syst. Bacteriol.* 42:315–323.

27. Liang, P., L. Averboukh, and A. B. Pardee. 1993. Distribution and cloning of eukaryotic mRNAs by means of differential display: refinements and optimization. *Nucleic Acids Res.* 21:3269–3275.

28. Liang, P. and A. B. Pardee. 1992. Differential display of eukaryotic messenger RNA by means of the polymerase chain reaction. *Science.* 257:967–971.

29. Lohmann, J., H. Schickle, and T. C. Bosch. 1995. REN display, a rapid and efficient method for nonradioactive differential display and mRNA isolation. *Biotechniques.* 18:200–202.

30. Marvin-Sikkema, F. D. and J. A. de Bont. 1994. Degradation of nitroaromatic compounds by microorganisms. *Appl. Microbiol. Biotechnol.* 42:499–507.

31. Men, A. E. and P. M. Gresshoff. 1998. Efficient cloning of DAF polymorphic markers from silver-stained polyacrylamide gels. *Biotechniques.* 24:593–595.

32. Michels, M. and E. P. Bakker. 1981. The mechanism of uncoupling by picrate in *Escherichia coli* K-12 membrane systems. *Eur. J. Biochem.* 116:513–519.

33. Morii, S., C. Fujii, T. Miyoshi, M. Iwami, and E. Itagaki. 1998. 3-Ketosteroid-delta1-dehydrogenase of *Rhodococcus rhodochrous*: sequencing of the genomic DNA and hyperexpression, purification, and characterization of the recombinant enzyme. *J. Biochem. (Tokyo).* 124:1026–1032.

34. Nilsson, J. R. 1995. pH-dependent effects of 2,4-dinitrophenol (DNP) on proliferation, endocytosis, fine structure and DNP resistance in *Tetrahymena. J. Eukaryot. Microbiol.* 42:248–255.

35. Patil, S. S. and V. M. Shinde. 1989. Gas chromatographic studies on the biodegradation of nitrobenzene and 2,4-dinitrophenol in the nitrobenzene plant wastewater. *Environ. Pollut.* 57:235–250.

36. Peschke, U., H. Schmidt, H. Z. Zhang, and W. Piepersberg. 1995. Molecular characterization of the lincomycin-production gene cluster of *Streptomyces lincolnensis* 78–11. *Mol. Microbiol.* 16:1137–1156.

37. Prousek, J. and M. Slosarikova. 1996. Fotokataliticka degradace 2,4,6-trinitrofenolu (kyselina pikrova) za pouziti vodnych polovodicovych disperznich smesi. *Chem. Listy.* 90:829–832.
38. Purwantini, E. and L. Daniels. 1998. Molecular analysis of the gene encoding F420-dependent glucose-6-phosphate dehydrogenase from *Mycobacterium smegmatis*. *J. Bacteriol.* 180:2212–2219.
39. Purwantini, E., T. P. Gillis, and L. Daniels. 1997. Presence of F420-dependent glucose-6-phosphate dehydrogenase in *Mycobacterium* and *Nocardia* species, but absence from *Streptomyces* and *Corynebacterium* species and methanogenic Archaea. *FEMS Microbiol. Lett.* 146:129–134.
40. Rajan, J., K. Valli, R. E. Perkins, F. S. Sariaslani, S. M. Barns, A. L. Reysenbach, S. Rehm, M. Ehringer, and N. R. Pace. 1996. Mineralization of 2,4,6-trinitrophenol (picric acid): characterization and phylogenetic identification of microbial strains. *J. Ind. Microbiol.* 16:319–324.
41. Rieger, P.-G., V. Sinnwell, A. Preuss, W. Francke, and H.-J. Knackmuss. 1999. Hydride-Meisenheimer complex formation and protonation as key reactions of 2,4,6-trinitrophenol biodegradation by *Rhodococcus erythropolis*. *J. Bacteriol.* 181:1189–1195.
42. Rivera-Marrero, C. A., M. A. Burroughs, R. A. Masse, F. O. Vannberg, D. L. Leimbach, J. Roman, and J. J. Murtagh, Jr. 1998. Identification of genes differentially expressed in *Mycobacterium tuberculosis* by differential display PCR. *Microb. Pathog.* 25:307–316.
43. Ross, I., P. Morgan, and P. Williams. 1999. The biodegradation of nitroaromatic waste. Presented at the Second International Symposium on Biodegradation of Nitroaromatic Compounds and Explosives. Leesburg, VA.
44. Russ, R., S. Ebert, and H.-J. Knackmuss. 1998. Transformation of picrate and 2,4-dinitrophenol by microorganisms. Abstract 98th Annu. Meet. Am. Soc. Microbiol. Atlanta, GA.
45. Sabater, J. F. and J. M. Zaragoza. 1993. A simple identification system for slowly growing *Mycobacteria*. II. Identification of 25 strains isolated from surface water in Valencia (Spain). *Acta Microbiol. Hung.* 40:343–349.
46. Shea, P. J., J. B. Weber, and M. R. Overcash. 1983. Biological activities of 2,4-dinitrophenol in plant-soil systems. *Residue Rev.* 87:1–41.
47. Shepard, B. D. and M. S. Gilmore. 1999. Identification of aerobically and anaerobically induced genes in *Enterococcus faecalis* by random arbitrarily primed PCR. *Appl. Environ. Microbiol.* 65:1470–1476.
48. Sidhu, H., S. D. Ogden, H. Y. Lung, B. G. Luttge, A. L. Baetz, and A. B. Peck. 1997. DNA sequencing and expression of the formyl coenzyme A transferase gene, *frc*, from *Oxalobacter formigenes*. *J. Bacteriol.* 179:3378–3381.
49. Simon, E. W. 1953. Mechanism of dinitrophenol toxicity. *Biol. Rev.* 28:453.
50. Stenlid, G. 1949. Effects of 2,4-dinitrophenol upon oxygen consumption and glucose uptake in young wheat roots. *Physiol. Plant.* 2:350.
51. Stratford, M. and P. A. Anslow. 1996. Comparison of the inhibitory action on *Saccharomyces cerevisiae* of weak-acid preservatives, uncouplers, and medium-chain fatty acids. *FEMS Microbiol. Lett.* 142:53–58.
52. Suen, W. C. and J. C. Spain. 1993. Cloning and characterization of *Pseudomonas* sp. strain DNT genes for 2,4-dinitrotoluene degradation. *J. Bacteriol.* 175:1831–1837.
53. Sutcliffe, I. C. 1998. Cell envelope composition and organisation in the genus *Rhodococcus*. *Antonie van Leeuwenhoek. J. Microbiol. Serol.* 74:49–58.

54. Tokiwa, H., R. Nakagawa, K. Horikawa, and A. Ohkubo. 1987. The nature of the mutagenicity and carcinogenicity of nitrated, aromatic compounds in the environment. *Environ. Health Perspect.* 73:191–199.

55. Tokiwa, H., N. Sera, A. Nakashima, K. Nakashima, Y. Nakanishi, and N. Shigematu. 1994. Mutagenic and carcinogenic significance and the possible induction of lung cancer by nitro aromatic hydrocarbons in particulate pollutants. *Environ. Health Perspect.* 102:107–110.

56. Valli, K., M. Chivukula, E. Armstrong, and S. Sariaslani. 1997. Mechanism of degradation of picric acid by *Nocardioides simplex*. Presented at the 97th Annu. Meet. Am. Soc. Microbiol. Miami Beach, FL.

57. Vorbeck, C., H. Lenke, P. Fischer, and H.-J. Knackmuss. 1994. Identification of a hydride-Meisenheimer complex as a metabolite of 2,4,6-trinitrotoluene by a *Mycobacterium* strain. *J. Bacteriol.* 176:932–934.

58. Welsh, J., K. Chada, S. S. Dalal, R. Cheng, D. Ralph, and M. McClelland. 1992. Arbitrarily primed PCR fingerprinting of RNA. *Nucleic Acids Res.* 20:4965–4970.

59. Whong, W. Z. and G. S. Edwards. 1984. Genotoxic activity of nitroaromatic explosives and related compounds in *Salmonella typhimurium*. *Mutat. Res.* 136:209–215.

60. Williams, J. G., A. R. Kubelik, K. J. Livak, J. A. Rafalski, and S. V. Tingey. 1990. DNA polymorphisms amplified by arbitrary primers are useful as genetic markers. *Nucleic Acids Res.* 18:6531–6535.

61. Wong, K. K. and M. McClelland. 1994. Stress-inducible gene of *Salmonella typhimurium* identified by arbitrarily primed PCR of RNA. *Proc. Natl. Acad. Sci. U.S.A.* 91:639–643.

62. Wyman, J. F., H. E. Guard, W. D. Won, and J. H. Quay. 1979. Conversion of 2,4,6-trinitrophenol to a mutagen by *Pseudomonas aeruginosa*. *Appl. Environ. Microbiol.* 37:222–226.

63. Wyman, J. F., M. P. Serve, D. W. Hobson, L. H. Lee, and D. E. Uddin. 1992. Acute toxicity, distribution, and metabolism of 2,4,6-trinitrophenol (picric acid) in Fischer 344 rats. *J. Toxicol. Environ. Health.* 37:313–327.

64. Yabannavar, A. V. and G. J. Zylstra. 1995. Cloning and characterization of the genes for *p*-nitrobenzoate degradation from *Pseudomonas pickettii* YH105. *Appl. Environ. Microbiol.* 61:4284–4290.

65. Yuk, M. H., E. T. Harvill, and J. F. Miller. 1998. The BvgAS virulence control system regulates type III secretion in *Bordetella bronchiseptica*. *Mol. Microbiol.* 28:945–959.

66. Zhang, J. P. and S. Normark. 1996. Induction of gene expression in *Escherichia coli* after pilus-mediated adherence. *Science.* 273:1234–1236.

CHAPTER 6

Microbial Degradation of Mononitrophenols and Mononitrobenzoates

Gerben J. Zylstra, Sang-Weon Bang, Lisa M. Newman, and Lynda L. Perry

CONTENTS

6.1 INTRODUCTION

Nitroaromatic compounds are found widely as environmental contaminants in the U.S. Historically, nitroaromatic compounds have been used widely in the chemical industry for the synthesis of pesticides and explosives. The degradation of complex nitro compounds is dealt with in detail in other chapters of this book. The degradation of simple nitroaromatic compounds such as nitrophenols and nitrobenzoates has received little attention recently compared with the more highly nitrated compounds, most likely since the simple nitroaromatic compounds are used in manufacturing processes as intermediates and are not necessarily released on a large scale. However, the fate and persistence of simple nitroaromatic compounds in the environment is of serious concern due to their increased water solubility compared with many of the more complex nitro compounds and, hence, the ability of simple nitroaromatic compounds to spread faster from a point source of contamination. This chapter will review what is known about the degradation of the three isomers of nitrophenol and the three isomers of nitrobenzoate.

6.2 INDUSTRIAL USES OF NITROPHENOLS AND NITROBENZOATES

Nitroaromatic compounds are utilized as the starting material for the synthesis of a wide variety of nitrogen-containing compounds. Final end uses fall into two classes: those compounds which still contain a nitro group (direct use) and those in which the product contains an amino group or other functional group derived from the nitro group. Nitrophenols provide prime examples of the many ways nitroaromatic compounds can be used in fashioning products for the explosives, pharmaceutical, and agricultural industries. Due to the intense yellow color of the phenolate anion and pH reactivity (p-nitrophenol, for instance, is colorless at pH 5.6 and the corresponding phenolate anion has a maximum yellow color at pH 7.6), nitrophenols are often used directly in titrations as indicators. Nitrophenols coupled to other substrates are often used in monitoring enzyme activity. For example, o-nitrophenyl, m-nitrophenyl, and p-nitrophenyl β-D-galactopyranosides are often used to measure β-galactosidase activity through release of the nitrophenol moiety and concomitant formation of the yellow color. In analogous fashion alkaline phosphatase activity is often measured through the use of p-nitrophenyl phosphate. Pharmaceutical uses of p-nitrophenol include the synthesis of the aspirin substitute acetaminophen, the antifungal nitrofungin, and the anthelmintic disophenol. Agricultural uses of nitrophenol esters include the organophosphate derivatives of p-nitrophenol, parathion (O,O-diethyl O-4-nitrophenyl phosphorothioate), and methylparathion (O,O-dimethyl O-4-nitrophenyl phosphorothioate), widely used as broad crop protection insecticides. The latter two compounds are readily hydrolized in soil and are thus a major source of p-nitrophenol found in the environment.

The nitrobenzoates also have uses in a variety of industries. For instance, p-nitrobenzoate ethyl ester can easily be reduced to p-aminobenzoate ethyl ester, also known as benzocaine, a local (usually surface) anesthetic. Reduction of m-nitrobenzoate to 5-aminosalicylic acid results in mesalamine, a compound used in the manufacture of light-sensitive paper and azo dyes. This compound is often used in the pharmaceutical world as an anti-inflammatory for the treatment of ulcerative colitis. Nitrobenzoates are often used in metal stripping solutions.

6.3 CATABOLIC PATHWAYS FOR THE DEGRADATION OF NITROPHENOLS

Many pure cultures of bacteria have been isolated for their ability to utilize nitrophenol isomers as the sole source of carbon and energy for growth. The metabolically productive pathways include those involving either initial reduction of the nitro group prior to removal from the aromatic ring or direct removal as nitrite via

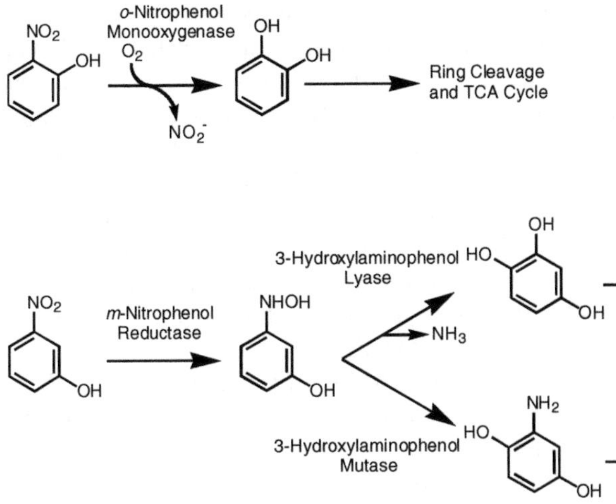

Figure 6.1　Catabolic pathways for the biodegradation of o-nitrophenol and m-nitrophenol. *Pseudomonas putida* B2[43,44] degrades o-nitrophenol through catechol (top). Two m-nitrophenol degradative pathways are known (bottom). *P. putida* B2 degrades m-nitrophenol through 1,2,4-trihydroxybenzene,[23] while *Ralstonia eutropha* JMP134 degrades m-nitrophenol through aminohydroquinone.[35,36] Demonstrated enzymatic reactions are labeled.

oxygenolytic cleavage. The simplest pathway for nitrophenol degradation is perhaps that for o-nitrophenol (Figure 6.1), involving a single enzyme for converting the substrate into a β-ketoadipate pathway intermediate. Zeyer and Kocher have shown that *Pseudomonas putida* B2 degrades o-nitrophenol via catechol with the release of nitrite.[43] The initial oxygenase enzyme was purified and shown to require 2 mol of NADPH as cofactor for the reaction.[44] This suggests that the enzyme initially removes the nitro group from o-nitrophenol to form o-benzoquinone with a further reduction to form catechol. The purified enzyme stoichiometrically converts o-nitrophenol to catechol with no detection of o-benzoquinone as an intermediate. The enzyme has a broad substrate range and oxidizes various alkyl- and halogen-substituted o-nitrophenols.[45] These latter compounds are not substrates for growth due to inability of the β-ketoadipate downstream pathway to handle the substituted catechols.

Two metabolic pathways have been described for m-nitrophenol degradation (Figure 6.1). In both of these pathways the initial enzymatic reaction is the NADPH-dependent partial reduction of the nitro group to a hydroxylamino moiety.[23,35] The two catabolic pathways diverge after the first step. In *P. putida* B2, m-hydroxylamino-phenol is converted to 1,2,4-trihydroxybenzene (1,2,4-benzenetriol, hydroxyquinol) via a lyase with release of ammonia.[23] *Ralstonia eutropha* JMP134, on the other hand, converts the m-hydroxylaminophenol to aminohydroquinone through the action of a mutase.[36] In the latter case the nitrogen is removed as ammonia at a later, as yet unidentified, step in the pathway. *R. eutropha* JMP134 also degrades 2-chloro-5-nitrophenol through the initial reduction and mutase steps and then dechlorination to form aminohydroquinone.

A number of bacterial strains have been isolated for the ability to degrade *p*-nitrophenol. Simpson and Evans[37] reported the first preliminary evidence of oxidative *p*-nitrophenol degradation with a *Pseudomonas* strain isolated from biological waste treatment. The strain utilized *p*-nitrophenol as the sole source of carbon and accumulated nitrite in the culture medium. 1,2,4-Trihydroxybenzene was suggested as an intermediate. Raymond and Alexander[33] reported that a *Flavobacterium* isolated from soil had the ability to utilize *p*-nitrophenol as a carbon and energy source with stoichiometric release of nitrite. 4-Nitrocatechol accumulated in the culture when resting cells were treated with 0.1% chloroform in the presence of *p*-nitrophenol. *Pseudomonas* strain 8P isolated from a microbial consortium previously adapted to parathion could metabolize *p*-nitrophenol as a carbon source.[25] Hydroquinone was tentatively identified as an initial metabolite.

Initial work by Spain and Gibson with a *Moraxella* species elucidated the entire metabolic pathway (Figure 6.2). *p*-Nitrophenol is degraded by an initial monooxygenase attack with the concomitant release of nitrite and the production of hydroquinone.[39] The reaction in crude cell extracts required 2 mol of NADPH for the oxidation of 1 mol of *p*-nitrophenol into 1 mol of hydroquinone produced. The data suggest that the initial monooxygenase converts *p*-nitrophenol into 1,4-benzoquinone at the expense of 1 mol of NADPH. The benzoquinone is then reduced into hydroquinone by a quinone reductase at the expense of an additional 1 mol of NADPH. However, the authors were not able to separate the monooxygenase activity from the quinone reductase activity, and, thus, whether two enzymes or one are responsible for these two enzymatic steps could not be determined. Both enzymatic reactions were associated with the particulate fraction of the cell extract. Subsequent degradation of hydroquinone by the *Moraxella* sp. proceeds by ring fission to form γ-hydroxymuconic semialdehyde.[38] The ring cleavage product is subsequently converted to maleylacetate by a dehydrogenase. Maleylacetate is then further reduced to β-ketoadipate and TCA cycle intermediates. Unexpectedly, a 1,2,4-trihydroxybenzene ring cleavage dioxygenase enzyme is coinduced with the hydroquinone

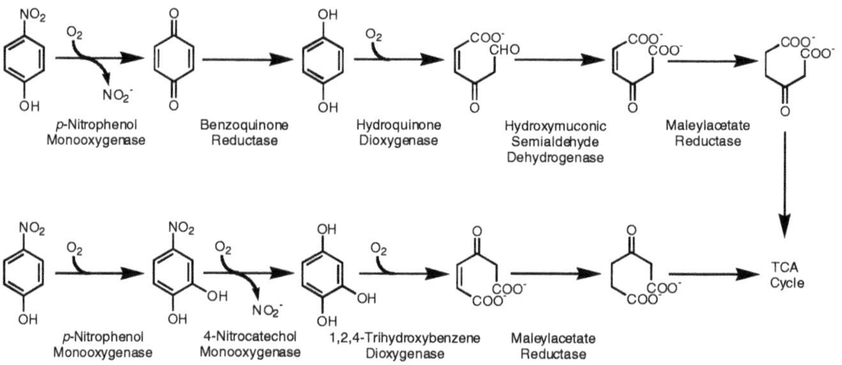

Figure 6.2 Catabolic pathways for the degradation of *p*-nitrophenol. A *Moraxella* sp. strain degrades *p*-nitrophenol through hydroquinone and β-ketoadipate.[38,39] *Arthrobacter* sp. strain JS443 degrades *p*-nitrophenol through 1,2,4-trihydroxybenzene and β-ketoadipate.[20]

dioxygenase. What role the enzyme might play in the degradation of p-nitrophenol is not clear and may simply be due to coinduction by related substrates.

Hanne et al.[18] showed that *Arthrobacter aurescens* TW17 also degrades p-nitrophenol through hydroquinone. In contrast, a *Nocardia* sp. converts p-nitrophenol to hydroquinone following exposure to p-nitrophenol,[18] but the same strain converts p-nitrophenol to 4-nitrocatechol when exposed to phenol, m-nitrophenol, or p-cresol. In the former case enzymes are specifically induced by p-nitrophenol for removal of the nitro group. In the latter case enzymes are induced by other phenols that hydroxylate the aromatic ring of p-nitrophenol. The ring-hydroxylation reaction is most likely due to nonspecific phenol hydroxylases induced by the phenolic compounds. A catabolic pathway for p-nitrophenol involving initial ring hydroxylation has been proposed for an *Arthrobacter* sp.[20] In this case (Figure 6.2), p-nitrophenol is first hydroxylated to 4-nitrocatechol by an initial hydroxylase. In a second monooxygenase catalyzed step, 1,2,4-trihydroxybenzene is formed with the concomitant release of nitrite. The benzenetriol is then cleaved into maleylacetate by a 1,2,4-trihydroxybenzene 1,2-dioxygenase. Maleylacetate is further metabolized via β-ketoadipate into TCA cycle intermediates.

The pathway proposed for *Arthrobacter* is apparently widespread among the Gram-positive bacteria. The degradation of 4-nitroanisole by two *Rhodococcus* strains involves the formation of p-nitrophenol by an o-demethylation reaction.[34] Oxygen uptake experiments indicate that p-nitrophenol degradation proceeds through 1,2,4-trihydroxybenzene via 4-nitrocatechol. Mitra and Vaidyanathan[24] purified a novel p-nitrophenol 2-hydroxylase from a *Nocardia* sp. strain. The hydroxylase catalyzes the hydroxylation of p-nitrophenol to form 4-nitrocatechol. A two-component p-nitrophenol monooxygenase was partially purified from *Bacillus sphaericus* JS905.[21] Based on the fact that this partially purified preparation could catalyze both of the initial monooxygenation steps in the pathway (p-nitrophenol to 1,2,4-trihydroxybenzene), the authors proposed that a single monooxygenase is involved in both steps of the pathway.

6.4 INDUCTION OF p-NITROPHENOL DEGRADATION

Spain and Gibson[38] noted that oxidation of p-nitrophenol in a *Moraxella* sp. is enhanced by pre-exposure to p-nitrophenol, indicating inducible gene expression of the enzymes involved in p-nitrophenol catabolism. However, for p-nitrophenol degradation by *P. putida* JS444, Nishino and Spain[30] reported that the ability to rapidly degrade p-nitrophenol is coupled to the number of cells that are present in the culture. More dilute suspensions of bacteria had a longer lag phase when exposed to p-nitrophenol. Hydroquinone accumulated transiently in the culture medium during the adaptation period. Addition of hydroquinone to the medium decreased the lag time, but only for the more concentrated cell suspensions. The authors suggested that induction of the appropriate enzymes for p-nitrophenol degradation is dependent upon the accumulation of hydroquinone or a factor whose release is stimulated by hydroquinone.

6.5 MOLECULAR BASIS FOR *p*-NITROPHENOL DEGRADATION

Very little has been done on determining the molecular basis for nitrophenol degradation. Initial studies have suggested that the genes responsible for *p*-nitrophenol degradation or, at least, an initial *p*-nitrophenol 4-monooxygenase gene are located on a plasmid in *A. aurescens* TW17.[18] The strain has multiple plasmids, and loss of the largest resulted in the inability to grow on *p*-nitrophenol. Prakash et al. have similarly shown that *P. cepacia* RKJ200 harbors a 50-kb plasmid carrying the genes for *p*-nitrophenol degradation and zinc resistance.[32] The genes for *p*-nitrophenol degradation have not been isolated yet from either of these organisms.

Recently, Bang and Zylstra[4,5] cloned the genes for *p*-nitrophenol degradation from *Pseudomonas* sp. strain ENV2030. The catabolic pathway for *p*-nitrophenol degradation by ENV2030 is the same as that shown for other Gram-negative bacteria (top pathway in Figure 6.2). The genes are arranged in at least three operons with two regulatory genes nearby (Figure 6.3). Interestingly, the genes encoding the first two enzymes in the pathway, *p*-nitrophenol monooxygenase (*pnpA*) and benzoquinone reductase (*pnpB*), are located next to each other but are divergently transcribed. The genes coding for the ring cleavage pathway (*pnpDEC*) are in an apparent operon along with several cotranscribed open reading frames with no known function and no homologs in the GenBank database. *p*-Nitrophenol monooxygenase (PnpA) is 30% identical to a family of monooxygenases from *Rhodococcus*,[6] *Comamonas*,[2] and *Escherichia coli*[13] that attack 3-hydroxyphenylpropionate. In contrast, benzoquinone reductase (PnpB) is 47% identical to a benzoquinone reductase recently identified in *Phanerochaete chrysosporium*[1] and is approximately 50% identical to proteins of unknown function found in the plants *Lithospermium erythrorhizon*[42] and *Prunus armeniaca* (GenBank accession AAD38143). Tryptophan repressor binding protein (WrbA) found in *E. coli*[41] is the most similar protein in the database to PnpB (57% identical), suggesting that WrbA may play some as yet unidentified

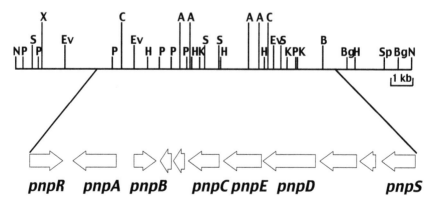

Figure 6.3 Restriction map of a 19-kb *Not*I fragment containing the genes for *p*-nitrophenol degradation: *pnpA*, gene for *p*-nitrophenol monooxygenase; *pnpB*, gene for benzoquinone reductase; *pnpC*, gene for hydroquinone 1,2-dioxygenase; *pnpD*, gene for hydroxymuconic semialdehyde dehydrogenase; *pnpD*, gene for maleylacetate reductase; *pnpR* and *pnpS*, genes for LysR-type regulatory proteins.

role in reducing quinones in *E. coli*. Hydroquinone 1,2-dioxygenase (PnpC) is about 40% identical to the 1,2,4-trihydroxybenzene 1,2-dioxygenases.[3,10,19] Hydroxymuconic acid semialdehyde dehydrogenase (PnpD) and maleylacetate reductase (PnpE) are found in a wide variety of different ring cleavage pathways, and, thus, PnpD and PnpE are highly similar to a wide variety of analogous enzymes in the database. Two genes, designated *pnpR* and *pnpS*, code for regulatory proteins in the LysR family. These two proteins are 44% identical to each other, 28% identical to the regulator PtxR from *Pseudomonas aeruginosa*,[17] and 25% identical to the regulator CatM from *A. calcoaceticus*.[28]

Transposon mutants have been identified where the transposon has inserted into either *pnpR* or *pnpS*, definitively identifying these genes as coding for positive regulatory elements involved in *p*-nitrophenol degradation.[4,5] Strains lacking *pnpR* do not metabolize *p*-nitrophenol, while strains lacking *pnpS* accumulate benzoquinone when grown on glucose in the presence of *p*-nitrophenol. Since hydroquinone spontaneously oxidizes into benzoquinone, these experiments cannot definitively determine whether PnpS controls expression of *pnpB* or *pnpDEC* or both. Thus, a model can be developed whereby *pnpR* controls the upper pathway and *pnpS* controls the lower pathway (Figure 6.4). In fact, a search for the LysR family DNA binding motif T(n){11}A forming an imperfect inverted repeat reveals seven probable candidates in the nucleotide sequence (Figure 6.4). These are in the positions predicted for control of each of the operons involved in *p*-nitrophenol degradation as well as in front of *pnpR* and *pnpS*, suggesting that the regulatory proteins control their own synthesis. Interestingly, the motif can also be found within *pnpR* and *pnpS* as well.

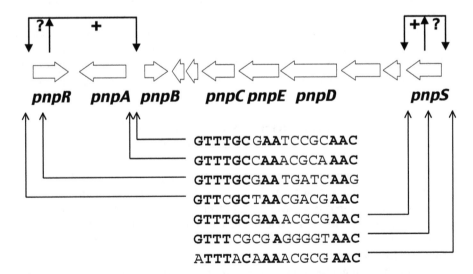

Figure 6.4 Proposed regulatory scheme for *p*-nitrophenol degradation in *Pseudomonas* sp. strain ENV2030. PnpR positively regulates *pnpA* and *pnpB,* while PnpS positively regulates the *pnpDEC* operon. The two regulatory proteins may also regulate their own synthesis. The sequence and location of potential PnpR and PnpS binding sites are shown.

Genes for *p*-nitrophenol degradation have been cloned from the analogous strain *P. putida* JS444 based on their hybridization to the *Pseudomonas* sp. strain ENV2030 genes (L. Perry and G. Zylstra, unpublished data). Nucleotide sequence analysis reveals that the genes are 90% identical with one exception: *P. putida* JS444 lacks the *pnpR* gene. Since PnpR in *Pseudomonas* sp. strain ENV2030 is absolutely required for induction of *pnpA*, some other regulator must be taking its place in *P. putida* JS444. Due to the similarities in the proteins it is highly likely that PnpS takes over the role of PnpR, but perhaps not very efficiently. This probably explains why *p*-nitrophenol degradation in the environment shows a long and unpredictable lag period. One could hypothesize that there are two populations of bacteria: one responds immediately to *p*-nitrophenol with *pnpR* and *pnpS,* while the other population with only *pnpS* needs a pathway intermediate to build up before the catabolic pathway is fully induced. Thus, the different population densities of the two bacterial genotypes would determine how quickly the natural population of bacteria responds to the presence of *p*-nitrophenol.

As mentioned above, *Arthrobacter* sp. strain JS443 degrades *p*-nitrophenol through 4-nitrocatechol and 1,2,4-trihydroxybenzene, a different catabolic pathway than that found in *Pseudomonas* sp. In order to clone the genes for this pathway, the 1,2,4-trihydroxybenzene dioxygenase was purified to homogeneity from *Arthrobacter* sp. strain JS443.[31] The corresponding gene was cloned using probes based on the N-terminal sequence. The deduced amino acid sequence shows 81% identity to 1,2,4-trihydroxybenzene dioxygenase derived from another *Arthrobacter* strain (BA-5–17) not known to be able to degrade *p*-nitrophenol.[26] On the other hand, the enzyme is 45 and 50% identical to 1,2,4-trihydroxybenzene dioxygenase involved in chlorophenol degradation from *Burkholderia cepacia* AC1100[10] and *Ralstonia pickettii* DTP0602. Since 1,2,4-trihydroxybenzene is a metabolite in several unrelated pathways, it is not surprising to find related genes for 1,2,4-trihydroxybenzene dioxygenase in other organisms. The close relatedness of 1,2,4-trihydroxybenzene dioxygenase in *Arthobacter* sp. strains JS443 and BA-5–17 suggests that the gene moved into *Arthrobacter* before being recruited for *p*-nitrophenol degradation.

6.6 CATABOLIC PATHWAYS FOR THE DEGRADATION OF NITROBENZOATES

The three isomers of nitrobenzoates, like the nitrophenols, can be degraded by a variety of different mechanisms. Early work in the 1950s and 1960s with Gram-negative and Gram-positive bacteria was inconclusive, and only in the last decade have the pathways for *m*-nitrobenzoate and *p*-nitrobenzoate been conclusively worked out. Cain and coworkers[7,9] suggested that the three isomers of nitrobenzoate are degraded through oxidative attack. The nitrite that is released would rapidly be converted to ammonia for use by the organism as a nitrogen source.[8] However, their data indicate that some reduction of the nitro group occurs. Although reduction proceeded at a considerable rate, the reaction was not considered metabolically fruitful. In the case of *o*-nitrobenzoate degradation, Warburg respirometry experiments with a *Nocardia*[9] showed that induced cells oxidized catechol, 4-nitrocatechol, and gentisate, but not anthranilate. Near endogenous levels of oxygen uptake were seen for anthranilate, protocatechuate, and salicylate. In the case of p-nitrobenzoate degradation, Warburg respirometry experiments showed that induced cells rapidly oxidized p-nitrosobenzoate, p-hydroxylaminobenzoate, p-hydroxybenzoate, proto-catechuate, and 4-nitrocatechol, but not p-aminobenzoate. This led the authors to conclude that the reduction reaction was accidental since it did not go to completion (to the amino) to induce the anthranilate or p-aminobenzoate degradation pathway. Once the Warburg respirometry data on the partial reduction products were ignored, the authors proposed that p-nitrobenzoate is degraded through p-hydroxybenzoate to protocatechuate and that o-nitrobenzoate is degraded through either protocate-chuate or catechol. Both pathways involve the oxidative release of nitrite into the culture medium. This conclusion is in contrast to the work in the same time period by Durham and colleagues with a $P.\ fluorescens$ in the case of p-nitrobenzoate degradation[11,12] and a $Flavobacterium$ sp. in the case of o-nitrobenzoate degrada-tion.[22] The authors provided convincing evidence, again based on respirometry experiments, that in their $Flavobacterium$ o-nitrobenzoate is degraded via o-nitrosobenzoate and o-hydroxylaminobenzoate, but does not involve anthranilate, catechol, salicylate, or protocatechuate. How the organism handles metabolism of the aromatic ring without going through catechol or protocatechuate remained a mystery, but based on current knowledge of metabolism of hydroxylamino com-pounds (see below) it is possible that metabolism proceeds through 2,3-dihydroxy-benzoate which was not included in the experiments. Similar work with the $P.\ fluorescens$ and p-nitrobenzoate degradation implicated p-aminobenzoate, p-hydroxy-benzoate, and protocatechuate as intermediates in the catabolic pathway. The path-ways proposed separately by Cain and Durham are shown in Figure 6.5. The pathway for o-nitrobenzoate degradation has not been rigorously established as of this writing.

In the last decade several groups have elucidated the exact steps (Figure 6.6) involved in the degradation of p-nitrobenzoate.[14–16] The pathway reconciles all of the data presented above: that p-nitrosobenzoate, p-hydroxylaminobenzoate, and proto-catechuate are actually involved in the pathway, but p-aminobenzoate is not. In the first enzymatic step of the pathway, p-nitrobenzoate undergoes two two-electron reduc-tions (2 mol of NAD(P)H) to p-hydroxylaminobenzoate. A single enzyme catalyzes the reaction as shown with purified p-nitrobenzoate reductase from $C.\ acidovorans$ NBA-10.[14] The p-hydroxylaminobenzoate is then converted to protocatechuate by a lyase reaction with the release of ammonia. No cofactors are needed for the lyase reaction by the purified enzyme.[15] This suggests that the pathway investigated by Cain in $Nocardia$[9] proceeds by the same reaction scheme. However, in the case of Durham's $P.\ fluorescens$[11,12] the simultaneous adaptation of p-nitrobenzoate grown cells to p-aminobenzoate and p-hydroxybenzoate suggests that a pathway involving the full reduction of the nitro group to the amino group can also be found in nature.

Initial work on m-hydroxybenzoate degradation by Cartwright and Cain[9] in $Nocardia$ showed that the nitro group was removed as nitrite in contrast to the degradation of o-nitrobenzoate and p-nitrobenzoate where the nitro group was removed as ammonia. The authors proposed a sequential reaction scheme where one monooxygenase reaction would remove the nitro group with the formation of m-hydroxybenzoate. A second monooxygenase reaction would then result in the

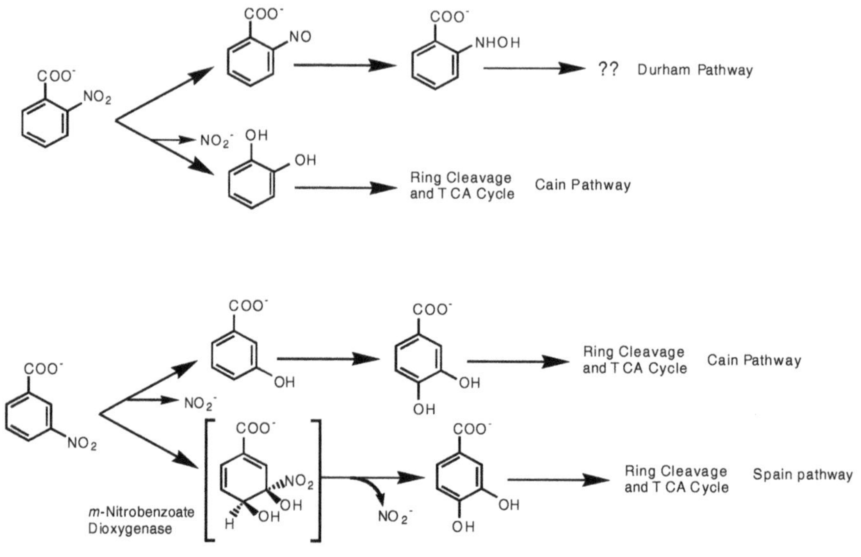

Figure 6.5 Proposed catabolic pathways for the degradation of o-nitrobenzoate and m-nitrobenzoate. Only the m-nitrobenzoate pathway involving the dioxygenase enzyme has been definitively demonstrated (see text for details). The Durham pathway for o-nitrobenzoate degradation through m-hydroxylaminobenzoate was suggested for a *Flavobacterium*,[22] while the Cain pathway for o-nitrobenzoate degradation through catechol was suggested for *Nocardia*.[9] The Cain pathway for m-nitrobenzoate degradation through m-hydroxybenzoate to protocatechuate was suggested for *Nocardia*,[9] while the Spain pathway for m-nitrobenzoate degradation through a proposed dihydrodiol to protocatchuate was proven for *Comamonas* sp. strain JS46.[27]

formation of protocatechuate which would then enter the β-ketoadipate pathway. However, respirometry experiments with m-nitrobenzoate grown cells showed significant oxygen uptake with protocatechuate, but not with m-hydroxybenzoate. Nevertheless, the authors still considered m-hydroxybenzoate to be an intermediate in the pathway. This dilemma was solved almost 40 years later by Nadeau and Spain.[27] These researchers showed with *Pseudomonas* sp. strain JS51 and *Comamonas* sp. strain JS46 that m-nitrobenzoate is subject to dioxygenase attack with the direct formation of protocatechuate and nitrite in a single reaction. That a dioxygenase is actually involved was confirmed using $^{18}O_2$ and showed that both atoms of molecular oxygen are incorporated into protocatechuate.

6.7 MOLECULAR BASIS FOR NITROBENZOATE DEGRADATION

Although molecular genetic tools would be quite useful in clarifying the exact pathways by which nitrobenzoates are degraded, not much has been done in this area. No genes have been cloned for either o-nitrobenzoate or m-nitrobenzoate

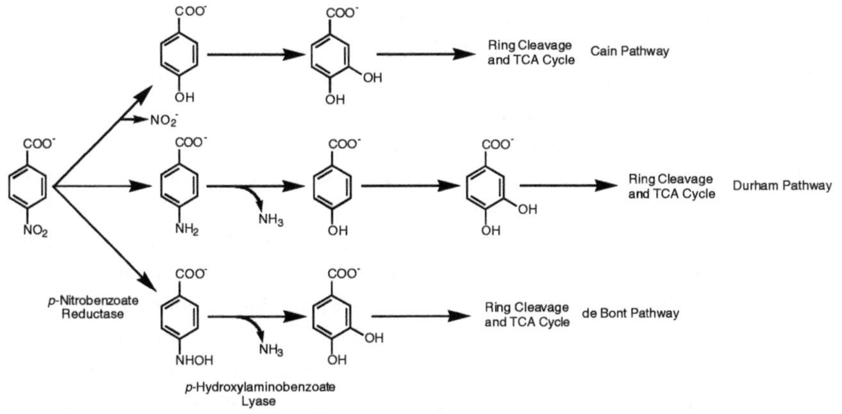

Figure 6.6   Proposed catabolic pathways for *p*-nitrobenzoate degradation. Only the de Bont pathway has been definitively demonstrated (see text for details). All three proposed catabolic pathways convert *p*-nitrobenzoate to protocatechuate as the ring cleavage substrate. The Cain pathway involving *p*-hydroxybenzoate as an intermediate was suggested for *Nocardia*,[9] while the Durham pathway involving *p*-aminobenzoate and *p*-hydroxybenzoate as intermediates was suggested for *P. fluorescens*.[11,12] The de Bont pathway involving *p*-hydroxylaminobenzoate as an intermediate was proven for *C. acidovorans* NBA-10.[14,15]

degradation. Yabannavar and Zylstra reported on the cloning of the genes for *p*-nitrobenzoate degradation from *R. pickettii* YH105.[40] Although this strain is capable of growing on either *p*-nitrobenzoate or *p*-aminobenzoate as a carbon source, it does so through independent pathways joining at protocatechuate.[40] Based on detection of metabolic intermediates from the cloned genes, *R. pickettii* YH105 degrades *p*-nitrobenzoate by the pathway proposed by de Bont and colleagues[14,15] (Figure 6.6). The nucleotide sequence shows two open reading frames (designated *pnbA* and *pnbB*) encoding the two enzymes needed for the conversion of *p*-nitrobenzoate to protocatechuate. These were confirmed by expression of each gene in *E. coli* or *R. pickettii*, followed by functional assays. In addition the deduced N-terminal sequence of the *p*-nitrobenzoate reductase matches the N-terminal sequence determined from the purified protein (C. Batie, personal communication). Immediately adjacent to the two genes for *p*-nitrobenzoate degradation and transcribed in the opposite direction is a gene coding for a regulatory protein in the LysR family. Analogous genes have been cloned from *Pseudomonas* sp. strain YH102, except in this case the gene for the *p*-nitrobenzoate reductase is not clustered with the genes required for regulation and the *p*-hydroxylaminobenzoate lyase.[29] In the case of YH102, mutants have been made of *pnbA* and *pnbR* by insertional inactivation. Interestingly, an insertional knockout of *pnbR* no longer has inducible *p*-nitrobenzoate reductase activity and the bacteria no longer grow on *p*-nitrobenzoate, indicating that *pnbA*, even though not clustered with *pnbB*, is still regulated by PnbR. An insertional mutant of *pnbB* results in the accumulation of *p*-hydroxylaminobenzoate, confirming its role in the degradation of *p*-nitrobenzoate.

Although the genes for *p*-nitrophenol degradation from *Pseudomonas* sp. strain ENV2030 and *P. putida* JS444 are almost identical, the same is not true for the genes for *p*-nitrobenzoate degradation from *R. pickettii* YH105 and *Pseudomonas* sp. strain YH102. The *pnbR* and *pnbB* genes show 49 and 64% identity between the two strains, respectively (Figure 6.7). In addition, Southern hybridization experiments (L. Newman and G. Zylstra, unpublished results) using the *pnb* genes from either YH102 or YH105 as probes show no or faint hybridization with *Pseudomonas* sp. strain 4NT[16] and *C. acidovorans* NBA-10[14] DNA, two organisms known to degrade *p*-nitrobenzoate by the same catabolic pathway as YH102 and YH105. In addition, analogous genes have been cloned and sequenced from *Pseudomonas* sp. strain TW3. These genes show a similar arrangement to what is seen in *Pseudomonas* YH102, but are 60% identical (M. A. Hughes and P. A. Williams, personal communication). This suggests that the genes for *p*-nitrobenzoate have existed for quite some time since they have widely diverged in the separate organisms. Searching for related genes in the GenBank database suggests that *pnbA* and *pnbB* are novel genes that have not simply evolved from genes encoding enzymes catalyzing similar reactions. There are no close analogs to the *Pseudomonas* sp. strain YH102 or *R. pickettii* YH105 *p*-hydroxylaminobenzoate lyase amino acid sequence in the GenBank database. The closest homolog to the *R. pickettii* YH105 *p*-nitrobenzoate reductase is a putative reductase in the *Streptomyces coelicolor* genome sequence (37% identical).

6.8 FUTURE DIRECTIONS

The future direction of research in the area of nitrophenol and nitrobenzoate degradation is obvious from the above discussion. Only a few examples of gene

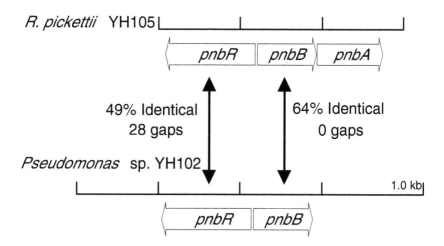

Figure 6.7 Comparison of the organization of the genes for *p*-nitrobenzoate degradation in two different Gram-negative organisms: *pnbA*, gene for *p*-nitrobenzoate reductase; *pnbB*, gene for *p*-hydroxylaminobenzoate lyase; *pnbR*, gene coding for a regulatory protein controlling *p*-nitrobenzoate degradation.

sequences are known from selected catabolic pathways and microorganisms. The sequence database needs to be expanded to include genes for all of the possible pathways for nitrophenol and nitrobenzoate degradation, as well as multiple examples from different microorganisms. The nucleotide and deduced amino acid sequences would be valuable in determining, for instance, how closely related *m*-nitrophenol reductase is to *p*-nitrobenzoate reductase. Such a comparison would answer the question as to whether the two reductase enzymes evolved from a common ancestor or evolved independently. The identification of the different gene sequences for *p*-nitrobenzoate degradation from different organisms aids in answering the evolutionary question since we now have an example of three different sequences that code for analogous enzymes. A related question is why there are two different pathways for *m*-nitrophenol degradation, both involving an initial reduction step but then diverging into two completely different pathways. How these two different pathways evolved and whether the two reductases are related to each other or perhaps evolved separately remain to be discovered. Although there is a large body of data on transport of polar aromatic compounds such as benzoate or *p*-hydroxybenzoate into the cell, it remains to be seen how polar compounds such as the nitrophenols and nitrobenzoates enter the cell. It is possible that the nitrophenols simply diffuse across the membrane, but for the nitrobenzoates there must be some mechanism of transport. Genes for this activity have not been identified yet, and physiological experiments have not been performed to show whether an active transport system is actually involved. Toxicity issues have not been addressed yet. *p*-Nitrophenol, for instance, is toxic to microorganisms and has even been used as a leather preservative because of its toxicity to fungi. Whether bacteria that degrade *p*-nitrophenol and other toxic nitro compounds have evolved mechanisms for protection against these compounds has not been determined. The complicated regulation of *p*-nitrophenol degradation needs to be explained — whether this is simply a coincidence due to the way that the genes evolved or perhaps that there may be an evolutionary advantage to why there are two regulatory genes and why *pnpA* and *pnpB* are divergently transcribed. Finally, it may be possible to create better nitro compound degraders by manipulating the sequences simply by rearranging the genes or perhaps by altering their regulation. The *p*-nitrophenol degradation pathway can perhaps provide the best example of this if the genes can be rearranged into a single operon and the competitiveness of the resulting organism compared with that of a wild-type strain. Alternatively, knowledge of the complexities of gene regulation for nitro compound degradation can be used to advantage through manipulation of the environment so that the natural consort of microorganisms can degrade the nitro compounds more efficiently.

REFERENCES

1. Akileswaran, L., B. J. Brock, J. L. Cereghino, and M. H. Gold. 1999. 1,4-Benzoquinone reductase from *Phanerochaete chrysosporium*: cDNA cloning and regulation of expression. *Appl. Environ. Microbiol.* 65:415–421.
2. Arai, H., T. Yamamoto, T. Ohishi, T. Shimizu, T. Nakata, and T. Kudo. 1999. Genetic organization and characteristics of the 3-(3-hydroxyphenyl)propionic acid degradation pathway of *Comamonas testosteroni* TA441. *Microbiology.* 145:2813–2820.

3. Armengaud, J., K. N. Timmis, and R. M. Wittich. 1999. A functional 4-hydroxysalicy-late/hydroxyquinol degradative pathway gene cluster is linked to the initial dibenzo-p-dioxin pathway genes in *Sphingomonas* sp. strain RW1. *J. Bacteriol.* 181:3452–3461.

4. Bang, S.-W. and G. J. Zylstra. 1996. Cloning and characterization of the genes involved in *p*-nitrophenol degradation by *Pseudomonas fluorescens* ENV2030. Abstr. Q166, p. 414. In Abstracts of the 96th General Meeting of the American Society for Microbiology, American Society for Microbiology, Washington, D.C.

5. Bang, S.-W. and G. J. Zylstra. 1997. Cloning and sequencing of the hydroquinone 1,2-dioxygenase, 2-hydroxymuconic semialdehyde dehydrogenase, and maleylacetate reductase genes from *Pseudomonas fluorescens* ENV2030. Abstr. Q383, p. 519. In Abstracts of the 97th General Meeting of the American Society for Microbiology, American Society for Microbiology, Washington, D.C.

6. Barnes, M. R., W. A. Duetz, and P. A. Williams. 1997. A 3-(3-hydroxyphenyl)propi-onic acid catabolic pathway in *Rhodococcus globerulus* PWD1: cloning and charac-terization of the *hpp* operon. *J. Bacteriol.* 179:6145–6153.

7. Cain, R. B. 1958. The microbial metabolism of nitroaromatic compounds. *J. Gen. Microbiol.* 19:1–14.

8. Cain, R. B. and N. J. Cartwright. 1960. Intermediary metabolism of nitrobenzoic acids by bacteria. *Nature.* 185:868–869.

9. Cartwright, N. J. and R. B. Cain. 1958. Bacterial degradation of the nitrobenzoic acids. *Biochem. J.* 71:248–261.

10. Daubaras, D. L., C. D. Hershberger, K. Kitano, and A. M. Chakrabarty. 1995. Sequence analysis of a gene cluster involved in metabolism of 2,4,5-trichlorophen-oxyacetic acid by *Burkholderia cepacia* AC1100. *Appl. Environ. Microbiol.* 61:1279–1289.

11. Durham, N. N. 1956. Bacterial oxidation of *p*-aminobenzoic acid by *Pseudomonas fluorescens*. *J. Bacteriol.* 72:333–336.

12. Durham, N. N. 1958. Studies on the metabolism of *p*-nitrobenzoic acid. *Can. J. Microbiol.* 4:141–148.

13. Fernandez, A., J. L. Garcia, and E. Diaz. 1997. Genetic characterization and expres-sion in heterologous hosts of the 3-3-(hydroxyphenyl)propionate catabolic pathway of *Escherichia coli* K-12. *J. Bacteriol.* 170:2573–2581.

14. Groenewegen, P. E., P. Breeuwer, J. M. van Helvoort, A. A. Langenhoff, F. P. de Vries, and J. A. de Bont. 1992. Novel degradative pathway of 4-nitrobenzoate in *Comamonas acidovorans* NBA-10. *J. Gen. Microbiol.* 138:1599–1605.

15. Groenewegen, P. E. J. and J. A. M. de Bont. 1992. Degradation of 4-nitrobenzoate via 4-hydroxylaminobenzoate and 3,4-dihydroxybenzoate in *Comamonas aci-dovorans* NBA-10. *Arch. Microbiol.* 158:381–386.

16. Haigler, B. E. and J. C. Spain. 1993. Biodegradation of 4-nitrotoluene by *Pseudomo-nas* sp. strain 4NT. *Appl. Environ. Microbiol.* 59:2239–2243.

17. Hamood, A. N., J. A. Colmer, U. A. Ochsner, and M. L. Vasil. 1996. Isolation and characterization of a *Pseudomonas aeruginosa* gene, *ptxR*, which positively regulates exotoxin A production. *Mol. Microbiol.* 21:97–110.

18. Hanne, L. F., L. L. Kirk, S. M. Appel, A. D. Narayan, and K. K. Bains. 1993. Degradation and induction specificity in actinomycetes that degrade *p*-nitrophenol. *Appl. Environ. Microbiol.* 59:3505–3508.

19. Hatta, T., O. Nakano, N. Imai, N. Takizawa, and H. Kiyohara. 1999. Cloning and sequence analysis of hydroxyquinol 1,2-dioxygenase gene in 2,4,6-trichlorophenol-degrading *Ralstonia pickettii* DTP602 and characterization of its product. *J. Biosci. Bioeng.* 87:267–277.

20. Jain, R. K., J. H. Dreisbach, and J. C. Spain. 1994. Biodegradation of *p*-nitrophenol via 1,2,4-benzenetriol by an *Arthrobacter* sp. *Appl. Environ. Microbiol.* 60:3030–3032.

21. Kadiyala, V. and J. C. Spain. 1998. A two-component monooxygenase catalyzes both the hydroxylation of *p*-nitrophenol and the oxidative release of nitrite from 4-nitro-catechol in *Bacillus sphaericus* JS905. *Appl. Environ. Microbiol.* 64:2479–2484.

22. Ke, Y.-H., L. L. Gee, and N. N. Durham. 1959. Mechanism involved in the metabolism of nitrophenylcarboxylic acid compounds by microorganisms. *J. Bacteriol.* 77:593–598.

23. Meulenberg, R., M. Pepi, and J. A. M. de Bont. 1996. Degradation of 3-nitrophenol by *Pseudomonas putida* B2 occurs via 1,2,4-benzenetriol. *Biodegradation.* 7:303–311.

24. Mitra, D. and C. S. Vaidyanathan. 1984. A new 4-nitrophenol 2-hydroxylase from a *Nocardia* sp. *Biochem. Int.* 8:609–615.

25. Munnecke, D. M. and D. P. Hsieh. 1974. Microbial decontamination of parathion and *p*-nitrophenol in aqueous media. *Appl. Microbiol.* 28:212–217.

26. Murakami, S., T. Okuno, E. Matsumura, S. Takenaka, R. Shinke, and K. Aoki. 1999. Cloning of a gene encoding hydroxyquinol 1,2-dioxygenase that catalyzes both intra-diol and extradiol ring cleavage of catechol. *Biosci. Biotechnol. Biochem.* 63:859–865.

27. Nadeau, L. J. and J. C. Spain. 1995. Bacterial degradation of *m*-nitrobenzoic acid. *Appl. Environ. Microbiol.* 61:840–843.

28. Neidle, E. L., C. Hartnett, and L. N. Ornston. 1989. Characterization of *Acinetobacter calcoaceticus catM*, a repressor gene homologous in sequence to transcriptional activator genes. *J. Bacteriol.* 171:5410–5421.

29. Newman, L. M. and G. J. Zylstra. 1997. Cloning and characterization of the genes involved in 4-nitrobenzoate degradation from *P. putida* YH102. Abstr. Q341, p. 512. In Abstracts of the 97th General Meeting of the American Society for Microbiology, American Society for Microbiology, Washington, D.C.

30. Nishino, S. F. and J. C. Spain. 1993. Cell density-dependent adaptation of *Pseudomonas putida* to biodegradation of *p*-nitrophenol. *Environ. Sci. Technol.* 27:489–494.

31. Perry, L. and G. J. Zylstra. 1999. Biochemical and molecular analysis of hydroxy-quinol 1,2-dioxygenase involved in *p*-nitrophenol degradation by *Arthrobacter* sp. strain JS443. Abstr. Q250, p. 581. In Abstracts of the 99th General Meeting of the American Society for Microbiology, American Society for Microbiology, Washington, D.C.

32. Prakash, D., A. Chauhan, and R. K. Jain. 1996. Plasmid-encoded degradation of *p*-nitrophenol by *Pseudomonas cepacia*. *Biochem. Biophys. Res. Commun.* 224:375–381.

33. Raymond, D. G. M. and M. Alexander. 1971. Microbial metabolism and cometabolism of nitrophenols. *Pestic. Biochem. Physiol.* 1:123–130.

34. Schafer, A., H. Harms, and A. J. Zehnder. 1996. Biodegradation of 4-nitroanisole by two *Rhodococcus* spp. *Biodegradation.* 7:249–255.

35. Schenzle, A., H. Lenke, P. Fischer, P. A. Williams, and H. J. Knackmuss. 1997. Catabolism of 3-nitrophenol by *Ralstonia eutropha* JMP134. *Appl. Environ. Microbiol.* 63:1421–1427.

36. Schenzle, A., H. Lenke, J. C. Spain, and H. J. Knackmuss. 1999. 3-Hydroxylami-nophenol mutase from *Ralstonia eutropha* JMP134 catalyzes a Bamberger rearrange-ment. *J. Bacteriol.* 181:1444–1450.

37. Simpson, J. L. and W. C. Evans. 1953. The metabolism of nitrophenols by certain bacteria. *Biochem. J.* 52:24.

38. Spain, J. C. and D. T. Gibson. 1991. Pathway for biodegradation of *p*-nitrophenol in a *Moraxella* species. *Appl. Environ. Microbiol.* 57:812–819.

39. Spain, J. C., O. Wyss, and D. T. Gibson. 1979. Enzymatic oxidation of *p*-nitrophenol. *Biochem. Biophys. Res. Commun.* 88:634–641.

40. Yabannavar, A. V. and G. J. Zylstra. 1995. Cloning and characterization of the genes for *p*-nitrobenzoate degradation from *Pseudomonas pickettii* YH105. *Appl. Environ. Microbiol.* 61:4284–4290.

41. Yang, W., L. Ni, and R. L. Somerville. 1993. A stationary-phase protein of *Escherichia coli* that affects the mode of association between the *trp* repressor protein and operator-bearing DNA. *Proc. Natl. Acad. Sci. U.S.A.* 90:5796–5800.

42. Yazaki, K., A. Bechthold, and M. Tabata. 1997. Molecular cloning and expression of a cDNA encoding a WrbA homologue from *Lithospermum erythrorhizon*. *Plant Physiol.* 115:1287.

43. Zeyer, J. and P. C. Kearney. 1984. Degradation of *o*-nitrophenol and *m*-nitrophenol by a *Pseudomonas putida*. *J. Agric. Food Chem.* 32:238–242.

44. Zeyer, J. and H. P. Kocher. 1988. Purification and characterization of a bacterial nitrophenol oxygenase which converts *ortho*-nitrophenol to catechol and nitrite. *J. Bacteriol.* 170:1789–1794.

45. Zeyer, J., H. P. Kocher, and K. N. Timmis. 1986. Influence of *para*-substituents on the oxidative metabolism of *o*-nitrophenols by *Pseudomonas putida* B2. *Appl. Environ. Microbiol.*

CHAPTER 7

The Role of Nitrate Ester Reductase Enzymes in the Biodegradation of Explosives

Richard E. Williams and Neil C. Bruce

CONTENTS

7.1 INTRODUCTION

Over the last 100 years, nitrate esters have found a variety of technological and medical applications, notably as high explosives and vasodilators. A wide variety of compounds have been synthesized — the structures of some are shown in Figure 7.1. Although useful therapeutically in low doses, nitrate esters and their metabolites are generally toxic at higher levels and also have significant environmental impact. In the case of nitroglycerin (glycerol trinitrate, GTN), mammalian toxicity data in Wendt et al.[57] suggest acute toxicity levels of 30 to 1300 mg/kg for GTN, and an LD_{50} of 1 mg/L for fish is quoted by Urbanski.[54] In practical terms, the main problems arise with wastewater streams from industrial plants manufacturing GTN, as well as from the need to dispose of munitions containing nitrocellulose and GTN when they reach the end of a limited shelf-life. GTN and related incompletely nitrated products are fairly soluble in water. Conventional chemical treatments of GTN waste streams include treatment with strong acid, strong alkali, or denitration with sodium sulfide. Products of this treatment may include partially denitrated GTN, glycidyl nitrate (2-nitrooxymethyl-oxirane), and glycidol (oxiranyl methanol), together with nitrite and nitrate, which can require further denitrifying treatment prior to final discharge. There is little evidence of biologically derived organic nitrate esters ($C-O-NO_2$) in the literature; an alkenyl nitrate that acts as an insect sex pheromone seems to be the

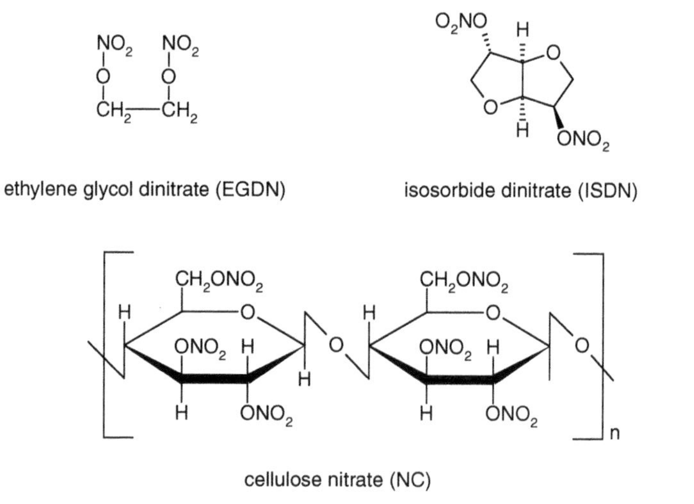

glycerol trinitrate (GTN)

pentaerythritol tetranitrate (PETN)

ethylene glycol dinitrate (EGDN)

isosorbide dinitrate (ISDN)

cellulose nitrate (NC)

Figure 7.1 Examples of nitrate esters.

sole example.[24] Some organic nitrates can form due to chemical interactions between NO_x and hydrocarbons in the atmosphere,[36] but multiple substituted nitrate esters have probably been newly introduced into the environment by man in the past 100 years. The way in which microorganisms have adapted over relatively short periods of time to tolerate or even derive nutrition from novel xenobiotic compounds is of particular interest. Characterization of nitrate ester-degrading systems also offers the prospect of utilizing either the organisms or their component enzymes in the identification and bioremediation of contaminated land.

7.2 BIOLOGICAL TRANSFORMATION OF NITRATE ESTERS

Nitrate esters are comparatively stable and are known to persist in the environment for significant periods of time. Degradation of these compounds by alkaline hydrolysis is possible with strong alkali (Table 7.1); however, given the existence of enzymes hydrolyzing sulfate and phosphate esters,[58] it might be expected that similar transformation of nitrate esters would occur in the environment. In fact, the bulk of the evidence points toward the reductive transformation of nitrate esters, resulting in the formation of alcohol and liberation of nitrite. As discussed by White and Snape,[58] nitrate esters are uncharged and therefore are less solvated than sulfate and phosphate esters, which might account for their stability to hydrolysis. Multiply-substituted nitrate esters have fairly low aqueous solubility and are lipophilic. However, their denitration products become steadily more soluble as nitrate groups are removed (Table 7.1). Interaction with sulfur occurs as a recurrent theme in biological interactions with nitrogen species. Thiol–nitrate interaction appears to be an important basis for mammalian effects of nitrates as nitric oxide donors.[53] The precise mechanisms involved remain controversial.

7.2.1 Mammalian Metabolism

The mammalian metabolism of organic nitrate esters has been extensively studied, as organic nitrates have been used as vasodilators for over 100 years. Despite much clarification in recent years, the mechanisms of nitrate ester action appear complex and are the subject of considerable conflicting data.[3,53] The pharmacological activity of nitrate esters is intimately related to the actions of nitric oxide, which is an important chemical messenger in many mammalian systems. Administration of GTN leads to vasodilation via the accumulation of cyclic GMP in vascular smooth muscle. The generally accepted model of action involves biotransformation of nitrate esters in the muscle tissue to form an activator of guanylate cyclase. It is the precise nature of this activator and the mechanism responsible for its formation that remains unclear. The existence of multiple enzymes transforming nitrate esters and the unstable nature of most postulated intermediates makes conclusive explanation of organic nitrate action extremely difficult.

At least two classes of enzymes have been implicated in mammalian nitrate ester biotransformation (Figure 7.2), both in hepatic and vascular tissues. Initial studies focused on glutathione-S-transferases, a group of enzymes mediating the detoxification

Table 7.1 Physical and Toxicological Properties of Nitrate Esters

Compound	Mammalian Toxicity (LD_{50} mg/kg)	Alkaline Hydrolysis Second-Order Rate Constant[18] at 30°C in 90% ethanol (10^{-5} $M^{-1}s^{-1}$)	Octanol/Water Partition Coefficient[61]	Aqueous Solubility at 20°C (g/L)
Glycerol trinitrate (GTN)	30–1,300	24,000	50.87	1.5[2]
Glycerol-1,2-dinitrate (1,2-GDN)	800–2,300		6.02	80
Glycerol-1,3-dinitrate (1,3-GDN)	500–800	4,500	5.05	80
Glycerol-1-nitrate (1-GMN)	900–6,000	2,150	0.26	700[57]
Glycerol-2-nitrate (2-GMN)	>5,000	920		700[57]
Ethylene glycol dinitrate (EGDN)		160		5[2]
Ethylene glycol nitrate (EGMN)		168		
Pentaerythritol tetranitrate (PETN)		34		2×10^{-3} [30]
Isosorbide dinitrate (ISDN)			15.06	
Isosorbide-5-nitrate (IS-5-N)			0.97	
Isosorbide-2-nitrate (IS-2-N)			0.65	

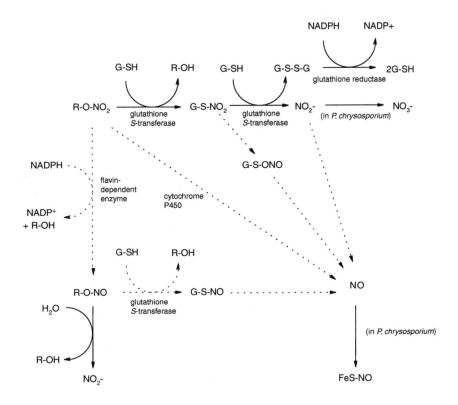

Figure 7.2 Eukaryotic metabolism of nitrate ester explosives.

of electrophilic xenobiotics by conjugation to glutathione (GSH). Glutathione-*S*-transferases (GSTs) catalyze the reduction of GTN to glycerol 1,2-dinitrate (1,2-GDN) and 1,3-GDN to liberate nitrite.[52] Given the known ability of nitric oxide to stimulate guanylate cyclase and the low vasodilator potency of nitrite, GST-mediated transformation of GTN is unlikely to be productive unless it somehow results in nitric oxide formation. Proposed schemes for nitric oxide production via GSTs have suggested the production of *S*-nitroso compounds as nitric oxide precursors. Such schemes also explain tolerance to GTN in terms of thiol depletion. *S*-nitroso compounds could arise from rearrangement of thionitrate ester intermediates produced by GST or from conversion of GTN to a nitrite ester prior to the action of GST.[62] Recent work has demonstrated that flavin mononucleotide (FMN) is capable of catalyzing the reduction of organic nitrate esters by NADPH and that thiols are able to participate in this chemistry,[62] enhancing nitric oxide production.

It is clear that enzymes other than GSTs might be responsible for the pharmacological action of GTN. Rat liver microsomes denitrated GTN in the presence of NADPH, with electron paramagnetic resonance (EPR) data suggesting the formation of a nitric oxide–heme iron complex.[41] The reaction was inhibited by oxygen and by classic inhibitors of cytochromes-P450. Conflicting data have been reported regarding the effect of P450 inhibitors on GTN transformation and action in intact blood vessel preparations.[3] It is hard to draw firm conclusions from the effects of inhibitors on uncharacterized enzymes, especially with the possibility of multiple transformation pathways.

7.2.2 Fungal Metabolism

Fungal metabolism of nitrate esters appears to be similar to that of mammals. It has, therefore, been studied both with a view to bioremediation and as a useful model of mammalian metabolism. GTN has a short half-life *in vivo*, prompting the search for alternative organic nitrate vasodilators more suitable for treatment of chronic conditions. Isosorbide mononitrates and glycerol mononitrates have longer half-lives than GTN,[9] but glycerol-2-nitrate (2-GMN) and 5-isosorbide mononitrate (5-ISMN) are difficult to synthesize chemically, making the search for biological routes to these potential drugs attractive. Screening of a range of microorganisms revealed that several fungi could denitrate isosorbide dinitrate (ISDN) to 2-ISMN and 5-ISMN, with varying regiospecificities.[38] This work does not appear to have been taken further. GTN is degraded by *Geotrichum candidum* into its mononitrate derivatives.[15] HPLC analysis showed the accumulation of large quantities of dinitrates prior to the formation of GMN. The evidence is consistent with sequential removal of nitrate groups. In contrast to *G. candidum*, *Phanerochaete chrysosporium* and *Sporotrichum paranense* were found to favor the production of 2-GMN.[16] The ratios of GDN and GMN isomers produced by *G. candidum* point to a regioselectivity toward removal of the C2 nitrate from GTN, but the C1 nitrate from 1,2-GDN. White and Snape[58] suggest that the difference in regiospecificity could arise if separate enzymes were responsible for the two denitrations.

During the metabolism of GTN, *P. chrysosporium* produced nitrite which was converted subsequently to nitrate. EPR spectra also suggested the formation of nitric oxide, which was initially present as an (Fe^{2+})–heme–NO complex and subsequently as a nonheme iron–sulfur–NO complex.[42] Analysis of the subcellular location of the enzyme activities responsible suggested the presence of a GST activity in the cytosol, together with both cytosolic and microsomal cytochrome-P450-like enzymes.[43] *P. chrysosporium* has been immobilized in a packed bed reactor that achieved 99% removal of GTN and complete denitration of a third of the initial 600 μM of GTN added, leaving a mixture of dinitrate and mononitrate esters.[5]

Nitrocellulose is a major component of gun propellants and poses a disposal problem since it has a limited shelf-life. To slow autocatalytic decomposition, stabilizers such as diphenylamine are present in the propellant mixture. These stabilizers are gradually consumed by nitration, resulting in unstable propellant which has been typically disposed of by open air burning/detonation. Increasing restrictions on these practices has driven the investigation of biodegradation methods. The success of these methods is limited by the bioavailability of nitrocellulose, which is insoluble in aqueous media. *Aspergillus fumigatus* grew in minimal media, with glucose as a carbon source and a nitrocellulose suspension as the sole nitrogen source, at a faster rate than would be supported by spontaneous hydrolysis of the nitrate ester linkage.[10] The biochemical basis for nitrocellulose degradation was not determined. A co-culture of the cellulolytic *Sclerotium rolfsii* and the denitrifying *Fusarium solani* were able to degrade nitrocellulose,[45] as measured by a gradual decrease in acetone-extractable material present in the culture medium. An observed increase in acetone-insoluble matter was attributed to growth of the fungi. The authors suggested that the significant drop in pH during growth probably prevented complete degradation, yet screening of a wider range of fungi[50] by the same protocol showed that substantial acidification only occurred with *S. rolfsii*, suggesting that there are additional obstacles to the degradation of nitrocellulose. An organism identified as *Penicillium corylophilum* Dierckx was isolated from a double base propellant containing nitrocellulose and GTN. This organism was able to partially degrade nitrocellulose with starch or xylan as a supplemental carbon source,[44] and to completely denitrate GTN in the presence of ammonium nitrate.[63] No biochemical basis for the activity has been demonstrated, but the involvement of hydroxyl radicals produced by lignin peroxidase in this organism was excluded.[49]

7.2.3   Bacterial Metabolism

In contrast to transformation by eukaryotic systems, the metabolism of nitrate esters by bacteria has only been characterized fairly recently. When reviewed 6 years ago by White et al.,[58] the clearest reports of bacterial degradation were that of glycerol trinitrate (GTN) by activated sludge[57] and ethylene glycol dinitrate (EGDN) by a strain of *Klebsiella oxytoca*.[51] In both cases nitrate groups were removed sequentially, but the degradation of the mononitrate was undetectable. The only suggestion of microbial degradation of pentaerythritol tetranitrate (PETN) arose from studies of the mammalian metabolism of PETN in rats. Tri-, di-, and mononitrated derivatives of ^{14}C-labeled PETN were produced on incubation with rat urine[26] — the transformation did not occur if the urine was filter sterilized or incubated at 4°C.

When propellant manufacturing plant wastewater with a GTN content of around 180 mg/L was treated in a pilot-scale (400-L) batch reactor, GTN was biodegraded to below detection limits. The reactor cycle involved an 8-h aerobic phase followed by a 5-h anoxic phase. The microbial community used the ethyl acetate present in the waste stream as a carbon source.[33] Aqueous GTN concentrations fell throughout the aerobic phase. The nitrate produced was then removed during an anoxic phase of the process. The authors suggested that GTN was also amenable to degradation under anoxic conditions. No analysis of the extent of denitration was reported.

GTN was completely mineralized under anaerobic conditions in sealed microcosms inoculated with digester sludge from a municipal waste treatment plant.[13] The mineralization proceeded via dinitrate and mononitrate intermediates, but no significant nitrate or nitrite accumulated. GTN was degraded faster when glucose was added as a carbon source. Denitration was regiospecific, and the formation of 1,3-GDN and 2-GMN was favored. Further work with batch and packed-bed reactors revealed complete denitration of GTN by both aerobic and anaerobic sludge.[5] Degradation under aerobic conditions was slow and required substantial addition of a supplementary carbon source.[4] A mixed culture, selectively enriched from aeration tank sludge of a treatment facility handling waste streams containing GTN, was able to use GTN as a sole carbon and nitrogen source under aerobic conditions.[1] Nitrogen was released predominantly as nitrite, although it was suggested that nitrite and nitrate could be consumed by denitrification processes.

Recently, the aerobic degradation of nitrate esters by pure cultures of bacteria has been investigated, and bacterial enzymes able to cleave nitrate esters have been characterized. Many of the bacterial isolates reported have been obtained from sites known to have been exposed to high levels of nitrate ester explosives, because at such sites one might expect selective pressure to have produced effective nitrate ester degraders. Screening of soil samples for bacteria with GTN tolerance and the ability to use GTN as the sole nitrogen source yielded strains identified as *Bacillus thuringiensis/cereus* and *Enterobacter agglomerans*.[29] In whole-cell assays, loss of GTN was correlated with production of nitrite. Further work showed that the former strain, deposited as *Bacillus* sp. ATCC 51912, was able to sequentially denitrate propylene glycol dinitrate (PGDN) to propylene glycol.[48] Selective enrichment of cultures for growth on GTN as a sole nitrogen source, with glycerol as a carbon source, led to the isolation of a *Pseudomonas* sp. from river sediment[59] and an *Agrobacterium radiobacter* strain from activated sewage sludge.[60] A consortium of *Arthrobacter ilicis* and *Agrobacterium radiobacter*, isolated from soil surrounding an explosives factory, denitrated EGDN and used it as sole nitrogen source for growth when supplemented with ethylene glycol as the carbon source.[34] Only the *Arthrobacter* sp. degraded EGDN; the role of the *Agrobacterium* sp. was apparently to consume the high levels of nitrate produced. Enrichment cultures from explosive-contaminated soil led to the isolation of an *E. cloacae* strain designated PB2, which was able to utilize PETN as a sole nitrogen source.[6] More recently, strains of *P. putida* and *P. fluorescens* capable of utilizing GTN as a sole nitrogen source were isolated from GTN-contaminated soil.[7,8] GTN degradation ability of the environmental isolates was compared to that of various strains from culture collections that exhibited some activity, but were not able to tolerate such high GTN concentrations. While the isolates from contaminated soil were more effective degraders, the ability of laboratory strains to transform GTN shows that GTN-degrading activity is widespread. The considerable variability in tolerance or growth on nitrate esters may arise from differences in nitrate/nitrite metabolism or toxicity of GTN metabolites, rather than in the initial transformation step.

7.3 BACTERIAL NITRATE ESTER REDUCTASES

In all bacteria where nitrate ester degradation has been characterized in detail, very similar enzymes have been identified. The enzymes catalyze the nicotinamide cofactor-dependent reductive cleavage of nitrate esters to give alcohol and nitrite. Purification of the PETN reductase from *E. cloacae* yielded a monomeric protein of around 40 kDa, which required NADPH for activity.[6] Similar enzymes were responsible for the nitrate ester-degrading activity in *A. radiobacter*[47] — "nitrate ester reductase" — and in the strains of *P. fluorescens* and *P. putida*[7] — "xenobiotic reductases." All utilize a noncovalently bound FMN as a redox cofactor. The *Agrobacterium* enzyme differs from the others in requiring NADH as the source of reducing equivalents. The genes that encode the nitrate ester reductases from

E. cloacae, A. radiobacter, P. putida, and *P. fluorescens* have now been cloned and sequenced[7,20,47] and are homologous. Thus, the enzymes form a novel family of nitrate ester reductases. The characteristics of these enzymes set them clearly apart from the glutathione- and iron-dependent enzymes implicated in mammalian and fungal transformation of nitrate esters.

Nitrate ester reductases might not be the only enzymes involved in bacterial transformation of nitrate esters. The enzyme(s) responsible for nitrate ester degradation by *Bacillus* sp. ATCC 51912 appeared to be cytosolic, but were membrane associated in *E. agglomerans*.[29] Enzyme assays carried out after dialysis of cell extracts from both strains indicated that diffusible cofactors were not needed for activity. This result is surprising, as it would seem to preclude a reductive denitration to release nitrite, instead favoring a hydrolytic mechanism releasing nitrate. Because the production of nitrite, not nitrate, was observed, the authors suggested that the reduction of nitrate to nitrite is catalyzed by nitrate reductase. Given that complete denitration of GTN required repeated addition of cell extract, it is conceivable that high molecular weight redox proteins are involved. Such a pool of reducing equivalents would be retained by a dialysis membrane, but would be rapidly depleted during *in vitro* experiments without a system for regeneration.

7.3.1 Extent of Activity

All reports of nitrate ester transformation by prokaryotic and eukaryotic systems involve sequential denitration steps, with the appearance of multiple partially denitrated products (Figure 7.3). Each step is slower than the previous step, and when complete denitration has been reported it is extremely slow. The findings correlate with the relative rates of alkaline hydrolysis (Table 7.1). A theoretical treatment of sequential GTN denitration by Smets et al. concluded that the use of GTN as a sole carbon and nitrogen source was thermodynamically feasible.[46] Cell extracts of *Bacillus* sp. ATCC 51912 converted GTN to glycerol and PGDN to propylene glycol, but the conversion was only achieved with the periodic addition of fresh cell extract to the reaction.[29,48] During growth of *Arthrobacter ilicis* on EGDN, the mononitrate ester accumulated only transiently, but denitration of ethylene glycol mononitrate could not be demonstrated *in vitro*.[34]

There are no conclusive reports of complete denitration of nitrate esters by purified nitrate ester reductases, nor by the organisms expressing them. Cells of *E. cloacae* PB2 and *Agrobacterium radiobacter* transformed PETN only as far as the dinitrate.[6,60] Cell yield of *E. cloacae* PB2 grown on PETN as the sole nitrogen source was 0.64 times that obtained after growth on equimolar amounts of NH_4Cl, which suggested that between two and three of the four PETN nitrate groups were supporting growth.[6] Comparison of the cell yield of the *A. radiobacter* strain grown on equimolar amounts of either GTN or sodium nitrite as the sole nitrogen source indicated that the strain utilized two of the three nitro groups of GTN. There was no observed breakdown of GMN to glycerol, and the organism could not use GTN as its sole carbon source. In extracts prepared from cells of *P. fluorescens* and *P. putida*,[8] nitrite was liberated at a rate of 1 mol per mole of GTN with limiting NADPH and up to 2 mol per mole with excess NADPH, which suggested that the

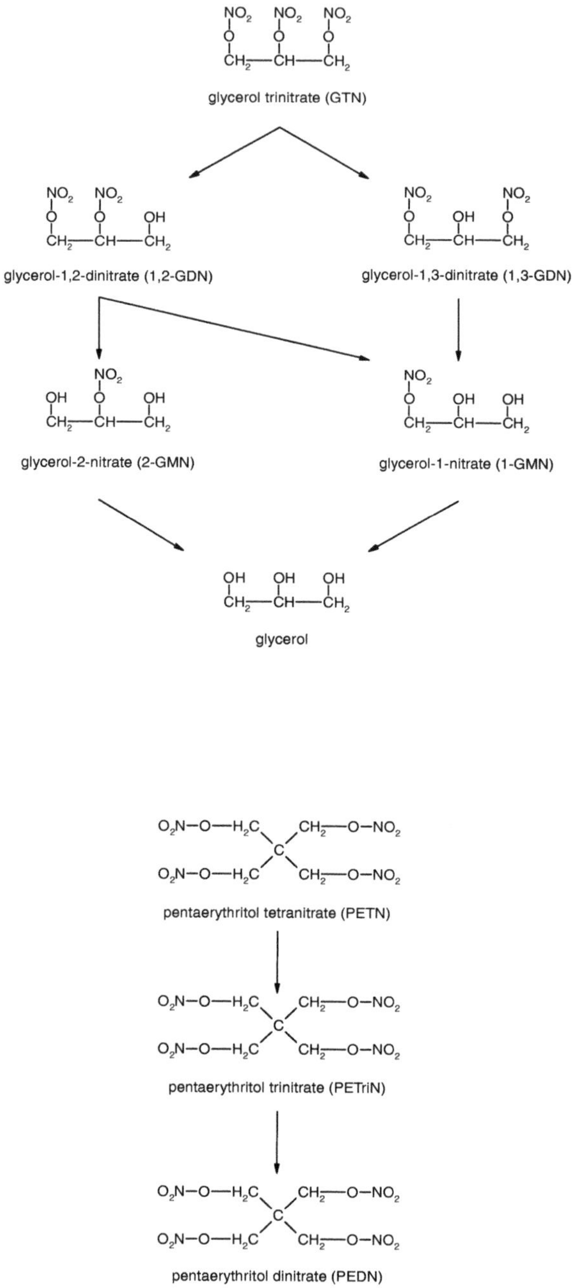

Figure 7.3   Bacterial transformation of GTN and PETN. Each step involves the oxidation of NADPH to NADP+ and the liberation of nitrite.

dinitrates and mononitrates are much poorer substrates than GTN. There was no significant conversion of mononitrates to glycerol, and neither strain could use GTN as a sole source of carbon and nitrogen. However, in the presence of large quantities of purified nitrate ester reductase from *P. putida* and excess NADPH, GTN was converted to GMN, and GMN levels gradually decreased. These results suggested that *P. putida* nitrate ester reductase was capable of catalyzing complete denitration, albeit at a very slow rate.[7,8]

7.3.2 Inducibility

When *Bacillus* sp. ATCC 51912 or *E. agglomerans* are grown in rich medium, addition of GTN does not enhance the expression of GTN-degrading activity.[29] PGDN-degrading activity in *Bacillus* sp. ATCC 51912 is also constitutive. Similarly, expression of PETN-degrading activity in *E. cloacae* was identical when cells were grown with PETN or NH_4NO_3 as nitrogen source.[6] The strains of *P. fluorescens* and *P. putida* showed little response to 1 m*M* GTN,[7,8] and mRNA levels of the gene responsible for the activity in *P. putida* were unchanged by addition of 0.9 m*M* GTN.[7] However, expression of GTN-degrading activity was 160-fold higher in *A. radiobacter* grown on GTN as the sole nitrogen source relative to cells grown in rich medium,[60] suggesting that expression of nitrate ester reductases may be regulated, although not necessarily by nitrate esters.

Although cleavage of nitrate esters makes nitrogen available to the organism, there is no reason to assume that the enzyme responsible is part of general nitrogen metabolism in the cell. Since nitrate esters have only entered the environment recently, it is likely that their biological breakdown is serendipitous rather than efficiently evolved. Strains found in treatment plants, water, and soil that have been exposed to nitrate esters for prolonged periods might well have been under selective pressure to constitutively express enzymes normally used for other purposes (and regulated accordingly). The fact that GTN-degrading enzymes purified from the strains of *P. putida* and *P. fluorescens* were present at up to 15% of the soluble cell protein[7,8] supports this reasoning.

7.3.3 Regiospecificity

GTN denitration can potentially produce two isomers of the dinitrate and mononitrate; both 1,2-GDN and 1-GMN have chiral centers. Inconsistencies in the isomeric composition of GTN degradation products have often been taken to suggest the existence of multiple degradation pathways[3] or the action of multiple enzymes with different regiospecificities.[58] However, it is clear that regiospecificity of GTN metabolism by bacterial nitrate ester reductases varies across species, despite significant similarity in enzyme structure. During growth with GTN as the sole nitrogen source, *A. radiobacter* produces mainly 1,3-GDN and less 1,2-GDN. Eventually 1-GMN and a small amount of 2-GMN accumulate in the culture broth. Denitration of the C2 nitrate group of GTN by resting cells was around ten times faster than the rate of denitration of the terminal nitrate groups.[60] Purified GTN reductase from this organism produced a similar ratio of GDN isomers.[47] Analysis of the products formed by the purified enzyme from *P. fluorescens* revealed a similar bias toward denitration at the 2-position of GTN and 1,2-GDN.[7] In contrast, denitration of GTN by the *P. putida* enzyme gave the 2:1 ratio of 1,2- to 1,3-GDN which would be expected if there was no regioselective bias, although there was a slight bias toward denitration of 1,2-GDN at the 1-position. Clearly, subtle differences in enzyme-substrate recognition influence regioselectivity, but the practical consequences of regioselective bias are limited.

7.3.4 Enzymes Related to Nitrate Ester Reductases

The enzymes form part of a growing family of flavoenzymes that are related in sequence and structure to the yeast Old Yellow Enzyme (OYE). This enzyme was the first protein in which it was demonstrated that the vitamin B2 derivative FMN was bound and involved in catalysis. In the 60 years since its discovery, the enzyme has been subjected to extensive physical characterization, yet the physiological role of OYE remains unknown. A large number of ligands and substrates for the oxidative and reductive half-reactions of OYE have been identified,[40] and the X-ray crystallographic structure of the enzyme in complex with some of them has been determined.[17] The majority of known substrates of the family possess an α,β-unsaturated carbonyl functional group, and the enzymes catalyze the NADPH-dependent reduction of the olefinic bond. Many phenolic compounds are competitive inhibitors, as are some steroids. The structures of some substrates and inhibitors of the family are shown in Figure 7.4.

Together with the archetype of the family OYE and the nitrate ester reductases, a few other members of the family have been characterized. Research into the degradation of morphine alkaloids by a strain of *P. putida* led to the identification of a two-stage transformation sequence from morphine via morphinone to hydromorphone. The second enzyme in the pathway, morphinone reductase,[19] is responsible

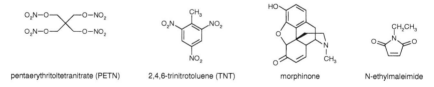

Figure 7.4 Substrates and inhibitors of the Old Yellow Enzyme family.

for the reduction of an α,β-unsaturated carbonyl functionality analogous to that of 2-cyclohexenone and other OYE substrates. The enzyme is a dimeric flavoprotein of similar subunit size to OYE, and steroids are potent competitive inhibitors. Another similar enzyme has been found in *P. syringae*.[37] A close relative of PETN reductase, *N*-ethylmaleimide reductase,[32] has been identified in *Escherichia coli*. No significant characterization of the enzyme is reported, but it should be noted that *N*-ethylmale-imide is a variation on the α/β-unsaturated carbonyl functional group. Many plants appear to contain a closely related enzyme, 12-oxophytodienoate reductase,[39] which catalyzes the reduction of an intermediate in wound response hormone synthesis.

7.3.5   Species Distribution and Evolution of Enzymes

With the growth in genomic sequencing projects, many uncharacterized relatives of the nitrate ester reductases are becoming apparent. Basic local alignment searches of the protein sequence of PETN reductase against completed and unfinished micro-bial genomes (http://www.ncbi.nlm.nih.gov/BLAST/unfinishedgenome.html) reveal several related enzymes in Gram-negative prokaryotes (Figure 7.5). There appears to be a strong level of sequence conservation across the family in Gram-negative

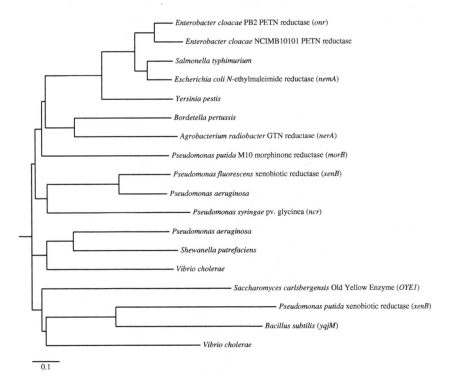

Figure 7.5   Phylogenetic relationship of prokaryotic nitrate ester reductases and related enzymes. The phylogeny is derived from the alignment of full-length protein sequences. Where no enzyme name is shown, the protein represented has not been characterized and is the conceptual translation of a genomic DNA sequence.

bacteria, yeast, and plants, although several completed bacterial genomes lack a close homolog. There is presumably some evolutionary significance here, although exactly what is hard to ascertain or prove. Alignment of deduced protein sequences from genes in the OYE family shows, as expected, that residues intimately involved in FMN binding and H bonding in the active site are well conserved.

7.3.6   Structural and Mechanistic Characterization of the Family

OYE operates by a Ping-Pong Bi-Bi kinetic mechanism,[40] with the enzyme-bound flavin being reduced by NADPH then reoxidized by substrate. Similar results have been obtained for PETN reductase.[20] The crystal structure of OYE in complex with the inhibitor *p*-hydroxybenzaldehyde[17] helped to rationalize the large quantity of biochemical data already gathered on OYE and gave fresh insight into the assumed structure–function relationships of the enzyme. The enzyme structures comprise a barrel shape formed from eight β-strands surrounded by eight α-helices, (Figure 7.6). There is a β-hairpin structure covering the base of the barrel, while the FMN is bound at the top of the barrel, with one face of the flavin accessible to the solvent, where it forms the bottom of the active site. Binding studies with a variety of substituted phenols suggest that it is the phenolate anion that is a ligand for the enzyme.[11] In the crystal structure, the phenolate oxygen forms two hydrogen bonds to His 191 and Asn 194, displacing a chloride ion bound in the empty oxidized

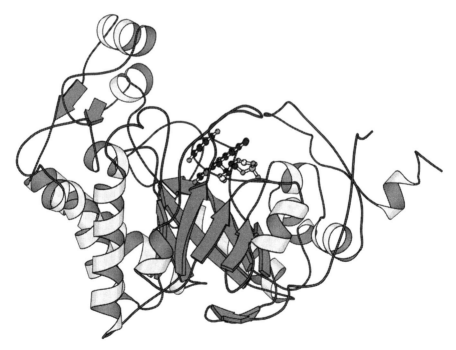

Figure 7.6   Ribbon diagram of the structure of Old Yelllow Enzyme with bound ligand. (Figure drawn using MOLSCRIPT[28] using coordinates from Reference 17 deposited as 1oyb in the Protein Databank [PDB].)

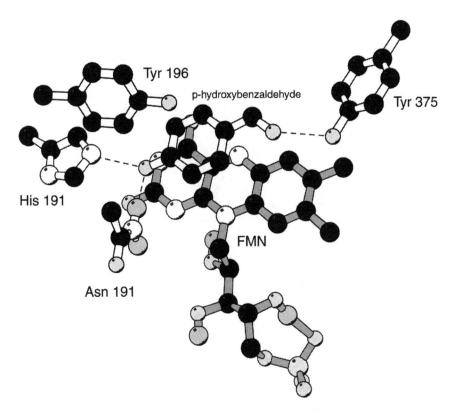

Figure 7.7   Active site of Old Yellow Enzyme in complex with *p*-hydroxybenzaldehyde. (Figure drawn using MOLSCRIPT[28] using coordinates from Reference 17 deposited as 1oyb in the Protein Databank [PDB].)

enzyme structure (Figure 7.7). Anchored in this position, the phenolic ring forms a π-π stacking interaction with the flavin. The inhibitor β-estradiol binds in a similar manner. If 2-cyclohexenone were to bind with the oxygen bound by His 191 and Asn 194, the β-carbon would be aligned above the flavin N5 atom, which correlates with earlier data suggesting hydride transfer from the flavin N5 atom to the β-carbon. Site-directed mutagenesis has implicated Tyr 196 in transfer of a proton to the α-carbon,[27] completing the saturation of the double bond.

The residues critical for activity in OYE are present in all the nitrate ester reductases, although Asn 194 is substituted by histidine in PETN reductase and close relatives. Given this high degree of homology and that PETN reductase is also active against cyclohexenone, it is reasonable to expect that the basic mechanism of OYE is also applicable to hydride transfer reactions catalyzed by nitrate ester reductases (Figure 7.8). OYE saturates nitrocyclohexene,[27] thus a nitro group will substitute for a carbonyl or phenolate functionality as a ligand for the His-Asn pair. Therefore, it is not surprising that nitrate esters are bound by nitrate ester reductases. There is no equivalent of the olefinic β-carbon in nitrate ester substrates, and it is not yet clear whether hydride transfer is involved in the reductive denitration of nitrate esters. It might well be the case that nitrate ester reductases catalyze both hydride transfer

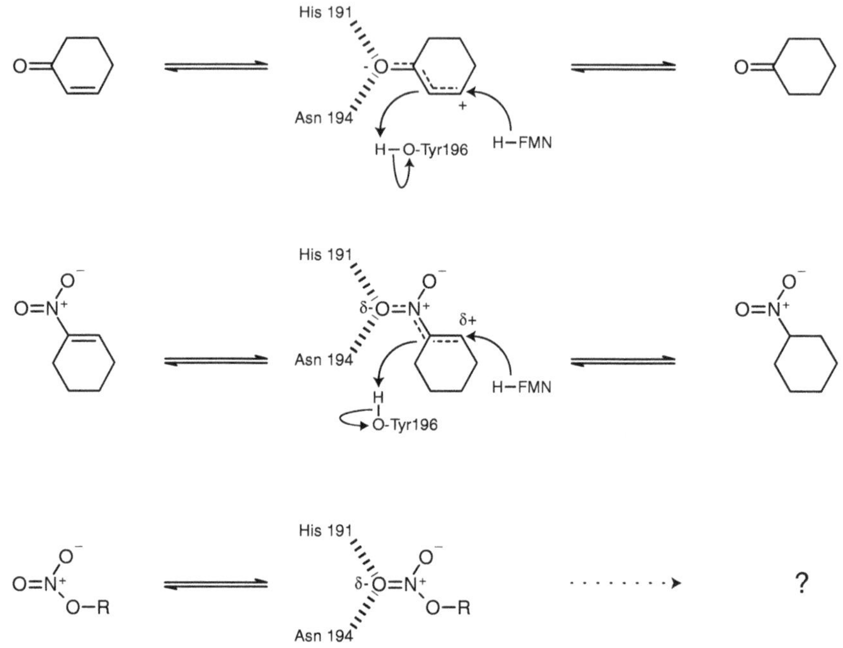

Figure 7.8   Mechanism of substrate reduction by nitrate ester reductases. 2-Cyclohexenone and nitrocyclohexene are reduced by similar hydride-transfer mechanisms. While binding of nitrate esters is likely to be similar, the mechanism of their reduction is, as yet, unclear.

and sequential electron transfer reactions, with nitrate ester reduction falling into the latter category.

The combined structural information implies that the active sites of the enzymes are formed by a small number of residues from loops at the top of the barrel. Structural variation in these loops is unlikely to affect the stability of the enzyme as a whole, suggesting that engineering of altered or enhanced specificity may be feasible.

7.3.7   Transformation of 2,4,6-Trinitrotoluene by Nitrate Ester Reductases

Although nitrate ester reductases have been isolated on the basis of their ability to liberate nitrite from PETN and GTN, it has become apparent that they might also share an ability to reduce trinitrotoluene (TNT). *Enterobacter cloacae* PB2 is able to grow very slowly on TNT as a sole nitrogen source,[21] and incubation of *E. cloacae* PETN reductase with NADPH and TNT resulted in the conversion of TNT to a mixture of reduced products.

Many flavin-containing enzymes act as nitroreductases, and hydroxylamino- and aminodinitrotoluenes were obtained both with PETN reductase and cloned *E. cloacae* nitroreductase.[12] However, additional strongly colored products, similar to those obtained with whole cells of picrate-utilizing *Rhodococcus erythropolis* and 4-nitrotoluene-utilizing *Mycobacterium* sp.,[55] are observed with PETN reductase[21] and the *P. fluorescens* nitrate ester reductase.[7] These colored products arise from hydride addition to the aromatic ring, yielding hydride– and dihydride–Meisenheimer complexes.[56] Thus, at least some nitrate ester reductases appear to reduce both nitro groups and the aromatic ring of TNT (Figure 7.9). Nitrite is formed by the action of these nitrate ester reductases on TNT,[21] but the nature of other end-products and the mechanism of their formation is as yet unknown. The inability of *E. cloacae* nitroreductase to liberate nitrite from TNT suggests that nitrite is more likely a product of the hydride addition pathway.

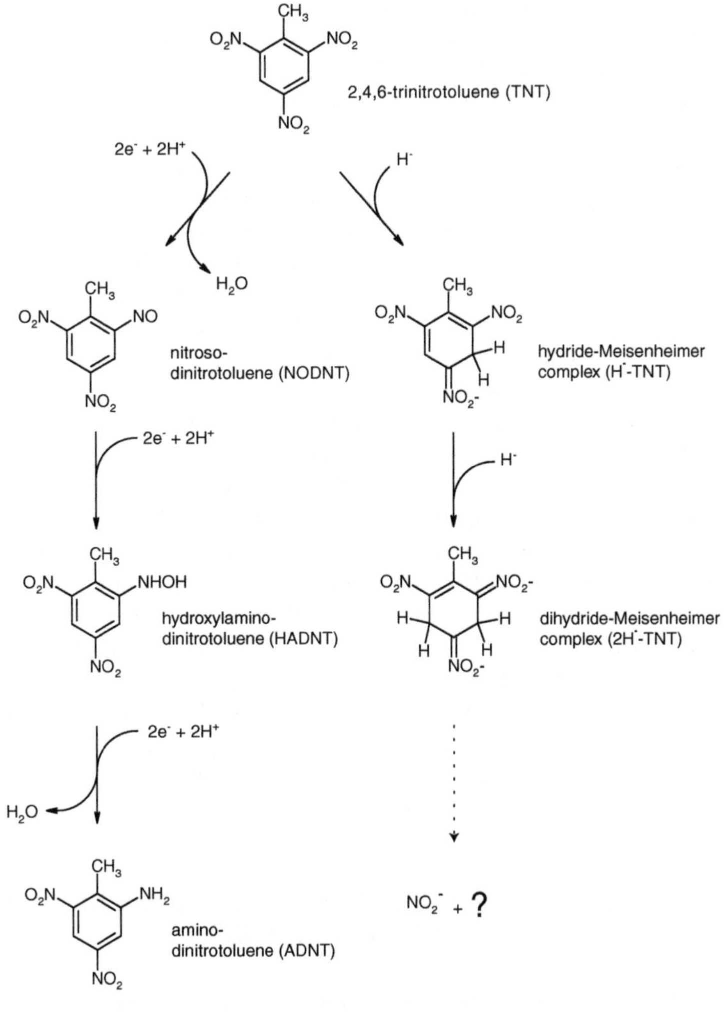

Figure 7.9 Products of TNT metabolism by PETN reductase. Each step involves the oxidation of NADPH to NADP+. Nitroso-dinitrotoluene has not been directly observed.

The white rot fungus *Phanerochaete chrysosporium*, mentioned earlier as a GTN degrader, is able to mineralize TNT.[25,31] Initial steps in the transformation involve a nitroreductase with activity against TNT,[35] while later stages are reliant on the expression of ligninolytic enzymes.[31] No hydride–Meisenheimer adducts of TNT have been reported. Thus, in contrast with the transformation of GTN and TNT by prokaryotic nitrate ester reductases, there is no evidence suggesting that a single enzyme is responsible for activity against both GTN and TNT in *P. chrysosporium*.

7.4 ENGINEERING NITRATE ESTER REDUCTASE ACTIVITY INTO PLANTS

There has been a considerable amount of interest recently in the use of plants for bioremediation of land contaminated with explosives,[14] and this subject is covered in detail elsewhere (see Chapter 10). A current problem with existing phytoremediation of TNT is that the explosives can be transformed to the more toxic and highly recalcitrant nitroso and hydroxyamino derivates.

Phytoremediation relies on the ability of plants to take up and, in some cases, metabolize the explosives. Unfortunately, whereas plants have many advantages for the remediation of contaminated land and water, they lack the catabolic versatility that enables microorganisms to mineralize such a wide diversity of xenobiotic compounds. This has raised the interesting question as to whether the impressive biodegradative capabilities of soil bacteria could be combined with the high biomass and stability of plants to yield an optimal system for *in situ* bioremediation of explosive residues in soil. To this end we have been engineering transgenic tobacco plants that express PETN reductase to degrade nitrate ester explosives and TNT.[22] The results were particularly promising because they demonstrated that seeds from transgenic tobacco plants that express PETN reductase were capable of germinating and growing in media containing levels of GTN and TNT that are toxic to the wild-type plant. The observation is particularly important in the case of TNT, since it implies that the products of TNT degradation by PETN reductase, which are as yet unidentified, are less toxic to plants than TNT and the aminodinitrotoluenes.[21,22] Previously, plants have been shown to denitrate GTN;[23] however, the denitration was enhanced by expression of PETN reductase in transgenic seedlings[22] (Figure 7.10) which also enhanced the denitration of GDN to GMN. While these results are compelling, it is necessary to determine if our transgenic approach to the phytoremediation of explosives can be performed with a plant that has the desired characteristics for field-scale applications.

7.5 AREAS FOR FUTURE RESEARCH ON THE NITRATE ESTER REDUCTASES

This chapter has sought to make the point that a good biochemical and genetic understanding of explosives metabolism has an important role to play in the development of processes for the bioremediation of explosive-contaminated land.

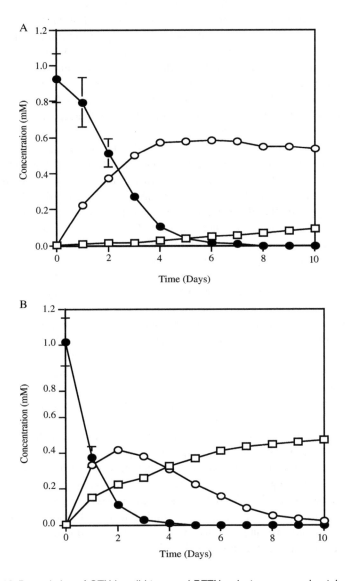

Figure 7.10 Degradation of GTN by wild-type and PETN reductase-expressing tobacco seed-lings. Wild-type (A) and PETN reductase-expressing (B) tobacco seedlings were incubated with 1 mM GTN in sterile water. Concentrations of GTN (●), GDN (○), and GMN (□) were determined by HPLC. Results shown are the mean and standard deviation of measurements on three independent flasks.

There is little doubt about the need to continue the elucidation of new degradative pathways, particularly if the diversity of available biocatalysts active against explo-sives is to increase. The selective pressure of environmental pollution is developing microorganisms that might be harnessed for explosives removal by biotechnolog-ical processes. Nevertheless, the fact that explosives persist in the environment emphasizes the failings of the existing catabolic activities in dealing with this problem. Screening natural diversity is unlikely to yield organisms that satisfy all the demands for the degradation of the most highly recalcitrant explosives, since these compounds have only been present in the environment for tens of years and microorganisms have therefore had little time to evolve enzymes suited to this task. The application of genetic engineering and biochemical techniques promises to improve and evolve natural biodegradative capabilities further. In particular, methods of directed evolution might hold the key to adapting explosive-degrading enzymes for bioremediation purposes.

The nitrate ester reductases are members of what is proving to be an important family of flavin oxidoreductases. Like all known members of the family, the true physiological role of nitrate ester reductases remains uncertain. Members of this family of enzymes display various chemistries with electrophilic substrates, and it has been suggested that the enzymes function as detoxification enzymes in antioxidant defense mechanisms. Further studies will be necessary to shed some light on the hypothesis. The recognition of steroid, alkaloid, nitrate ester, and nitroaromatic substrates acted upon by this family of enzymes is particularly intriguing. We have recently embarked on a crystallographic study of PETN reductase and related enzymes with a view to understanding the structures of the enzyme–substrate complexes and subsequent reaction mechanisms.

Metabolic engineering could also have a considerable influence on the development of bioremediation systems in the near future; although the field is still a very young one, much has been accomplished and this is beginning to have an impact on environmental biotechnology. The combination of increasing commercial interest and advances in understanding the genetic and biochemical basis of metabolic pathways is now producing a more rational approach to engineering metabolic pathways. The continual development of new genetic tools is making the manipulation of particular genes or chromosomes increasingly easier. Recently, attention has focused on the use of transgenic plants for phytoremediation. Innate biodegradative abilities of plants are less impressive than those of adapted bacteria and fungi, but these disadvantages are balanced by the large amounts of plant biomass which can easily be sustained in the field. This raises the exciting prospect for the future possibility of combining the impressive biodegradative abilities of certain bacteria with the high biomass and stability of plants to yield an optimal system for *in situ* bioremediation of organic pollutants in soil.

REFERENCES

1. Accashian, J.V., R.T. Vinopal, B.-J. Kim, and B.F. Smets. 1998. Aerobic growth on nitroglycerin as the sole carbon, nitrogen, and energy source by a mixed bacterial culture. *Appl. Environ. Microbiol.* 64:3300–3304.
2. Alm, A., O. Dalman, I. Frolen-Lindgren, F. Hulten, T. Karlsson, and M. Kowalska. 1978. Analysis of Explosives. FOA Report No. C-20267-D1. National Defence Research Institute, Stockholm.

3. Bennett, B.M., B.J. McDonald, R. Nigam, and W.C. Simon. 1994. Biotransformation of organic nitrates and vascular smooth muscle cell function. *Trends Pharmacol. Sci.* 15:245–249.

4. Bhaumik, S., C. Christodoulatos, B.W. Brodman, and N. Pal. 1998. Biodegradation of glycerol trinitrate by activated sludge: cosubstrate requirements, inhibition, and kinetics. *J. Environ. Sci. Health A.* 33:547–571.

5. Bhaumik, S., C. Christodoulatos, G.P. Korfiatis, and B.W. Brodman. 1997. Aerobic and anaerobic biodegradation of nitroglycerin in batch and packed bed bioreactors. *Water Sci. Technol.* 36:139–146.

6. Binks, P.R., C.E. French, S. Nicklin, and N.C. Bruce. 1996. Degradation of pentaerythritol tetranitrate by *Enterobacter cloacae* PB2. *Appl. Environ. Microbiol.* 62:1214–1219.

7. Blehert, D.S., B.G. Fox, and G.H. Chambliss. 1999. Cloning and sequence analysis of two *Pseudomonas* flavoprotein xenobiotic reductases. *J. Bacteriol.* 181:6254–6263.

8. Blehert, D.S., K.L. Knoke, B.G. Fox, and G.H. Chambliss. 1997. Regioselectivity of nitroglycerin denitration by flavoprotein nitroester reductases purified from two *Pseudomonas* species. *J. Bacteriol.* 179:6912–6920.

9. Bogaert, M.G. 1994. Clinical pharmacokinetics of nitrates. *Cardiovasc. Drugs Ther.* 8:693–699.

10. Brodman, B.W. and M.P. Devine. 1981. Microbial attack of nitrocellulose. *J. Applied Polym. Sci.* 26:997–1000.

11. Brown, B.J., Z. Deng, P.A. Karplus, and V. Massey. 1998. On the active site of old yellow enzyme — role of histidine 191 and asparagine 194. *J. Biol. Chem.* 273:32753–32762.

12. Bryant, C., L. Hubbard, and W.D. McElroy. 1991. Cloning, nucleotide-sequence, and expression of the nitroreductase gene from *Enterobacter cloacae*. *J. Biol. Chem.* 266:4126–4130.

13. Christodoulatos, C., S. Bhaumik, and B.W. Brodman. 1997. Anaerobic biodegradation of nitroglycerin. *Water Res.* 31:1462–1470.

14. Cunningham, S.D. and D.W. Ow. 1996. Promises and prospects of phytoremediation. *Plant Physiol.* 110:715–719.

15. Ducrocq, C., C. Servy, and M. Lenfant. 1989. Bioconversion of glyceryl trinitrate into mononitrates by *Geotrichum candidum*. *FEMS Microbiol. Lett.* 65:219–222.

16. Ducrocq, C., C. Servy, and M. Lenfant. 1990. Formation of glyceryl 2-mononitrate by regioselective bioconversion of glyceryl trinitrate — efficiency of the filamentous fungus *Phanerochaete chrysosporium*. *Biotechnol. Appl. Biochem.* 12:325–330.

17. Fox, K.M. and P.A. Karplus. 1994. Old yellow enzyme at 2Å resolution — overall structure, ligand-binding, and comparison with related flavoproteins. *Structure.* 2:1089–1105.

18. Fraser, R.T.M. 1968. Stability of nitrate esters. *Chem. Ind.* 1117–1118.

19. French, C.E. and N.C. Bruce. 1994. Purification and characterization of morphinone reductase from *Pseudomonas putida* M10. *Biochem. J.* 301:97–103.

20. French, C.E., S. Nicklin, and N.C. Bruce. 1996. Sequence and properties of pentaerythritol tetranitrate reductase from *Enterobacter cloacae* PB2. *J. Bacteriol.* 178:6623–6627.

21. French, C.E., S. Nicklin, and N.C. Bruce. 1998. Aerobic degradation of 2,4,6-trinitrotoluene by *Enterobacter cloacae* PB2 and by pentaerythritol tetranitrate reductase. *Appl. Environ. Microbiol.* 64:2864–2868.

22. French, C.E., S.J. Rosser, G.J. Davies, S. Nicklin, and N.C. Bruce. 1999. Biodegradation of explosives by transgenic plants expressing pentaerythritol tetranitrate reductase. *Nature Biotechnol.* 17:491–494.

23. Goel, A., G. Kumar, G.F. Payne, and S.K. Dube. 1997. Plant cell biodegradation of a xenobiotic nitrate ester, nitroglycerin. *Nature Biotechnol.* 15:174–177.

24. Hall, D.R., P.S. Beevor, D.G. Campion, D.J. Chamberlain, A. Cork, R.D. White, A. Almestar, and T.J. Henneberry. 1992. Nitrate esters — novel sex-pheromone components of the cotton leafperforator, *Bucculatrix thurberiella*. *Tetrahedron Lett.* 33:4811–4814.

25. Hawari, J., A. Halasz, S. Beaudet, L. Paquet, G. Ampleman, and S. Thiboutot. 1999. Biotransformation of 2,4,6-trinitrotoluene with *Phanerochaete chrysosporium* in agitated cultures at pH 4.5. *Appl. Environ. Microbiol.* 65:2977–2986.

26. King, S.Y. and H.L. Fung. 1984. Rapid microbial degradation of organic nitrates in rat excreta. Re-examination of the urinary and fecal metabolite profiles of pentaerythritol tetranitrate in the rat. *Drug Metab. Dispos.* 12:353–357.

27. Kohli, R.M. and V. Massey. 1998. The oxidative half-reaction of old yellow enzyme — the role of tyrosine 196. *J. Biol. Chem.* 273:32763–32770.

28. Kraulis, P.J. 1991. MOLSCRIPT: a program to produce both detailed and schematic plots of protein structure. *J. Appl. Crystallogr.* 24:946–949.

29. Meng, M., W.-Q. Sun, L.A. Geelhaar, G. Kumar, A.R. Patel, G.F. Payne, M.K. Speedie, and J.R. Stacy. 1995. Denitration of glycerol trinitrate by resting cells and cell extracts of *Bacillus thuringiensis/cereus* and *Enterobacter agglomerans*. *Appl. Environ. Microbiol.* 61:2548–2553.

30. Merril, E.J. 1965. Solubility of pentaerythritol tetranitrate-1,2–^{14}C in water and saline. *J. Pharm. Sci.* 54:1670–1671.

31. Michels, J. and G. Gottschalk. 1995. Pathway of 2,4,6-trinitrotoluene (TNT) degradation by *Phanerochaete chrysosporium*, in *Biodegradation of Nitroaromatic Compounds*, J.C. Spain, Ed., Plenum Press, New York, chap. 9.

32. Miura, K., Y. Tomioka, H. Suzuki, M. Yonezawa, T. Hishinuma, and M. Mizugaki. 1997. Molecular cloning of the *nemA* gene encoding *N*-ethylmaleimide reductase from *Escherichia coli*. *Biol. Pharm. Bull.* 20:110–112.

33. Pesari, H. and D. Grasso. 1993. Biodegradation of an inhibitory nongrowth substrate (Nitroglycerin) in batch reactors. *Biotechnol. Bioeng.* 41:79–87.

34. Ramos, J.L., A. Haidour, E. Duque, G. Pinar, V. Calvo, and J.M. Oliva. 1996. Metabolism of nitrate esters by a consortium of two bacteria. *Nature Biotechnol.* 14:320–322.

35. Rieble, S., D.K. Joshi, and M.H. Gold. 1994. Aromatic nitroreductase from the basidiomycete *Phanerochaete chrysosporium*. *Biochem. Biophys. Res. Commun.* 205:298–304.

36. Roberts, J.M. 1990. The atmospheric chemistry of organic nitrates. *Atmos. Environ.* 24A:243–287.

37. Rohde, B.H., R. Schmid, and M.S. Ullrich. 1999. Thermoregulated expression and characterization of an NAD(P)H-dependent 2-cyclohexen-1-one reductase in the plant pathogenic bacterium *Pseudomonas syringae* pv glicinea. *J. Bacteriol.* 181:814–822.

38. Ropenga, J.S. and M. Lenfant. 1988. Bioconversion of isosorbide dinitrate by various microorganisms. *Appl. Microbiol. Biotechnol.* 27:358–361.

39. Schaller, F. and E.W. Weiler. 1997. Molecular cloning and characterization of 12-oxophytodienoate reductase, an enzyme of the octadecanoid signaling pathway from *Arabidopsis thaliana*. *J. Biol. Chem.* 272:28066–28072.

40. Schopfer, L.M. and V. Massey. 1991. Old Yellow Enzyme, in *A Study of Enzymes*, S.A. Kuby, Ed., CRC Press, Boca Raton, FL, chap. 10.

41. Servent, D., M. Delaforge, C. Ducrocq, D. Mansuy, and M. Lenfant. 1989. Nitric oxide formation during microsomal hepatic denitration of glyceryl trinitrate — involvement of cytochrome P-450. *Biochem. Biophys. Res. Commun.* 163:1210–1216.

42. Servent, D., C. Ducrocq, Y. Henry, A. Guissani, and M. Lenfant. 1991. Nitroglycerin metabolism by *Phanerochaete chrysosporium* — evidence for nitric oxide and nitrite formation. *Biochim. Biophys. Acta.* 1074:320–325.

43. Servent, D., C. Ducrocq, Y. Henry, C. Servy, and M. Lenfant. 1992. Multiple enzymatic pathways involved in the metabolism of glyceryl trinitrate in *Phanerochaete chrysosporium*. *Biotechnol. Appl. Biochem.* 15:257–266.

44. Sharma, A., S.T. Sundaram, Y.Z. Zhang, and B.W. Brodman. 1995a. Biodegradation of nitrate esters. Degradation of nitrocellulose by a fungus isolated from a double-base propellant. *J. Appl. Polym. Sci.* 55:1847–1854.

45. Sharma, A., S.T. Sundaram, Y.Z. Zhang, and B.W. Brodman. 1995b. Nitrocellulose degradation by a coculture of *Sclerotium rolfsii* and *Fusarium solani*. *J. Ind. Microbiol.* 15:1–4.

46. Smets, B.F., R.T. Vinopal, D. Grasso, K.A. Strevett, and B.-J. Kim. 1995. Nitroglycerin biodegradation: theoretical thermodynamic considerations. *J. Energet. Mater.* 13:385–398.

47. Snape, J.R., N.A. Walkley, A.P. Morby, S. Nicklin, and G.F. White. 1997. Purification, properties and sequence of glycerol trinitrate reductase from *Agrobacterium radiobacter*. *J. Bacteriol.* 179:7796–7802.

48. Sun, W.-Q., M. Meng, G. Kumar, L.A. Geelhaar, G.F. Payne, M.K. Speedie, and J.R. Stacy. 1996. Biological denitration of propylene glycol dinitrate by *Bacillus* sp. ATCC 51912. *Appl. Microbiol. Biotechnol.* 45:525–529.

49. Sundaram, S.T., Y.Z. Zhang, A. Sharma, and B.W. Brodman. 1998. Screening for the involvement of the hydroxyl radical in the biodegradation of glyceryl trinitrate by *Penicillium corylophilum* Dierckx. *Waste Manage.* 17:437–441.

50. Sundaram, S.T., Y.Z. Zhang, A. Sharma, K. Ng, and B.W. Brodman. 1995. Screening of mycelial fungi for nitrocellulose degradation. *J. Appl. Polym. Sci.* 58:2287–2291.

51. Tan-Walker, R. L. B. 1987. Techniques for Analysis of Explosive Vapours. Ph.D. thesis. University of London, U.K.

52. Taylor, I.W., C. Ioannides, and D.V. Parke. 1989. Organic nitrate reductase — reassessment of its subcellular-localization and tissue distribution and its relationship to the glutathione transferases. *Int. J. Biochem.* 21:67–71.

53. Thatcher, G.R.J. and H. Weldon. 1998. NO problem for nitroglycerin: organic nitrate chemistry and therapy. *Chem. Soc. Rev.* 27:331–337.

54. Urbanski, T. 1984. *Chemistry and Technology of Explosives*. Pergamon Press Ltd., Oxford.

55. Vorbeck, C., H. Lenke, P. Fischer, and H.J. Knackmuss. 1994. Identification of a hydride-Meisenheimer complex as a metabolite of 2,4,6-trinitrotoluene by a *Mycobacterium* strain. *J. Bacteriol.* 176:932–934.

56. Vorbeck, C., H. Lenke, P. Fischer, J.C. Spain, and H.J. Knackmuss. 1998. Initial reductive reactions in aerobic microbial metabolism of 2,4,6- trinitrotoluene. *Appl. Environ. Microbiol.* 64:246–252.

57. Wendt, T.M., J.H. Cornell, and A.M. Kaplan. 1978. Microbial degradation of glycerol nitrates. *Appl. Environ. Microbiol.* 36:693–699.

58. White, G.F. and J.R. Snape. 1993. Microbial cleavage of nitrate esters: defusing the environment. *J. Gen. Microbiol.* 139:1947–1957.

59. White, G.F., J.R. Snape, and S. Nicklin. 1996a. Bacterial biodegradation of glycerol trinitrate. *Int. Biodeterior. Biodegrad.* 38:77–82.

60. White, G.F., J.R. Snape, and S. Nicklin. 1996b. Biodegradation of glycerol trinitrate and pentaerythritol tetranitrate by *Agrobacterium radiobacter*. *Appl. Environ. Microbiol.* 62:637–642.
61. Wildfeuer, A., H. Laufen, and M. Leitold. 1985. Pharmacology of isosorbide dinitrate after transdermal application. *Arzneim. Forsch. Drug Res.* 35(II):1289–1291.
62. Wong, P.S.Y. and J.M. Fukuto. 1999. Reaction of organic nitrate esters and *S*-nitrosothiols with reduced flavins: a possible mechanism of bioactivation. *Drug Metab. Dispos.* 27:502–509.
63. Zhang, Y.Z., S.T. Sundaram, A. Sharma, and B.W. Brodman. 1997. Biodegradation of glyceryl trinitrate by *Penicillium corylophilum* Dierckx. *Appl. Environ. Microbiol.* 63:1712–1714.

CHAPTER 8

Anaerobic Transformation
of TNT by *Clostridium*

Farrukh Ahmad and Joseph B. Hughes

CONTENTS

8.1 INTRODUCTION

Trinitrotoluene (TNT) contamination occurs in the environment predominantly in the surface and shallow subsurface soil.[54] The presence of TNT contamination in easily accessible surficial soils makes it amenable to treatment by either *in situ* or *ex situ* engineered biological systems. However, components of secondary explosive formulations other than TNT, such as the relatively polar dinitrotoluene (DNT) impurities of TNT and the poorly sorbing explosive compound 1,3,5-trinitrohexahydro-1,3,5-triazine (RDX), have been found in shallow groundwater aquifers.[9,10,71]

The electrophilic nature of the nitro groups of TNT makes them prone to reductive attack, even under oxidative conditions.[59,64] The complete reduction of a single nitro group involves a six-electron transfer resulting in the formation of an amino group. This reduction pathway proceeds sequentially via the potentially stable two- and four-electron transfer intermediates of nitroso and hydroxylamino groups, respectively (Figure 8.1). The reduction pathway does not preclude odd-numbered electron transfers that lead to the formation of highly unstable anion radicals that are rarely observed in natural systems.[17] Despite their thermodynamic stability (when compared to nitro groups), nitroso and hydroxylamino groups are chemically reactive and can undergo a variety of reactions, including rapid abiotic condensation reactions with each other to form toxic azoxy compounds.[17,40] The complete reduction of polynitroaromatics such as TNT gets progressively slower with the reduction of each nitro group. This phenomenon is largely due to the destabilization of the aromatic ring with each reduction and the subsequent decrease in the electrophilic nature of the remaining nitro substituents. In biological systems, the complete reduction of TNT to 2,4,6-triaminotoluene (TAT) has been observed only under strictly anaerobic conditions.[59]

Over the past decade, efforts focusing on the cleanup of contamination resulting from herbicides[39] and explosives[23,58] has stimulated an interest in anaerobic systems due to their ability to completely reduce polynitroaromatics. A substantial amount

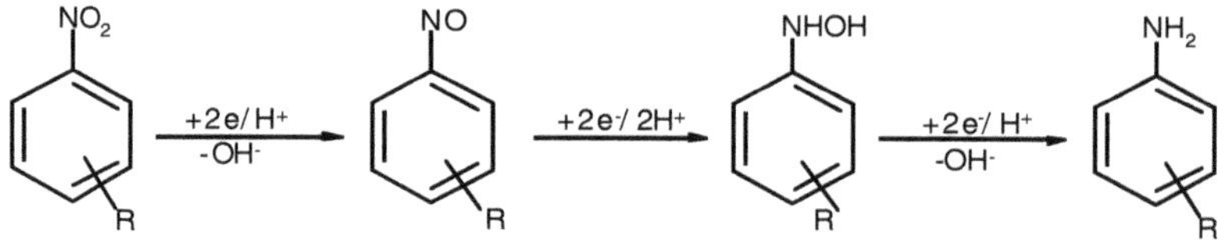

Figure 8.1 Reduction pathway of a nitro group showing the relatively stable two-, four-, and six-electron addition products of nitroso, hydroxylamino, and amino groups, respectively. The unstable one- and three-electron transfer anion radical intermediates are not shown.

of research has been conducted with mixed anaerobic cultures, which forms the basis of two commercially available biological processes. Both of these processes attain the complete reduction of TNT, but provide little insight into the biochemistry employed by the organisms involved. The recent isolation of clostridia from a commercially available anaerobic consortium capable of reducing TNT[63] has revitalized research on the ability of clostridia to reduce nitroaromatics. Clostridia have long been known to possess enzymes that are capable of rapid nitro group reduction ("nitroreductase" activity). Interestingly, the rates of nitro group reduction achieved in studies with pure clostridial cultures, cell extracts, and purified enzymes have been at least an order of magnitude faster than the commercially available anaerobic consortia (based on an equal degree of reduction). On the other hand, evidence for the complete reduction of polynitroaromatics by clostridial enzymes has been inconclusive; the enzymes isolated so far have instead shown an accumulation of the chemically reactive hydroxylamino compounds. Discrepancies such as these warrant further investigation of TNT biotransformation by clostridia so that the true fate of TNT with these organisms can be determined and, subsequently, their role in the nitro group reduction by anaerobic fermentative consortia can be assessed.

Clostridia are generally classified as Gram-positive, endospore-forming obligate anaerobes that are incapable of the dissimilatory reduction of sulfate.[3,14,15] With close to 100 species,[14,15] the genus *Clostridium* is one of the largest genera among prokaryotes.[3] Clostridia are ubiquitous in nature due to their fermentative diversity and their ability to form resistant spores.[26,30] They have been isolated from soils, decomposing biological materials, and the lower gut of mammals. In addition, they have been widely studied because some proteolytic species (e.g., *C. tetani*) produce strong toxins and other species (e.g., *C. acetobutylicum*) are of industrial importance due to their solvent producing ability. Clostridia are known to possess strong nitroreductase properties that have been extensively investigated with two purified enzymes and cell cultures.[4,5,16,18,51,55,58] Furthermore, clostridia have been isolated from an anaerobic consortium that was used to transform nitroaromatic compounds in the presence of a suitable carbohydrate carbon source.[23,63] Therefore, saccharolytic clostridia, typically nonpathogenic organisms, are especially well suited for the study of biotransformations of TNT in anaerobic engineered systems.

8.2 OVERVIEW OF CLOSTRIDIAL METABOLISM

This section provides only a brief overview of clostridial metabolic diversity and metabolic pathways. Many facets of this topic have been the subject of extensive reviews: namely, acid production or acidogenesis,[47] solvent production or solventogenesis,[8,37,38] proteolytic and purinolytic fermentations,[3] and acetate production via multiple substrates or homoacetogenesis.[47] Readers are guided to these publications for detailed discussions on specific topics. The purpose of this section is to provide background material on the broader metabolic capabilities of clostridia that are relevant to their nitroreductase activity.

8.2.1   Fermentative Metabolism

Clostridia are unable to completely oxidize an electron donor due to the lack of an electron-transport phosphorylation system and are limited to substrate-level phosphorylation (SLP) for energy generation. SLP is the formation of intermediates containing high-energy phosphoryl bonds (from the reaction of an inorganic phosphate with an activated organic substrate) that are enzymatically coupled to the production of ATP.[26] Consequently, fermentative organisms including clostridia suffer from an energy limitation and must rely on a high throughput of substrate for active growth (also known as the "Pasteur Effect"). As a result, anaerobic fermenters relying on SLP for growth tend to overproduce reducing power in the form of reduced carriers (e.g., NADH or reduced ferredoxin),[26,30] which can be partially recovered for biosynthesis in the form of NADPH.[37] The reoxidation of the reduced carriers (hereafter termed "dissipation of reducing power") is essential for maintaining a constant supply of energy in fermentative organisms because these carriers are present in limited quantities within a cell.

In pure cultures, fermentative pathways serve two main purposes: the generation of ATP and the dissipation of reducing power. As mentioned earlier, energy is obtained from SLP reactions that are mediated by dehydrogenses (e.g., glyceraldehyde 3-phosphate dehydrogenase of the Embden–Meyerhoff–Parnas glycolytic pathway) and from reactions mediated by kinases that convert high-energy intermediates to organic acids. In clostridia, reducing power in excess of biosynthesis requirements can be dissipated by the evolution of hydrogen gas (via the ferredoxin/hydrogenase system), the formation of a second high-energy intermediate such as butyryl-CoA from acetyl-CoA, or the generation of reduced fermentation end products such as solvents. In saccharolytic fermentations, the reduction of an unrelated electrophilic compound such as TNT can provide an additional means for reoxidizing reduced electron carriers that are needed for energy generation via SLP.

Clostridia demonstrate enormous fermentative diversity by utilizing a wide variety of substrates (Table 8.1) and are commonly classified by the type of substrate

Table 8.1   Fermentative Diversity of Some Common Clostridia that Have Been Utilized in Nitroreduction Studies[3,8,37,38,47]

Species	Substrate(s)	Product(s)
Clostridium acetobutlicum	Starch, disaccharides, simple sugars	Acetate, butyrate, acetone, butanol, ethanol
C. bifermentans	Glucose, proteins, amino acids	Acetate, isocaproate, valerate, isovalerate, other acids
C. clostridiiforme	Hexoses, xylose	Acetate, lactate
C. kluyveri	Ethanol, propanol, succinate, acetate+carbon dioxide	Butyrate, caproate
C. paraputrificum	Hexose, disaccharide, starch, steroids	Acetate, propionate, lactate
C. pasteuranium	Hexoses, disaccharides	Acetate, butyrate
C. perfringens	Hexoses, disaccharides, starch	Acetate, butyrate, lactate, ethanol
C. thermoaceticum	Glucose, fructose, xylose, C-1 compounds	Acetate

they ferment. They are generally divided into four categories: (1) saccharolytic, (2) proteolytic, (3) both saccharolytic and proteolytic, and (4) possessing specialized metabolism (e.g., purine fermenters such as *C. acidiurici* and ethanol fermenters such as *C. kluyveri*). However, it should be noted that regardless of the substrate, fermentation pathways usually proceed through common high-energy intermediates such as acetyl-CoA and butyryl-CoA.[30] Clostridia are also casually classified by the type of fermentation end products they produce, e.g., solventogenic (alcohol and acetone producers), acidogenic (organic acid producers), and acetogenic (acetate producers). Some saccharolytic clostridia such as *C. acetobutylicum* are typically acidogenic, but undergo a switch in their metabolism to solventogenesis under conditions of stress (Figure 8.2).

8.2.2 Autotrophic Metabolism

All clostridia are heterotrophs. However, some are capable of autotrophic metabolism.[47,62] Examples of such species are *C. aceticum* and *C. thermoaceticum*. These organisms can derive carbon for cell growth from the reduction of carbon dioxide via carbon monoxide to acetate. This autotrophic pathway (Figure 8.3) is mediated by the enzyme carbon monoxide dehydrogenase (CO-DH). The reducing power for the pathway is directly generated via the uptake of molecular hydrogen by hydrogenase or reduced ferredoxins formed from the fermentation of an electron donor.[47]

8.3 ENZYMES CATALYZING TRANSFORMATION OF NITRO SUBSTITUENTS

Over the years, interest in the nitroreductase capability of clostridial enzymes has been driven by several pragmatic reasons. Originally, when 2,4-dinitrophenol (DNP) was implicated as an uncoupler of phosphorylation,[4] researchers discovered that DNP could undergo photochemical reduction in ferredoxin-rich systems such as chloroplasts of photosynthetic plants.[6,48,75] This discovery led del Campo et al.[18] to explore dark reactions that could also reduce DNP. They found that DNP could be effectively reduced to 2,4-diaminophenol by H_2 (established by H_2 uptake stoichiometry) in the presence of the hydrogenase/ferredoxin system of *C. pasteurianum*. Later in an unrelated study, medical researchers attempting to halt the protein synthesis of a rapidly growing culture of *C. acetobutylicum* were unable to do so using chloramphenicol,[55] a broad-spectrum antibiotic in common use at the time. Earlier, chloramphenicol, a nitroaromatic compound, had been shown to be an effective antibiotic against a wide variety of clostridia[42] via a mechanism that involved the inhibition of protein synthesis. Hence, the nitroreduction of this antibiotic by clostridia served as a potential mechanism for antibiotic resistance and became a cause for concern owing to the pathogenic nature of certain proteolytic clostridia. More recently, attempts to remediate soil contaminated with di- and polynitroaromatics have prompted an interest in anaerobic biotransformation; these types of compounds undergo complete reduction only under strictly anaerobic conditions.[59] Therefore, the enzyme systems of clostridia, a common soil anaerobe, are

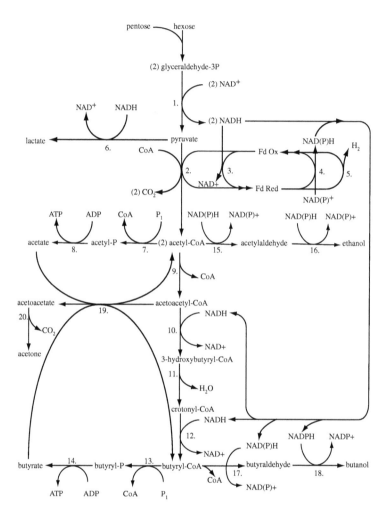

Figure 8.2 Saccharolytic fermentative pathway of the butyric acid clostridia, *C. acetobutylicum.* Reactions for both the acidogenic phase (acetate and butyrate products) and the solventogenesis phase (acetone, butanol, and ethanol products) are shown. Enzyme numbering is as follows: (1) glyceraldehyde-3-phosphate dehydrogenase, (2) Pyruvate-ferredoxin oxidoreductase, (3) NADH-ferredoxin oxidoreductase, (4) NADPH-ferredoxin oxidoreductase, (5) hydrogenase, (6) lactate dehydrogenase, (7) phosphate acetyltransferase, (8) acetate kinase, (9) thiolase (acetyl-CoA acetyltransferase), (10) 3-hydroxybutyryl-CoA dehydrogenase, (11) crotonase, (12) butyryl-CoA dehydrogenase, (13) phosphate butyltransferase, (14) butyrate kinase, (15) acetaldehyde dehydrogenase, (16) ethanol dehydrogenase, (17) butyraldehyde dehydrogenase, (18) butanol dehydrogenase, (19) acetoacetyl-CoA:acetate/butyrate:CoA transferase, and (20) acetoacetate decarboxylase. Enzymes not shown include granulose (glycogen) synthase and granulose phosphorylase which become active during the stationary phase. (Adapted from Jones, D.T. and D.R. Woods. 1986. *Microbiol. Rev.* 50:484.)

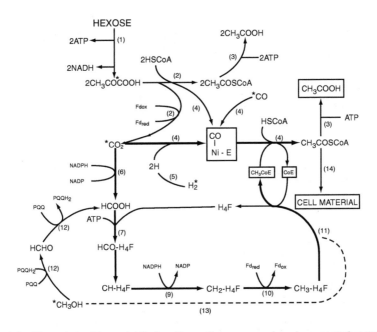

Figure 8.3 The autotrophic acetyl-CoA pathway (heavy arrows) and connected metabolism of hexoses, methanol, and carbon monoxide in *C. thermoaceticum.* Symbols key: H₄F, tetrahydrofolate; CoE, corrinoid enzyme; CO-Ni-E, carbon monoxide dehydrogenase (CO-DH) with CO moiety bound to nickel; PQQ, pyrroloquinoline quinone; Fd, ferredoxin. Enzymes or reaction sequences are numbered as follows: (1) glycolysis, (2) pyruvate-ferredoxin oxidoreductase, (3) phosphotransacetylase and acetate kinase, (4) CO-DH, (5) hydrogenase, (6) formate dehydrogenase, (7) formyl-H₄ folate synthetase, (8) methenyl-H₄ folate cyclohydrolase, (9) methylene-H₄ folate dehydrogenase, (10) methylene-H₄ folate reductase, (11) transmethylase, (12) methanol dehydrogenase, (13) methanol cobamide methyltransferase, and (14) anabolism. (From Ljungdahl, L. et al. 1989. *Clostridia Biotechnology Handbooks, Vol. 3.* Plenun Press, New York. With permission.)

being investigated for their ability to rapidly transform compounds like TNT to potentially less harmful products.

The reduction of nitro groups by clostridia has been mainly attributed to the gratuitous activity of oxidoreductases that have the well-defined metabolic function of reducing electron carriers with an extremely low midpoint redox potential (these enzymes are also referred to as "ferredoxin-reducing enzymes" and include hydrogenase, pyruvate-ferredoxin oxidoreductase, and NADH-ferredoxin oxidoreductase). Table 8.2 presents a comparative overview of studies that have utilized clostridial crude cell extracts or purified cell components (e.g., enzymes or electron carriers) to successfully reduce the nitro groups of various compounds. The clostridial enzymes conclusively identified as having nitroreductase activity, namely, hydrogenase and CO-DH, are different from the classical Type I (oxygen-insensitive) and Type II (oxygen-sensitive) nitroreductases purified from aerobic and facultative bacteria. Classical bacterial nitroreductases are flavin mononucleotide (FMN) requiring flavoproteins that operate at a suitably high midpoint redox that allows them to

Table 8.2 Reduction of Compounds Containing Nitro Substituents by Crude Cell Extracts or Purified Components from Clostridia.

Species	Biological Component			Electron Donor/Reductant	Electron Carrier	pH	Starting Compound	Intermediates	Final Product	Ref.
	Extract	Enzyme	Electron Carrier							
C. acetobutylicum	X			PYR	MV/FD	7.0	Chloramphenicol	NR	NR	O'Brien and Morris[55]
	X			PYR	MV	7.0	2/3-NP	NR	NR	(*)
	X			PYR	MV	7.0	2/3/4-NBA	NR	NR	
	X			PYR	MV	7.0	2NBD	NR	NR	
	X			PYR	MV	7.0	NF	NR	NR	
	X			PYR	MV	7.0	3NAn	NR	NR	
	X			PYR	MV	7.0	24DNP	NR	NR	
	X			H₂	—	7.8	TNT	DHA6NT	Aminophenol	Hughes et al.[34]
				H₂	—	7.8	24DNT, 26DNT	HANTs, HAATs	DHATs	Hughes et al.[35]
C. kluyveri	X			H₂/NADH	—	8.5/11.5	pNB	—	pHAB	Angermaier et al.[4] and
		X		H₂	FD	8.5	pNB	—	pHAB	Angermaier and Simon[5]
C. pasteuranium		X		H₂	FD/FMN/BV	8.0	24DNP	NR	NR (6e⁻/NO₂)	del Campo et al.[18] (*, **)
	X			H₂	FD	7.5	TNT	4HADNT	TAT	McCormick et al.[51]

Organism		Electron donor	Electron carrier	pH				Reference
	X	Dithionite	FD/MV	7.0	Metronidazole	NR	NR (4e⁻/NO₂)	Lindmark and Müller[46] (*, **)
	X	H₂	FD/FLV	8.0	Metronidazole	NR	NR	Chen and Blanchard[16] (*)
	X	H₂	FD	8.5	pNB	—	pHAB	Angermaier et al.[4] and Angermaier and Simon[5]
	X	H₂	MV	8.0	DA6NT	—	DA6HAT	Preuss et al.[58]
C. sporogenes	X	H₂	FD	8.5	pNB	—	pHAB	Angermaier and Simon[5]
C. thermoaceticum	X	CO	MV	8.0	DA6NT	—	DA6HAT	Preuss et al.[58]

Note: Electron carriers: FD = ferredoxin; FLV = flavodoxin; FMN = flavin mononucleotide; BV = benzyl viologen; and MV = methyl viologen.

Electron donors: PYR = pyruvate; H₂ = hydrogen gas; NADH = reduced nicotinamide adenine dinucleotide; and CO = carbon monoxide.

Compounds: DA6HAT = 2,4-diamino-6-hydroxylaminotoluene; DA6NT = 2,4-diamino-6-nitrotoluene; DHA6NT = 2,4-dihydroxylamino-6-nitrotoluene; DAP = diaminophenol; DNP = dinitrophenol; 24DNT = 2,4-dinitrotoluene; 26DNT = 2,6-dinitrotoluene; HAATs = hydroxylamino-aminotoluenes; HANTs = hydroxylaminonitrotoluenes; 4HADNT = 4-hydroxylamino 2,6-dinitrotoluene; NAn = nitroaniline; NBA = nitrobenzoic acid; NBD = nitrobenzaldehyde; NF = nitrofurantoin; NP = nitrophenol; pHAB = para-aminobenzoate; pNB = para-nitrobenzoate; TAT = 2,4,6-triaminotoluene; and TNT = 2,4,6-trinitrotoluene.

Other: NR = not reported; * = loss of reactant monitored; and ** = hydrogen uptake monitored.

accept electrons from either NADH or NADPH (for a more detailed review of classical bacterial nitroreductases, see Bryant and McElroy[13] or Chapter 7). In contrast, clostridial hydrogenase and CO-DH have highly reducing iron-sulfur redox centers that allow transfer of electrons to electron carriers at a redox near that of the hydrogen electrode. Conversely, recent research suggests that other transformations such as the rearrangement of hydroxylamino compounds to aminophenols[34] can also be catalyzed by clostridial enzymes. Such enzymes can potentially divert the partially reduced intermediates of nitroaromatic compounds away from the reduction pathway catalyzed by nitroreductases.

8.3.1   The Role of the Hydrogenase/Ferredoxin System in Nitro Group Reduction

Clostridial hydrogenases and ferredoxins are proteins that contain highly reducing inorganic iron-sulfur centers that allow them to participate in electron transfer reactions at extremely low redox potentials.[77] In clostridia, the iron-sulfur centers of hydrogenase are believed to contain three iron-sulfur clusters of the general form [4Fe-4S] (Figure 8.4).[1,7] Unlike the hydrogenases of aerobic hydrogen bacteria, the hydrogenases purified from clostridia are usually "bidirectional."[1,77] This term means that they can catalyze the transfer of electrons to and from an electron carrier as shown by the following equation:

Hydrogenase

$$(\text{Electron Carrier})_{\text{reduced}} + 2\text{H}^+ \rightleftharpoons \text{H}_2 + (\text{Electron Carrier})_{\text{oxidized}}$$

Exposure to oxygen irreversibly inactivates hydrogenase, whereas exposure to carbon monoxide reversibly inhibits hydrogenase activity.[1]

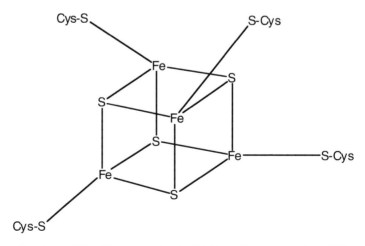

Figure 8.4   Structure of [4Fe-4S] nonheme iron/acid labile sulfur cluster in clostridial ferredoxin, hydrogenase, and carbon monoxide dehydrogenase.

In obligate anaerobes, hydrogenases interact with an electron carrier that has a suitably low midpoint redox potential and can be reduced by other oxidoreductases. In clostridia, this role is fulfilled by ferredoxins. Clostridial ferredoxins generally possess two [4Fe-4S] clusters that can each cycle between +3 and +2 oxidation states (excluding charge contributed by bound cysteine residues), thereby allowing each ferredoxin molecule to transfer one or two electrons.[52,77] The midpoint redox potentials of clostridial ferredoxins are in the range of −390 to −434 mV[77] based on the particular species from which they are extracted. In saccharolytic fermentations, ferredoxins can be reduced by oxidoreductases involved in the pyruvate phosphoroclastic reaction[60] or directly by NADH via the enzymatic action of NADH-ferredoxin oxidoreductase. *In vitro*, electron transfer with hydrogenase can be maintained when ferredoxins are replaced with other biological or nonbiological electron carriers such as flavodoxins and redox dyes, respectively.[1,18] Table 8.3 compares the midpoint redox potentials for a number of electron carriers. Together, hydrogenase and ferredoxin form an electron transport system that plays a critical part in clostridial metabolism. During fermentation, clostridia as well as other fermentative anaerobes use this system to dissipate excess reducing power by reducing protons to liberate hydrogen gas. On the other hand, clostridia such as *C. thermoaceticum* use the system to acquire reducing power that is used in the autotrophic synthesis of acetyl-CoA from carbon dioxide.

Clostridia exhibit an interesting regulatory mechanism under iron-limiting or iron-deprived conditions. Under such conditions, the formation of ferredoxin ceases; the iron of existing ferredoxin molecules is recycled for use in the production of essential enzymes,[65] and the formation of a small flavoprotein, flavodoxin, is induced.[49] Flavodoxins require FMN as cofactor, lack iron and labile sulfide, and have a low enough midpoint redox potential to substitute for ferredoxins as electron carriers. It should be noted, however, that flavodoxins are single electron carriers. Therefore, the change in electron carrier under iron-limiting conditions can result in different reduction kinetics, which may help to explain some of the variation in research findings regarding the reduction of TNT by clostridia.

The ability of the hydrogenase/ferredoxin system to reduce nitro groups was first observed by del Campo et al.[18] when they reduced DNP using hydrogen gas as the electron donor in the presence of purified hydrogenase and ferredoxin from *C. pasteurianum*. They also showed that spinach ferredoxin, benzyl viologen, and FMN were suitable replacements for the ferredoxin electron carrier. O'Brien and Morris,[55] using crude cell-free extracts of *C. acetobutylicum* and pyruvate as the electron donor, expanded earlier findings to include various other nitroaromatic

Table 8.3   Midpoint Redox Potentials (E′$_o$) of Clostridial Electron Carriers that Have Been Shown to Mediate Nitroreduction

Electron Carrier	E′$_o$ (mV)	Ref.
Methyl viologen	−440	46, 55, 58
Ferredoxin	−410	4, 5, 16, 18, 46, 55
Benzyl viologen	−385	18
Flavodoxin	−370	16
Flavin mononucleotide (FMN)	−190	18

compounds, including the antibiotic chloramphenicol. In addition, using ferre-doxin-free extracts, they were able to demonstrate that ferredoxin or a similar electron carrier was essential for the nitro reduction. Furthermore, they demon-strated the inability of FMN, flavin adenine dinucleotide (FAD), NADH, and NADPH to interact with the hydrogenase/ferredoxin system, contradicting some of the earlier findings of del Campo et al. In a nonenzymatic reaction, Lindmark and Müller[46] were able to partially reduce the antibiotic metronidazole, a nitro-imidazole (2-methyl-5-nitroimidazole-1-ethanol), with electron carriers that had been previously reduced using dithionite. However, rates of the nonenzymatic reduction were much slower than rates with the hydrogenase/electron carrier system in the earlier studies. Building on the findings of the earlier studies, Chen and Blanchard[16] developed a hydrogenase-linked electron carrier assay using the reduction of metronidazole in the H_2/hydrogenase/ferredoxin or flavodoxin system. They went on to note that this system was more sensitive to the concentration of electron carrier as compared to that of the electron donor.

The first experiment that utilized the H_2/hydrogenase/ferredoxin system to reduce TNT was performed by McCormick et al.[51] They catalyzed the formation of TAT via 4-hydroxylamino-2,6-dinitrotoluene (4HADNT) using ferredoxin-rich cell extracts of *C. pasteurianum*. The study was the first to monitor the formation of the partially reduced hydroxylamino intermediates. They used thin layer chromatogra-phy followed by the application of various dyes for separating and visualizing the reaction products. Earlier studies had only monitored the loss of nitroaromatic reactants and the stoichiometric consumption of electron donor to postulate the identity of the reduced product. Prediction of reduced products based on hydrogen uptake assays can lead to an overestimation of the degree of reduction, especially when other ferredoxin-oxidizing enzymes (e.g., clostridial NADPH-ferredoxin oxidoreductase) are present in a system.

The accumulation of hydroxylamino products in two separate studies conducted with purified hydrogenase suggests that the hydrogenase/ferredoxin system is inef-ficient at achieving the complete reduction of nitro substituents to amino substituents. Angermaier and Simon[4,5] used purified components of the H_2/hydrogenase (*C. kluyveri*)/ferredoxin (*Clostridium* sp. La1) system to partially reduce *para*-nitroben-zoate (pNB) to *para*-hydroxylaminobenzoate (pHAB) via a dianion radical that was detected using electron paramagnetic resonance (EPR) spectroscopy. The presence of the dianion radical intermediate confirmed that odd-numbered electron reductions do occur in these systems.[17] They obtained similar results with cell extracts of *C. kluyveri*. However, they noted that during H_2 uptake studies with clostridial cell cultures there was a slow uptake of the third mole of hydrogen needed for the complete reduction to an amino substituent. Angermaier and Simon concluded that in the H_2/hydrogenase/ferredoxin system, the pHAB product must be a poor substrate for the reoxidation of the reduced ferredoxin. On a similar note, Preuss et al.[58] discovered that 2,4-diamino-6-hydroxylaminotoluene (DA6HAT) accumulated as the final product when 2,4-diamino-6-nitrotoluene (DA6NT) was incubated with purified hydrogenase from *C. pasteurianum* and methyl viologen as the electron carrier. In the same study they were able to catalyze the complete reduction of DA6NT to TAT with intact cells of *C. pasteurianum*. Based on these findings, Preuss et al. suggested that a second enzyme might be involved in the complete reduction of the nitro group of DA6NT to an amino group.

A recent study by Hughes et al.,[34] performed with cell extracts of *C. acetobutylicum* also showed partial reduction of a nitroaromatic compound. Hughes et al. provided rigorous proof of the accumulation of an aminophenolic product (2-hydroxylamino-4-amino-5-hydroxy-6-nitrotoluene) formed from TNT via a partially reduced 2,4-dihydroxylamino-6-nitrotoluene (DHA6NT). The results suggested that the DHA6NT underwent a Bamberger rearrangement[17,53,68,70] to form the aminophenolic product (Figure 8.5). A similar, biologically mediated rearrangement has been reported to occur with hydroxylaminobenzene in aerobic cultures of *Pseudomonas pseudoalcaligenes*.[53,70] Unlike transformations in clostridia, the reduction of the nitro group followed by the Bamberger rearrangement in *P. pseudoalcaligenes* was part of a productive catabolic sequence. It should be noted, however, that aminophenol formation was not detected when clostridial cell extracts were incubated with 2,4-dinitrotoluene (DNT) and 2,6-DNT as starting compounds.[35]

The transient or permanent presence of hydroxylamino products has repeatedly been detected in studies involving the reduction of nitro compounds with clostridial hydrogenase/ferredoxin systems, ever since these compounds have been monitored. Studies with purified hydrogenase/ferredoxin systems suggest the involvement of other factors during the complete reduction observed in active clostridial cell cultures. Further studies are needed in order to elucidate whether these factors are biological (other enzymes or other reduced electron carriers) or environmental (culture conditions). Difficulties encountered with complete reduction of nitroaromatic compounds by the hydrogenase/ferredoxin system may facilitate the occurrence of competing biologically mediated reactions such as Bamberger rearrangements that can form products amenable to aerobic mineralization by other organisms.[53,70]

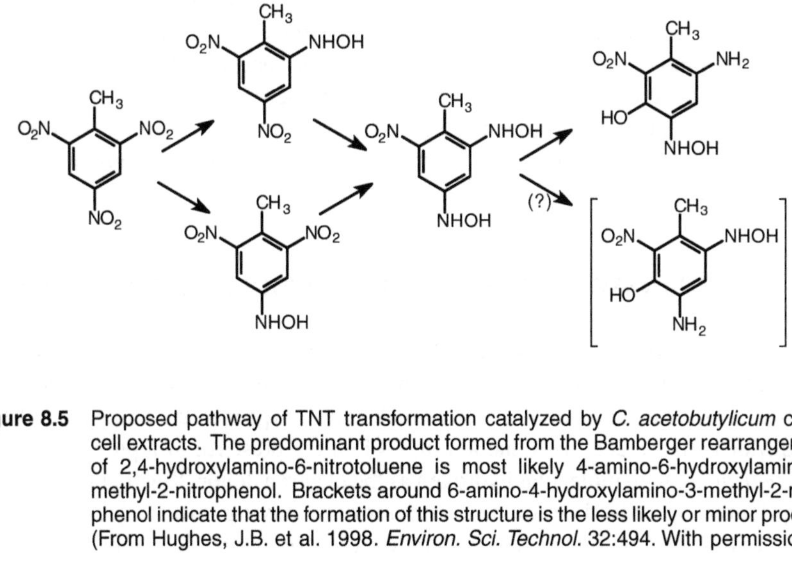

Figure 8.5 Proposed pathway of TNT transformation catalyzed by *C. acetobutylicum* crude cell extracts. The predominant product formed from the Bamberger rearrangement of 2,4-hydroxylamino-6-nitrotoluene is most likely 4-amino-6-hydroxylamino-3-methyl-2-nitrophenol. Brackets around 6-amino-4-hydroxylamino-3-methyl-2-nitrophenol indicate that the formation of this structure is the less likely or minor product. (From Hughes, J.B. et al. 1998. *Environ. Sci. Technol.* 32:494. With permission.)

8.3.2 Other Enzymes Catalyzing Reductions of Nitro Substituents

In studies with nitroaromatic compounds, two other clostridial enzymes besides hydrogenase have been implicated for their nitroreductase activity. In a study that focused on the nitroreductase activity of the hydrogenase/ferredoxin system, Agermaier et al.[4] made the unusual observation that the nitroreductase activity of cell extracts of *C. kluyveri* was NADH dependent and was maximum at a pH of 11.5. The observation was surprising because hydrogenase cannot interact directly with the NAD⁺/NADH couple and typically has a much lower optimal pH.[1] The authors ruled out several enzymes by performing multiple enzymatic assays on fractions collected from the separation of the extract on a DEAE-Sepharose column. They concluded that there was a high probability that the enzyme responsible for the nitroreductase activity was butyryl-CoA dehydrogenase, an NADH-dependent enzyme that participates in the formation of butyryl-CoA from crotonyl-CoA in butyric acid clostridia. It should be noted that the authors did not rule out NADH-ferredoxin oxidoreductase, another ferredoxin-reducing enzyme common in clostridia.

Preuss et al.[58] reported results similar to their work with the H₂/hydrogenase/methyl viologen system when they employed the CO/CO-DH/methyl viologen system to reduce DA6NT. An accumulation of DA6HAT was also observed in the CO/CO-DH/methyl viologen system. The CO-DH had been partially purified from *C. thermoaceticum*. A more recent study employing purified CO-DH showed that an electron carrier was unnecessary to produce dihydroxylaminonitrotoluenes from TNT[31] when CO was used as the electron donor. The highly reducing redox center of CO-DH contains nickel atoms, iron atoms, and [4Fe-4S] clusters.[47] The CO electron donor directly forms a complex with the nickel in the redox center. This complex is essential to the homoacetogenic pathway of *C. thermoaceticum* (Figure 8.3).

Incubations of nitroaromatic compounds with purified clostridial enzymes in the presence of a suitable electronic carrier have resulted in the rapid accumulation of hydroxylamino compounds. From these studies, it appears that the initial reduction of nitroaromatic compounds is the fortuitous property of enzymes involved in the fermentative and homoacetogenic metabolic pathways of clostridia. However, the complete reduction of a nitro substituent to an amino substituent might involve other factors besides the nitroreductases evaluated to date.

8.4 TRANSFORMATION OF TNT BY WHOLE CELL SYSTEMS

O'Brien and Morris[55] discovered that the nitro substituent of chloramphenicol could be readily reduced by actively growing cultures of *C. acetobutylicum* using pyruvate as the electron donor. McCormick et al.[51] expanded the findings of O'Brien and Morris to include the reduction of TNT in the presence of a pure culture of *C. pasteurianum* with molecular hydrogen as the electron donor. Together these findings implicated ferredoxin-reducing enzymes as having nitroreductase activity (Figure 8.2) and prompted the exploration of various practical applications. Such applications included the biocatalysis of optically active aliphatic hydroxylamines to their corresponding amines,[11] and the treatment of soil contaminated with nitroaromatic explosive compounds. The recent identification of *Clostridium* as the predominant organism in a commercially available anaerobic consortium that had rapidly reduced TNT- and RDX-contaminated soil,[63] has regenerated an interest in remediation studies with pure cultures of this organism. Consequently, a variety of such studies utilizing different species, nutrients (synthetic media/suspensions vs. complex media/active cultures), growth phases (active cells vs. resting cells), and parent nitroaromatic compounds have been conducted. Table 8.4 provides a listing of such pure culture biotransformation studies. An evaluation of trends evident among the studies follows.

8.4.1 Cometabolic Nitro Substituent Reduction

Clostridia have diverse nutritional requirements for growth (e.g., salts, trace metals, vitamins, amino acids, etc.). In the laboratory, such requirements are best and most conveniently met by complex media such as peptone-yeast extract-glucose (PYG) or chopped meat-carbohydrate (CMC) media.[3] Because of the intricate nutritional requirements of pure clostridial cultures, it is inherently difficult to design traditional nutrient limitation studies[12] to evaluate the role of nitroaromatic compounds as a sole source of carbon and/or nitrogen for growth. Results can be further confounded if internal changes in the distribution of a nutrient occurs while growth of the pure culture continues unhampered. For example, the observation that an iron-deprived culture of *C. pasteurianum* showed little change in growth or in the level of its iron-containing enzymes while its ferredoxin levels plummeted, indicated that internal cycling of iron from ferredoxin to essential enzymes[65] was occurring. Therefore, attempts to simplify the media in transformation studies involving nitroaromatic compounds have inevitably led to drastic reductions in growth.[21,33,45,67] Furthermore, the slow growth has been accompanied by correspondingly slower rates of reduction and often with partial reduction of the parent nitroaromatic compound,[33,67] especially when resting cells with low acidogenic activity[67] or synthetic media supplemented with inadequate carbon sources[45,67] have been used. Hence, in lieu of effective nutrient limitation studies with clostridia, other lines of evidence must be examined to determine whether nitroaromatic compounds can serve as primary nutritional and/or energy substrates or whether their metabolism is strictly cometabolic.

Transformation studies performed with ^{14}C-labeled nitroaromatic compounds have failed to show the generation of radiolabeled carbon dioxide, indicating that parent nitroaromatic compounds are not mineralized by pure clostridial cultures.[33,34,41,45,66] In addition, mass balances performed in radiolabeled studies have shown that the majority of the radioactivity remains in solution,[41,66] suggesting that components of the nitroaromatic compounds are not incorporated into biomass. It should be pointed out, however, that transformation studies with labeled nitrogen in the nitro group have yet to be performed.

Given the plethora of nitroreduction data for ferredoxin-reducing enzymes, it comes as no surprise that clostridial cells, either in cultures (actively growing cells) or suspensions (resting cells), can carry out at least the initial reductions of nitroaromatic

Table 8.4   Reduction of Nitroaromatic Compounds by Intact Cells of Clostridia

Clostridial Species	Nitroaromatic or Intermediate Transformed	Ref. (listed chronologically)
C. acetobutilicum	Chloramphenicol	O'Brien and Morris[55]
	TNT	Khan et al.[41]
	TNT, DA6NT	Ederer et al.[21]
	TNT, DHA6NT	Hughes et al.[33]
	TNT	Hughes et al.[34]
	24DNT, 26DNT	Hughes et al.[35]
C. bifermentans	TNT, RDX	Regan and Crawford[63]
	TNT	Shin and Crawford[66]
	TNT, DA6NT, TAT	Lewis et al.[45]
	TNT	Shin et al.[67]
	TNT, DA6NT	Ederer et al.[21]
C. clostridiiforme	1NPy, pNBA, 1,3DNPy, 1,6DNPy	Rafii et al.[61]
C. kluyveri	pNB	Angermaier et al.[4]
	pNB	Angermaier and Simon[5]
C. leptum	1NPy, pNBA, 1,3DNPy, 1,6DNPy	Rafii et al.[61]
C. paraputrificum	1NPy, pNBA, 1,3DNPy, 1,6DNPy	Rafii et al.[61]
C. pasteuranium	TNT	McCormick et al.[51]
	pNB	Angermaier et al.[4]
	pNB	Angermaier and Simon[5]
	TNT, DA6NT	Preuss et al.[58]
	pNP, mNP, 2,4DNP, pNB, pNA	Gorontzy et al.[24]
C. perfringens	6NC	Manning et al.[50]
C. sordellii	TNT, DA6NT	Ederer et al.[21]
Clostridium species	1NPy, pNBA, 1,3DNPy, 1,6DNPy	Rafii et al.[61]
C. sporogenes	pNB	Angermaier and Simon[5]
	TNT, DA6NT	Ederer et al.[21]
Clostridium species W1	pNP, mNP, 2,4DNP, pNB, pNA	Gorontzy et al.[24]
C. thermoaceticum	TNT, DA6NT	Preuss et al.[58]

Note: Compound abbreviations: DA6NT = 2,4-diamino-6-nitrotoluene; DHA6NT = 2,4-dihydroxylamino-6-nitrotoluene; DAP = diaminophenol; 2,4DNP = 2,4-dinitrophe-nol; 1,3DNPy = 1,3-dinitropyrene; 1,6DNPy = 1,6-dinitropyrene; 24DNT = 2,4-dinitrotoluene; 26DNT = 2,6-dinitrotoluene; mNP = meta-nitrophenol; 6NC = 6-nitrochrysene; pNA = *para*-nitroaniline; pNB = *para*-nitrobenzoate; pNBA = *para*-nitrobenzoic acid; pNP = *para*-nitrophenol; pNPy = *para*-nitropyrene; RDX = Research Department Explosive (hexahydro-1,3,5-trinitro-*s*-triazine); TAT = 2,4,6-triaminotoluene; and TNT = 2,4,6-trinitrotoluene.

compounds without any prior acclimation.[41,55] This observation suggests that the initial reduction of nitro groups is the gratuitous activity of clostridial enzymes that are already present in significant concentrations prior to exposure of the organism to nitroaromatic compounds. The absence of an acclimation phase, together with the lack of mineral-ization evidence, provide ample proof for the cometabolism of nitroaromatic com-pounds under fermentative conditions. Although the reduction of nitroaromatic compounds provides no nutritional benefit to clostridia, the reduction serves to reoxidize reduced electron carriers needed for energy production via SLP.

8.4.2   Significance of Acidogenic Conditions in the Reduction of Nitro Substituents

The reduction of nitroaromatic compounds has been catalyzed most rapidly and effectively by actively growing cultures of fermentative acidogenic clostridia.[21,25,33,41,45,66] Such organisms usually dissipate excess reducing power (i.e., in excess of biosynthesis requirements) generated in SLP by employing one or more of three reaction pathways. The first of these dissipation pathways during clostridial fermentations is the reduction of protons via the ferredoxin/hydrogenase system to liberate hydrogen gas. Under acidogenic conditions, the generation of molecular hydrogen serves as the only means of maintaining the redox balance offset by the production of a large quantity of partially oxidized fermentation products. To initiate this reaction, ferredoxin is reduced by either pyruvate-ferredoxin oxidoreductase or NADH-ferredoxin oxidoreductase. The second pathway is common to all butyric acid clostridia that constitute greater than 50% of all clostridia isolated to date. In this pathway butyryl-CoA is generated from acetyl-CoA via several NADH-dependent reactions. Butyric acid clostridia can utilize the first two pathways simultaneously, as indicated by the fact that high levels of hydrogenase activity can be maintained under acidogenic conditions.[2,29,37] The third pathway is present in species that can induce enzymes capable of producing solvents or "solventogenesis," thereby moving the organism away from acidogenic metabolism. A common example of such a species is the acetone-, ethanol-, and butanol-producing *C. acetobutylicum,* which undergoes a switch in its metabolism from producing acetate and butyrate to producing the more reduced solvents. (Figure 8.2.) Although it has been widely shown that the onset of solventogenesis occurs in batch cultures of *C. acetobutylicum* during the late exponential growth phase, the exact mechanism for the transformation is unclear. However, low extracellular pH, the difference between intracellular and extracellular pH, and intracellular undissociated butyric acid concentration have been cited as important factors in triggering solventogenesis under nonnutrient-limiting conditions.[8,37] Morphological changes have also been associated with the onset of solventogenesis.[36,37] It should also be noted that the first and third reaction pathways for dissipating reducing power are mutually exclusive; a large drop in hydrogenase activity occurs with the onset of solventogenesis.[2,29,37,38]

The metabolism of *C. acetobutylicum* is ideally suited for studies on the role of the different pathways responsible for dissipation of reducing power, because this species is capable of utilizing any one of the three pathways. Also, the metabolism of this species is perhaps the best studied of all clostridia, owing to its industrial importance as an alternative means of producing solvents. Studies that are carefully designed to inhibit one or more dissipation pathways at key branch points can help to reveal which enzymes are incapable of nitroreductase activity. One study that attempted such an approach was reported by Khan et al.[41] When solventogenesis was induced by carbon monoxide, a common inhibitor for hydrogenase activity, TNT was reduced to hydroxlamino compounds much more slowly than in acidogenic

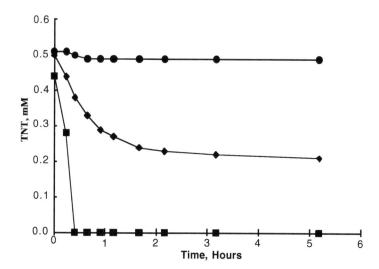

Figure 8.6   The transformation of TNT by cell cultures of *C. acetobutylicum* in acidogenic (■), solventogenic (♦), and stationary (●) phases of growth. (From Khan, T.A. et al. 1997. *J. Ind. Microbiol. Biotechnol.* 18:198. With permission.)

controls (Figure 8.6). The result suggested that enzymes involved in solvent production and the acetyl-CoA to butyryl-CoA pathway are incapable of effective reduction of nitroaromatic groups, thereby reestablishing the importance of hydrogenase and acidogenic enzymes in nitroaromatic reductions.

A few researchers have attempted nitroaromatic compound reduction with stationary phase or resting cells.[41,67] Clostridial cells in the stationary phase generally have low acidogenic activity because their metabolism is diverted to the accumulation of granulose before the onset of sporulation.[37] Such experiments have not demonstrated reduction of TNT beyond the monohydroxylamino stage.

8.4.3   Incomplete Nitro Substituent Reduction and the Formation of Aminophenols

Recent studies involving the reduction of nitroaromatic compounds,[25,33,34,41,45,66,67] especially those conducted with radiolabeled TNT in the presence of active clostridial cultures,[33,41,45,66,67] have conclusively demonstrated the formation of large quantities of polar products other than TAT (Figure 8.7). In one study, the polar product formed was identified as an aminophenol[34] formed by a Bamberger-type rearrangement[17,64,68] of a dihydroxylamino intermediate of TNT. Furthermore, in an earlier study with active cultures, a reasonable mass balance could not be established between the aminonitrotoluene reduction intermediates and the completely reduced TAT product,[51] indicating the potential existence of alternative pathways.

Studies that have revealed aminophenol formation have been performed with active clostridial cultures grown in batch systems without any pH control. Since the

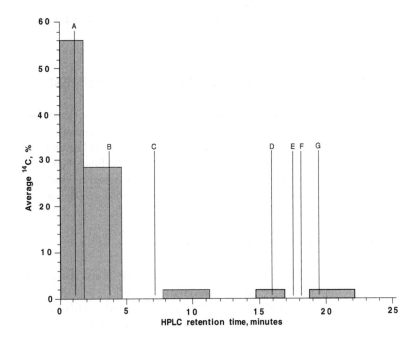

Figure 8.7   The distribution of radioactivity initially added as ^{14}C-radiolabeled TNT in HPLC fractions of a sample collected after the complete transformation of TNT by an acidogenic culture of *C. acetobutylicum*. Corresponding HPLC retention times of standards are labeled as (A) TAT, (B) DA6NT, (C) TNT, (D) 2HADNT, (E) 4HADNT, (F) 2ADNT, and (G) 4ADNT. Note that the UV spectra of authentic standards of TAT, DA6NT, and TNT did not match those of observed polar products. (From Khan, T.A. et al. 1997. *J. Ind. Microbiol. Biotechnol.* 18:198. With permission.)

pH in such cultures drops substantially over time due to acidogenic activity, it is difficult to discern whether the Bamberger rearrangement is biologically mediated or abiotic because of the low pH.[64,68] However, more recent studies conducted with extracts prepared from acidogenic cells at neutral pH have duplicated the results obtained from growing cultures,[34] indicating that the Bamberger rearrangement is enzyme catalyzed. Further evidence supporting this finding is the fact that rearrangement cannot be reproduced from arylhydroxylamino starting compounds in the absence of cell extract or active cell cultures.[34]

Although the formation of polar products other than TAT has been observed in a number of studies, TAT has not been detected as an intermediate in such studies.[33,34,41] Incubations of cell cultures with TAT have not produced any other polar products,[45] indicating that TAT is a dead end product. Similar results have been obtained with mixed anaerobic cultures.[32] These results suggest that it is possible that more than one type of aminophenol may be forming from more than one hydroxylamino precursor, depending on the degree of reduction achieved in a system. Therefore, it can be inferred from these findings that the formation of aminophenols is a biologically mediated pathway that competes with the complete reduction of TNT to TAT. Further studies are needed to determine the factors controlling the degree of TNT reduction and the extent of competing reaction pathways such as aminophenol formation.

8.5 PRACTICAL CONSIDERATIONS AND FUTURE RESEARCH

Bioremediation processes based on clostridial fermentations require additional research for a better understanding of the fate of the nitroaromatic contaminant, better predictability of process performance, and greater control of undesired competing reactions. The additional research can vary from the fundamental (e.g., analytical methods or biochemical, microbiological, or toxicological studies) to more applied engineering considerations (e.g., process selection). A brief discussion of specific research considerations follows.

8.5.1 Analytical Methods Development

Biotransformation studies with TNT have traditionally relied heavily on reverse phase high performance liquid chromatographic (RP-HPLC) separations followed by ultraviolet (UV) spectral analysis[23,24,39,41,44] to separate and identify the anticipated aminonitrotoluene reduction products. With the recent discovery of polar products, including aminophenols, in clostridial systems, it is imperative that new methods be developed using analytical systems (e.g., capillary electrophoresis (CE) followed by MS compound confirmation) that are more suitable for non-UV-absorbing or polar compounds. In addition, owing to their inherent instability, the hydroxylamino intermediates produced by clostridia also require better analytical methods for quantification. The hydroxylamino intermediates are often confused with their corresponding amino analogs because they have similar UV spectra. However, hydroxylamino intermediates elute slightly earlier than amino intermediates on RP-HPLC columns owing to their greater polarity. Pre-column derivatization procedures used in conjunction with RP-HPLC/diode array detection may help to resolve this problem by stabilizing the hydroxylamino intermediates and giving them distinct UV spectra. Appropriate analytical methods, together with radiolabeled studies, can establish better material balances to determine the true fate of nitroaromatic contaminants in clostridial systems.

8.5.2 Biochemical Pathway Studies

Although the enzymes and reactions of certain species of clostridia have been extensively studied, little is known about their potential for the biotransformation of xenobiotic compounds. With the exception of hydrogenase and CO-DH, no enzymes of clostridia have been thoroughly evaluated for nitroreductase activity. Furthermore, the identity of the enzyme responsible for Bamberger rearrangement has yet to be determined. Therefore, it is necessary to purify and conduct biotransformation studies with various enzymes from the acidogenic and homoacetogenic pathways, not only to screen them for their nitroreductase activities, but also to determine rate constants for the formation of competing biotransformation products.

The elucidation of the nitroreductase activity of enzymes involved in fermentative pathways in clostridia can be further aided by metabolic engineering techniques applied to the known genome sequence of a common strain of *C. acetobutylicum*, ATCC 824. The genomic of this nonpathogenic and saccharolytic clostridium was mapped as part of the Department of Energy (DOE) Microbial Genome Project due to interest in this organism's solvent producing ability. The genetic sequence of the pathogenic *C. difficile* is also being mapped. In addition, the genes of several enzymes from the fermentative pathway of *C. acetobutylicum* ATCC 824 have been independently identified[74] and the remainder can be identified using various recombinant DNA methods. Once an enzyme's genetic sequence has been identified, its nitroreductase activity can be evaluated using enzyme assays following overexpression and purification from an expression vector, or directly by employing mutagenesis techniques (site directed or transposon).

Conventional research conducted with the hydrogenase/electron carrier system has largely failed to achieve the complete reduction of polynitroaromatic compounds such as TNT. At this point, it is unclear whether complete reduction of TNT is infeasible with the hydrogenase/electron carrier system or is a function of the type and concentration of electron carrier available. The addition of supplemental electron carriers[55] has yielded better rates and extents of nitro group reduction. Additional studies to determine the effects of iron limitation (resulting in varying levels and types of natural electron carriers) on the degree of reduction, as well as the reoxidizability of reduced carrier at different stages in the reduction pathway, are needed to attain better control of the reduction pathway.

8.5.3 Microbiological Studies

Microbiological studies are needed to correlate the findings of clostridial pure culture and enzyme studies with undefined mixed culture studies. Currently available anaerobic mixed culture processes rely on saccharolytic fermentation conditions to reduce polynitroaromatic compounds.[23–25,39,64] Under such conditions, the only means by which acidogenic organisms can maintain a redox balance is by the production of molecular hydrogen. A fermentative system producing hydrogen favors the reduction of nitroaromatic compounds because of a high level of expression of ferredoxin-reducing enzymes. Solventogenic organisms can divert some of their electron flow away from reduced ferredoxin by producing the more reduced solvent compounds. Therefore, studies employing methods such as enzyme assays or phylogenetic probing and surface protein assays for the enumeration of specific organisms capable of acidogenic fermentation, including clostridia, may be useful for determining the potential nitroreductase activity of an undefined anaerobic consortium.

Research conducted with clostridial cultures and extracts has demonstrated the rapid nitroreductase activity of these organisms. In the laboratory, clostridia are enriched by first producing harsh conditions, such as heat treatment to 80°C or ethanol treatment to induce spore formation and also to kill nonspore-forming organisms. Since clostridia are the only fermentative endospore-forming anaerobes, they can be enriched from spores by supplying a suitable fermentation substrate under anaerobic conditions or by supplying an excess of the fermentation substrate to drive the system anaerobic by utilizing the oxygen scavenging activities of aerobic spore formers. Hence, the development of field-scale methods for the selection and enrichment of clostridia in mixed anaerobic cultures can help to develop startup conditions for non-inoculated *in situ* bioremediation applications. Conversely, such methods can be used to develop an inoculum rich in clostridial endospores that can be used for bioaugmentation.

8.5.4 Toxicological Studies

Although the toxicity and mutagenicity of TNT have been widely investigated,[73,76] limited information is available on the toxicity and mutagenicity of amino and hydroxylamino intermediates of TNT. Some studies have demonstrated the mutagenicity of aminonitrotoluenes.[72] Owing to the inherent unstable nature of the hydroxylamino intermediates in the environment, they cannot be used in long-term chronic toxicity tests. However, when *C. acetobutylicum* cell extracts are incubated with TNT solution mutagenicity is maximum when the 2,4-dihydroxylamino-6-nitrotoluene concentration is maximal[56] (Figure 8.8). This finding is in agreement with the finding that 2,6-DNT is highly mutagenic via covalent binding to DNA and proteins after the activation by acetylation or sulfation (sulfate ion activation) of its hydroxylamino intermediate in the rat liver.[57] Hydroxylamino intermediates can also

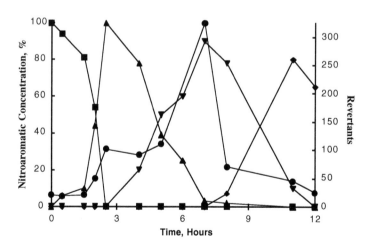

Figure 8.8 Ames mutagenicity testing results of the transient TNT metabolites during transformation by *C. acetobutylicum* crude cell extracts. The concentration of nitroaromatic compounds are in percent of initial TNT(■), sum of monohydroxylaminoditrotoluenes (▲), 2,4-dihydroxylamino-6-nitrotoluene (▼), and the proposed Bamberger rearrangement product (◆). Concurrent Ames Test results are the Revertants (●). (From Padda, R.S. et al. Submitted for publication. With permission.)

undergo abiotic condensation reactions with nitroso intermediates to form highly toxic azoxy compounds.[17,40] Aminophenolic products show only a fraction of the mutagenicity displayed by the hydroxylamino intermediates of TNT.[56] This evidence suggests that the goal of any anaerobic remediation process should be to carry on TNT treatment past the hydroxylamino intermediates.

8.5.5 Process Selection

The selection of a process for the biodegradation of nitroaromatic compounds by anaerobic fermenters such as clostridia must take into account two crucial factors for the proper operation of the process: pH control for acidogenic activity and the limitation of sorptive interactions of biodegradation intermediates with soil. Currently available mixed culture processes do not take such factors into account.

For acidogenic cultures, pH could be maintained more easily in continuous flow systems because such systems provide a means of removal of the acidified fermentation end products from the culture. Also, when clostridia capable of solvent production are involved, it becomes essential to avoid a significant drop in pH so that solventogenesis is not triggered.

The interactions of nitroaromatic compounds and their amino intermediates with soil have been vigorously investigated in recent years.[19,20,22,27,28,43,69] Nitroaromatic compounds undergo sorption via cation exchange interactions at siloxane groups of soil mineral media,[27,28] whereas amino compounds covalently bind to soil natural organic matter.[22,43] No sorption information on hydroxylamino compounds is available; however, their strong interactions with DNA and proteins in mutagenicity studies suggest that their sorption behavior may be similar to that of amino compounds. Interactions of hydroxylamino compounds with soil and natural organic matter warrant further investigation.

In addition to the above considerations, the incorporation of an aerobic stage after the anaerobic treatment is another line of investigation that should be pursued. Aerobic cultures of *P. pseudoalcaligenes* catalyze the ring cleavage of an aminophenol formed from the rearrangement of hydroxylaminobenzene.[52,70] The incorporation of such an aerobic stage to cleave aminophenols formed in clostridial systems could offer a pathway to the mineralization of nitroaromatic contaminants.

REFERENCES

1. Adams, M. W., L. E. Mortenson, and J. S. Chen. 1981. Hydrogenase. *Biochim. Biophys. Acta.* 594:105–176.
2. Andersch, W., H. Bahl, and G. Gottschalk. 1983. Level of enzymes involved in acetate, butyrate, acetone, and butanol formation by *Clostridium acetobutylicum. Eur. J. Appl. Microbial. Biotechnol.* 18:327–332.
3. Anderseen, J. R., H. Bahl, and G. Gottschalk. 1989. Introduction to the physiology and biochemistry of the genus *Clostridium.* In N. P. Minton and D. J. Clarke (eds.), *Clostridia* (Biotechnology Handbooks, Vol. 3). Plenum Press, New York.

4. Angermaier, L., F. Hein, and H. Simon. 1981. Investigations on the reduction of aliphatic and aromatic nitro compounds by *Clostridium* species and enzyme systems. In H. Bothe and A. Trebst (eds.), *Biology of Inorganic Nitrogen and Sulfur*, p. 266–275. Springer, Berlin.

5. Angermaier, L. and H. Simon. 1983. On the reduction of aliphatic and aromatic nitro compounds by *Clostridia*, the role of ferredoxin and its stabilization. *Hoppe-Seyler's Z. Physiol. Chem.* 364:961–975.

6. Arnon, D. I. 1965. Ferredoxin and photosynthesis. *Science.* 149:1460–1470.

7. Beinert, H. 1990. Recent developments in the field of iron-sulfur proteins. *FASEB J.* 4:2483–2491.

8. Bennett, G. N. and D. J. Petersen. 1993. Cloning and expression of *Clostridium acetobutylicum* genes involved in solvent production. In M. Sebald (ed.), *Genetics and Molecular Biology of Anaerobic Bacteria*. Springer-Verlag, New York.

9. Bradley, P. M., F. H. Chapelle, J. E. Landmeyer, and J. G. Schumacher. 1994. Microbial transformation of nitroaromatics in surface soils and aquifer materials. *Appl. Environ. Microbiol.* 60:2170–2175.

10. Bradley, P. M., F. H. Chapelle, J. E. Landmeyer, and J. G. Schumacher. 1997. Potential for intrinsic bioremediation of a DNT-contaminated aquifer. *Ground Water.* 35:12–17.

11. Braun, H., F. P. Schmittchen, A. Schneider, and H. Simon. 1991. Microbial reduction of n-allylhydroxylamines to n-allylamines using *Clostridia*. *Tetrahedron.* 47:3329–3334.

12. Bruhn, C., H. Lenke, and H. J. Knackmuss. 1987. Nitrosubstituted aromatic compounds as nitrogen source for bacteria. *Appl. Environ. Microbiol.* 53:208–210.

13. Bryant, C. and W. D. McElroy. 1991. Nitroreductases. In F. Muller (ed.), *Chemistry and Biochemistry of Flavoenzymes, Vol. II.* CRC Press, Boca Raton, FL.

14. Cato, E. P., W. L. George, and S. M. Finegold. 1986. Genus *Clostridium*. In P. H. A. Sneath, N. S. Mair, M. E. Sharpe, and J. G. Holt (eds.), *Bergey's Manual of Systematic Bacteriology, Vol. 2*, p. 1141–2000, Williams & Wilkins, Baltimore, MD.

15. Cato, E. P. and E. Stackenbrandt. 1989. Taxonomy and phylogeny, In N. P. Minton and D. J. Clarke (eds.), *Clostridia* (Biotechnology Handbooks, Vol. 3). Plenum Press, New York.

16. Chen, J. S. and D. K. Blanchard. 1979. A simple hydrogenase-linked assay for ferredoxin and flavodoxin. *Anal. Biochem.* 93:216–222.

17. Corbett, M. D. and B. R. Corbett. 1995. Bioorganic chemistry of the arylhydroxylamine and nitrosoarene functional groups. In J. C. Spain (ed.), *Biodegradation of Nitroaromatic Compounds*. Plenum Press, New York.

18. del Campo, F. F., J. M. Ramirez, A. Paneque, and M. Losada. 1966. Ferredoxin and the dark and light reduction of dinitrophenol. *Biochem. Biophys. Res. Commun.* 22:547–553.

19. Daun, G., H. Lenke, M. Reuss, and H. J. Knackmuss. 1998. Biological treatment of TNT-contaminated soil. 1. Anaerobic cometabolic reduction and interaction of TNT and metabolites with soil components. *Environ. Sci. Technol.* 32:1956–1963.

20. Drzyzga, O., D. Bruns-Nagel, T. Gorontzy, K. H. Blotevogel, D. Gemsa, and E. von Low. 1998. Incorporation of ^{14}C-labeled 2,4,6-trinitrotoluene metabolites in different soil fractions after anaerobic and anaerobic-aerobic treatment of soil/molasses mixtures. *Environ. Sci. Technol.* 32:3529–3535.

21. Ederer, M. M., T. A. Lewis, and R. L. Crawford. 1997. 2,4,6-Trinitrotoluene (TNT) transformation by clostridia isolated from a munition-fed bioreactor: comparison with non-adapted bacteria. *J. Ind. Microbiol. Biotechnol.* 18:82–88.

22. Elovitz, M. S. and E. J. Weber. 1999. Sediment-mediated reduction of 2,4,6-trinitrotoluene and fate of the resulting (poly)amines. *Environ. Sci. Technol.* 33:2617–2625.
23. Funk, S. B., D. J. Roberts, D. L. Crawford, and R. L Crawford. 1993. Initial-phase optimization for bioremediation of munition compound-contaminated soils. *Appl. Environ. Microbiol.* 59:2171–2177.
24. Gorontzy, T., J. Küver, and K.-H. Blotevogel. 1993. Microbial transformation of nitroaromatic compounds under anaerobic conditions. *J. Gen. Microbiol.* 139:1331–1336.
25. Gorontzy, T., I. Raber, D. Bruns-Nagel, O. Drzyzga, and K.-H. Blotevogel. 1997. Transformation of the explosive TNT by clostridia. In *In Situ and On-Site Bioremediation, Volume 2.* Battelle Press, Columbus, OH.
26. Gottschalk, G. 1979. *Bacterial Metabolism*, p. 182–192. Springer-Verlag, New York.
27. Haderlein, S. B. and R. P. Schwarzenbach. 1993. Adsorption of substituted nitrobenzenes and nitrophenols to mineral surfaces. *Environ. Sci. Technol.* 27:316–326.
28. Haderlein, S. B., K. W. Weissmahr, and R. P. Schwarzenbach. 1996. Specific adsorption of nitroaromatic explosives and pesticides to clay minerals. *Environ. Sci. Technol.* 30:612–622.
29. Hartmanis, M. G. N. and S. Gatenbeck. 1984. Intermediary metabolism in *Clostridium acetobutylicum*: levels of enzymes involved in the formation of acetate and butyrate. *Appl. Environ. Microbiol.* 47:1277–1283.
30. Holland, K. T., J. S. Knapp, and J. G. Shoesmith. 1987. *Anaerobic Bacteria*, p. 42–45 and 97–100. Blackie & Son Limited, Glasgow.
31. Huang, S., P. A. Lindahl, C. Wang, G. N. Bennett, F. B. Rudolph, and J. B. Hughes. 2000. 2,4,6-Trinitrotoluene reduction by carbon monoxide dehydrogenase from *Clostridium thermoaceticum*. *Appl. Environ. Microbiol.* 66:1474–1478.
32. Hawari, J., A. Halasz, L. Paquet, E. Zhou, B. Spencer, G. Ampleman, and S. Thiboutot. 1998. Characterization of metabolites in the biotransformation of 2,4,6-trinitrotoluene with anaerobic sludge: role of triaminotoluene. *Appl. Environ. Microbiol.* 64:2200–2206.
33. Hughes, J. B., C. Y. Wang, R. Bhadra, A. Richardson, G. N. Bennett, and F. Rudolph. 1998. Reduction of 2,4,6-trinitrotoluene by *Clostridium acetobutylicum* through hydroxylamino intermediates. *Environ. Toxicol. Chem.* 17:343–348.
34. Hughes, J. B., C. Y. Wang, K. Yesland, A. Richardson, R. Bhadra, G. N. Bennett, and F. Rudolph. 1998. Bamberger rearrangement during TNT metabolism by *Clostridium acetobutylicum*. *Environ. Sci. Technol.* 32:494–500.
35. Hughes, J. B., C. Y. Wang, and C. Zhang. 1999. Anaerobic biotransformation of 2,4-dinitrotoluene and 2,6-dinitrotoluene by *Clostridium acetobutylicum*: a pathway through dihydroxylamino intermediates. *Environ. Sci. Technol.* 33:1065–1070.
36. Jones, D. T., V. D. Westhuizen, S. Long, R. Allcock, S. J. Reid, and D. R. Woods. 1982. Solvent production and morphological changes in *Clostridium acetobutylicum*. *Appl. Environ. Microbiol.* 43:1434–1439.
37. Jones, D. T. and D. R. Woods. 1986. Acetone-butanol fermentation revisited. *Microbiol. Rev.* 50:484–524.
38. Jones, D. T. and D. R. Woods. 1989. Solvent production. In N. P. Minton and D. J. Clarke (eds.), *Clostridia* (Biotechnology Handbooks, Vol. 3). Plenum Press, New York.
39. Kaake, R. H., D. J. Roberts, T. O. Stevens, R. L. Crawford, and D. L. Crawford. 1992. Bioremediation of soils contaminated with the herbicide 2-sec-butyl-4,6-dinitrophenol (dinoseb). *Appl. Environ. Microbiol.* 58:1683–1689.

40. Kaplan, D. L. and A. M. Kaplan. 1982. Thermophilic biotransformation of 2,4,6-trinitrotoluene under simulated composting conditions. *Appl. Environ. Microbiol.* 44:757–760.

41. Khan, T. A., R. Bhadra, and J. Hughes. 1997. Anaerobic transformation of 2,4,6 TNT and related nitroaromatic compounds by *Clostridum acetobutylicum*. *J. Ind. Microbiol. Biotechnol.* 18:198–203.

42. Kiser, J. S., G. C. de Mello, D. H. Richard, and J. H. Williams. 1950. Chemotherapy of experimental clostridial infections. *J. Infect. Dis.* 28:76–80.

43. Knicker, H., D. Bruns-Nagel, O. Drzyzga, E. von Low, and K. Steinbach. 1999. Characterization of ^{15}N-TNT residues after an anaerobic/aerobic treatment of soil/molasses mixtures by ^{15}N NMR spectroscopy. 1. Determination and optimization of relevant NMR spectroscopic parameters. *Environ. Sci. Technol.*, 33:343–349.

44. Lewis, T. A., M. M. Ederer, R. L. Crawford, and D. L. Crawford. 1997. Microbial transformation of 2,4,6-trinitrotoluene. *J. Ind. Microbiol. Biotechnol.* 18:89–96.

45. Lewis, T. A., S. Goszcsynski, R. L. Crawford, R. A. Korus, and W. Admassu. 1996. Products of anaerobic 2,4,6 trinitrotoluene (TNT) transformation by *Clostridium bifermentans*. *Appl. Environ. Microbiol.* 62:4669–4674.

46. Lindmark, D. G. and M. Müller. 1976. Antitrichomonad action, mutagenicity, and reduction of metronidazole and other nitroimidazoles. *Antimicrob. Agents Chemother.* 10:476–482.

47. Ljungdahl, L. G., J. Hugenholtz, and J. Wiegel. 1989. Acetogenic and acid-producing clostridia. In N. P. Minton and D. J. Clarke (eds.), *Clostridia (Biotechnology Handbooks, Vol. 3)*. Plenum Press, New York.

48. Losada, M., J. M. Ramirez, A. Paneque, and F. F. del Campo. 1965. Light and dark reductions of nitrate in a reconstituted chloroplast system. *Biochim. Biophys. Acta.* 109:86–96.

49. Ludwig, M. L. and C. L. Luschinsky. 1991. Structure and redox properties of clostridial flavodoxin. In F. Muller (ed.), *Chemistry and Biochemistry of Flavoenzymes, Vol. III*. CRC Press, Boca Raton, FL.

50. Manning, B. W., W. L. Campbell, W. Franklin, K. B. Delclos, and C. E. Cerniglia. 1988. Metabolism of 6-nitrochrysene by intestinal microflora. *Appl. Environ. Microbiol.* 54:197–203.

51. McCormick, N. G., F. E. Feeherry, and H. S. Levinson. 1976. Microbial transformation of 2,4,6-trinitrotoluene and other nitroaromatic compounds. *Appl. Environ. Microbiol.* 31:949–958.

52. Mortenson, L. E. and G. Nakos. 1973. Bacterial ferredoxins and/or iron-sulfur proteins as electron carriers. In W. Lovenberg (ed.), *Iron-Sulfur Proteins Vol. 1*. Academic Press, New York.

53. Nishino, S. F. and J. C. Spain. 1993. Degradation of nitrobenzene by a *Pseudomonas pseudoalcaligenes*. *Appl. Environ. Microbiol.* 59:2520–2525.

54. Noyes, R. 1996. *Chemical Weapons Destruction and Explosive Waste/Unexploded Ordinance Remediation*, p. 102–141. Noyes Publications, New Jersey.

55. O'Brien, R. W. and J. G. Morris. 1971. The ferredoxin-dependent reduction of chloramphenicol by *Clostridium acetobutylicum*. *J. Gen. Microbiol.* 67:265–271.

56. Padda, R. S., C. Y. Wang, J. B. Hughes, and G. N. Bennett. Mutagenicity of trinitrotoluene and its metabolites formed during anaerobic degradation by *Clostridum acetobutylicum* ATCC 824. Submitted for publication.

57. Parkinson, A. 1996. Biotransformation of xenobiotics. In C. D. Klaassen (ed.), *Casarett & Doull's Toxicology: The Basic Science of Poisons*. McGraw-Hill, New York.

58. Preuss, A., J. Fimpel, and G. Diekert. 1993. Anaerobic transformation of 2,4,6 trinitrotoluene (TNT). *Arch. Microbiol.* 159:345–353.

59. Preuss, A. and P. G. Rieger. 1995. Anaerobic transformation of 2,4,6-trinitrotoluene. In J. C. Spain (ed.), *Biodegradation of Nitroaromatic Compounds*. Plenum Press, New York.

60. Rabinowitz, J. C. 1993. The *Clostridium pasteurianum* ferredoxin gene. In M. Sebald (ed.), *Genetics and Molecular Biology of Anaerobic Bacteria*. Springer-Verlag, New York.

61. Rafii, F., W. Franklin, R. H. Heflich, and C. E. Cerniglia. 1991. Reduction on nitroaromatic compounds by anaerobic bacteria isolated from the human gastrointestinal tract. *Appl. Environ. Microbiol.* 57:962–968.

62. Ragsdale, S. W. and H. G. Wood. 1985. Acetate biosynthesis by acetogenic bacteria. *J. Biol. Chem.* 260:3970–3977.

63. Regan, K. M. and R. L. Crawford. 1994. Characterization of *Clostridium bifermentans* and its biotransformation of 2,4,6 trinitrotoluene and 1,3,5-triaza-1,3,5-trinitrocyclohexane (RDX). *Biotechnol. Lett.* 16:1081–1086.

64. Rieger, P. G. and H. J. Knackmuss. 1995. Basic knowledge and perspectives on biodegradation of 2,4,6-trinitrotoluene and related nitroaromatic compounds in soil. In J. C. Spain (ed.), *Biodegradation of Nitroaromatic Compounds*. Plenum Press, New York.

65. Schonheit, P., A. Brandis, and R. K. Thauer. 1979. Ferredoxin degradation in growing *Clostridium pasteurianum* during periods of iron deprivation. *Arch. Microbiol.* 120:73–76.

66. Shin, C. Y. and D. L. Crawford. 1995. Biodegradation of trinitrotoluene (TNT) by a strain of *Clostridium bifermentans*. In R. W. Hinchee, J. Fredrickson, and B. C. Alleman (eds.), *Bioaugmentation for Site Remediation*. Battelle Press, Columbus, OH.

67. Shin, C. Y., T. A. Lewis, and D. L. Crawford. 1997. 2,4,6 Trinitrotoluene (TNT) biodegradation by *Clostridium bifermentans*. In *In Situ and On-Site Bioremediation: Volume 2*. Battelle Press, Columbus, OH.

68. Shine, H. J. 1967. The rearrangements of phenylhydroxylamines, In *Aromatic Rearrangements*, p. 182–190. Elsevier, Amsterdam.

69. Shen, C. F., S. R. Guiot, S. Thiboutot, G. Ampleman, and J. Hawari. 1998. Fate of explosives and their metabolites in bioslurry treatment processes. *Biodegradation*. 8:339–347.

70. Spain, J. C. 1995. Bacterial degradation of nitroaromatic compounds under aerobic conditions. In J. C. Spain (ed.), *Biodegradation of Nitroaromatic Compounds*. Plenum Press, New York.

71. Spanggord, R. J., T. Mill, T. W. Chou, W. R. Mabey, J. H. Smith, and S. Lee. 1980. Environmental fate studies on certain munition wastewater constituents. Final Report Phase I: Literature Review. Prepared by SRI International for the U. S. Army Medical Research and Development Command, Fort Detrick, MD, Contract No. DAMD17–78-C-8061.

72. Spanggord, R. J., K. E. Mortelmans, A. F. Griffing, and V. F. Simmon. 1982. Mutagenicity in *Salmonella typhimurium* and structure activity relationships of waste water components emanating from the manufacture of trinitrotoluene. *Environ. Mutagen.* 4:163–179.

73. Tan, E. L., C. H. Ho, W. H. Griest, and R. L. Tyndall. 1992. Mutagenicity of trinitrotoluene and its metabolites formed during composting. *J. Toxicol. Environ. Health.* 36:165–172.

74. Walter, K. A., R. V. Nair, J. W. Cary, G. N. Bennett, and E. T. Papoutsakis. 1993. Sequence and arrangement of genes encoding the two butyrate-formation enzymes of Clostridium acetobutylicum ATCC 824. *Gene*. 134:107–111.

75. Wessels, J. S. C. 1965. Mechanism of the reduction of organic nitro compounds by chloroplasts. *Biochim. Biophys. Acta*. 109:357–371.

76. Won, W. D., L. H. di Salvo, and J. Ng. 1976. Toxicity and mutagenicity of 2,4,6-trinitrotoluene and its microbial metabolites. *Appl. Environ. Microbiol*. 31:576–580.

77. Yoch, D. C. and R. P. Carithers. 1979. Bacterial iron-sulfur proteins, *Microbiol. Rev.* 43:384–421.

CHAPTER **9**

Fungal Degradation of Explosives: TNT and Related Nitroaromatic Compounds

Wolfgang Fritsche, Katrin Scheibner, Angela Herre, and Martin Hofrichter

CONTENTS

9.1 INTRODUCTION

The production and disposal of trinitrotoluene (TNT) and its derivatives left many sites polluted. The contamination of soil, surfacewater, and groundwaters by

TNT, its metabolites, and related compounds represent a worldwide environmental problem: these compounds exhibit considerable toxicity to humans, animals, and microorganisms.[61] The problem is especially severe in Germany, because the former armament plants were destroyed after World War II in the course of demilitarization without environmental regards.[4] TNT and its primary transformation products are still detectable 50 years later in soil. Continual leaching of the contaminants from these soils has resulted in the contamination of groundwater. Consequently, these sites must be remediated.

Because incineration, the only available proven technology for elimination of explosives in Western Europe, is expensive, bioremediation represents an alternative technology. The biological decomposition of TNT and related compounds is limited by the fact that no microorganisms able to use TNT as the sole source of carbon and energy have been identified to date. The bulky and electronegative nitro substituents of TNT present steric constraints and reduce the electron density of the aromatic ring, impeding the electrophilic attack by oxygenases. The high chemical energy potential of TNT is thus unavailable to microorganisms.

During the past 20 years, there has been a significant increase in research on biotransformation of nitroaromatic compounds as illustrated by several reviews.[19,26,46,47,59] Many microorganisms have enzyme systems that catalyze the reduction of one aromatic nitro group to the amino group. In contrast, only the white-rot fungi producing the extracellular ligninolytic enzyme system are capable of the oxidative destruction and mineralization of the aromatic nucleus of TNT.[5,14,39,42] However, mineralization, measured by the formation of $^{14}CO_2$ from ring-labeled ^{14}C-TNT, does not mean that these fungi are able to use TNT as a carbon and energy source. TNT mineralization is a cometabolic process where carbohydrates are utilized as actual growth substrates for the degrading fungi.[5,14,39,48,51] The ligninolytic fungi attack a broad spectrum of recalcitrant and toxic environmental chemicals including polyaromatic hydrocarbons (PAHs), polychlorinated biphenyls (PCBs), chlorinated phenols, and pesticides, e.g., DDT.[2,6,35,40] These capacities led to the proposal of using white-rot fungi for bioremediation of persistent and toxic chemicals in the environment.[7]

The degradation of TNT and related compounds by *Phanerochaete chryso-sporium* (syn. *Sporotrichum pulverulentum*) has been studied extensively.[4,14,23,38,48,53] The fungus has become the most widely studied model organism in both basic and applied research on degradation of lignin and xenobiotics. The fungus offers rapid growth in chemically defined liquid media, high metabolic activity, and asexual reproduction with easy to handle conidia spores.[13] The metabolism of TNT by *P. chrysosporium* was presented in two excellent reviews by Stahl and Aust,[53] and Michels and Gottschalk.[39] There are a number of other ligninolytic wood-decaying and litter-decomposing fungi that can offer advantages for the remediation of munition wastes. This chapter presents an overview of recent research on TNT metabolism by fungi, particularly the less studied white-rot and litter-decomposing fungi, that are well adapted to soil. In this regard, the chapter focuses on new results on the enzymatic oxidation of TNT metabolites by manganese peroxidase, which was first reported by Scheibner et al.[42] Special emphasis is placed on degradation processes in liquid media and soils, which seem to be very different. The transition from water to soil is characterized by fundamental differences in chemical and physical conditions.

9.2 DISCOVERING FUNGI WITH THE CAPABILITY TO MINERALIZE TNT

Fungi used for bioremediation of contaminated soils have to be adapted to growth in soils. Moreover, they have to tolerate higher TNT concentrations (>100 ppm) and be able to compete with the indigenous microflora. It has been estimated that there are about 1.5 million fungal species. They colonize a wide range of habitats where they are found as saprophytes in soil and litter and as parasites or symbionts in living plants and animals. Figure 9.1 shows habitats and ecological groups of fungi that are of interest for bioremediation.

Mineralization of xenobiotic compounds is the most desired aim of bioremediation. Therefore, we performed an extensive screening program to examine the ability of fungi to mineralize TNT.[43] We screened 91 fungal strains belonging to 32 genera of different ecophysiological and taxonomic groups under cometabolic conditions. Litter-decomposing basidiomycetes were evaluated for the first time. A high number of such fungi that grow in the organic soil horizons have ligninolytic activities. Selected results of the screening program are summarized in Table 9.1. All fungal strains tested metabolized TNT rapidly by forming aminodinitrotoluenes (4- and 2-ADNT). Micromycetes produced higher amounts of ADNTs than wood- and litter-decaying basidiomycetes did. Only a few wood- and litter-decaying basidiomycetes catalyzed significant mineralization, as determined by formation of $^{14}CO_2$ from uniformly ring-labeled ^{14}C-TNT. Up to 30% of the applied TNT was mineralized.[43] We have investigated the mineralization mechanism in two ligninolytic species of the order Agaricales, family Strophariaceae: the litter decomposer *Stropharia rugosoannulata* and the white-rot wood decayer *Nematoloma frowardii*.[41,42]

Most previous research on TNT metabolism by ligninolytic fungi has focused on *Phanerochaete chrysosporium*. Recently, additional species of white-rot fungi, including *Phlebia radiata*, have been investigated by Van Aken et al.[56,57,58] *P. radiata*, as well as *Phanerochaete chrysosporium*, are corticoid basidiomycetes of the order Meruliales. *Bjerkandera adusta*, recognized by Field et al.[15] as an effective degrader of xenobiotic compounds, has also been studied by Spreinat et al.[49] Generally speaking, it is assumed that ligninolytic macrofungi are able to mineralize TNT.

9.3 METABOLISM OF NITROAROMATIC COMPOUNDS BY LIGNINOLYTIC AND NON-LIGNINOLYTIC FUNGI

9.3.1 Initial Attack by Reduction

All of the fungi investigated so far by various research groups reduce at least one nitro group of TNT. Therefore, TNT disappears completely by reduction in fungal cultures. ADNTs (4- and 2-ADNT) are the predominant metabolites. The *para*-nitro group of TNT is usually reduced preferentially. Intermediates of the stepwise reduction

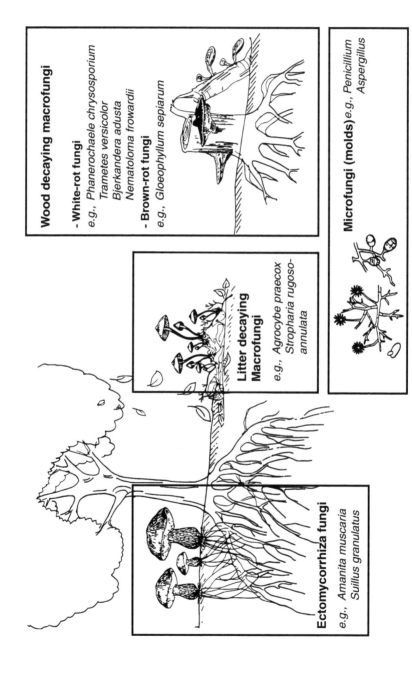

Figure 9.1   Ecophysiological groups of fungi.

Table 9.1

Fungal strain	Growth in TNT Contaminated Soil	TNT Metabolized %	HADNTs Formed %	4-/2-ADNTs Formed %	$^{14}CO_2$ Released %
Wood-Decaying Basidiomycetes					
Laetiporus sulphureus	+/-	100	0	9/5	0.5
Fomes fomentarius	-	93	3	13/10	8
Hypholoma fasciculare	+	96	0	6/4	17
Nematoloma frowardii	n.d.	100	0	n.d.	22
Kuehneromyces mutabilis	+	100	0	8/6	24
Trametes versicolor	+/-	100	0	9/4	28
Litter-Decomposing Basisiomycetes					
Paxillus involutus	-	100	0	12/6	2
Agaricus aestivalis	-	100	0	5/3	9
Agrocybe praecox	+	100	0	10/7	20
Stropharia rugosoannulata	+	100	0	12/6	36
Micromycetes (Molds)					
Aspergillus terreus	+	100	0	18/13	<1
Fusarium solani	-	96	4	23/18	1
Penicillium frequentans	+	100	9	21/17	<2
Cunninghamella elegans	+	100	0	12/10	<2
Control (uninoculated)	0	0	0	0	0

Note: The transformation was determined after 6 days (TNT concentration 0.25 mM). The mineralization was determined after 64 days. Uniformly ring-labeled ^{14}C-TNT was added to a final concentration of 4.75 µCi/l corresponding to 0.1 mM unlabeled TNT.

Source: Adapted from Scheibner et al.[43] and Scheibner.[44]

are nitroso-dinitrotoluenes (NsTs) and hydroxylaminodinitrotoluenes (HADNTs).[23] The subsequent formation of further transformation products depends on the fungal species, the culture conditions, and the incubation time. Investigations of the physiology of *P. chrysosporium* have differentiated between ligninolytic and non-ligninolytic conditions, i.e., the presence or absence of peroxidases.[5,39] The production of lignin peroxidase is suppressed in media containing nitrogen and carbon in excess.[55] However, this strict distinction applies only for *P. chrysosporium* and its lignin peroxidase. Manganese peroxidase, the second major ligninolytic enzyme of *P. chrysosporium* and other ligninolytic fungi, is regulated by other factors to be discussed below.

The temporal onset and time course of TNT transformation are crucial parameters in the detection of intermediates. Hawari et al.[23] classified several TNT metabolites into primary and secondary products, as summarized in Table 9.2. The initial steps of TNT reduction to monoaminodinitrotoluenes are followed by their successive transformation to secondary azoxy-, azo-, and hydrazoproducts. The formation of the primary metabolites (HADNTs and ADNTs) is coupled with the rearrangement to their corresponding aminophenols (Bamberger rearrangement) and with the formation of azoxy compounds. Chemical condensation of nitroso- and hydroxylaminoderivatives results in the formation of azoxy derivatives such as 4,4'-azoxy-2,2', 4,4'-tetranitrotoluene.[39] Fungal cultures also catalyze the formation of diaminonitrotoluenes (DANTs),[2,5,23,51] whereas they do not seem to reduce 4-amino-2,6-dinitrotoluene directly to 2,4-diamino-6-nitrotoluene.[39]

Table 9.2   Metabolites of TNT Transformations by *Phanerochaete chrysosporium* under Non-Ligninolytic and Ligninolytic Conditions

Abbreviations	Metabolites	Ref.
	Primary Phase Metabolites	
2-NsT	2-Nitroso-4,6-dinitrotoluene	23
4-NsT	4-Nitroso-2,6.dinitrotoluene	23
2-HADNT	2-Hydroxylamino-4,6-dinitrotoluene	23
4-HADNT	4-Hydroxylamino-2,6-dinitrotoluene	23, 38, 39
2-ADNT	2-Amino-4,6-dinitrotoluene	23, 38, 49
4-ADNT	4-Amino-2,6-dinitrotoluene	23, 38, 39, 49
2A-5OH-4,6-DNT	2-Amino-5-hydroxy-4,6-dinitrotoluene	23
4A-5OH-2,6-DNT	4-Amino-5-hydroxy-2,6-dinitrotoluene	23
2,4-DANT	2,4-Diamino-6-nitrotoluene	49
4,4'-Az	4,4'-Azoxy-2,2',4,4'-tetranitrotoluene	23, 39
	Secondary Phase Metabolites	
	2-N-Acetylamido-4,6-dinitrotoluene and p-isomer	23
	2-Formylamido-4,6-dinitrotoluene and p-isomer	23
	4-N-Acetylamino-2-amino-6-nitrotoluene	23
	4-N-Formylamido-2-amino-6-nitrotoluene	23, 39
	4-N-Acetylhydroxy-2,6-dinitrotoluene	23
	4-N-Acetoxy-2,6-dinitrotoluene	23
	4-N-Acetylamido-2-hydroxylamino-6-nitrotoluene	23
	Azo and hydrazo dimers	23

Although the products of fungal reduction of TNT are well known, the enzymatic mechanisms are still under discussion. Two processes have been proposed. Stahl and Aust demonstrated that the reduction of TNT is dependent on the presence of living intact mycelia.[50-53] They proposed the involvement of a membrane-bound redox system correlated with the proton secretion system used by the fungus to maintain the pH of its microenvironment at 4.5. The results were confirmed by experiments with metabolic inhibitors of membrane redox systems.[50,51] According to experiments conducted by Michels and Gottschalk (1995) with cell free extracts of the same fungus (*P. chrysosporium*), the reductive activity is intracellular and depends on NADPH.[39] However, the enzymatic reactions involved in the fungal reduction steps require more detailed investigations. Other ligninolytic white-rot fungi should be used to broaden the understanding.

To date, all investigations support the conclusion that the ligninolytic enzyme system is not involved in the initial reduction steps. Molds which do not produce lignin peroxidase and manganese peroxidase are able to catalyze the same reductive reactions (Table 9.1).

9.3.2 Mineralization of TNT by Ligninolytic Fungi

Ligninolytic fungi produce a variety of extracellular enzymes involved in the attack and modification of wood, other plant materials, and humic substances. The function of the ligninolytic enzyme system is the selective degradation of lignin inside the lignocellulose complex to make cellulose and hemicelluloses, the actual carbon and energy sources of the fungi, accessible. Therefore, lignin biodegradation can be basically considered as a cometabolic process.[12] Ligninolysis is probably brought about by the synergistic action of different ligninolytic enzymes: manganese peroxidase (MnP), lignin peroxidase (LiP), and laccase. Both peroxidases are heme-containing glycoproteins requiring hydrogen peroxide for function. Laccase is a copper-containing phenol oxidase which uses molecular oxygen as the electron acceptor. Multiple forms of these enzymes are secreted by ligninolytic fungi. In addition, the fungi produce complementary enzymes, e.g., glyoxal oxidase, and aryl alcohol oxidase that generate hydrogen peroxide.[20,21] Different groups of fungi can be defined based on the enzyme patterns of ligninolytic fungi[21,22] (Figure 9.2).

Almost all strains of *P. chrysosporium* produce LiP and MnP. Both peroxidases have the ability to catalyze one-electron oxidations resulting in the formation of highly reactive free radical intermediates.[1,2,40] Detailed investigation of LiP with respect to the degradation of xenobiotics has also led to a certain reassessment of MnP. This enzyme oxidizes Mn^{2+} into Mn^{3+} which is stabilized by chelating organic acids produced by the fungi, e.g., oxalate. Chelated Mn^{3+} is a strong oxidant and small enough to diffuse into the lignocellulose complex (Figure 9.3). MnP production is regulated by the manganese concentration (Mn^{2+}) in the culture medium.[3]

The two ligninolytic fungi used in our investigations are *S. rugosoannulata* and *N. frowardii*. *S. rugosoannulata* produces MnP and laccase. *N. frowardii* synthesizes three ligninolytic enzymes: MnP, LiP, and laccase. The presence of Mn^{2+} in the culture medium increases MnP activity considerably (Figure 9.4). Furthermore, it is apparent from Figure 9.4 that the increase in MnP activity correlates with the

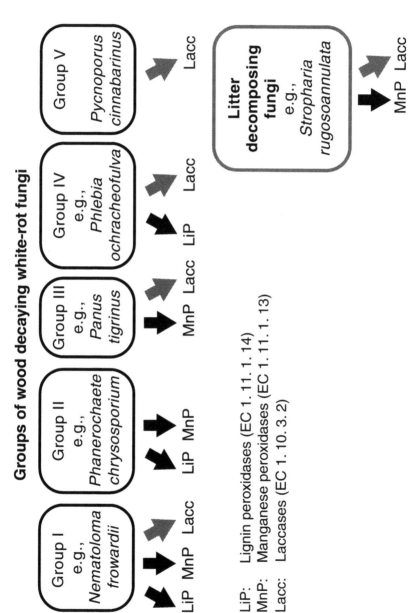

Figure 9.2 Ligninolytic enzymes produced by fungi. The categories of white-rot fungi are modified from Hatakka.[21,22]

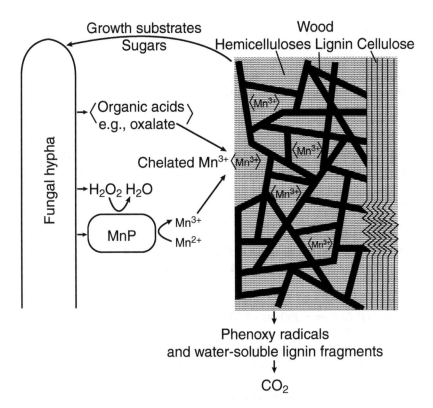

Figure 9.3   Generalized mechanism for the cometabolic degradation of lignocellulose by wood-decaying fungi able to produce MnP. (Adapted from Fritsche, W. 1999. *Mikrobiologie*. Spektrum-Verlag, Heidelberg.)

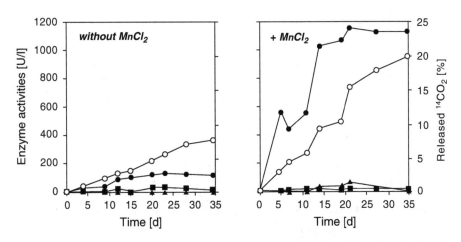

Figure 9.4   Effect of manganese addition on the production of MnP and the mineralization of ring-labeled ^{14}C-TNT in liquid cultures of *N. frowardii*. Symbols apply to both figures.[44] MnP (●), Laccase (■), LiP (▲), [14]CO₂ (○).

increase of TNT mineralization. More than 20% of ^{14}C-U-ring-labeled TNT applied was degraded to $^{14}CO_2$ within 30 days. Due to this manganese effect we focused our further experiments on MnP.

TNT metabolites produced by the selected fungi were investigated in order to elucidate possible substrates of the oxidative reactions of MnP. Comparison of TNT metabolism by the ligninolytic fungus *S. rugosoannulata* with that of the non-ligninolytic mold *Penicillium frequentans* demonstrates that both fungi produce the same metabolites: hydroxylaminodinitrotoluenes and aminodinitrotoluenes (Figure 9.5). However, in the experiment with the ligninolytic fungus the mass balance was different due to the liberation of CO_2. As shown in Figure 9.6 and Table 9.1, the reduced metabolites of TNT are dead end products of molds, whereas they represent intermediates in the metabolism of ligninolytic fungi that can undergo oxidative reactions.

9.4 OXIDATION OF REDUCED TNT METABOLITES TO CO₂ BY MANGANESE PEROXIDASE: ENZYMATIC COMBUSTION

The oxidative steps of TNT mineralization by ligninolytic fungi were investigated by using cell free preparations of MnP obtained from *S. rugosoannulata* and *N. frowardii*.[27,29] The *in vitro* system is able to mineralize a mixture of reduction products from ^{14}C-U-ring-labeled TNT as well as ADNTs and 2,6-DANT. MnP catalyzed the oxidation of 2,6-DANT at a rate that is nearly tenfold higher than that of ADNTs.[41,42] Recently, Van Aken et al. published similar results for the oxidation of 4-HADNT, ADNTs, and DANTs by MnP from *Phlebia radiata*.[56–58] Thus, a broad spectrum of reduction products of TNT are substrates of the MnP system.

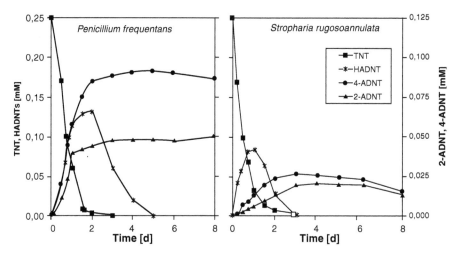

Figure 9.5 Comparison of the metabolism of TNT by *Penicillium frequentans* and the litter-decomposing fungus *S. rugosoannulata*. Symbols apply to both figures: TNT (■), HADNT (*), 4-ADNT (●), 2-ADNT (▲).[44]

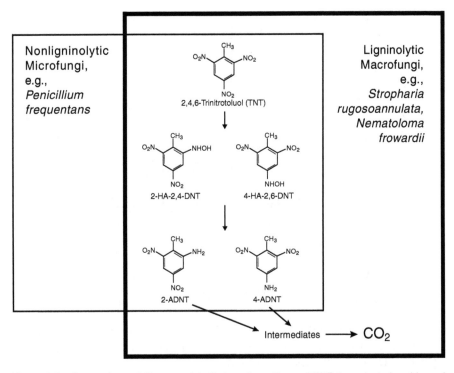

Figure 9.6 Comparison of the cometabolic transformations of TNT by selected molds and ligninolytic macrofungi.

The transformation and mineralization of reduced derivatives of TNT by MnP are significantly enhanced in the presence of cysteine or reduced glutathione (GSH) (Figure 9.7). At a concentration of 1 mM GSH, nearly 40% of the [^{14}C]-2-ADNT was degraded to $^{14}CO_2$. In the absence of GSH or cysteine, the MnP system mineralized only 5% of the applied [^{14}C]-2-ADNT. The concentration of thiols has a strong influence on the mineralization rate.[29,57] The Mn^{3+}- and thiol-mediated oxidation of lignin model compounds has been previously reported.[16,37,60] It was proposed that a reactive complex of Mn(III)-GSH or Mn(II)-GS• is the ultimate oxidant. Our findings demonstrate that the MnP-GSH system is able to mineralize different aromatic and aliphatic compounds, including pentachlorophenol and glyoxylate.[29] Very recently, the involvement of unsaturated fatty acids, e.g., linoleic acid, in the MnP-catalyzed mineralization of lignins has been demonstrated.[31] The role of other sources of reactive radicals which are involved in the enzymatic oxidation of reduced derivatives of TNT should be investigated. MnP alone was always capable of effecting a low, but significant mineralization. That means that thiols and unsaturated fatty acids are only enhancers of the MnP-catalyzed mineralization process.

About 40% of the applied [^{14}C]-TNT was mineralized (Figure 9.7). But what happens with the residual radioactivity in the reaction mixture? A special HPLC column for the separation of organic acids was used to analyze the residual radioactivity. The result of this analysis is shown in Figure 9.8. Radioactivity was mostly eluted from the column at the same time as glyoxylate, oxalate, formate, malonate,

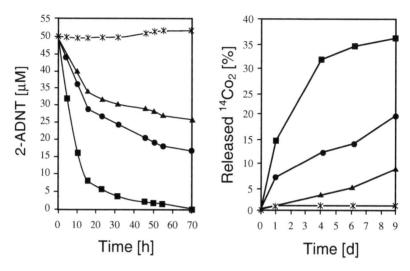

Figure 9.7 Degradation and mineralization of 2-amino-4,6-dinitrotoluene by MnP of *S. rugo-soannulata* with and without mediating thiols.[44] Symbols apply to both figures: Control (∗), MnP/Mn²⁺ (▲), MnP/Mn²⁺/L-cysteine (●), MnP/Mn²⁺/GSH (■).

and acetate, indicating that probably similar compounds are formed. Mass balance for released $^{14}CO_2$ and organic acid-like substances was about 80% of the applied [^{14}C]-2-ADNT after 9 days of incubation.

MnP of ligninolytic fungi mineralizes uniformly ring-labeled reduced TNT metabolites to CO_2. This observation strongly suggests the cleavage of the aromatic nucleus to CO_2. The proposed pathway is given in Figure 9.9. It is important to mention that dioxygen plays a role in the mineralization process by reacting spontaneously with the metabolites formed through the action of Mn^{3+}. The concept of enzymatic combustion was adopted from Kirk et al.[33,34] They wrote in an excellent review on microbial lignin degradation:[33] "It is this nonspecific subsequent oxidation of lignin that leads us to refer the process as enzymatic combustion."

In contrast to the MnP system, the understanding of the enzymatic steps in TNT conversion by LiP is limited. Michels and Gottschalk[39] reported that LiP catalyzed the oxidation of early TNT intermediates — HADTs — leading to the formation of the corresponding nitrosodinitrotoluenes and coupling products (azoxy-tetranitrotoluenes). Direct mineralization, however, was not observed. Furthermore, the authors found that LiP is not capable of oxidizing the main TNT reduction products — ADNTs. Based on the known reactions of MnP, the role and actual function of LiP in TNT mineralization should be clarified and reassessed.

9.5 FUNGAL BIOREMEDIATION OF TNT FROM CONTAMINATED SOILS

The best approach of soil bioremediation would be natural attenuation by the indigenous microflora. However, the continued contamination of soils by TNT and its derivatives 50 years after World War II indicates a low intrinsic bioremediation

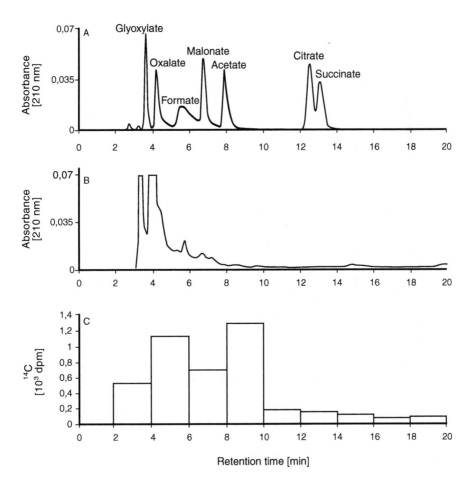

Figure 9.8 Analysis of residual activity in samples of [^{14}C]-2-ADNT, which was treated with MnP of *S. rugosoannulata* for 72 h: (A) low-molecular-weight organic acid standards, (B) chromatogram of the reaction mixture, and (C) distribution of residual activity. The radioactivity of the organic acids represents about 40% of the total radioactivity. (Adapted from Scheibner, K. 1998. Ph.D. Thesis. Friedrich-Schiller-Universität, Jena.)

capacity. A natural degradative microflora has not evolved because TNT and its co-contaminants are not substrates for a productive metabolism and do not support growth. In fact the toxicity of nitroaromatic compounds inhibits growth of *Phanerochaete chrysosporium*.[46] However, in many contaminated soils the explosives are in crystalline form and are not very water soluble. Therefore, their bioavailability is low. Nuggets, crystals, and the heterogeneous distribution of TNT are reasons for large standard deviations in experiments with soil samples from contaminated field sites.

Recently, a variety of bioremediation techniques for treating TNT-contaminated soils has been reviewed (see Chapter 14). The application of ligninolytic macrofungi is one of the strategies. The principle of fungal remediation technology is a combination

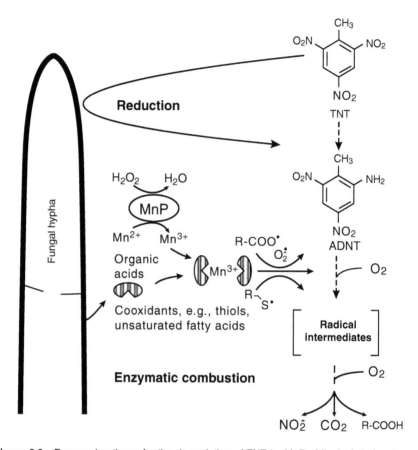

Figure 9.9 Proposed pathway for the degradation of TNT by MnP of ligninolytic fungi.

of bioaugmentation and biostimulation. The inoculation with fungi is combined with biostimulation by addition of growth substrates such as straw or wood chips.

P. chrysosporium has been the fungus of choice for many soil experiments.[14,48,53] Laboratory-scale experiments suggested that bioremediation of TNT-contaminated soils by *P. chrysosporium* is feasible. It is important, however, to distinguish between disappearance of TNT by transformation to bound residues and mineralization. In various systems, about 85% of the initial TNT disappeared over periods of 30 to 90 days, but mineralization did not exceed 5 to 20%. Recently, Dutta et al.[11] demonstrated TNT mineralization under non-ligninolytic conditions by *P. chrysosporium*. Soil highly contaminated with TNT was diluted with malt extract medium. The production of LiP is suppressed in media containing carbon and nitrogen sources in excess. In contrast, MnP may be synthesized.[23] One of the advantages of *P. chrysosporium* is the capability to grow in stirred reactors. Most of the basidiomycetes are sensitive to agitation. Loomis et al. demonstrated the feasibility of encapsulating *P. chrysosporium* in calcium alginate pellets as a potential method of delivering this fungus to toxic wastes.[36] This interesting method seems to be without significance in bioremediation of TNT-contaminated soils.

9.5.1   Bioremediation in Static Piles

The applicability of the litter-decomposing basisiomycete *S. rugosoannulata* for soil bioremediation was proven in a model experiment in a static pile. The process in static piles is similar to the composting process; however, the soil is not mixed repeatedly because the fungi are sensitive to agitation. The limiting factor in soil bioremediation is the mass transfer of the pollutant and nutrients to microbial cells. In static piles, the mycelia migrate actively from the inoculum into the polluted soil. Generally, the biopile processes require prolonged incubation periods, but the additional time can be offset by the reduced cost of mixing.

We conducted experiments to examine the bioremediation potential of the litter-decomposing fungus *S. rugosoannulata* at the former ammunition plant "Tanne" in the Harz mountains in Germany.[24,25] A static pile ($1 \times 2 \times 0.5$ m) of TNT-contaminated soil was inoculated with the fungus grown on wheat straw. The ratio of wheat straw plus fungus to soil was 1:3 v/v. A control pile contained a straw layer without the fungus. The piles were built on a wooden platform sealed with plastic foil. The leachate water from the open piles was collected in reservoir tanks. Growth of the fungus from the straw layer into the soil was noticable 1 week after inoculation. The fungus actively colonized the whole pile, and after 3 to 6 months fruiting bodies were developed on the pile surface. During the first week of incubation the TNT concentration decreased from about 80 to 20 mg of TNT per kilogram of soil and further down to 5 mg/kg after 300 days (Figure 9.10). The high standard deviations in the TNT analysis could be explained by the heterogenous TNT distribution in the soil. A decreasing standard error in the inoculated pile indicates the dissolution of crystals of TNT. TNT was degraded much more slowly in the control pile without the fungus. The extremely low concentrations of TNT and its metabolites in the leachate indicated the effectiveness of the remediation process. The fate of TNT and the mass balance was investigated in static microcosms described below.

Spreinat et al.[49] described similar results from soil of the polluted area of Tanne, using *Bjerkandera adusta*. They measured a decrease of TNT from 4000 mg/kg of soil dry weight to about 100 mg/kg in a period of 60 days. The pilot plant experiment is in operation.

9.5.2   Fate of ^{14}C-labeled TNT in Bioremediated Soil: A Laboratory Microcosm Study

Static microcosms were used to determine the fate of TNT in inoculated and straw supplemented soils. The carbon transfer from [^{14}C]-TNT was investigated by using a battery of soil columns.[32,45,54] It has to be emphasized that the physiological and chemical conditions in the microcosms differed from those in the static pile. The microcosms were aerated by humidity saturated air at room temperature.

Mass balance and distribution of radioactivity are shown in Figure 9.11. The mineralization was relatively low. Only 3% of the total radioactivity applied was converted to CO_2 within 30 weeks. Instead, the labeled carbon became associated with the soil humus fractions. A phase of physico-chemical adsorption was

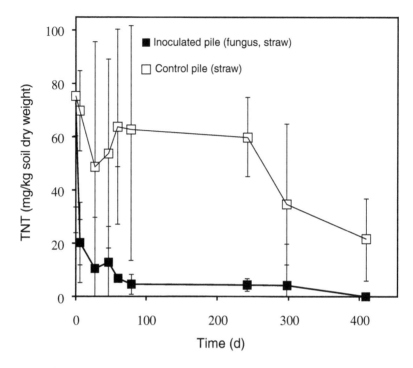

Figure 9.10 Effect of inoculation by *S. rugosoannulata* on TNT concentration in an authentically contaminated soil in a static pile. (Adapted from Herre, A. et al. 1998. *TerraTech.* 4:52).

followed by incorporation of radioactivity into the fractions of fulvic acids, humic acids, and humins (a total of 86% after 30 weeks). The analysis of the labeled metabolites in the water and methanol extractable fractions indicates the transformation of TNT to aminodinitrotoluenes and traces of diaminonitrotoluenes. The radioactivity in the aqueous extracts decreased to about 2% and in the methanolic extract to 13% of total ^{14}C applied within 30 weeks of incubation. The results were similar to those of Comfort et al. who have demonstrated that 34.6% of the applied TNT could not be extracted from solid soil particles at the end of their long-term experiments.[8]

The disappearance of TNT and its reduced metabolites during the composting process seems to be largely due to the reactivity of the intermediates of humic acid precursors. We assume that the gradual formation of ^{14}C-labeled humic substances is an indication of the covalent coupling of TNT metabolites to the organic soil matrix (humus). Ligninolytic enzymes, particulary laccases,[18] may be involved in these reactions. Dawel et al.[9] investigated the coupling of 2,4-diamino-6-nitrotoluene (2,4-DANT) in an enzymatic model system containing laccase and the humic constituent guaiacol (Figure 9.12). The first reaction step is a laccase-mediated oxidative dimerization of guaiacol followed by the spontaneous oxidation to a diphenoxyquinone. The second step is the nucleophilic addition of 2,4-DANT to the quinone structure. It is well known that phenolic and amino compounds

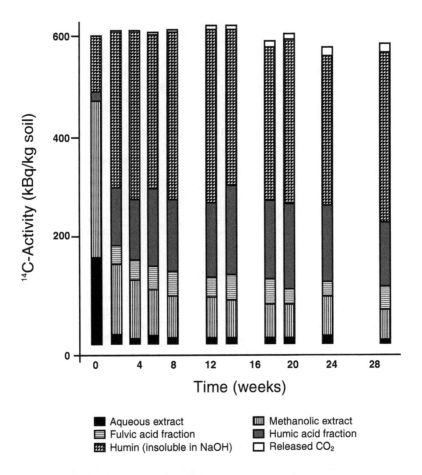

Figure 9.11 Transformation and immobilization of TNT in static soil microcosms filled with 1 kg soil and straw inoculated with *S. rugosoannulata* (2:1 v/v). The soil was artificially contaminated by ring-labeled ^{14}C-TNT. (Adapted from Scheibner, K. et al. 1998. Biologische Vertahren zur Bodensanierung: 4. Statusseminar des Verburdvorhabens Biologishe Sanierung von Rüstungsaltlasten.)

have the tendency to form bound residues.[18] Generally speaking, the nature and stability of bound immobilized metabolites of TNT is unclear. Analytical procedures are required to differentiate between sorption and covalent binding during the composting process.

The results of our soil microcosm experiments seem to be contradictory to those in liquid cultures which demonstrated high TNT mineralization. The comparison of the conditions in liquid media and soil shows considerable differences, mainly concerning the energy and carbon sources (Figure 9.13). In soil, there is an excess of aromatic compounds derived from the degradation of straw or wood chips. In addition, an extensive humus matrix comprises a huge number of reactive compounds. Ligninolytic fungi are responsible for the degradation of plant biomass,

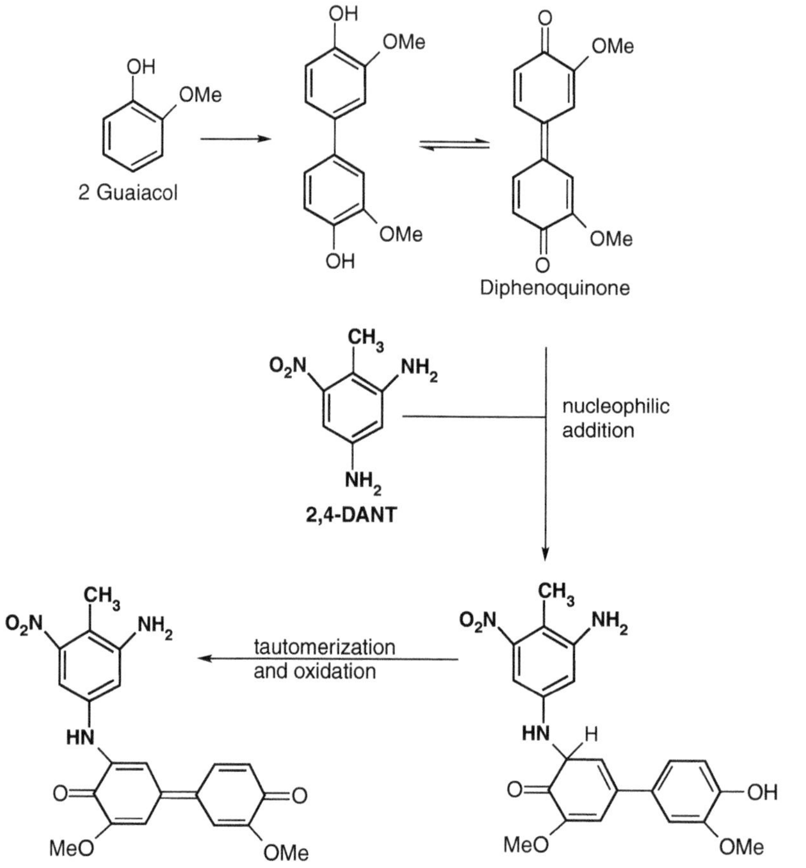

Figure 9.12 Proposed reactions of laccase-mediated coupling of 2,4-diamino-6-nitrotoluene to guaiacol, a model for humification. (Adapted from Dawel, G. et al. 1997. *Appl. Environ. Microbiol.* 63:2560.)

including lignin, and TNT. It has to be assumed that there is a high chemical affinity between the fungal metabolites of TNT and the reactive aromatic residues of lignocellulose degradation, e.g., phenolic and quinoid structures. Therefore, TNT metabolites very probably react with the aromatic lignin fragments and are not available to be mineralized by peroxidases of the ligninolytic system. The natural processes of humus formation (humification) are based on similar phenomena.[62] Repolymerization of aromatic degradation products of plant biomass occurs faster than the respective mineralization and is the actual reason for the existence of humus. Nevertheless, there is a continuous turnover of humic substances in nature. Thus, once a pollutant is covalently coupled to humic substances, its long-term release is accompanied by simultaneous mineralization.[9] Therefore, the investigation of the turnover of TNT over a long time is necessary to evaluate the structure of products that might be released.

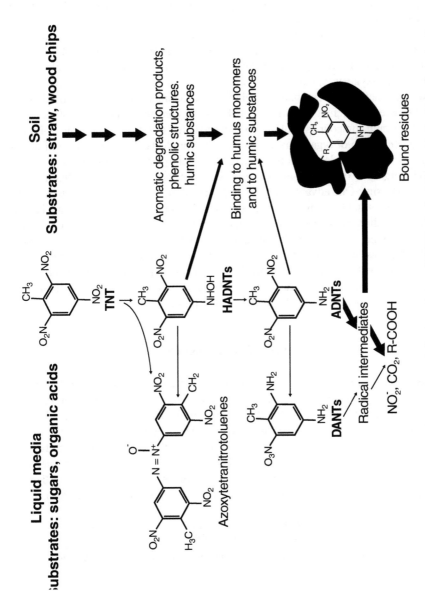

Figure 9.13 Proposed pathways for the cometabolic degradation of TNT by ligninolytic fungi in liquid media and soils.

9.6 DETOXIFICATION OF ARSENIC-CONTAINING ORGANIC COMPOUNDS BY MANGANESE PEROXIDASE

MnPs peroxidase, produced by ligninolytic fungi, is able to oxidize a wide variety of organic pollutants. The Chemical Weapons Convention (CWC), which was ratified by many countries, sets an explicit timetable for signatory countries to destroy their chemical weapon stockpiles and to remediate contaminated sites. Therefore, there is an increasing interest in biological methods to destroy chemical warfare agents. Organo-arsenic compounds were produced during World War I and II as chemical warfare agents, e.g., alkylarsines, chlorovinylarsines, and phenylarsines. These extremely toxic nerve gases have to be eliminated. Preliminary experiments have been carried out to test the potential of fungal MnP to attack arsenic-containing agents (Table 9.3).[20,28] Organo-arsenic compounds could not be detected after enzymatic treatment. This indicates a conversion of organo-arsenic to inorganic products. The mechanism of degradation as well as the degradation products are currently under investigation.

The highly toxic organo-arsenic agents do not inactivate enzymes like they do microorganisms. One advantage of the application of MnP is its enormous stability. The enzyme is stable for weeks in sterile filtered solutions. Organic solvents including ethanol, dioxane, and acetonitrile, at concentrations up to 40 vol%, are tolerated by the enzyme without significant loss in activity. A simple technique for MnP production has been developed and is currently being optimized.[28] Incineration is the accepted technology for the remediation of chemical weapons. However, due to the high cost of incineration, research should be focused on alternative technologies. Application of enzymes may be an alternative process.

9.7 CONCLUSIONS

A broad spectrum of ligninolytic fungi, including white-rot and litter-decomposing macrofungi, are able to mineralize TNT after they reduce it to aminonitro derivatives. Peroxidases are involved in the mineralization process, which is based on the enzymatic combustion of the aromatic nucleus. The magnitude of enzymatic mineralization is on the order of 10 to 50%, depending on the experimental conditions.

The mineralization potential for TNT is limited in soil. Results from the literature and our own investigation presented in this chapter indicate that mineralization ranges between 3 and 20%. The reason for the relatively low mineralization seems

Table 9.3   Degradation of Highly Toxic Organo-Arsenic Compounds, Produced as Chemical Warfare Agents, by MnP

Compound	Structure	Degradation (%)	Reaction Time (h)
Lewisite I	$Cl-CH=CH-AsCl_2$	99	0.5
Lewisite II	$(Cl-CH=CH)_2-AsCl$	99	0.5
Lewisite III	$(Cl-CH=CH)_3-As$	97	1
Dick	$C_2H_5-AsCl_2$	95	2
Clark 1	$C_6H_5-AsCl-C_6H_5$	100	24
Pfiffikus	$C_6H_5-AsCl_2$	91	120

Source: Laboratory experiments adapted from Haas et al.[20] and Hofrichter et al.[28]

to be that the reduced metabolites interact rapidly with soil components. Radicals formed by fungal enzymes react with humic substances. Humification is part of the natural carbon cycle, and it is not surprising that derivatives of nitroaromatic compounds are incorporated into the soil organic matrix. Therefore, xenobiotic compounds that cannot be removed by extraction with nonpolar or polar solvents were termed "bound residues."[32] Such compounds represent novel molecular species formed from the intermediates of fungal metabolism and may form complexes with organic constituents in soil. Owing to the difficulties in characterization of their actual chemical structures and the enormous heterogeneity, the environmental risk will have to be assessed by ecotoxicological tests.

Ligninolytic fungi stabilize TNT and its biotransformation products with the soil humic compounds in three major steps: (1) the transformation of nitroaromatic compounds, (2) the formation of phenolic and quinoid molecules from plant material, and (3) the covalent coupling of both reactants by laccases and peroxidases. It can be assumed that small molecules, the macromolecular organic matter, and clay minerals are involved in these coupling and binding processes. Humification of nitroaromatic explosives represents a complex chemical process. Knowledge of the stoichiometric relations between fungal transformation of nitroaromatic compounds and the formation of humic substance-like precursors from plant material is essential for the manipulation of humification. High priority in future research should be given to the nature and amount of additives that provide the matrix for the binding process.

Humified aromatic residues of explosives are involved in the long-term turnover of humic substances in nature. Therefore, the remobilization of humified nitroaromatics has to be examined. Ligninolytic macrofungi probably play a significant part in humus turnover. The peroxidases of the fungi are not only involved in the polymerization process, but also in the decomposition of humus. The capability to degrade humic compounds could offer a new way to clarify the long-term fate of bound residues in short-term tests. The white-rot fungus *N. frowardii* with its MnP system may be one of the appropriate candidates for this experiment.[30]

Based on the considerations described previously, we suggest that ligninolytic fungi have the potential for effective soil bioremediation. Despite the recognized potential, the development of fungal applications has been hampered by inconsistent treatment conditions. The growth of fungi and their biodegradation rates are relatively low in comparison to bacteria. However, the ability of ligninolytic fungi to degrade a wide range of persistent xenobiotics, e.g., TNT, RDX, HMX, PAHs, PCP, and PCBs, provides a strong rationale for research and development of field applications. These fungi provide treatment systems that will operate *in situ* in large areas of contaminated soils (landfarming in combination with bioaugmentation and biostimulation by straw and wood chips). However, the potential toxicity of products remains to be evaluated.

ACKNOWLEDGMENT

Research in the authors' laboratory has been supported by the Bundesministerium für Bildung und Forschung (BMF grants 1450821A2, 0327051D), as well as by the Fonds der Chemischen Industrie.

REFERENCES

1. Aust, S. D. 1997. Factors affecting biodegradation by white-rot fungi. In: *In Situ and On-Site Bioremediation*. Papers 4th Symp. Int. In Situ and On-Site Biorem. New Orleans 1997, Vol. 2, 537–544. Batelle Press, Columbus, OH.

2. Barr, D. P. and S. D. Aust. 1994. Mechanisms white rot fungi use to degrade pollutants. *Environ. Sci. Technol.* 28:78A–87A.

3. Bonnarme, P. and T. W. Jeffries. 1990. Mn(II) regulation of lignine peroxidases and manganese-dependent peroxidases from lignin degrading white rot fungi. *Appl. Environ. Microbiol.* 56:210–217.

4. Braedt, M., H. Hörseljau, F. Jacobs, and F. Knolle. 1998. Die Sprengstoffabrik "Tanne" in Clausthal-Zellerfeld: Geschichte und Perspektive einer Harzer Rüstungssaltlast. Papierflieger, Clausthal-Zellerfeld.

5. Bumpus, J. A. and M. Tatarko. 1994. Biodegradation of 2,4,6-trinitrotoluene by *Phanerochaete chrysosporium*: identification of initial degradation products and the discovery of a TNT metabolite that inhibits lignin peroxidases. *Curr. Microbiol.* 28:185–190.

6. Bumpus, J. A. and S. D. Aust. 1987. Biodegradation of DDT (1,1,1-trichloro-2,2-bis(4-chlorophenyl)ethane) by the white-rot fungus *Phanerochaete chrysosporium*. *Appl. Environ. Microbiol.* 53:2001–2008.

7. Bumpus, J. A., M. Tien, D. Wright, and S. D. Aust. 1985. Oxidation of persistent environmental pollutants by a white-rot fungus. *Science.* 228:1434–1436.

8. Comfort, S. D., P. J. Shea, L. S. Hundal, Z. Li, B. L. Woodbury, J. L. Martin, and W. L. Powers. 1995. TNT transport and fate in contaminated soil. *J. Environ. Qual.* 24:1174–1182.

9. Dawel, G., M. Kästner, J. Michels, W. Poppitz, W. Günther, and W. Fritsche. 1997. Structure of a laccase-mediated product of coupling of 2,4-diamino-6-nitrotoluene (2,4-DANT) to guaiacol, a model for coupling of 2,4,6-trinitrotoluene (TNT) metabolites to humic organic soil matrix. *Appl. Environ. Microbiol.* 63:2560–2565.

10. Dec, J., K. L. Shuttleworth, and J.-M. Bollag. 1990. Microbial release of 2,4-dichlorophenol bound to humic acid or incorporated during humification. *J. Environ. Qual.* 19:546–551.

11. Dutta, S. K., M. M. Kackson, L. H. Hou, D. Powell, and H. E. Tatem. 1998. Non-ligninolytic TNT mineralization in contaminated soil by *Phanerochaete chrysosporium. Bioremediation J.* 2:97–103.

12. Erikson, K. E., R. A. Blanchette, and P. Ander. 1990. *Microbial and Enzymatic Degradation of Wood and Wood Compounds*. Springer-Verlag, Berlin.

13. Fernando, T. and S. D. Aust. 1994. Biodegradation of toxic chemicals by white rot fungi. In: Chaudhry, G. R. (ed.). *Biological Degradation and Bioremediation of Toxic Chemicals*, p. 386–402. Chapman & Hall, London.

14. Fernando, T., J. A. Bumpus, and S. D. Aust. 1990. Biodegradation of TNT (2,4,6-trinitrotoluene) by *Phanerochaete chrysosporium. Appl. Environ. Microbiol.* 56:1666–1671.

15. Field, J. A., E. de Jong, G. Feijoo-Costa, and J. A. M. de Bont. 1993. Screening for ligninolytic fungi applicable to the biodegradation of xenobiotics. *Trends Biotechnol.* 11:44–49.

16. Forrester, I. T., A. C. Grabski, R. R. Burgess, and G. F. Leatham. 1988. Manganese, Mn-dependent peroxidases, and the biodegradation of lignin. *Biochem. Biophys. Res. Commun.* 157:992–999.

17. Fritsche, W. 1999. *Mikrobiologie*. Spektrum-Verlag, Heidelberg.

18. Gianfreda, L., F. Xu, and J.-M. Bollag. 1999. Laccases: a useful group of oxidoreductive enzymes. *Bioremediation J.* 3:1–25.

19. Gorontzy, T., O. Drzyzga, M. W. Kahl, D. Bruns-Nagel, J. Breitung, E. von Löw, and K.-H. Blotevogel. 1994. Microbial degradation of explosives and related compounds. *Crit. Rev. Microbiol.* 20:265–284.

20. Haas, R., K. Scheibner, and M. Hofrichter. 1998. Verfahren zum Abbau arsenorganischer Verbindungen. Deutsches Patentamt, Aktenzeichen 19826607.3, 16. June 1998.

21. Hatakka, A. 1994. Lignin modifying enzymes from selected white-rot fungi: production and role in lignin degradation. *FEMS Microbiol. Rev.* 13:125–135.

22. Hatakka, A., T. Vares, and T. Lundell. 1999. Fungal enzyme systems in lignin biodegradation, In: Steinbüchel, A. (ed.). *Biochemical Principles and Mechanisms of Biosynthesis and Biodegradation of Polymers.* p. 273–283. Wiley-VCH, Weinheim.

23. Hawari, J., A. Halasz, S. Beaudet, L. Paquet, G. Ampleman, and S. Thiboutot. 1999. Biotransformation of 2,4,6-trinitrotoluene with *Phanerochaete chysosporium* in agitated cultures at pH 4,5. *Appl. Environ. Microbiol.* 65:2977–2986.

24. Herre, A., K. Scheibner, and W. Fritsche. 1998. Bioremediation von 2,4,6-Trinitrotoluol-kontaminiertem Boden durch Pilze auf einem Rüstungsaltlastenstandort. *TerraTech.* 4:52–55.

25. Herre, A., J. Michels, K. Scheibner, and W. Fritsche. 1997. Bioremediation of 2,4,6-trinitrotoluene (TNT) contaminated soil by a litter decaying fungus. In: *In Situ and On-Site Bioremediation.* Papers 4th Int. Symp. In Situ and On-Site Biorem. New Orleans 1997. Vol. 2, 493–498. Batelle Press, Columbus, OH.

26. Higson, F. K. 1992. Microbial degradation of nitroaromatic compounds. Adv. *Appl. Microbiol.* 37:1–19.

27. Hofrichter, M. and W. Fritsche. 1997. Depolymerization of low-rank coal by extracellular fungal enzyme systems. II. The ligninolytic enzymes of the coal-humic-depolymerizing fungus *Nematoloma frowardii* b19. *Appl. Microbiol. Biotechnol.* 47:566–571.

28. Hofrichter, M., K. Scheibner, R. Haas, J. Nüske, and W. Fritsche. 1999. Die Mangan-Peroxidase ligninolytischer Pilze: Ein innovatives biokatalytisches System zur Eliminierung von Umwelt-Schadstoffen. In: Heiden, S. et al. (eds.). *Biotechnologie im Umweltschutz. Bioremediation: Entwicklungsstand — Anwendungen — Perspektiven.* p. 108–116. Schmidt-Verlag, Berlin.

29. Hofrichter, M., K. Scheibner, I. Schneegass, and W. Fritsche. 1998. Enzymatic combustion of aromatic and aliphatic compounds by manganese peroxidase from *Nematoloma frowardii. Appl. Environ. Microbiol.* 64:399–404.

30. Hofrichter, M., K. Scheibner, I. Schneegaß, D. Ziegenhagen, and W. Fritsche. 1998. Mineralization of synthetic humic substances by manganese peroxidase from the white-rot fungus *Nematoloma frowardii. Appl. Microbiol. Biotechnol.* 49:584–588.

31. Kapich, A., M. Hofrichter, T. Vares, and A. Hatakka. 1999. Coupling of manganese peroxidase mediated lipid peroxidation with destruction of non-phenolic model compounds and ^{14}C-labeled lignins. *Biochem. Biophys. Res. Commun.* 259:212–219.

32. Kästner, M., S. Streibich, M. Beyrer, H. H. Richnow, and W. Fritsche. 1999. Formation of bound residues during microbial degradation of [^{14}C]anthracene in soil. *Appl. Environ. Microbiol.* 65:1834–1842.

33. Kirk, T. K. and R. L. Farrel. 1987. Enzymatic "combustion": the microbial degradation of lignin. *Annu. Rev. Microbiol.* 41:465–505.

34. Kirk, T. K., E. Schulze, W. J. Connors, L. F. Lorenz, and J. G. Zeikus. 1978. Influence of culture parameters on lignin metabolism by *Phanerochaete chrysosporium. Arch. Microbiol.* 117:277–285.

35. Lamar, R. T. 1992. The role of fungal lignin-degrading enzymes in xenobiotic degradation. *Curr. Opin. Biotechnol.* 3:261–266.
36. Loomis, A. K., A. M. Childress, D. Daigle, and J. W. Bennett. 1997. Alginate encapsulation of the white-rot fungus *Phanerochaete chrysosporium*. *Curr. Microbiol.* 34:127–130.
37. McEldoon, J. P. and J. S. Dordick. 1991. Thiol and Mn^{2+}-mediated oxidation of veratryl alcohol by horseradish peroxidase. *J. Biol. Chem.* 266:14288–14293.
38. Michels, J. and G. Gottschalk. 1994. Inhibition of the lignin peroxidase of *Phanerochaete chrysosporium* by hydroxylamino-dinitrotoluene, an early intermediate in the degradation of 2,4,6-trinitrotoluene. *Appl. Environ. Microbiol.* 60:187–194.
39. Michels, J. and G. Gottschalk. 1995. Pathway of 2,4,6-trinitrotoluene (TNT) by *Phanerochaete chrysosporium,* In: Spain, J.C. (ed.). *Biodegradation of Nitroaromatic Compounds.* p. 135–149. Plenum Press, New York.
40. Paszczynski, A. and R. L. Crawford. 1995. Potential for bioremediation of xenobiotic compounds by the white-rot fungus *Phanerochaete chrysosporium*. *Biotechnol. Prog.* 11: 368–379.
41. Scheibner, K. and M. Hofrichter. 1998. Conversion of aminonitrotoluenes by fungal manganese peroxidase. *J. Basic Microbiol.* 38:63–71.
42. Scheibner, K., M. Hofrichter, and W. Fritsche. 1997. Mineralization of 2-amino-4,6-dinitrotoluene by manganese peroxidase of the white-rot fungus *Nematoloma frowardii*. *Biotechnol. Lett.* 19:835–839.
43. Scheibner, K., M. Hofrichter, A. Herre, J. Michels, and W. Fritsche. 1997. Screening for fungi intensively mineralizing 2,4,6-trinitrotoluene. *Appl. Microbiol. Biotechnol.* 47:452–457.
44. Scheibner, K. 1998. Abbau von 2,4,6-Trinitrotoluol durch Pilze unter besonderer Berücksichtigung ligninolytischer Basidiomyceten und des Mangan-Peroxidase-Systems. Ph.D. thesis. Friedrich-schiller-Universität, Jena.
45. Scheibner, K., A. Herre, and W. Fritsche. 1998. Abbau von 2,4,6-Trinitrotoluol durch Basidiomyceten: Einfluß des Mangan-Peroxidase-Systems und Schadstoffbilanzierung im Boden. In: Tagungsband zum Statusseminar des Forschungsverbundes "Biologische Verfahren zur Bodensanierung; 4. Statusseminar des Verbundvorhabens Biologische Sanierung von Rüstungsaltlasten" 7. Mai 1998, Clausthal-Zellerfeld, p. H 1–37. Umweltbundesamt Berlin, Projektträger des BMBF.
46. Spain, J. C. (ed). 1995. *Biodegradation of Nitroaromatic Compounds.* Plenum Press, New York.
47. Spain, J. C. 1995. Biodegradation of nitroaromatic compounds. *Annu. Rev. Microbiol.* 49:523–555.
48. Spiker, J. K., D. L. Crawford, and R. L. Crawford. 1992. Influence of 2,4,6-trinitrotoluene (TNT) concentration on the degradation of TNT in explosive contaminated soils by the white-rot fungus *Phanerochaete chrysosporium*. *Appl. Environ. Microbiol.* 58:3199–3202.
49. Spreinat, A., H.-J. Abken, K. Tölle, and G. Gottschalk. 1998. Sanierung von Rüstungsaltlasten durch Pilze. *TerraTech.* 3:46–49.
50. Stahl, J. D. and S. D. Aust. 1993. Metabolism and detoxification of TNT by *Phanerochaete chrysosporium*. *Biochem. Biophys. Res. Commun.* 192:477–482.
51. Stahl, J. D. and S. D. Aust. 1993. Plasma membrane-dependent reduction of 2,4,6-trinitrotoluene by *Phanerochaete chrysosporium*. *Biochem. Biophys. Res. Commun.* 192:471–476.
52. Stahl, J. D. and S. D. Aust. 1995. Properties of a transplasma membrane redox system of *Phanerochaete chrysosporium*. *Arch. Biochem. Biophys.* 320:369–374.

53. Stahl, J. D. and S. D. Aust. 1995. Biodegradation of 2,4,6-trinitrotoluene by the white rot fungus *Phanerochaete chrysosporium*, In: Spain, J.C. (ed.). *Biodegradation of Nitroaromatic Compounds*. p. 117–134. Plenum Press, New York.

54. Stegmann, R., S. Lotter, and J. Heerenklage. 1991. Biological treatment of oil-contaminated soils in bioreactors. In: Hinchee, R. E. and R. F. Olfenbuttel (eds.). *On-Site Bioreclamation*. p. 188–208. Butterworth-Heinemann, Boston.

55. Tien, M. and T. K. Kirk. 1988. Lignin peroxidase of *Phanerochaete chrysosporium. Methods Enzymol.* 161:238–249.

56. Van Aken, B., L. M. Godefroid, C. M. Peres, H. Naveau, and S. N. Agathos. 1999. Mineralization of ^{14}C-U-ring labeled 4-hydroxylamino-2,6-dinitrotoluene by manganese-dependent peroxidase of the white-rot basidiomycete *Phlebia radiata. J. Biotechnol.* 68:159–169.

57. Van Aken, B., M. Hofrichter, K. Scheibner, A. I. Hatakka, H. Naveau, and S. N. Agathos. 1999. Transformation and mineralization of 2,4,6-trinitrotoluene (TNT) by manganese peroxidase from the white-rot basidiomycete *Phlebia radiata. Biodegradation.* 10:83–91.

58. Van Aken, B., K. Skubisz, H. Naveau, and S. N. Agathos. 1997. Biodegradation of 2,4,6-trinitrotoluene (TNT) by the white-rot basidiomycete *Phlebia radiata. Biotechnol. Lett.* 19:813–817.

59. Walker, J. E. and D. L. Kaplan. 1992. Biological degradation of explosives and chemical agents. *Biodegradation.* 3:369–385.

60. Wariishi, H., K. V. Valli, V. Reganathan, and M. H. Gold. 1989. Thiol-mediated oxidation of nonphenolic lignin model compounds by manganese peroxidase of *Phanerochaete chrysosporium. J. Biol. Chem.* 264:14185–14191.

61. Won, W. D., L. H. Di Salvo, and J. Ng. 1976. Toxicity and mutagenicity of 2,4,6-trinitrotoluene and its microbial metabolites. *Appl. Environ. Microbiol.* 31:575–580.

62. Ziechmann, W. 1994. *Humic Substances. A Study about Their Theory and Reality.* Wissenschaftsverlag, Mannheim.

CHAPTER 10

Phytoremediation and Plant Metabolism of Explosives and Nitroaromatic Compounds

Joel G. Burken, Jacqueline V. Shanks, and Phillip L. Thompson

CONTENTS

10.1 INTRODUCTION

The development of methods to remediate soils and groundwaters contaminated with 2,4,6-trinitrotoluene (TNT), hexahydro-1,3,5-trinitro-1,3,5-triazine (RDX), octohydro-1,3,5,7-tetranitro-1,3,5,7-tetraazocine (HMX), and related explosive-manufacturing byproducts has been an intense research interest for several decades. As discussed in other chapters, the ability of microorganisms to metabolize explosives has been studied extensively, although complete biomineralization of TNT has not been observed.[75,76] Bioremediation has been used to treat contaminated media at facilities in the U.S. and Europe. Only since the mid-1990s has phytoremediation (the use of plants for the restoration of contaminated soils, groundwater, or surface water) emerged as a potential low cost technology for remediation of explosives-contaminated soils and groundwater.[26] Prior to that time, studies of plant–explosives interactions had been limited to toxicity evaluations and investigating potential foodchain contamination due to explosives accumulation in plant tissue.[20,33,35,36,58,69] However, the observation of rapid TNT disappearance in aquatic plant systems[67,87] has spurred investigations of explosives phytoremediation using aquatic or terrestrial plants. Such investigations range from pathway analysis to field demonstrations. The investigations have targeted an improved understanding of plant metabolism allowing for assessment of the true potential of phytoremediation. The goal of this chapter is to present an overview and discussion of explosives phytoremediation, with emphasis on TNT because it has been the explosive most frequently studied in plant systems.

10.2 CONTAMINANT ATTENUATION MECHANISMS IN PHYTOREMEDIATION

Contaminant attenuation mechanisms involved in phytoremediation are complex and not limited to the direct metabolism of contaminants by plants. In fact, the introduction of plants in contaminated areas has the potential to yield several indirect contaminant attenuation mechanisms that assist in the cleanup, or management, of contaminated media. Examples of indirect attenuation mechanisms in phytoremediation include the metabolism of contaminants by plant-associated microbes and plant-induced changes to the contaminated environment (alteration of redox conditions, shifts in microbial communities, and modification of site hydrodynamics). Thus, the term phytoremediation encompasses a range of processes beyond direct plant–contaminant interactions that must be considered in the evaluation of phytoremediation. In certain cases, the indirect mechanisms may dominate contaminant attenuation. In such cases, the technology may be best described as "plant-assisted" remediation.

Several terms that relate to the specific attenuation mechanism have been used to better describe specific applications of phytoremediation. Such terms include phytoextraction, phytodegradation, phytovolatilization, and rhizodegradation.[68] Phytoextraction describes the simple case where a contaminant enters the plant (i.e., uptake) and is stored within the plant tissue in a recoverable form. The process often occurs with heavy metals and may be important for certain organic chemicals that are resistant to plant metabolism.

Phytodegradation and phytovolatilization refer to processes beyond uptake and storage of contaminants. In the case of phytodegradation, contaminants are metabolized in plant tissues following uptake. Phytodegradation has been studied extensively to understand the fate of herbicides in crop plants. Information on herbicide phytodegradation has been extended to cell cultures[12] to noncrop species,[48,57,73] including hybrid poplar trees.[16,79] The overall metabolic process involved in phytodegradation is, in some ways, analogous to human metabolism of xenobiotic chemicals, thus a "green liver" conceptual model is often used to describe phytodegradation.[46] This conceptual model is described in more detail below. Phytovolatilization is the release of contaminants from plant tissues to the atmosphere after uptake. The process has been observed for contaminants such as trichloroethylene (TCE), which has been detected in the off-gas from plant leaves in the laboratory[18] and in the field.[22]

Rhizodegradation describes processes that differ significantly from the preceding classifications of phytoremediation. In rhizodegradation, the contaminant is transformed by microbes in the rhizosphere (i.e., the microbe-rich zone in intimate contact with a vascular root system). In the rhizosphere, soil redox conditions, organic content, moisture, and other soil properties are manipulated by the activity of plant roots. Rhizodegradation is responsible for the enhanced removal of total petroleum hydrocarbons from soil by deep-rooted trees[19] and other annual species.[70]

A number of the listed degradation processes can be active in phytoremediation systems. Figure 10.1 is a schematic of the phytoremediation processes that can impact the removal of organic compounds from soil and groundwater. Although direct and indirect attenuation processes involved in phytoremediation have been identified, many questions remain with respect to mechanisms, ultimate fate, and kinetics. Trapp and McFarland[82] compiled a review of applicable information and basic mathematical descriptions of xenobiotic uptake, metabolism, and fate in plants. The U.S. Environmental Protection Agency (EPA) has also published a handbook on phytoremediation design and processes.[29] A compilation of phytoremediation projects and abstracts is also available as an EPA report.[30]

10.2.1 The "Green Liver" Conceptual Model of Contaminant Metabolism

The specific interactions of a pollutant with soil, water, and plants will vary depending on the chemical properties of the contaminant, the physiological properties of the introduced plant species, and the contaminated medium. Collectively, these properties will determine whether a contaminant is subject to phytoextraction, phytodegradation, phytovolatilization, or rhizodegradation. In each case the process of phytoremediation begins with contaminant transport to the plant. In terrestrial species, transport of contaminants to the plant is dominated by the uptake of water by the roots, and distribution within the plant relies on xylem or phloem transport (Figure 10.1). For wetland plant species, contaminants can enter through the roots or can partition from the water column into plant tissues directly (Figure 10.2). Uptake in terrestrial plants has been studied for many plant-contaminant combinations, and quantitative models to predict uptake rates have been developed.[82] Such

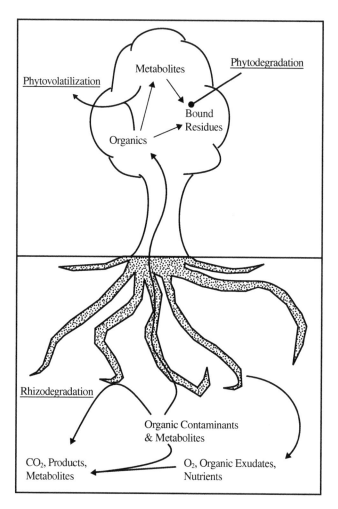

Figure 10.1   Phytoremediation mechanisms in a terrestrial, deep-rooted plant. Rhizodegrada-tion, phytovolatilization, and phytodegradation pathways are the potential degra-dation and removal mechanisms for organic compounds.

models, which are based on flow in the transpiration stream and organic partition relationships, have been used in the development of Transpiration Stream Concen-tration Factors (TSCFs) for various compounds.[13,17,41] TSCF relationships are devel-oped strictly for terrestrial plants, and TSCFs (or similar relationships) have not been developed for aquatic plants where diffusion and partitioning are thought to play a larger role in uptake (Figure 10.2).

The green liver model is often used to describe the fate and disposition of organic contaminants (xenobiotic compounds) within plants.[64] The transformation of xeno-biotic compounds by plants differs greatly from metabolism carried out by micro-organisms, although there are certainly many similarities in plant and microbial biochemistry (e.g., cytochrome P-450). A primary difference is that the plants are

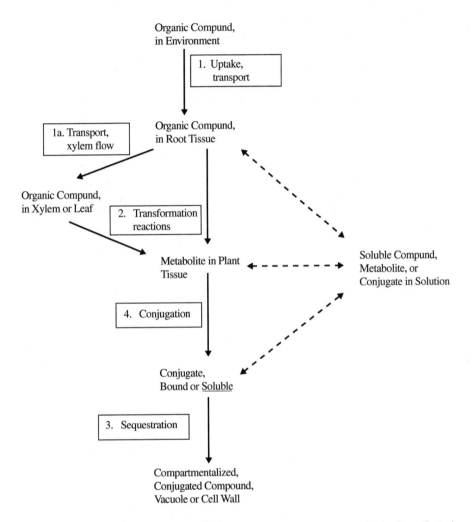

Figure 10.2 Schematic of the green liver model for metabolism of xenobiotics in plants. Dotted lines show diffusion/transport possible in aquatic plants.

not heterotrophs and, therefore, do not have an extensive array of catabolic pathways for use of organic compounds as carbon or nitrogen sources. Plants rely on photosynthates as the energy supply to respiring cells, and, thus, the need to metabolize other organic compounds does not exist. The metabolism of foreign compounds in plant systems is generally considered to be a "detoxification" process that is similar to the metabolism of xenobiotic compounds in humans,[46] hence the name green liver. This concept evolved from studies of plant exposure to herbicides.[37,38] During detoxification, xenobiotic compounds are metabolized in three stages: transformation, conjugation, and sequestration (Figure 10.2).

Initial transformation processes can include reduction, oxidation, and hydrolysis reactions. Such reactions are enzyme catalyzed and result in the formation of compounds that are amenable to subsequent conjugation reactions. Oxidation reactions are the predominant transformation reactions in the metabolism of pesticides; however, reduction reactions have been found for certain nitroaromatic compounds.[47] Hydroxylation plays an important role, because the hydroxyl group is a suitable site for conjugation reactions. Cytochrome P-450-catalyzed hydroxylations are key steps in detoxification of herbicides by various crop species.[47] Transformation of TNT could include oxidation of the aromatic ring or the methyl group and the reduction of nitro groups to nitroso, hydroxyl-amino, or amino derivatives.

The modified xenobiotic compound undergoes conjugation with an organic molecule of plant origin. The conjugate formation plays a key role in the metabolism of xenobiotic compounds by plants because it leads to a reduction in toxicity to the plant.[25,82] Functional groups associated with TNT are resistant to conjugation reactions without transformation; however, reduced forms of aryl nitro groups, particularly aryl amines, have the potential to undergo rapid conjugation. Examples of compounds used in the conjugation of xenobiotic amino groups include glucose and malonate, which are added to the transformation product by glucosyltransferases and malonyltransferases, respectively.[25]

Following conjugation, the resulting compound can undergo various sequestration processes. One common sequestration mechanism is the storage of conjugates in plant organelles such as vacuoles,[25] where they are no longer capable of interfering with cell function. In other cases, xenobiotic conjugates are incorporated into biopolymers such as lignin where they are characterized as "bound" (i.e., nonextractable) residues. Wetland plants can excrete the conjugates, which results in "storage" outside of the plant (Figure 10.2).

10.2.2 Plant Metabolism of TNT

Numerous studies have reported TNT "disappearance" from aqueous solutions in the presence of terrestrial or aquatic plants.[3,5,7,8,20,27,33,36,50–56,58,65,69,75,79,85,86,87] During such studies TNT cannot be recovered stoichiometrically from plant tissue because of phytodegradation. Preliminary evidence indicates that RDX is taken up by plants, but is not transformed as rapidly as TNT. Generally, RDX is stable in solution and accumulates in plant tissues. A diverse array of plants (or plant systems) has been tested for this ability. Representative aquatic and terrestrial species are listed in Tables 10.1 and 10.2.

In the earliest published study of terrestrial plants exposed to TNT, Palazzo and Leggett[58] reported that TNT disappearance resulted in the formation of traces of aminated reduction products similar to those produced by microbial metabolism (i.e., aminated nitrotolutenes). At the time, it was unknown whether the aminated derivatives were the result of microbial activity (rhizodegradation) or metabolism by the plant itself. Harvey et al.[36] analyzed tissues from plants that were exposed to ^{14}C-TNT and found labeled compounds that were not TNT or the reduction products and suggested that acid hydrolyzable conjugates were formed in plants. However, neither of the above studies was designed to directly determine the role of plant vs. microbial metabolism in the disappearance of TNT.

Table 10.1 Wetland Species Tested in Laboratory Studies with Explosives or Nitroaromatic Compounds

Plant Species	Common Name	Compounds investigated	Ref.
Myriophyllum aquaticum	Parrot feather	TNT	42
		TNT, RDX, HMX	63
		TNT	85
		RDX, TNT	2
M. spicaticum	Eurasian water milfoil	TNT	59
		TNT	42[a]
		RDX, TNT	2
M. braziliense	Parrot feather	RDX, TNT, HMX, TNB	74
Scriptus validus	Bulrush	RDX, TNT	63
Cannaceae sp.	Canna	RDX, TNT	63
Elodea canadensis	Elodea	RDX, TNT	2, 63
Salviniaceae sp.	Fern	RDX, TNT	63
Bamusa pynaea	Bamboo	RDX, TNT	63
Sagittarria	Arrowhead	RDX, TNT	2, 63
Egeria densa	Egeria	RDX, TNT	2
Vallisneria americana	Vallisineria	RDX, TNT	2
Potamogeton crispus	Curlyleaf pondweed	RDX, TNT	2
P. pectinatus	Sago pondweed	RDX, TNT	2
Heteranthera dubia	Water star-grass	RDX, TNT	2
Eleocharis parvula	Dwarf spikerush	RDX, TNT	2
P. nodosus	American pondweed	RDX, TNT	2
Ceratophyllum demersum	Coontail	RDX, TNT	2
Alisma subcodatum	Water-plantain	RDX, TNT	2
Carex vulpinoidea	Fox sedge	RDX, TNT	2
Scirpus cyperinus	Wool-grass	RDX, TNT	2
E. obtusa	Blunt spikerush	RDX, TNT	2
Phalaris arundinacea	Reed canary grass	RDX, TNT	2
Typha angustifolia	Narrowleaf cat-tail	RDX, TNT	2
Nitella	Stonewort	TNT	87

[a] Conducted with organ tissue cultures.

Recently, research with axenic plant tissues and intact plants has provided detailed insight into the fate of TNT in plant systems, with emphasis on how plant metabolism contributes to explosives disappearance and on the determination of metabolic products. This approach was able to differentiate between the contributions of plants and microbes during TNT disappearance. In 1997, Hughes et al.[42] provided rigorous mass balances that confirmed the ability of plants to take up and transform TNT. Three different plant systems (*Catharanthus roseus* hairy root cultures, axenic *Myriophyllum* plants, and native *Myriophyllum*) were exposed to uniformly labeled ^{14}C-TNT, and the fate of ^{14}C was evaluated. TNT was completely transformed in all systems containing viable plant tissue and mineralization did not occur, which is consistent with the green liver model. The hairy root cultures allow rapid lab studies

Table 10.2 Terrestrial Species Tested in Laboratory Studies with Explosives or Nitroaromatic Compounds

Plant Species	Common Name	Compounds Investigated	Ref.
Phaseolus vulgaris	Bush bean	TNT, RDX	36, 65
Medicago sativa	Alfalfa	TNT	33, 65
Allium schoenoprasum	Allium	TNT	33
Populus sp. deltoides × nigra	Hybrid poplar	TNT, RDX	78–81
Catharanthus roseus	Periwinkle	TNT	7,[a] 42[a]
Lupinus angusifolius		TNT	65
Trifolium repens	Clover	TNT	65
Phacelia seicea		TNT	65
Triticum aestivum	Wheat	TNT	65, 71[a]
Aleopecurus pratensis		TNT	65
Bromus inermis		TNT	65
Festuca rubra	Red fescue	TNT	65
Lolium perenne		TNT	65
Phleum sp.		TNT	65
Zea mays	Maize	RDX	21
Glycien max	Soybean	RDX	21
Sorghum sudanese	Sorghum/sudangrass	RDX	21
T. aestivum	Wheat	RDX	21
Cyperus esculentus	Yellow nutsedge	RDX	50
Z. mays	Corn	RDX	50
Lacuta sativa	Lettuce	RDX	50
Lyopersicum esculentum	Tomato	RDX	50

[a] Conducted with organ tissue cultures.

and grow nearly as fast to high biomass yields as cell suspensions. In contrast, they have inherent genetic stability, resulting in reproducible, stable growth and metabolic characteristics.[28,72]

The above results were the first to demonstrate conclusively that plants were capable of metabolizing TNT, but they provided little information on the products. Products of metabolism were characterized into four classes: aminonitrotoluenes (2-amino-4,6-dinitrotoluene and 4-amino-2,6-dinitrotoluene) observed in the media; unidentified ^{14}C-labeled soluble products; extractable plant-associated ^{14}C fraction that could not be identified as reduction products (i.e., aminonitrotoluenes, hydroxyl-aminonitrotoluenes, or axozynitrotoluenes); and bound residues (i.e., a non-extractable plant-associated material that could be quantified after combustion of the plant tissue). The aminodinitrotoluenes detected were the result of aryl nitro group reduction (as is commonly observed in microbial studies), but their concentrations (typically less than 10% of the total) were much less than would be present if stoichiometric conversion was taking place. The soluble products and extractable plant-associated fractions accounted for the majority of the products. Soluble products comprised 30 to 50% of the ^{14}C, while extractable plant-associated ^{14}C was as much as 70% of the total. Later studies[85] revealed that the bound fraction increased with time, with a commensurate decrease in levels of extractable plant-associated metabolites.

Additional studies focused on ^{14}C distribution in microbe-free hairy root cultures during the initial stages of TNT metabolism by plants.[7] Plant tissues converted TNT

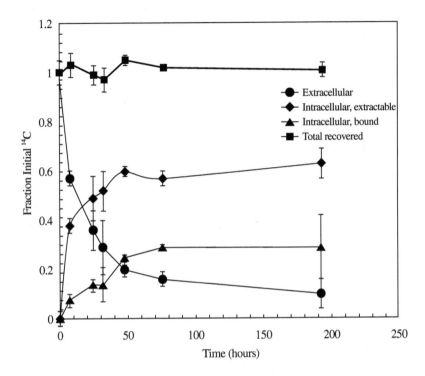

Figure 10.3   Mass balance of ¹⁴C in hairy root cultures of *C. roseus* over an 8-day period. Initial TNT concentration was 25 mg/L.[7] (Adapted from Bhadra, R. et al. 1999. *Phytoremediation and Innovation Strategies for Specialized Remedial Applications*. Batelle Press, Columbus, OH.)

rapidly to an unidentifiable intracellular extractable fraction, which was subsequently converted to an intracellular bound residue (Figure 10.3). Levels of identifiable nitroaromatic compounds (TNT and the aminated nitrotoluenes) decreased to below 5% of the initial ¹⁴C added within 75 h (Figure 10.4). The maximum combined level of the aminated nitrotoluenes at any time during the experiment was 12%.[7]

Two important conclusions can be drawn from the results of the studies described above. First, plants clearly utilized an intrinsic capacity to metabolize explosives. Second, the distribution of transformation products, conjugates, and bound residues is consistent with a green liver model of xenobiotic compound metabolism. The studies did not, however, address plant transformation pathways or identify the intermediates and products of metabolism.

The following sections are provided to summarize the current state of understanding of the metabolism of TNT by plants. The sections are organized as stages in the green liver model presented previously. It is important to note that most studies have focused on the identification of metabolic products or the characteristics of products (i.e., bound residue), and little information exists regarding the enzymes involved in plant metabolism, the regulation of their activity, or their kinetic properties.

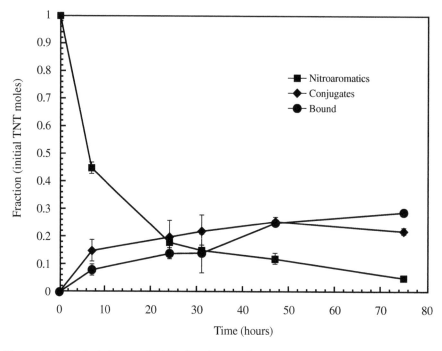

Figure 10.4   Mole balance of TNT, 2-amino-4,6-dinitrotoluene, 4-amino-2,6-dinitrotoluene, conjugate metabolites, and bound fractions expressed as fraction of initial TNT. The nitroaromatics group represents the sum of TNT, 2-amino-4,6-dinitrotoluene, and 4-amino-2,6-dinitrotoluene. The conjugates represent the sum of TNT-1, TNT-2, 2A-1, 4A-1.[7] (Adapted from Bhadra, R. et al. 1999. *Phytoremediation and Innovation Strategies for Specialized Remedial Applications.* Batelle Press, Columbus, OH.)

10.2.3 Initial TNT Transformation

TNT transformation is common among a variety of plant species and their tissues (Tables 10.1 and 10.2). Direct conjugation reactions are not likely to occur, since the TNT molecule does not contain functional groups amenable to conjugation. Thus, transformation is required to initiate plant metabolism. Based on the structure of TNT, two initial transformation processes are possible. First is the reduction of one or more nitro groups, yielding hydroxylamino or amino groups. Second is oxidation of either the methyl group or the ring directly. Interestingly, both forms of initial transformation take place in plants.

10.2.3.1 Reduction

In most of the studies listed in Table 10.1, TNT transformation by plants has been accompanied by the appearance of low levels of its monoamino derivatives, 2-amino-4,6-dinitrotoluene and 4-amino-2,6-dinitrotoluene.[7,42,56,58,59,61,63,65,85] Diamino TNT derivatives are seldom detected, although 2,4-diamino-6-nitrotoluene may have been observed at a maximum of 0.4% of the TNT taken up by hybrid poplars,[78] but

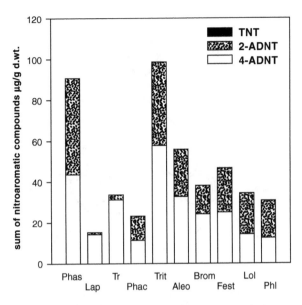

Figure 10.5 Nitroaromatic compounds extracted from the roots of plants, cultivated in soil supplemented with 100 mg of TNT per kilogram of soil. Results represent the average of six replicates. (From Scheidemann, P. et al. 1998. *Plant Physiol.* 152:242. With permission.)

chemical confirmation was not conducted. Quantitatively, the majority of reduction products observed are isolated from the roots of plants (terrestrial and aquatic) (Figure 10.5) where their concentrations can exceed that of TNT.[33,52,65,79,85]

Two studies have reported the formation of azoxy compounds in the aquatic weed *Myriophyllum*.[8,59] Trace quantities of 2,2′-azoxy-tetranitrotoluenes were detected in the medium of *Myriophyllum* incubated with TNT.[59] In another study, two binuclear metabolites (characterized by mass spectrometry and ^{1}H-NMR) different from available standards of 4,4′-azoxy-tetranitrotoluenes and 2,2′-azoxy-tetranitrotoluenes were detected in the medium at levels of 5.6% of the initial level of TNT.[8] The complete structure of the metabolites was not determined.

The low levels of aminated reduction products in plant systems may result from the high reactivity of hydroxylamino intermediates (Figure 10.6) with oxygen.[43] Hydroxylaminoderivatives of TNT are reasonably stable under anaerobic conditions *in vitro*, but decompose rapidly when exposed to oxygen.[43] Hydroxylaminonitrotoluenes are central intermediates in the bacterial reduction of TNT,[62] as discussed by Ahmad and Hughes (Chapter 8), and 4-hydroxylamino-2,6-dinitrotoluene has been detected in studies where *Myriophyllum* was exposed to TNT.[58] In plant systems where oxygen is present, the reactivity of hydroxylamines has the potential to divert further reductive metabolism to more stable amino compounds. Ferredoxin NADP$^+$ reductase isolated from spinach leaves catalyzed the nearly stoichometric conversion of TNT to 4-hydroxylamino-2,6-dinitrotoluene.[32] The hydroxylamines are postulated as a key intermediate in the transformation pathway of TNT (Figure 10.6), but conclusive research is needed.

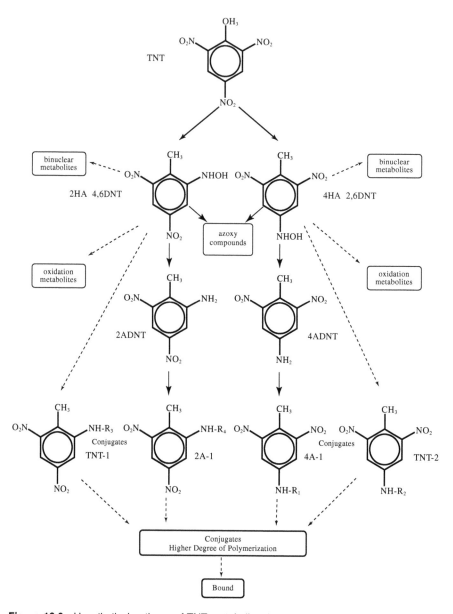

Figure 10.6   Hypothetical pathway of TNT metabolism in plants.

It should be noted that Rivera et al.[63] have proposed that reductive metabolism by plants yields 2,4,6-triaminotoluene (TAT) that subsequently undergoes ring cleavage. TAT was not detected in these studies, and it is doubtful that TAT would be produced in large quantities in plants. TAT formation requires strongly reducing conditions ($E_h \leq -200$ mV) (62) not present in plant cells. If TAT were produced, accumulation would be prevented by its autooxidation (catalyzed by oxygen and metal ions such as Mn^{2+}) followed by polymerization.[62]

10.2.3.2 Oxidation

Until recently, oxidative metabolism of TNT by plants had not been investigated, because the products are highly polar and analysis is difficult. Oxidative metabolism of herbicides is well documented as the most common initial reaction leading to activation or deactivation of the parent herbicide.[23,24] Cytochrome P-450 enzymes play a central role in chemical transformation of herbicides, usually to compounds with reduced phytotoxicity.[24,37,38,64] Plant species vary in their abilities to transform herbicides, which suggests that there would be considerable species variability in both the production and the phytotoxicity of oxidized compounds.

To date, oxidative metabolism of TNT has been reported only once.[8] Products of oxidative metabolism of TNT were isolated and characterized in the aquatic plant *M. aquaticum*.[8] Six oxidation metabolites (Table 10.3) were isolated and characterized by spectroscopic methods. The compounds include 2-amino-4,6-dinitrobenzoic acid, 2,4-dinitro-6-hydroxy-benzyl alcohol, 2-*N*-acetoxyamino-4,6-dinitrobenzaldehyde, and 2,4-dinitro-6-hydroxytoluene. The detection of these compounds provides clear evidence of initial oxidative transformation of TNT by plants via oxidation of the ring-substituted methyl group or through aromatic hydroxylation. Oxidation of the methyl group of TNT can also result from abiotic and light-catalyzed reactions.[34] Care was taken in the plant studies cited above to eliminate photooxidation as a confounding variable. The *M. aquaticum* plant system likely contained small amounts of microbes (it was not strictly axenic). However, the amount of oxidized products was over 30% of the initial ^{14}C label, which suggests a high activity of plant oxidative processes. Oxidized products would be expected to be lower if oxidation were solely catalyzed by microbes. Additional testing of different plant species for the presence of the oxidation products is needed to elucidate the oxidative pathways.

The oxidation products appear to follow formation of the hydroxylamino compounds (Figure 10.6). When *M. aquaticum* was exposed to either 2-ADNT or 4-ADNT, none of the oxidation products was detected. Although not shown in Figure 10.4, it is possible that the oxidation products are conjugated. The fate and turnover of the oxidation products has not been examined.

The details of the pathways for oxidative metabolism can only be speculated upon because no mechanistic studies exist. The formation of 2-amino-4,6-dinitrobenzoic acid could be mechanistically similar to the multistep hydroxylation observed in the metabolism of the herbicide oxadiazon in rice[44] and peanut[11] plants. The metabolites 2-*N*-acetoxyamino-4,6-dinitrobenzaldehyde and 2,4-dinitro-6-hydroxy-benzyl alcohol could be modified intermediates or byproducts of the multistep hydroxylation of the methyl group. Both compounds also exhibit additional modifications, such as the acetoxy addition at the aryl amino group and aromatic hydroxylation, respectively. The formation of 2,4-dinitro-6-hydroxy-benzyl alcohol and 2,4-dinitro-6-hydroxytoluene could result from the ring hydroxylation of TNT. In both cases, the elimination of a nitro group or a reduced form of the nitro group would be required. Aromatic hydroxylation of the herbicide 2,4-D is known to occur in plants; however, the elimination of ring-substituted chlorine is a minor reaction.[38]

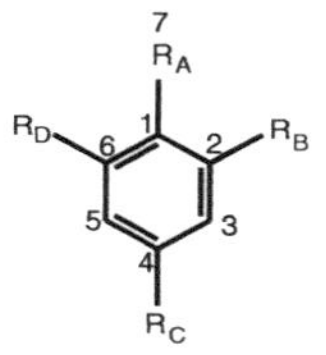

Table 10.3 TNT Phytotransformation Metabolites

Reduction Metabolites

Analyte	R_A	R_B	R_C	R_D
4-Amino-2,6-dinitrotoluene	CH_3	NO_2	NH_2	NO_2
2-Amino-4,6-dinitrotoluene	CH_3	NH_2	NO_2	NO_2
2,4-Diamino-6-nitrotoluene	CH_3	NH_2	NH_2	NO_2
2-Hydroxylamino-4,6-dinitrotoluene	CH_3	$NHOH$	NO_2	NO_2

Oxidation Metabolites

Analyte	R_A	R_B	R_C	R_D
2-Amino-4,6-dinitrobenzoic acid	$COOH$	NH_2	NO_2	NO_2
2,4-Dinitro-6-hydroxy-benzyl alcohol	CH_2OH	NO_2	NO_2	OH
2-N-Acetoxyamino-4,6-dinitrobenzaldehyde	CHO	$NHOCOCH_3$	NO_2	NO_2
4-N-Acetoxyamino-2,6-dinitrobenzaldehyde	CHO	NO_2	$NHOCOCH_3$	NO_2
2,6-Dinitro-4-hydroxytoluene	CH_3	NO_2	OH	NO_2
2,4-Dinitro-6-hydroxytoluene	CH_3	NO_2	NO_2	OH

Conjugation Metabolites

Analyte	R_A	R_B	R_C	R_D
TNT-1	CH_3	NO_2	NH-R_1[a]	NO_2
4A-1	CH_3	NO_2	NH-R_2[a]	NO_2
TNT-2	CH_3	NH-R_3[a]	NO_2	NO_2
2A-1	CH_3	NH-R_4[b]	NO_2	NO_2

[a] R_1, R_2, R_3, and R_4 are six carbon sugars.

Studies to determine the specific enzymes responsible for the oxidative transformations have not been performed. From the current state of understanding of herbicide metabolism in plants, the most plausible enzyme candidates for the oxidative metabolism of TNT are cytochrome P-450 enzymes. Cytochrome P-450 enzymes are localized primarily in the microsomes (endoplasmic reticulum) of plant cells[37,64] and require molecular oxygen as a second substrate and NADPH or NADH as cofactors.[37,64] With the advances in identifying the genes for cytochrome P-450 enzymes in plants, the details of the pathway structure of oxidative metabolism will be more easily elucidated. These reactions are particularly interesting because oxidation products of TNT have rarely been observed in microbial studies and little is known about their long-term fate in aquatic environments.

10.2.4 Conjugation of TNT-transformation Products.

Harvey et al.[36] first suggested that acid hydrolyzable conjugates could be formed during the metabolism of TNT in plants. Since that report, several TNT-derived conjugates have been isolated[7] (Table 10.3). The underlying basis for identifying metabolites as "conjugated products" is the isolation of ^{14}C-labeled compounds from plant tissues and confirming an increase in molecular weight by mass spectroscopy. Four unique TNT-derived conjugates have been isolated from *C. roseus* hairy root cultures. Interestingly, two have UV-VIS spectroscopic characteristics similar to 2-ADNT (TNT-2, 2A-1); the others have UV-VIS spectroscopic characteristics similar to that of 4-ADNT (TNT-1, 4A-1). The conjugates were originally thought to be formed by conjugations of the amino groups of 2-ADNT and 4-ADNT, as supported by a range of spectroscopic techniques and the ability to chemically or enzymatically hydrolyze the compounds with the concomitant appearance of an individual aminodinitrotoluene isomer. From the mass spectral evidence, it is likely that at least a six-carbon unit from the plant intracellular milieu was involved in conjugate formation. The resulting mass balance and distribution of the initial TNT added to the cell culture is displayed in Table 10.4.

Table 10.4 Mass Balance Estimate for TNT Fate in *Myriophyllum aquaticum* 12 Days after Amendment

Metabolite	Percent of Initial TNT (mol%)
2,4,6-Trinitrotoluene (TNT)	10
2-Amino-4,6-dinitrotoluene	2.3
4-Amino-2,6-dinitrotoluene	5.3
2-Amino-4,6-dinitrobenzoic acid	4.4
2,4-Dinitro-6-hydroxy-benzyl alcohol	8.1
2-N-Acetoxyamino-4,6-dinitrobenzaldehyde	7.8
2,4-Dinitro-6-hydroxytoluene	15.6
Conjugates	2
Binuclear metabolites	5.6
Bound	33
Total fraction accounted for	94
Unknown	6

A time course of TNT metabolism in hairy root cultures of *C. roseus* (Figure 10.4) reveals that conjugates within plant tissues were a significant fraction of the TNT metabolized (15 to 26% over a 75-h period). Conjugates appear to be "gateway" intermediates for bound residues.[49] The hypothesis is supported by the turnover of individual conjugates and accumulation of unknown conjugate products in studies of TNT-, 2-ADNT-, or 4-ADNT-amended hairy roots.[86]

An interesting observation from additional study suggested that hydroxylamino intermediates were involved in a portion of the conjugation process. When hairy roots were exposed to either 2-ADNT or 4-ADNT, the conjugate profile differed from the profile obtained if TNT was the parent compound.[86] Instead of two conjugates from either aminodinitrotoluene, only one identifiable conjugate was detected. The absence of certain conjugates in aminodinitrotoluene amended systems suggests that conjugation can occur prior to complete reduction of the nitro group. This hypothesis is supported by spectroscopic examination of the conjugates from TNT that were not formed in hairy root cultures exposed to aminodinitrotoluene. Also, unique unidentified conjugates were detected in monoamino-amended roots that were not detected in TNT-amended roots. Thus, the distribution and levels of TNT-derived conjugates could be dependent on several factors.[86] Formation of TNT conjugates is an important finding and contributes to the understanding of the fate of TNT in plant systems, particularly because it is the intermediate step between initial transformation and the sequestration of metabolites as bound residue.[7]

10.2.5 Sequestration of Conjugates

The large molecules formed during sequestration and the inability to resolve the contribution of a TNT-derived molecule within a large chemical structure using spectroscopic methods complicate the chemical characterization of sequestered conjugates. In some cases, sequestration has been characterized as a decrease in extractability of conjugates over time. For example, studies of ^{14}C binding in *Myriophyllum* demonstrated that the level of plant-extractable metabolites was reduced from approximately 40% after 5 days to 3% after 21 days.[84] A recent study by Sens et al.[71] examined the incorporation of ^{14}C from ^{14}C-labeled TNT into the different cell wall fractions of wheat and beans. Most of the ^{14}C label in the cell wall was covalently bound in the lignin fraction and not the pectin, cellulose, or protein fractions. Gel permeation chromatography showed that the molecular weights of the TNT metabolites in lignin from wheat and beans are roughly between 1000 to 80,000 g/mol.

Formation of bound residues from nitrogen-containing aromatic compounds is well established,[46] and the incorporation of nitrogen-containing compounds in plant tissues is often faster than for nonnitrogen-containing aromatic compounds.[15,46] The formation process and conformation of such bound residues can affect the residue's bioavailability and toxicity.[1,46] For example, non-ruminant mammals cannot degrade lignin to release constituents of the polymer. Bakke et al.[1] found that sheep and rats almost exclusively excreted ^{14}C ingested as bound residues in corn and sorghum that had taken up ^{14}C-atrazine. Generally, bound residues are considered to be nontoxic or of lesser toxicity than parent compounds or intermediates. However, the toxicity of bound residues originating from nitroaromatic compounds is not understood.

10.2.6 Mass Balance on All Stages of Metabolism

A complete mass balance for all identified metabolites has not been performed for any plant species. A preliminary estimate (Table 10.4) indicates that after 12 days of exposure to ^{14}C-radiolabeled TNT, *M. aquaticum* concentrated most of the radiolabel as oxidation products and bound residues. Altogether, 94% of the label was recovered, but the chemical characteristics are unknown for 43% of the TNT added initially (the bound residues plus the 10% undetermined). As indicated by Sens et al.,[71] the chemical structure of the bound residues will be extremely difficult to determine.

Hairy roots tissue cultures convert TNT to a variety of uncharacterized products (Figures 10.3 and 10.4). A large fraction of the metabolic products are found in the polar solvent front during HPLC analyses. Preliminary results indicate that they are not oxidation products,[7] and they have yet to be identified.

Because TNT degradation leads primarily to bound residues, future toxicity studies should focus on the toxicity of plant biomass. Identifying all products and intermediates is unlikely because conjugates are poorly defined and quite possibly species specific. There is also limited utility in conducting toxicity tests on low levels of pure compounds when the concern will be the overall effect of the mixture of compounds. Certainly the identification of intermediates and products is crucial to understanding pathways and their regulation, but evaluating the toxicity of all intermediates is not a high priority in the application of phytoremediation systems.

10.3 DESIGN VARIABLES AND CONSIDERATIONS

Plants, both aquatic and terrestrial, can transform TNT (Tables 10.1 and 10.2), and several features of the transformation pathways may be similar (Figure 10.6). Aquatic and terrestrial plant systems will have distinct roles in phytoremediation due to differing interactions with contaminants, as illustrated in Figures 10.1 and 10.2. This section serves to highlight some of the phytoremediation design considerations for the two systems.

10.3.1 Aquatic Species

Conclusive evidence that plant-mediated processes can contribute to the disappearance of TNT was demonstrated with tissue cultures of the terrestrial plant *C. roseus*[42] and axenic cultures of the aquatic plant *M. aquaticum*. Under conditions of low TNT/plant biomass ratios, TNT disappears to below detection limits. For example, studies with hairy root tissue cultures showed that TNT does not accumulate in the plant, even after exposure of up to 80 mg of TNT per liter.[7] A number of laboratory studies with wetland plants have provided similar results.[56,59,63]

Tables 10.1 and 10.2 indicate that the ability to remove TNT seems to be ubiquitous across a wide range of plant species and tissue types. Enzyme studies support this conclusion, and TNT disappearance is observed in aquatic sediment-water systems.[83,87] Plant nitroreductases were initially suggested as a catalytic mechanism for the initial disappearance of TNT;[55,87] however, the enzymatic pathways have yet to be validated. More recent work has shown that ferredoxin NADP[+] reductase isolated from spinach leaves and other plant enzymes can also transform TNT.[32]

The disappearance of TNT is known to be dependent on plant biomass concentration.[59] Several groups have reported an initial bulk rate constant (**k**) for the disappearance of TNT in media using a pseudo-first-order dependence of extracellular concentration with time. This rate constant can be estimated by Equation 10.1.

$$-\frac{d[X]}{dt} = \mathbf{k}[P][X] \tag{10.1}$$

where **k** is the rate coefficient, $[P]$ is biomass concentration ($g_{fresh\ weight}$/mL), and $[X]$ is the extracellular concentration of TNT (mg/L). Rate coefficients are on the order of 1 mL/($g_{fresh\ weight}$/h).[59,86] Rate coefficients can be estimated quickly from experiments during which biomass concentration is relatively constant in fully grown plants or in mature tissue. A sample of rate coefficients, as determined in laboratory work by Pavlostathis et al.[59] for the removal of TNT by *M. spicatum,* is shown in Table 10.5.

The pseudo-first order assumption is appropriate at low TNT/biomass ratios,[54] but not at higher TNT concentrations. This concept is illustrated by experiments in which TNT is added at various concentrations to stationary phase (mature, nongrowing) hairy root cultures at a biomass concentration of 100 $g_{fresh\ weight}$/L (Figure 10.7). For initial levels of 50 mg/L or lower, TNT disappeared completely from the aqueous medium within 90 h.[54] At higher initial TNT concentrations, however, TNT remained in the medium. Because the physiological conditions of hairy root cultures and temperature were controlled in the study, results indicate that high initial TNT levels interfere with plant activity. Therefore, transformation is linked to metabolism and is an active plant process. The hypothesis is consistent with the green liver model and is also supported for other measures of plant metabolic activity. Lower TNT disappearance rates were observed in plants that experience light deprivation.[59] The rates became progressively lower as light deprivation continued. The same research group performed studies with a metabolic inhibitor. Two reactors containing *M. spicatum* plants were dosed with TNT, and removal rates were nearly identical. One reactor was inactivated with azide and did not remove TNT from solution, whereas

Table 10.5 Rate Constants for TNT Removal in Laboratory Wetlands Containing Eurasian Water Milfoil

Initial TNT Concentration (μM)	First Order Rate Constant (h^{-1})	Plant-Normalized Rate Constant (mL/g/h)
5.8 ± 0.1	0.142 ± 0.028	4.26 ± 0.84
22.7 ± 0.3	0.080 ± 0.016	2.40 ± 0.48
49.0 ± 0.5	0.065 ± 0.030	1.95 ± 0.90
127.4 ± 1.2	0.017 ± 0.005	0.51 ± 0.15
259.5 ± 2.4	0.013 ± 0.002	0.39 ± 0.06
500.4 ± 2.4	0.029 ± 0.013	0.87 ± 0.39

Source: Adapted from Pavlostathis, S.G. et al. 1998. *Environ. Toxicol. Chem.* 17:2266.

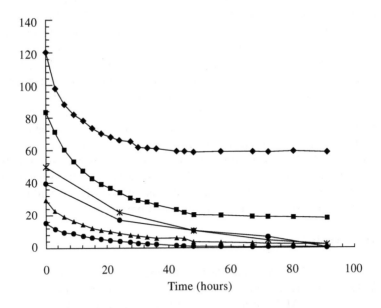

Figure 10.7 TNT disappearance in aqueous media and *C. roseus* hairy root cultures. TNT was added to stationary phase hairy root cultures. Biomass concentration was 100 $g_{fresh\ weight}$/L.[54]

TNT removal in the untreated reactor was relatively unchanged. It should be noted that the untreated reactors were not axenic. Any microbial activity would have also been terminated in the azide-amended reactor. Linking TNT removal and degradation to plant metabolism is an important finding. Proof that TNT removal is an active process in plants dispels hypotheses that TNT removal is a potentially reversible sorption process and provides substantial support for the green liver model.

Kinetic rate coefficients have been calculated for the removal of RDX and HMX from aqueous solutions containing wetland species. The explosives RDX and HMX are more recalcitrant than TNT to transformation by plants. In studies with *C. roseus* hairy roots and *Myriophyllum*, RDX disappeared at a much lower rate than TNT.[10] Most of the ^{14}C label accumulated intracellularly as RDX in *C. roseus* hairy roots (mass balances were not performed on *Myriophyllum*).[10] In contrast, *C. roseus* hairy roots removed HMX only at levels slightly higher than controls, and *Myriophyllum* did not remove HMX from solution.[10] Initial results indicate that plants are not able to transform RDX and HMX to products susceptible to conjugation and subsequent sequestration reactions.

Research completed at the U.S. Army Corps of Engineers Waterways Experiment Station (Vicksburg, MS) investigated the uptake and disappearance of TNT and RDX by submersed and emergent plant species. The initial tests were conducted as screening tests to evaluate plant species for use at the Iowa Army Ammunition Plant (IAAP) (Middletown, IA) and the Milan Army Ammunition Plant (MAAP) (Milan, TN).[2,3] These studies produced results of a qualitative nature, comparing the disappearance in reactors with a variety of plants. The qualitative findings do not ascertain the exact role of the plants in the disappearance of the RDX and TNT, but, while not definitive, results were consistent with other bodies of work. TNT did not accumulate in plant tissues, and 2- and 4-ADNT (low concentrations) were the only transformation products found in plant extracts. RDX was readily translocated, and accumulated in aerial plant tissues. Arrowhead aerial portions accumulated RDX up to 200 µg/g$_{dry\ weight}$.[3] The lack of mass balances makes it difficult to evaluate the extent to which RDX was transformed by any aquatic species. The results of the screening study[3] were used to select American pondweed, coontail, and common arrowhead for more intensive study[4] and possible use in future phytoremediation. A rather consistent finding in most aquatic studies listed in Tables 10.1 and 10.2 is that a major portion of TNT ended up in an unidentified form, which was usually in the plant tissues and partially nonextractable. Subsequent work investigating product identification and transformation mechanisms has led to insight regarding pathways beyond the initial ADNT products, as discussed previously.

10.3.2 Terrestrial Species

Uptake of TNT has been documented for many species of terrestrial plants. An initial study characterized the uptake of TNT for yellow nutsedge (*Cyperus esculentus*).[58] TNT and ADNT isomers were present in the plant, and 4-ADNT was the most prevalent compound. Formation of the 4- and 2-ADNT was proposed to occur within the plant. Transformation within the plant is consistent with findings stated above for aquatic species.

The fate of TNT in blando brome, bush bean, and wheat has been studied.[20,36] ^{14}C-TNT was used in both hydroponic and soil experiments. Following 60 days of exposure, bush bean accumulated up to 77% of the TNT-related label in root tissues and less than 20% of that fraction was recovered as TNT. The remainder consisted of 4-ADNT, 2-ADNT, a highly polar unknown, and nonextractable residues. About 15% of the added ^{14}C was recovered in the stem and identified as 4-ADNT, 2-ADNT, and a number of highly polar components — which upon acid hydrolysis yielded 4- and 2-ADNT. Accumulation in leaf tissues reached as high as 13% of the added [14]C. Overall, greater than 80% of ^{14}C was conjugated or transformed in the tissues of each plant species. Results indicate that some of the TNT transformation products could be transported as conjugates, which is consistent with the green liver model for plants and axenic cultures discussed earlier, including acid hydrolysis liberation of the aminodinitrotoluene isomers. Terrestrial and aquatic plants seem to have similar metabolic processes. Therefore, the same TNT transformation pathways (Figure 10.6) could be involved in both terrestrial and aquatic plant species.

Other studies that involved testing TNT with a variety of terrestrial plants have led to similar conclusions. Screening studies have consistently revealed that 2- and 4-ADNT along with low concentrations of TNT accumulate in root tissues.[33] Scheidemann et al.[65] tested 11 plant species at three different soil TNT concentrations. *Phaseolus vulgaris* was the only species to survive 500 mg of TNT per kilogram of soil, and high levels of nitroaromatic compounds were found in roots (460 µg/g$_{dry\ weight}$). When grown in soil contaminated to 100 mg/kg, *P. vulgaris* and *Triticum aestivum* contained the highest levels of TNT metabolites (91 and 98 $\mu g/g_{dry\ weight}$, respectively). Tests of rhizosphere degradation revealed that four of the six cultivars of *T. aestivum* tested significantly reduced the TNT concentration in the rhizosphere soil.[64] The results indicate that these species have a beneficial impact on TNT-contaminated soil; however, the mechanisms were not fully determined.

In evaluating uptake and degradation of TNT in hydroponic systems, Gorge et al.[33] determined that levels of nitrotoluenes were much higher in *Allium schoeno-prasum* than in *Medicago sativa* grown in hydroponic solutions containing 0.1 to 10 mg of TNT per liter. The majority of nitrotoluenes detected in the plant tissue was found in the root tissues, and the predominant nitrotoluenes were 2- and 4-ADNT with very little TNT remaining. Root concentrations were up to 20 and 30 times higher than shoot concentrations. From predictive relationships previously developed by Briggs et al.[13] and by Burken and Schnoor,[17] TNT should be more mobile than observed by Gorge et al.[33] and should reach the stem and leaves. The lack of mobility and the inability to isolate TNT from root tissues indicate that TNT is rapidly transformed in the root tissues. Transformation products appear to be a "gateway" into the formation of bound residues.

Thompson et al.[79] studied the uptake of ^{14}C-TNT by poplar trees (*Populus deltoides* × *nigra*, DN34) and found that uptake rates are dependent on the media used. TNT was taken up rapidly from hydroponic solutions by poplar hybrids, while sorption to soil retarded uptake. Aging of the contaminated soil further reduced the uptake rate of the ^{14}C-TNT (Figure 10.8). Uptake after approximately 20 days of aging was reduced by nearly 80%. Mass balances based on ^{14}C revealed that up to 75% of the ^{14}C label remained in root tissues with up to 10% eventually being translocated to the leaves. Transformation of TNT to 4-ADNT, 2-ADNT, and possibly 2,4-diamino-6-nitrotoluene occurred within the plant. Trees also transformed TNT to at least four additional unidentified polar metabolites (as indicated by their HPLC retention times). Only one of the detected polar products was translocated to leaf tissues. Polar metabolites were also detected and were subsequently identified in *Myriophyllum* degradation of TNT.[8] TNT metabolism in terrestrial species occurs in the root tissues, and transformation pathways are similar to those for aquatic species (Figure 10.6).

As expected, uptake and transport of RDX were very different from TNT in terrestrial plants.[36] Bush beans grown in soil containing 60 ppm RDX removed less than 16% of the RDX from the soil after 60 days, and uptake was greater than 60% from a hydroponic solution containing 10 ppm RDX after 7 days. After uptake, RDX and metabolites remained in plant tissues, with the majority remaining as RDX. Polar metabolite production increased with time. After 60 days, only 21 to 51% of the radiolabel in plant extracts remained as RDX, 8 to 30% existed as unknown polar compounds, and 20 to 50% was no longer extractable.[36] Rapid uptake from the hydroponic solutions resulted in lower levels of transformation. Aerial tissues of bush beans grown in 10 ppm hydroponic solution reached foliar RDX concentrations of approximately 97 ppm. These results indicate that some transformation does occur in the roots of bush beans.

Results of RDX uptake and transformation in and by hybrid poplars produced similar findings to those obtained with bush beans. Following uptake, RDX was

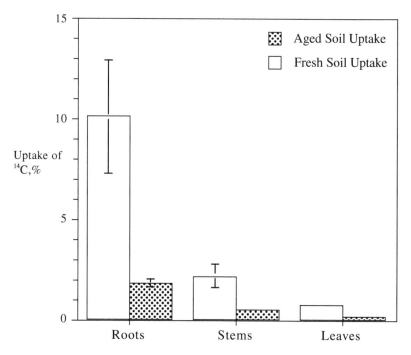

Figure 10.8   Soil spiked with [U-^{14}C]TNT was aged for 270 days. (n = 6 for "soil" and n = 2 for aged soil. Error bars indicate ±1 standard deviation.) The uptake after 22 days of exposure to aged TNT soil was significantly lower than that for trees exposed for 20 days to freshly spiked IAAP soil.[78]

readily translocated to the leaves, as indicated by the fact that 60% of the assimilated ^{14}C was found in the leaf tissues. No additional metabolites were detected, and only 15% of the ^{14}C was in a bound form.[81] This trend of bioaccumulation of RDX vs. the transformation and sequestration of TNT is depicted in Figure 10.9.

Accumulation of RDX has also been characterized in garden vegetables and agricultural crops,[50] where bioaccumulation was observed in corn, lettuce, tomato, and cyperus. Other work examined the effect of plant species.[61] Exposure to contaminated irrigation water alone did not cause bioaccumulation of RDX in radish roots, tomato fruit, or corn kernels, but it did result in accumulation of RDX in the leafy portions of lettuce, yellow nutsedge, and corn stover. Increasing soil RDX concentration increased the uptake for all species tested. Tomato fruit did not accumulate RDX to detectable levels when the concentration was less than 7.68 mg of RDX per kilogram of soil, and RDX was nondetectable in corn kernels when RDX concentration in the soil was less than 50.3 mg/kg. Soil type and total organic matter also affected uptake. Soils low in clay content contributed to higher uptake rates, while increased organic content of low-clay soils significantly decreased plant uptake rates.[61] Plant uptake of RDX by maize (*Zea mays*), soybean (*Glycien max*), sorghum/sudangrass (*Sorghum sudanese*), and wheat (*T. aestivum*) was similar to other crop species.[21] RDX concentrations in plant tissues increased linearly with increasing

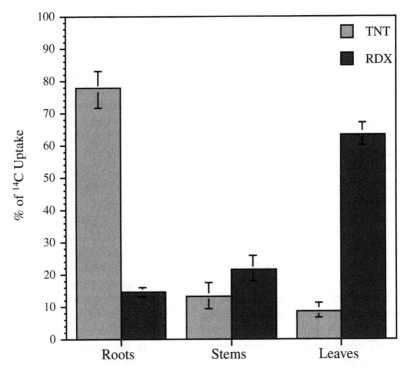

Figure 10.9 After the uptake of [U-^{14}C]TNT and [U-^{14}C]RDX from hydroponic solution (pH 7.0 and 20°C) for 2 days, the distribution of radiolabel between poplar tissues was in sharp contrast for the two compounds. The majority of the label remained in the roots for [U-^{14}C]TNT, whereas the leaves contained the bulk of [U-^{14}C]RDX-related label.[78] (n = 3 for each group and error bars represent ±1 standard deviation.)

RDX levels in hydroponic solution and soils. Furthermore, soil properties have a significant effect on plant uptake, limiting availability and transport to roots.

The consistent identification of RDX in the aboveground (edible) portions of the terrestrial plants is cause for concern in the application of phytoremediation. RDX in groundwater could potentially be translocated to and accumulate in aboveground tissues. Tissues could then be eaten by grazing animals, introducing RDX into foodchains. One interesting observation was decreasing mass balances for RDX over time in laboratory studies.[77] The use of ^{14}C compounds limits the possibilities to explain the decline in mass balance, which was not observed with TNT and has not been noted in earlier studies with atrazine (a triazine compound like RDX).[14,16,78] Because terrestrial species have limited pathways for compounds to exit the tissues, the formation of a volatile compound has been entertained as a possibility. The biodegradation of RDX by bacteria leads to a variety of small molecules that might be volatile (see Chapter 11). Similar reactions could take place in plant leaves by photolytic degradation. To date, this is only a hypothesis that has not been tested.

10.3.3 Toxicity

Toxicity reduction is obviously a concern in the successful implementation of phytoremediation. Toxicity of explosive compounds has been evaluated in screening studies and detailed laboratory experiments, and TNT can be toxic to plants.[27,69,75] The level of phytotoxicity depends upon several factors, including plant species, stage of growth, availability of TNT, and other environmental factors. *M. aquaticum* showed chlorosis, or yellowing of leaves, after exposure to TNT at concentrations of 5.9 μM (1.3 mg/L).[59] TNT was phytotoxic to the aquatic plant duckweed[20] above 1 mg/L (4.4 μM) and to yellow nutsedge[58] at 5 mg/L (22 μM). Growth and transpiration by hybrid poplar trees were significantly reduced at TNT concentrations greater than 5 mg/L.[80] Most of the TNT studies with hairy root cultures have been performed with TNT added to stationary phase cultures (cultures that are metabolically active but not undergoing growth). TNT, however, does inhibit growth of hairy root cultures. The growth rate of hairy roots decreases shortly after exposure to TNT amendment in the active growth phase (Figure 10.10). The decreased growth rate was similar at the TNT levels tested (20 to 40 mg/L). Interestingly, TNT disappearance rates (normalized to biomass concentration) in actively growing hairy root cultures were not appreciably different from those in metabolically active, stationary phase hairy roots (data not shown).[54] Germination and growth of tall fescue (*Festuca arundinacea Schreb.*) were affected by TNT concentrations greater than 30 mg/L.[59] In the same study, germination was not significantly affected by the maximum concentration of 4-ADNT tested (15 mg/L), although growth was reduced. Switchgrass and smooth bromegrass

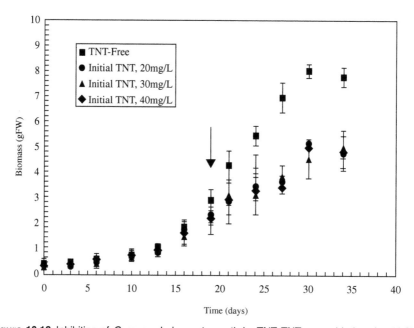

Figure 10.10 Inhibition of *C. roseus* hairy root growth by TNT. TNT was added at day 19.[54]

germination and growth were capable of withstanding soil solution concentrations less than 15 mg/L of TNT.[60]

Plants grown in soil have a higher threshold to TNT levels than aquatic or hydroponically grown plants. *Daturia innoxia* and *Lycopersicon peruvianum* were able to grow well in soils contaminated with 750 mg of TNT per kilogram, although both plants were negatively affected by concentrations in the 1000 mg/kg range. Scheidemann et al.[65] determined that *P. vulgaris* tolerated soil TNT concentrations of up to 500 mg/kg. The reason that toxicity is lower in soil systems than in hydroponic systems is most likely a decrease in the bioavailability of TNT in soil, which subsequently limits exposure of the plants to TNT. In examining toxicity in soil systems, the distribution coefficient (K_d) plays a key role in bioavailability and should be considered.

Based on the results described above, TNT is toxic to plants within the range of 1 to 30 mg/L in hydroponic systems. The mechanism of toxicity appears to be similar among the species tested. The exact mechanism of phytotoxicity of TNT is unknown, although it seems likely that the metabolic fate of TNT contributes to the overall effect on plant growth (detoxification through metabolism). Interestingly, germination and growth of wild-type seeds of tobacco were inhibited at 0.05 mM TNT (11 mg/L), but seeds from transgenic plants that expressed a pentaerthritol tetranitrate reductase (PETN) were able to germinate and grow.[31] Initial TNT concentration plays a role in toxicity, as transgenic plants expressing PETN could not grow in media containing 0.25 mM TNT (57 mg/L).[31] More research is needed on the metabolic fate of TNT in plants to elucidate the mechanisms of phytotoxicity.

RDX toxicity is understood even less than TNT toxicity. However, the limited number of studies to date indicates that RDX is less toxic. RDX toxicity to maize and wheat grown in hydroponic solutions was estimated to be 21 mg/L (the highest concentration tested) with the no observed adverse effect level (NOAEL) being 13 mg/L RDX ($p < 0.05$).[21] Sorghum or soybean plants showed no toxic effects for RDX concentrations up to 21 mg/L and after 30 days of exposure. Similarly, RDX at 21 mg/L (14 days) did not affect growth or transpiration rates in hybrid poplars.

10.3.4 Terrestrial Species Field Studies

Terrestrial phytoremediation systems have not been implemented at field scale to date. However, screening and sampling of numerous plant species growing on explosive-contaminated soils has yielded a body of knowledge for understanding TNT uptake in the natural environment. Two of the screening studies were performed on plants growing at the IAAP site and the Joliet Army Ammunition Plant site (JAAP, Joliet, IL). RDX accumulated in the tissues of many of the plants tested. Species that tested positive for RDX uptake included black locust (*Robinia pseudoacacia*), red cedar (*Juniperus virginiana*), smartweed, bromegrass (*Bromus inermis*), pigweed (*Amaranthus* sp.), reed canary grass (*Phalaris arundinacea*), Canadian goldenrod (*Solidago canadensis*), and ragweed (*Ambrosia artemislilfolia*).[66] Black locust growing in soil contaminated with 114 mg/kg RDX had 38.6 mg/kg in leaf tissues; red cedar growing in soil with 42 mg/kg RDX had 17 mg/kg in leaf tissues. Given that the K_d is 1.0 L/kg, one might expect higher concentrations in the leaves, but perhaps the lower concentrations are a result of photodegradation or some other unidentified mechanism. Only common ragweed (*A. artemislilfolia*) tested positive for TNT uptake or the presence of 2- or 4-ADNT metabolites. No leaf tissues samples contained detectable levels of TNT or known metabolites. Schneider et al.[66] determined the potential for field-level, foodchain contamination of plant tissue by TNT, 4-ADNT, 2-ADNT, 2,4-diamino-6-nitrotoluene, 2,6-diamino-6-nitrotoluene, and unidentified metabolites in carrots (*Daucus corota*), radishes (*Raphunus sativus*), kale (*Brassica oleracea*), lettuce (*Lactuca sativa*), and lamb lettuce (*Valerianella locusta*). Field sampling was consistent with laboratory studies with respect to TNT uptake and transformation, revealing that TNT and identifiable metabolites do not translocate from root tissues. Results of field and lab studies were partially in agreement for RDX translocation, although lower concentrations were observed in the field than predicted from earlier lab studies. Hypotheses for the lower concentrations in the field include plant metabolism occurring over longer periods in the field, photolytic pathways in plant tissues, or increased microbial activity (rhizodegradation) in the field setting.

10.4 CASE STUDY: IOWA ARMY AMMUNITION PLANT

The Iowa Army Ammunition Plant (IAAP), in Middletown, IA is a 20,000-acre site (31.25 mi^2) of which 8000 acres are used for both munitions production and testing. The IAAP is on the Comprehensive Environmental Response and Compensation Liability Act (CERCLA) National Priority List. From 1943 to 1975, the U.S. Army generated large volumes of explosives-laden wastewater at its Line 1 and Line 800 production facilities at the IAAP. In 1948, a concrete impoundment was constructed near Line 1 to contain explosives wastewater in a lagoon having an area of about 3.6 acres. The 5-acre Line 800 Pink Water Lagoon Area is surrounded by an earthen berm and at one time was also used for sludge disposal.

The lagoon basins and underlying groundwater have been contaminated with TNT and RDX. The improper disposal of wastes from detonator manufacturing has also left trace amounts of heavy metals such as lead and vanadium in the contaminated areas. The action or cleanup levels were driven by the explosives contamination. The soil cleanup levels were set at 47 mg/kg$_{soil}$ for TNT and 1.3 mg/kg$_{soil}$ for RDX. For RDX and TNT, the cleanup levels for groundwater and surface water at the site were set at 2 μg/L, the U.S. EPA health advisory level for drinking water.

Preliminary studies investigating the potential of phytoremediation and active remediation at the site were coordinated between the U.S. Army Corps of Engineers, Waterways Experiments Station, and The University of Iowa.[41,45] Areas of IAAP were too grossly contaminated to support plant life or any phytoremediation processes. Approximately 100,000 yd^3 of soil were excavated from the Line 1 (24,000 yd^3) and Line 800 (76,000 yd^3) areas in early 1997. Excavation was expected to diminish the threat to local groundwater supplies, but the U.S. Army Corps of Engineers anticipated residual contamination equal to or less than the action levels. As a safety margin, residual contamination would require additional treatment.

Studies were then conducted to assess the potential for using phytoremediation with poplars as a final polishing step for soil and with wetlands to treat residual contamination in surfacewater and groundwater.

At IAAP a full-scale wetlands treatment system is currently entering the third year of operation. The process of selecting a lagoon treatment approach, obtaining regulatory approval, and eventually construction were important factors in the type of phytoremediation used. A number of studies, including those discussed previously, were incorporated into the IAAP planning and investigation work.

The research of Thompson et al.[78–81] evaluated the potential use of poplar trees to remediate TNT and RDX contamination at the IAAP. This work determined the feasibility and impacts of terrestrial phytoremediation at the IAAP site. Goals of that work were to

- Evaluate the intrinsic ability of hybrid poplar trees to translocate and transform TNT and RDX from water and soil
- Determine the toxicity of RDX and TNT on the trees
- Determine how sorption to IAAP soil controls the mobility and bioavailability of TNT, RDX, and any transformation products

The above goals were achieved in laboratory studies. In summary of the results, the potential use of terrestrial vegetation was restricted by competition with soil sorption because it would take too long to reach cleanup goals (Figure 10.11). The

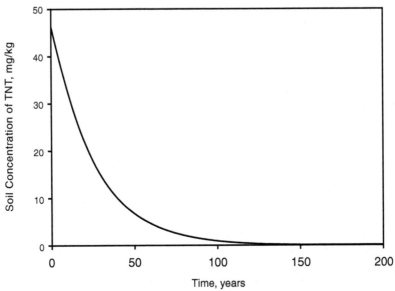

Contaminated area = 1 acre × 2 m deep = 8094 m³   Porewater volume = 0.3 * 8094 m³ = 2428 m³ =
Uniform contamination of 47 mg/kg TNT   TSCF = 0.46
Bulk density = 1.5 kg/L   Trees per acre = 2000
Porosity = 0.3   Transpiration per tree per year = 20 gal/d * 180 d
Total mass of TNT = 5.7 × 10⁷ mg TNT   Season = 13,626 L/yr (a liberal estimate)

Figure 10.11 Model of time required for phytoremediation of soil contaminated at an initial TNT concentration of 47 ppm, assuming that plant uptake is the removal mechanism.

translocation of RDX to leaf tissues, where it primarily remained as the parent compound, also limited the use of poplars.[81] Because of these reasons and the prevalence of saturated soils at the site, the use of poplars was not suggested.

The work presented by Larson et al. and Best et al.[3,4,50,51] was also related to investigating the potential of phytoremediation, but focused on aquatic-based systems. The goals of their work were to

- Quantify the ability of aquatic and wetland plants to treat TNT and RDX containing surface water or groundwater
- Investigate the transport and fate of the explosives in plants
- Determine the degradation byproducts and distribution in potential phytoremediation species and sediments

The research was successful at identifying plants with the ability to remove contaminants; however, the fates and transport and the metabolic pathways were not determined. Initial screening of wetland plants revealed plants with the capability of removing TNT and RDX from aqueous systems, as summarized above. Plants with the greatest potential for use in wetland systems were the submerged plants coontail and American pondweed and the emergent plants common arrowhead, reed canary grass, and fox sedge. The studies indicated that the prescribed action concentration of RDX (2 ppb) was not expected to be reached under complete-mixed conditions.[4]

In lab studies, TNT was not detected in plant tissues, and TNT degradation products were also below detection limits. However, related lab studies did detect both aminodinitrotoluene isomers from TNT.[3] RDX accumulated in a number of plants, reaching concentrations up to 1000 $\mu g/g_{dry\ weight}$.[3] Prior to the initial excavation, RDX was also detected in plant tissues from the IAAP site. RDX concentrations in arrowhead roots (7 $mg/kg_{dry\ weight}$) and in reed canary grass shoots (10 $mg/kg_{dry\ weight}$)[66] were cause for concern going into implementation at the IAAP.

Following excavation of Line 1 and Line 800 sites at the IAAP, the excavation pits were not backfilled. Pits were converted to constructed wetlands to phytoremediate the collected groundwater, to provide ecological enhancement to the area, and simply to avoid the costly process of backfilling the area from another location. Sediment was obtained from nearby Stump Lake to introduce indigenous species. Following the addition of sediment, the wetlands filled with infiltrating groundwater.

RDX from the soils contaminated the groundwater that filled the wetland and resulted in a rebound effect in the RDX concentration (Figure 10.12). The seasonal swings in RDX concentration were observed in the Line 1 and Line 800 impoundment wetlands. Water was not released from the wetland that contained RDX above the 2 ppb action level. The cause of the observed increase in RDX during the winter months has not been delineated at the site. Possible reasons for the increase include decreases in plant activity, biomass concentration, transformation rates, physical-chemical reaction rates, and photolytic reaction rates. Also, with less precipitation, concentrations in the infiltrating groundwater would most likely be increased. Residual soil and sediment concentrations of RDX are greater than 1.3 mg/kg of soil. This residual concentration, while not widespread, is capable of contaminating groundwater and surface water at greater than 2 $\mu g/L$, the U.S.

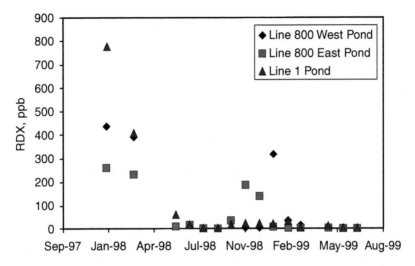

Figure 10.12 RDX concentrations in wetlands treating residual groundwater contamination at Line 1 and Line 800 at IAAP.

EPA Health Advisory Level and the action level at the IAAP. A similar seasonal trend was seen in work at MAAP.[74]

As stated above, bioaccumulation of RDX in plant tissues was a concern at the site. Sampling of the plant material took place in June 1999. Duplicate testing revealed that RDX and TNT were both below detection limits in the plant tissues tested.[45,53] Detection limits were determined to be lower than 1 mg/kg$_{dry\ weight}$. Plant species tested included water plantain, parrot feather, and coontail. Trinitrobenzene was detected in some plant tissues at <30 mg/kg$_{dry\ weight}$.

The results described previously provide insight to the application of the overall process and mechanisms involved. For example, RDX did not accumulate in plant tissues as expected. Simple modeling based upon laboratory research indicates that RDX should have been detectable in plant tissues. The plant concentrations measured in lab studies ranged to 200,000 μg RDX per kilogram$_{dry\ weight}$.[3] The lack of detection in the current wetlands and the relatively low levels detected in tissues (<10 mg/kg$_{dry\ weight}$) prior to excavation of high RDX concentrations[66] suggest that additional mechanisms were acting to remove RDX from plant tissues or to degrade it before uptake. One distinct difference between the field and lab studies is the light exposure for the plants. Sunlight might cause photolytic reactions in the leaves, and the photolytic reactions may result in RDX degradation to a volatile metabolite.

The IAAP treatment wetlands are providing effective treatment for both TNT and RDX. Further treatment of the surface water and groundwater at the site is planned. Brush Creek is to be diverted to the Line 1 wetland in 2000. Brush Creek flows through the Line 1 site and the residual soil contamination and commonly contains RDX at concentrations near 10 ppb. Groundwater is possibly going to be extracted and treated in the Line 800 wetland. The groundwater in question contains RDX at concentrations near 20 ppb. Integrating other treatment systems is also under consideration, potentially including a permeable iron treatment system or other phytoremediation applications. Terrestrial systems utilizing upland poplars or willows to control groundwater movement are being considered by the U.S. Army Corps of Engineers for certain applications. Plant and aqueous concentrations of the nitroaromatic compounds and metabolites are to undergo continued monitoring.

10.5 SUMMARY

Phytoremediation of explosives and nitroaromatic compounds is a promising technology; however, limitations and concerns exist. Phytoremediation systems are subject to toxicity at highly contaminated sites. Toxicity varies among plant species, is dependent on soil type, and is affected by other contaminants that may be present. Plants can also accumulate explosives such as RDX and metabolites of TNT and RDX that are yet unidentified. Many of the unidentified compounds are thought to be bound residues. Bound residues of other compounds, including nitroaromatic pesticides, are nontoxic, but little is known concerning bound residues resulting from RDX, TNT, HMX, or related compounds. Terrestrial systems also appear to be constrained by the bioavailability of the compounds in question. The use of terrestrial plants appears to be limited to stabilization of residual contamination and would prove to be prohibitively slow when active removal and degradation are the goals in cleaning a highly contaminated site. Wetlands systems have shown potential in laboratory-, pilot-, and full-scale systems. Questions still exist regarding the final fate of the contaminants and the long-term ecological impacts of potential metabolites. Interestingly, laboratory studies have identified oxidized compounds that may be precursors to ring cleavage.

A commonly overlooked benefit of phytoremediation systems is the concurrent ecological benefits offered. Utilizing natural systems such as phytoremediation can speed the reestablishment of biological systems and return contaminated sites to a more natural habitat. This "holistic" approach has not been considered in the comparisons of technologies to date. Many other remediation approaches, such as excavation/incineration, thermal treatments, or capping, are devastating to existing ecosystems. These technologies can strip the soil of any living organisms and destroy soil organic matter and nutrients in order to remove a contaminant at parts per million or parts per billion levels. The long-term goal of remediation is generally to remove potential health hazards and improve the ecological health of the site, not devastate or remove it. The potential ecological improvement, simplicity in concept, and aesthetic advantages are responsible, in part, for the strong public acceptance of phytoremediation applications. Such public acceptance is lacking for many other technologies.

Given the limitations and benefits of phytoremediation applications, there appears to be a place in the remediation sector for this technology. Treatment of mildly contaminated sites and of residual soil and water contamination appears to be a niche that is well suited to phytoremediation and phytotreatment technologies. Questions still remain that must be answered through research if broad-based acceptance and application are to develop. Critical issues for the future are presented in the following section.

10.6 FUTURE RESEARCH

Knowledge of plant metabolism of nitroaromatic compounds is important for natural attenuation processes and phytoremediation applications. Currently, implementation of phytoremediation to treat RDX and TNT in full-scale systems is ahead of lab studies and the scientific knowledge base. The results observed at IAAP and MAAP in full- and pilot-scale systems were not altogether predicted by laboratory studies, even though the same species, sediments, and waters were used. An evaluation of the literature suggests several specific areas where additional research is needed. Future laboratory studies should examine the role of intermediate hydroxylamines and unknown polar products, determine the mechanisms of phytotoxicity, perform complete mass balances on plant systems, examine more plant systems, and provide insight about integrated systems at the field scale. Generally, plant processes for other explosives, RDX and HMX, and other nitroaromatic compounds are less characterized or not characterized at all. Such compounds need additional research because of their high solubility.

In full-scale systems the combined effects of photolysis, plant, microbial, and sediment interaction appear to remove explosives at rates in excess of what existing laboratory studies predict. The fate of the contaminants is also more favorable than predicted, with little or no RDX accumulating in plant tissues. *In situ* studies would allow better understanding and characterization of the integrated natural systems such as wetlands. Such understanding will allow improved engineering and utilization of plants to treat nitroaromatic and explosive compounds. In the end, plants do not act as axenic cultures in the applied natural systems. Certainly understanding each mechanism in the whole system has scientific merit, but may not improve the application or optimize the design.

Characterization of phytocatalytic pathways, (i.e., identification of intermediates, reactions, enzymes, and genes) enables quantitative modeling of the fate and disposition of nitroaromatic compounds. It may speed the development of *in situ* analytical tools for the realistic assessment of nitroaromatic fate in the environment. Analysis of pathway structure may be the route to assessing intrinsic kinetics of the fate of TNT and the bioavailability of metabolic products. Plant metabolic engineering could be a novel way to lead to optimized plant traits in the remediation of TNT and related compounds.[39] Metabolic engineering may lead to enhanced growth of plant species in the presence of contaminant, enhanced rates of transformation, and reduction of recalcitrant products. The work described by Williams and Bruce in Chapter 7 is an excellent example of early metabolic engineering. Identification and/or engineering enzymes with enhanced catalytic capabilities will be central to plant metabolic engineering, as well as to opening up avenues in the future for plant catabolism of nitroaromatic compounds. In plant metabolic engineering, the indigenous plant transformation pathways will compete with engineered ones, and thus quantitative understanding of indigenous pathways will aid in design of trangenic phytoremediation systems.

Continued efforts to attain complete confirmation of degradation byproducts have recently led to elucidation of a number of TNT metabolites and insight to the mechanisms that produce them. Future research also needs to address toxicity concerns related to bound residues and unidentified metabolic products that may accumulate in plant tissues. Regulatory agencies require such information if phytoremediation is to become widely used. The biological endpoint of contaminants is of great concern and should be considered. For example, investigating the toxicity and biological effect of the plant tissue as a whole may be advantageous (e.g., toxicity of leaf tissues to worms). Identifying all products may prove to be futile and would still require toxicologic studies for each identified product. Research evaluating and determining ecological impacts of explosives phytoremediation could result in the proliferation of plant-assisted remediation.

ACKNOWLEDGMENTS

The authors extend thanks to J. Hughes, R. Bhadra, M. Barman, J. Lauritzen, M. Stone, M. Vanderford, D. Wayment, R. Williams, E. Best, J. Miller, and D. Werner. Our appreciation to contributors from the U.S. Army Corps of Engineers, Kevin Howe, Don Moses, and Jackson Kiker for their large contribution to this work, and to Craig Just and M. Subramanian for their contributions to this chapter and critical reviews.

This material is based, in part, upon work supported by DSWA Project No. 01–97–1-0020, by the Texas Advanced Technology Program under Grant No. 003604-045, by NSF Young Investigator Award BES-9257938 to J.V.S., and USEPA Athens Research Laboratory under Grant 8L-0644-NTEX.

REFERENCES

1. Bakke, J.E., R.H. Shimabukuro, K.L. Davison, and G.L. Lamoureux. 1972. Sheep and rat metabolism of the insoluble ^{14}C residues present in ^{14}C treated sorghum. *Chemosphere*. 1:21–24.
2. Best, E.H.P., S.L. Sprecher, H.L. Fredrickson, M.E. Zappi, and S.L. Larson. 1997. Screening submersed plant species for phytoremediation of explosives contaminated groundwater from the Milan army ammunition plant, Milan, Tennessee. Technical Report EL-97-24. U.S. Army Engineer Waterways Experiment Station, Vicksburg, MS.
3. Best, E.H.P., M.E. Zappi, H.L. Fredrickson, S.L. Sprecher, S.L. Larson, and M. Ochman. 1997. Screening of aquatic and wetland plant species for the phytoremediation of explosives-contaminated groundwater from the Iowa Army Ammunition *Plant. Ann. N.Y. Acad. Sci.* 829:179–194.
4. Best, E.H.P., J.L. Miller, H.L. Fredrickson, S.L. Larson, M.E. Zappi, and T.H. Streckfuss. 1998. Explosives removal from groundwater of the Iowa Army Ammunition Plant in continuous-flow laboratory systems planted with aquatic and wetland plants. Technical Report EL-98–13. U.S. Army Engineer Waterways Experiment Station, Omaha, NE.

 5. Best, E.H.P., S.L. Sprecher, S.L. Larson, H.L. Fredrickson, and D.F. Bader. 1999. Environmental behavior and fate of explosives from groundwater from the Milan Army Ammunition Plant in aquatic and wetland plant treatments. Mass balances of TNT and RDX. *Chemosphere.* 38:3383–3396.
 6. Bhadra, R. and J.V. Shanks. 1997. Transient studies of nutrient uptake, growth, and indole alkaloid accumulation in heterotrophic cultures of hairy roots of *Catharanthus roseus. Biotechnol. Bioeng.* 55:527–534.
 7. Bhadra, R., D. Wayment, J.B. Hughes, and J.V. Shanks. 1999. Confirmation of conjugation processes during TNT metabolism by axenic plant roots. *Environ. Sci. Technol.* 33:446–452.
 8. Bhadra, R., D. Wayment, J.B. Hughes, and J.V. Shanks. 1999. Characterization of oxidation products of TNT metabolism in aquatic phytoremediation systems of *Myriophyllum aquaticum. Environ. Sci. Technol.* 33:3354–3361.
 9. Bhadra, R., D. Wayment, J.B. Hughes, and J.V. Shanks. 1999. Conjugate formation during the metabolism of 2,4,6-trinitrotoluene in plant roots. p 121–125. In A. Leeson and B.C. Alleman (eds.), *Phytoremediation and Innovation Strategies for Specialized Remedial Applications.* Batelle Press, Columbus, OH.
10. Bhadra, R., R. Williams, S. Barman, M.B. Stone, J.B. Hughes, and J.V. Shanks. 2000. Fate of RDX and HMX in axenic plant roots. *Chemosphere.* In press.
11. Bingham, S.W., R.L. Shaver, and C.L. Guyton. 1980. Peanut uptake and metabolism of ^{14}C oxadiazon from soil. *J. Agric. Food Chem.* 28:735–740.
12. Bockers, M., C. Rivero, B. Thielde, T. Janowski, and B. Schmidt. 1994. Translocation and metabolism of 3,4-dichloroaniline in soybean and wheat plants. *J. Bioscience.* 49:719–726.
13. Briggs, G.G., R.H. Bromilow, and A.A. Evans. 1982. Relationships between lipophicity and root uptake and translocation of non-ionized chemicals by barley. *Pestic. Sci.* 13:495–504.
14. Burken, J.G. 1993. Masters thesis. University of Iowa, Iowa City.
15. Burken, J.G. 1996. Ph.D. thesis. University of Iowa, Iowa City.
16. Burken, J.G. and J.L. Schnoor. 1997. Uptake and metabolism of atrazine by hybrid poplar trees. *Environ. Sci. Technol.* 31:1399–1406.
17. Burken, J.G. and J.L. Schnoor. 1998. Predictive relationships for uptake of organic contaminants by hybrid poplar trees. *Environ. Sci. Technol.* 32:3379–3385.
18. Burken, J.G. and J.L. Schnoor. 1999. Distribution and volatilization of organic contaminants following uptake by hybrid poplar trees. *Int. J. Phytoremed.* 1:139–152.
19. Carman, E., T. Crossman, and E. Gatliff. 1998. Phytoremediation of no. 2 fuel-oil contaminated soil. *J. Soil Contam.* 7:455–466.
20. Cataldo, D.A., S. Harvey, R.J. Fellows, R.M. Bean, and B.D. McVeety. 1989. An evaluation of the environmental fate and behavior of munitions material (TNT, RDX) in soil and plant systems: TNT. U.S. DOE Contract 90-012748, Pacific N.W. Laboratories, Richland, WA.
21. Chen, D. 1993. M.S. thesis. Univerisity of Illinois, Urbana-Champaign.
22. Compton, H.R., D.M. Harosi, S.R. Hirsch, and J.G. Wrobel. 1998. Pilot-scale use of trees to address voc contamination. p. 245. In G.B. Wickramanayake and R.E. Hinchee (eds.), *Bioremediation and Phytoremediation, Chlorinated and Recalcitrant Compounds.* Battelle Press, Columbus, OH.
23. Cole, D.J. 1983. Oxidation of xenobiotics in plants. p. 199–254. In D.H. Hutson and T.R. Roberts (eds.), *Progress in Pesticide Chemistry and Toxicology.* John Wiley & Sons, New York.

24. Cole, D.J. 1994. Detoxification and activation of agrochemicals in plants. *Pest. Sci.* 42:209–222.
25. Coleman, J.O.D., M.M.A. Blake-Kalff, and T.G.E. Davies. 1997. Detoxification of xenobiotics by plants: chemical modification and vacuolar compartmentation. *Trends Plant Sci.* 2:144–151.
26. Cunningham, S.D., T.A. Anderson, A. Paul Schwab, and F.C. Hsu. 1996. Phytoremediation of soils contaminated with organic pollutants. p. 55–113. In D.L. Sparks (ed.), *Advances in Agronomy.* Vol. 56. Academic Press, New York.
27. Dacre, J. and D.H. Rosenblatt. 1974. Mammalian toxicology and toxicity to aquatic organisms of four important types of waterborne munitions pollutants — an extensive literature evaluation. AD-A778725. TR-7403. U.S. Army Medical Bioengineering Research and Development Laboratory, Fort Detrick, Fredrick, MD.
28. Doran, P.M. (ed.). 1997. *Hairy Roots: Culture and Applications.* Harwood Academic Publishers, Amsterdam.
29. EPA, U.S. 1999. Introduction to Phytoremediation. EPA/600/R-99/107. U.S. EPA, Cincinnati, OH.
30. EPA, U.S. 1999. Phytoremediation Resource Guide. EPA/542/B-99/003. U.S. EPA, Cincinnati, OH.
31. French, C.E., S.J. Rosser, G.J. Davies, S. Nicklin, and N.C. Bruce. 1999. Biodegradation of explosives by transgenic plants expressing pentaerythritol tetranitrate reductase. *Nature Biotechnol.* 17:491–494.
32. Goheen, S.C., J.A. Campbell, S.K. Roach, Y. Shi, and M.M. Shah. 1999. Degradation products after digestion of TNT using ferredoxin NADP$^+$ reductase. Second International Symposium on Biodegradation of Nitroaromatic Compounds and Explosives 1999. Leesburg, VA.
33. Gorge, E., S. Brandt, and D. Werner. 1994. Uptake and metabolism of 2,4,6 TNT in higher plants. *Environ. Sci. Pollut. Res.* 1:229–233.
34. Gorontzy T., O. Drzyzga, M.W. Kahl, D. Bruns-Nagel, J. Breitung, E. von Loew, and K.H. Blotevogel. 1994. Microbial degradation of explosives and related compounds. *Crit. Rev. Microbiol.* 20:265–284.
35. Harvey, S.D., R.J. Fellows, D.A. Cataldo, and R.M. Bean. 1990. Analysis of 2,4,6-trinitrotoluene and its transformation products in soils and plant tissues by high performance liquid chromatography. *J. Chromatogr.* 518:361–374.
36. Harvey, S.D., R.J. Fellows, D.A. Cataldo, and R.M. Bean. 1991. Fate of the explosive hexahydro-1,3,5-trinitro-1,3,5-triazine (RDX) in soil and bioaccumulation in bush bean hydroponic plants. *Environ. Toxicol. Chem.* 10:845–855.
37. Hatzios, K.K. and D. Penner (eds.) 1982. *Metabolism of Herbicides in Higher Plants.* CRC Press, Boca Raton, FL.
38. Hatzios, K.K. 1991. Biotransformation of herbicides in higher plants. p. 141–185. In R. Grover and A.J. Cessna (eds.), *Environmental Chemistry of Herbicides.* CRC Press, Boca Raton, FL.
39. Hooker, B.S. and R.S. Skeen. 1999. Transgenic phytoremediation blasts onto the scene. *Nature Biotechnol.* 17:428.
40. Howe, K., Moses, D., and Kiker, J. 1999. U.S. Army Corps of Engineers. Personal communications.
41. Hsu, F.C., R.L. Marxmiller, and A.Y. Yang. 1990. Study of root uptake and xylem translocation of cinamethylin and related compounds in detopped soybean roots using a pressure chamber technique. *Plant Physiol.* 93:1573–1578.

42. Hughes, J.B., J.V. Shanks, M. Vanderford, J. Lauritzen, and R. Bhadra. 1997. Transformation of TNT by aquatic plants and plant tissue cultures. *Environ. Sci. Technol.* 31:266–271.

43. Hughes, J.B. 1999. Unpublished data.

44. Ishizuka, K., H. Hirata, and K. Fukunaga. 1975. Absorption, translocation, and metabolism of oxadiazon in rice plants. *Agric. Biol. Chem.* 39:1431–1446.

45. Just, C.L. 1999. University of Iowa. Personal communications.

46. Klein, W. and I. Scheunert. 1982. Bound pesticide residues in soil. plants and food with particular emphasis on the application for nuclear techniques. IEIA-SM-263/38. International Atomic Energy Agency.

47. Komoβa, D., C. Langerbartels, and H. Sandermann. 1995. Metabolic processes for organic chemicals in plants. p. 69–103. In S. Trapp, S. and J. McFarland (ed.), *Plant Contamination — Modeling and Simulation of Organic Chemical Processes.* CRC Press, Boca Raton, FL.

48. Lamoureux, G.L., R.H. Shimabukuro, H.R. Swanson, and D.S. Frear. 1970. Metabolism of atrazine in excised sorghum leaf sections. *J. Agric. Food Chem.* 18:81–87.

49. Lamoureaux, G.L. and D.G. Rusness. 1986. Xenobiotic conjugation in higher plants. p. 62. In G.D. Paulson (ed.), *Xenobiotic Conjugation Chemistry.* American Chemical Society, Washington, D.C.

50. Larson, S.L. 1997. Fate of explosive contaminants in plants. *Ann. N.Y. Acad. Sci.* 829:195–201.

51. Larson, S.L., C.A. Weiss, B.L. Escalon, and D. Parker. 1999. Increased extraction efficiency of acetonitrile/water mixtures for explosives determination in plant tissues. *Chemosphere.* 38:2153–2162.

52. Larson, S.L., R.P. Jones, B.L. Escalon, and D. Parker. 1999. Classification of explosives transformation products in plant tissue. *Environ. Toxicol. Chem.* 18:1270–1276.

53. Larson, S.L. 1999. Waterways Experiments Station. Personal communication.

54. Lauritzen, J. 1998. M.S. thesis. Rice University, Houston.

55. Medina, V.F. and S.C. McCutcheon. 1996. Phytoremediation: modeling of TNT and its breakdown products. *Remediation.* 6:31–45.

56. Medina, V.F., R. Rivera, S. Larson, and S.C. McCutcheon. 1996. Phytoremediation: modeling of TNT and its breakdown products. *Soil Groundwater Cleanup.* Feb/Mar:19–24.

57. Montgomery, M.L. and V.H. Freed. 1964. Metabolism of triazine herbicides by plants. *J. Agric. Food Chem.* 12:11–14.

58. Palazzo, A.J. and D.C. Leggett. 1986. Effect and disposition of TNT in terrestrial plants. *J. Environ. Qual.* 15:49–52.

59. Pavlostathis, S.G., K.G. Comstock, M.E. Jacobson, and F.M. Saunders. 1998. Transformation of 2,4,6-trinitrotoluene by the aquatic plant *Myriophyllum spicatum. Environ. Toxicol. Chem.* 17:2266–2273.

60. Peterson, M., G. Horst, P. Shea, and S. Comfort. 1998. Germination and seedling development of switchgrass and smooth bromegrass exposed to 2,4,6-trinitrotoluene. *Environ. Pollut.* 99:53–59.

61. Price, R.A., J.C. Pennington, S.L. Larson, D. Naumann, and C.A. Hayes. 1997. Plant uptake of explosives from contaminated soil and irrigation water at the former Nebraska Ordanance Plant, Mead, Nebraska. Technical Report EL-97–11. U.S. Army Engineer Waterways Experiment Station, Kansas City, MO.

62. Rieger, P.G. and H.J. Knackmuss. 1995. Basic knowledge and perspectives on biodegradation of 2,4,6,-trinitrotoluene and related nitroaromatic compounds. p. 1–18. In J.C. Spain (ed.), *Biodegradation of Nitroaromatic Compounds.* Plenum Press, New York.

63. Rivera R., V.F. Medina, S.L. Larson, and S.C. McCutcheon. 1998. Phytotreatment of TNT-contaminated groundwater. *J. Soil Contam.* 7:511–529.

64. Sandermann, J.H. 1994. Higher plant metabolism of xenobiotics: the 'green liver' concept. *Pharmacogenetics.* 4:225–241.

65. Scheidemann, P., A. Klunk, C. Sens, and D.J. Werner. 1998. Species dependent uptake and tolerance of nitroaromatic compounds by higher plants. *Plant Physiol.* 152:242–247.

66. Schneider, K.J., J. Oltmanns, T. Radenberg, T. Schneider, and D. Pauly-Mundegar. 1996. Uptake of nitroaromatic compounds in plants. Implications for risk assessment of ammunition sites. *Environ. Sci. Pollut. Res. Int.* 3:135–138.

67. Schnoor, J.L., L.A. Licht, S.C. McCutcheon, N.L. Wolfe, and L.H. Carreira. 1995. Phytoremediation of organic and nutrient contaminants. *Environ. Sci. Technol.* 29:318–323.

68. Schnoor, J.L. 1997. Phytoremediation. Technology Evaluation Report TE-97-01. National Environmental Technology Applications Center, Pittsburgh, PA.

69. Schott, C.D. and E.G. Worthley. 1974. The toxicity of TNT and related wastes to an aquatic flowering plant: *Lemna perpusilla Torr.* Edgewood Arsenal. Technical Report EB-TR-74016. Edgewood Arsenal, Aberdeen Proving Ground, MD.

70. Schwab, A.P. and M.K. Banks. 1994. Biologically mediated dissipation of polyaromatic hydrocarbons in the root zone. In T.A. Anderson and J.R. Coats (eds.), *Bioremediation through Rhizosphere Technology.* American Chemical Society, Washington, D.C.

71. Sens, C., P. Scheidemann, and D. Werner. 1999. The distribution of ^{14}C TNT in different biochemical compartments of the monocotyledenous *Triticum aestivum.* *Environ. Pollut.* 104:113–119.

72. Shanks, J.V. and J. Morgan. 1999. Plant hairy root culture. *Curr. Opin. Biotechnol.* 10:151–155.

73. Shimabukuro, R.H. and H.R. Swanson. 1969. Atrazine metabolism, selectivity, and mode of action. *J. Agric. Food Chem.* 17:199–205.

74. Sikora, F.J., L.L. Behrends, W.D. Phillips, H.S. Coonrod, E. Bailey, and D.F. Bader. 1997. A microcosm study on remediation of explosives-contaminated groundwater using constructed wetlands. In R.K. Bajpai and M.E. Zappi (eds.), *Bioremediation of Surface and Subsurface Contamination.* New York Academy of Sciences, New York.

75. Smock, L.A., D.L. Stoneburger, and J.R. Clark. 1976. The toxic effects of trinitrotoluene (TNT) and its primary degradation products on two species of algae and the fathead minnow. *Water Res.* 10:537–543.

76. Spain, J. C. 1995. Biodegradation of nitroaromatic compounds. *Annu. Rev. Microbiol.* 49:523–555.

77. Spain, J. C. 1995. Biodegradation of nitroaromatic compounds under aerobic conditions, p. 19–36. In J.C. Spain (ed.), *Biodegradation of Nitroaromatic Compounds.* Plenum Press, New York.

78. Thompson, P.L. 1997. Ph.D. thesis. University of Iowa, Iowa City.

79. Thompson, P.L., L. Ramer, and J.L. Schnoor. 1998. Uptake and transformation of TNT by hybrid poplar trees. *Environ. Sci. Technol.* 32:975–980.

80. Thompson, P.L., L. Ramer, A. Guffey, and J.L. Schnoor. 1998. Decreased transpiration in poplar tress exposed to 2,4,6-trinitrotoluene. *Environ. Toxicol. Chem.* 17:902–906.

81. Thompson, P.L., L. Ramer, and J.L. Schnoor. 1999. Hexahydro-1,3,5-trinitro-1,3,5-triazine translocation in poplar trees. *Environ. Toxicol. Chem.* 18:279–284.

82. Trapp, S. 1995. Model for uptake of xenobiotics into plants. p. 107–152 In S. Trapp and J.C. McFarlane (eds.), *Plant Contamination: Modeling and Simulation of Organic Chemical Processes.* CRC Press, Boca Raton, FL.

83. van Beelen P. and D.R. Burris. 1995. Reduction of the explosive 2,4,6-trinitrotoluene by enzymes from aquatic sediments. *Environ. Toxicol. Chem.* 14:2115–2123.

84. Vanderford, M. 1996. M.S. thesis. Rice University, Houston.

85. Vanderford, M., J.V. Shanks, and J.B. Hughes. 1997. Phytotransformation of trinitro-toluene (TNT) and distribution of metabolic products in *Myriophyllum aquaticum.* *Biotechnol. Lett.* 19:277–280.

86. Wayment, D.G., R. Bhadra, J. Lauritzen, J. Hughes, and J.V. Shanks. 1999. A transient study of formation of conjugates during TNT metabolism by plant tissues. *Int. J. Phytoremed.* 1:227–239.

87. Wolfe, N.L., T.Y. Ou, L. Carriera, and D. Gunnison. 1994. Alternative methods for biological destruction of TNT: a preliminary feasibility assessment of enzymatic degradation. Technical Report IRRP-94–3. U. S. Army Corps of Engineers Waterways Experiment Station, Vicksburg, MS.

CHAPTER 11

Biodegradation of RDX and HMX: From Basic Research to Field Application

Jalal Hawari

CONTENTS

11.1 INTRODUCTION

Of the nearly 20 different energetic compounds used in conventional munitions by the military today, hexahydro-1,3,5-trinitro-1,3,5-triazine (RDX) and octahydro-1,3,5,7-tetranitro-1,3,5,7- tetrazocine (HMX) (Figure 11.1) are the most powerful and commonly used. RDX and HMX are acronyms that stand for Royal Demolition Explosive and High Melting Explosive, respectively. The release of these chemicals, which has resulted in severe soil and groundwater contamination during most of the 20th century, has been associated with various commercial and military activities including manufacturing, waste discharge, testing and training, ordnance demilitarization, and open burning/open detonation (OB/OD).[35,66,99] It has been estimated that during RDX manufacturing up to 12 mg/L can be discharged into the environment in process wastewaters.[48] In general, data on the extent of contamination by energetic chemicals are scarce.[23,24] Some military agencies either have no data on the environmental fate and impact of energetic chemicals or such data are kept classified. In Canada, environmental problems associated with the manufacturing of energetic chemicals and their use in training is currently under extensive investigation by the Department of National Defence and private manufacturers. There are two main producers (ICI and Expro) of energetic chemicals in Canada, but there are no accessible data on the level and extent of contamination. An estimated 103 training (air and land force bases) and three OB/OD sites are listed by the Canadian Department of National Defence as possible sources of environmental contamination by RDX, HMX, 2,4,6-trinitrotoluene (TNT), and other chemicals.[98] Recent characterization studies on such sites by Jenkins et al.[49] and Thiboutot et al.[98] revealed that the levels of contamination varied drastically. At some sites the level of contamination did not exceed 10 mg/kg for any of the three explosives. At other antitank firing ranges, where a melt-cast explosive (30% HMX and 70% TNT) was used, HMX concentrations ranged from 1640 mg/kg near one target to 2.1 mg/kg at a distance of 15 m from the target.[49,98]

In the past, energetic chemicals received much less attention in the environmental community than petroleum products and other industrial chemicals. This is despite the fact that environmental contamination by explosives is a worldwide problem as a result of military activities beginning in World War I.[2] Recently there has been a buildup of public and governmental interest in the environmental behavior of explosives. Several reviews have been published that address the physicochemical properties,[77,96] biodegradation,[30,52] and toxicity[77] of explosives such as RDX and HMX. In

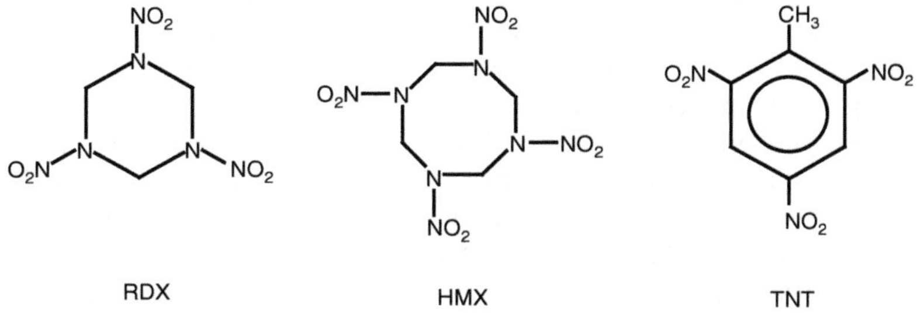

Figure 11.1 Structures of RDX, HMX, and 2,4,6-trinitrotoluene (TNT).

general, nitroamines are toxic[37,75,96,102,105] and are considered as class C possible carcinogens. They also have adverse effects on the central nervous system in mammals. The toxicity and persistence of nitroamines necessitates that their fate in the environment be understood and that contaminated soil and groundwater be remediated. Bioremediation has always been desirable because other physicochemical techniques such as incineration (OB/OD),[28,66] adsorption by granular activated carbon followed by alkaline hydrolysis,[42-44] and advanced photooxidation (UV/O₃) are either costly (up to $800/ton for the incineration) or hazardous due to the emission of toxic byproducts. Recently, we tested a flooded soil biopile[31] and a pilot-scale bioreactor[34] for the treatment of RDX and HMX. As far as we are aware, no *in situ* bioremediation technologies of the two energetic chemicals have been used in the field.

Whereas there is extensive information concerning the biodegradation and metabolic pathways of polynitroaromatics such as TNT,[26,39,59,72,90] there is little information regarding the nonaromatic cyclic nitroamines. This chapter will attempt to describe cyclic nitroamine biodegradation under both aerobic and anaerobic conditions, discuss their potential breakdown products and metabolic pathways, and explore the current state of bioremediation technologies. This chapter will also identify the potential problems that will be encountered in the extrapolation of laboratory research to design field applications.

11.2 BIODEGRADATION PATHWAYS

11.2.1 Chemical Properties and Decomposition Patterns

Several chemical studies[11,12,65,81,100,109] have shown that once the nonaromatic cyclic nitramine RDX or HMX undergoes a change in its molecular structure, the ring collapses to produce small nitrogen-containing (N_2O, NO_2, NH_3) and small carbon-containing (HCHO, HCOOH, and CO_2) products. The behavior of RDX and HMX is distinguishable from that of the aromatic compound, TNT, which is (bio)transformed under several aerobic and anaerobic conditions to produce stable

intermediates (amines, acetyl derivatives, azo and azoxy compounds) while maintaining its stable aromatic ring structure.[39,40] For instance, bond dissociation energies
of N–N, C–N, and C–H bonds in RDX are relatively low (48, 85, and 94 kcal/mol
in the case of RDX[65]). Once a bond, more likely the weak –N–N– bond in either
RDX and HMX, is cleaved the remaining chemical bonds become much weaker,
causing a rapid molecular decomposition involving ring cleavage of the molecule.[65]
Therefore, following the cleavage of the –N–NO$_2$ bond in RDX whether through
chemical or biological means, the inner –C–N– bond β to the cleaved –N–NO$_2$ bond
becomes destabilized (2 kcal/mol), resulting in rapid ring cleavage (Figure 11.2).

Thermolysis of either RDX or HMX in the gas phase[12,65,81,109] and the condensed
phase[100] leads to decomposition through a concerted mechanism to release methylenenitramine (CH$_2$=NNO$_2$) units and/or a less ordered bond fissioning (N-N and C-
H cleavage) to generate free radicals (I) (Figure 11.2). Bimolecular elimination (E2)
of HNO$_2$ via alkaline hydrolysis also leads to the production of the cyclohexenyl
intermediate derivative (II).[21,44] The primary intermediates generated can continue
to decompose to yield HCN, HNO$_2$, NO$_2$, HCHO, and N$_2$O (Figures 11.2 and 11.3).

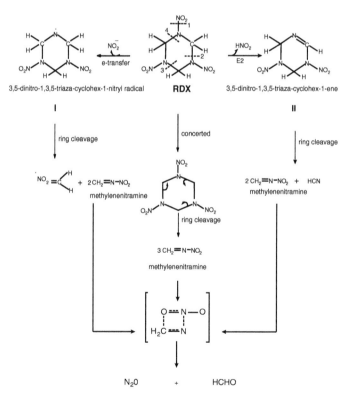

Figure 11.2   Thermal degradation routes of RDX showing secondary products from the prime
cleavage of N-N, C-H bonds and concerted elimination of methylenenitramine
(CH$_2$=NNO$_2$).[81,109]
The numbered dotted lines in RDX show the bond dissociation energies (BDEs)
and order of bond cleavage in the molecule during thermolysis. BDEs are 49,
28, 16, and 2 kcal/mol for the bonds marked 1, 2, 3, and 4, respectively.[3,65]

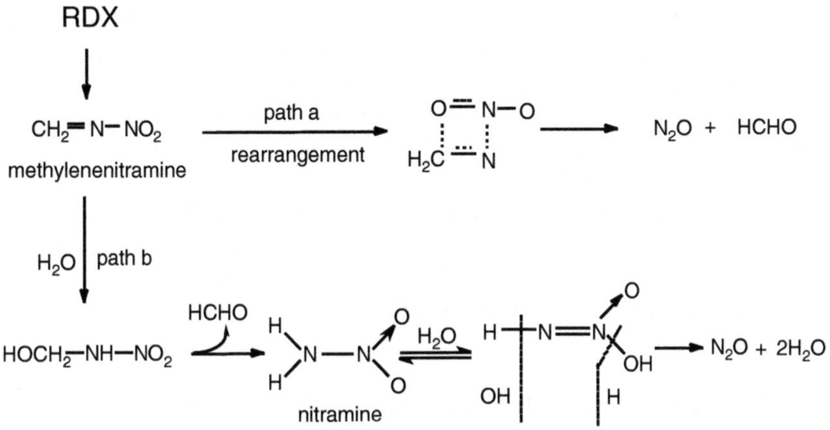

RDX

Figure 11.3 Potential reaction routes of methylenenitramine, $CH_2=NNO_2$ (*path a*), and hydroxymethylnitramine, $HOCH_2NHNO_2$ (*path b*), to nitrous oxide and HCHO.[3,95,109]

These products, particularly N_2O and HCHO, were also produced from RDX and HMX during photolysis[17] and reduction by zero valent metals,[1,86] suggesting that once the cyclic nitramine is attacked, ring cleavage occurs to produce similar distribution of products. Thus the detection of both N_2O and HCHO during biodegradation of RDX and HMX may indicate the occurrence of ring cleavage patterns similar to those described in Figure 11.2.

All RDX decomposition routes summarized in Figure 11.2 (bimolecular elimination: E_2 of HNO_2 under alkaline conditions, concerted decomposition under thermal conditions, and electron transfer followed by denitration) lead to the formation of methylenenitramine, $CH_2=NNO_2$. Methylenenitramine, which acts as a monomeric unit of RDX and HMX, is unstable and undergoes spontaneous decomposition to produce HCHO and N_2O (Figure 11.3, *path a*).[3,56,109] In water, methylenenitramine gives hydroxymethylnitramine ($HOCH_2NHNO_2$), which exists as an equilibrated mixture with its dissociated products HCHO and nitramine (NH_2NO_2). Nitramines also decompose in water to give N_2O (Figure 11.3, *path b*).[56,57,99] The previous detailed account of RDX chemical decomposition (Figures 11.2 and 11.3) is provided as a paradigm for RDX biodegradation, because very little information is available regarding metabolites. The comparison is appropriate because the decomposition routes and products described in Figure 11.2 can possibly be produced enzymatically. For instance, a one-electron reduction of the $-NO_2$ group via a nitroreductase can lead to the formation of an anion radical. If the resulting intermediate loses a nitrite ion, then the resulting free radical decomposes to give methylenenitramine (Figure 11.2). Furthermore, a hydrolase enzyme capable of breaking the inner $-C-N-$ bonds of the cyclic nitramine would decompose in a manner similar to the decomposition routes described in Figures 11.2 and 11.3.

The involvement of free radicals as intermediates in the breakdown of cyclic nitroamines is important in the case of fungi where free radicals are expected to form by reaction with the peroxidase system. The argument is supported by the unusually high levels of mineralization (75%) obtained when RDX was incubated with *Phanerochaete chrysosporium*.[25] The results provide strong evidence for a free radical chain mechanism in the degradation of RDX.

Finally, the nitroso compounds, which are frequently observed during the biodegradation of RDX and HMX,[6,34,63,64,106,107] are also formed chemically during their thermolysis[11,12] and after reduction by Fe(0).[1,86] Nitroso derivatives are energetically similar to their nitro counterparts and are expected to yield ring cleavage patterns similar to those discussed for their precursors RDX and HMX. The knowledge gained from the above chemical studies will be exploited in subsequent sections to aid discussion of new metabolic pathways for the biodegradation of RDX and HMX.

11.2.2   Early Attempts to Describe the RDX Degradation Pathway

11.2.2.1 McCormick Pathway

The only detailed biotransformation pathway ever published for cyclic nitramines is that described by McCormick et al.[63] for RDX transformation by a municipal anaerobic sludge. They proposed a pathway based on the sequential reduction of RDX to hexahydro-1-nitroso-3,5-dinitro-1,3,5-triazine (MNX), hexahydro-1,3-dinitroso-5-nitro-1,3,5-triazine (DNX), and hexahydro-1,3,5-trinitroso-1,3,5-triazine (TNX). The nitroso compounds undergo further transformation to unstable hydroxylamine derivatives, HOHN-RDX, that subsequently undergo ring cleavage to eventually yield HCHO, CH_3OH, NH_2NH_2, and $(H_3C)_2NNH_2$. McCormick et al.[63] provided insufficient experimental data to validate their pathway, including close to two dozen hypothetical structures (Figure 11.4). The authors included methylenedinitramine (III, $CH_2(NHNO_2)_2$) and nitramine (NH_2NO_2) in their pathway (Figure 11.4), but neither nitrous oxide (N_2O) nor carbon dioxide (CO_2) was shown despite previous chemical evidence that both methylenedinitramine (III) and nitramine decompose in water to produce N_2O.[56,57,99]

RDX biodegradation in a municipal anaerobic sludge gave N_2O and CO_2 as major end products.[41] Interestingly, such products are frequently formed during the thermal,[12,109] chemical,[1,86] photolytic,[17] and hydrolytic[21] cleavage of RDX. Furthermore, the inclusion of hydrazine, 1,1-dimethylhydrazine, and 1,2-dimethylhydrazine in the pathway described by McCormick et al.[63] is confusing because it gives the reader a false impression of their presence as major ring cleavage products. In fact, McCormick et al.[63] had to concentrate the RDX culture medium a thousandfold to be able to see only traces of these hydrazines. Under the same conditions, HMX did not produce hydrazines of any sort. In a subsequent study, the authors reported that hydrazines are unstable under both aerobic and anaerobic conditions.[64] Even if hydrazines were indeed RDX biodegradation products they would not withstand the severe thermal conditions used by McCormick et al.[63] to prepare the treated sample for analysis. In a more recent study, Hawari et al.[41] did not detect hydrazine or 1,2-dimethylhydrazine during RDX biodegradation by a municipal anaerobic sludge. Only traces (picogram levels) of 1,1-dimethylhydrazine were detected using a sensitive SPME/GC/MS analysis. The

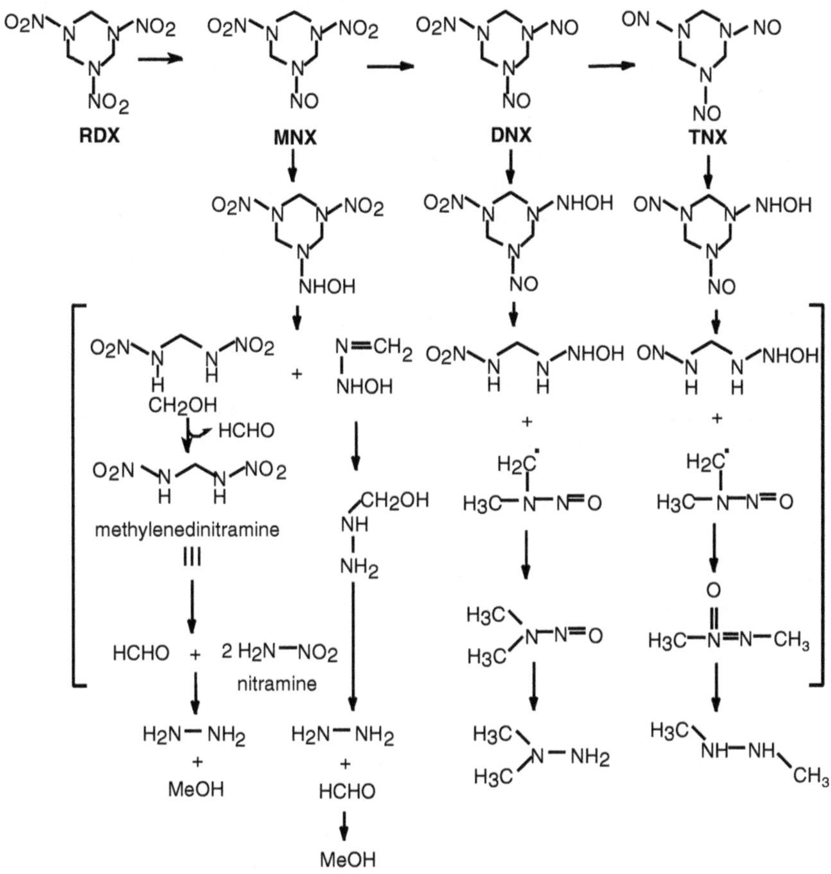

Figure 11.4 A comprehensive overview of RDX biotransformation routes as postulated by McCormick et al.[63] Square brackets indicate that none of the products were observed.

authors could not confirm if 1,1-dimethylhydrazine was a genuine RDX metabolite. Whether hydrazines are RDX degradation products or not is a controversial issue and can be resolved in the future by using ring labeled ^{15}[N]-RDX.

Finally, McCormick et al.[63] reported that ring cleavage occurred only via the primary hydroxylamino derivatives of RDX, HOHN-RDX, rather than through RDX itself (Figure 11.4). Other mechanisms including the enzymatic cleavage of the inner –N–C–N– or the outer N-NO$_2$ bonds in RDX are strong possibilities. Anaerobic municipal sludge is a rich source of microbial communities and is bound to contain unlimited numbers of bacterial strains and other catalytic enzymes such as hydrolases and nitroreductases. Despite these limitations, the McCormick pathway (Figure 11.4) remains the most detailed pathway ever proposed to describe the biodegradation of cyclic nitramines. The subsequent discussion will examine recent experimental findings and further literature reports[6,18,54,106,107] in an attempt to clarify the current understanding of the biodegradation pathways of both RDX and HMX.

11.2.2.2 Other Attempts to Describe RDX Degradation Pathways

Recently, Binks et al.[6] reported two new intermediate products from the treatment of RDX with the aerobic isolate *Stenotrophomonas maltophilia* PB1. The first product was identified by GC/MS (EI) as the chloride salt of methylene-*N*-nitroamino-*N'*-acetoxyamine (m/z 171 Da) and the second as methylene-*N*-(hydroxymethyl)-hydroxylamine-*N'*-(hydroxymethyl)-nitramine (m/z 167 Da). The authors did not elaborate on the degradation pathway. The first product was presumed to originate from the impurity, 1-acetylhexahydro-3,5-dinitro-1,3,5-triazine, which is known to be present with RDX. Interestingly, the reaction of methylenedinitramine (III) with HCl would give a salt with the same molecular mass (m/z at 171 Da) as observed by Binks et al.[6] More recently, Bier et al.[5] reported the formation of III (m/z 136 Da) during photolysis of RDX in the presence of a Fenton reagent. The intermediate was detected as the dihydrogenphosphate salt of III (IV, x = $H_2PO_4^-$) (Figure 11.5), which suggests that the product observed by Binks et al.[6] at m/z 171 Da (136 from VIII + 35 from Cl) was probably the chloride salt of III. Apparently, III is a breakdown product of both RDX and HMX since its formation is feasible under varied conditions of photolysis,[5,17] reduction with zero valent metals,[46] and biological degradation.[6,41,63] On the other hand, metabolism of RDX in laboratory animals produces several small carbon-containing fragments including CO_2, bicarbonate, and HCOOH as end products. No large metabolites such as the nitroso derivatives (MNX, DNX, and TNX) or the ring cleavage product III have been identified.[78,79]

On the other hand, hydrolysis,[21] photolysis,[5,17] thermolysis,[109] and zero valent metal reduction[46] of RDX all produce similar product distributions with some resemblance to those obtained via the McCormick pathway.[63,64] The above discussion emphasizes that once a change, whether biological or chemical, is initiated at the cyclic nitramine (RDX and HMX), the formation of breakdown products and degradation pathways are essentially the same. Exact product distribution, however, will depend on the actual chemical and enzymatic components of the system used to

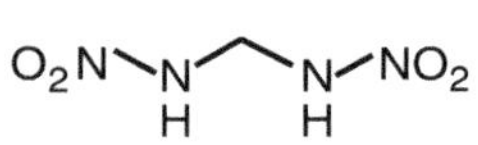

methylenedinitramine

III

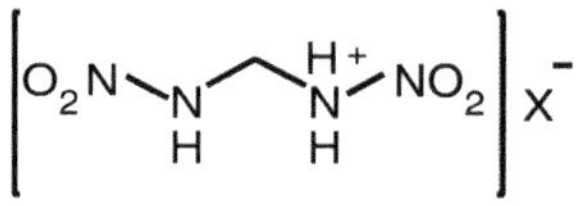

$(x = Cl^-, H_2PO_4^-)$

IV

Figure 11.5 Representative structures of methlynedinitramine (III) as a potential breakdown product of cyclic nitramine and its salt derivative (IV).

degrade the nitramine. The above discussion suggests that RDX and HMX biodegradation are not necessarily restricted to the transformation routes described by McCormick et al.,[63] wherein ring cleavage proceeds exclusively via the hydroxylamino derivatives of RDX, HOHN-RDX (Figure 11.4). Attempted biodegradation of trinitroso RDX in anaerobic municipal sludge showed that the compound was less reactive than RDX. Its disappearance rate was lower than that of RDX by at least a factor of 2, and it produced ten times less N_2O than RDX under the same conditions (Hawari et al., unpublished data). Other important routes such as concerted decomposition to eliminate $CH_2=NNO_2$, hydrolytic cleavage to generate hydroxymethylnitramine $HOCH_2NHNO_2$ or methylenedinitramine $O_2NNHCH_2NHNO_2$ (III), and/or N-NO_2 bond fissioning to generate the free radical (I) or the cyclohexenyl derivative (II) (Figure 11.2) should also be considered. We would like to emphasize here that the degradation patterns marked by ring cleavage for the nonaromatic compound, RDX (Figure 11.2), are not necessarily applicable to the more stable aromatic compound, TNT.

11.3 ANAEROBIC DEGRADATION

11.3.1 Biodegradation in Liquid Cultures

Information on the biodegradation of cyclic nitramines in soil and wastewater under anaerobic conditions has been increasing steadily over the last two decades.[7,28,30,34,63,64,82,83,106,107] RDX and HMX can be degraded by either anaerobic sludge,[63] electron-accepting agents (under nitrate-reducing, sulfidogenic, and methanogenic conditions),[8,9,27,54,60] specific isolates,[9,54,104,106] or mixed microbial cultures.[28,36,45,107] However, it seems that there has been no further progress with respect to characterizing the degradation pathways of nitramines since the early work of McCormick et al.[63] (Figure 11.4).

Kitts et al.[54] used three soil isolates (*Providencia rettgeri, Citrobacter freundii,* and *Morganella morganii*) of the Enterobacteriaceae family to degrade both RDX and HMX under O_2-depleting conditions. They identified only the nitroso derivatives (MNX, DNX, and TNX) as metabolites. They did not identify the carbon-containing (HCHO and HCOOH) or nitrogen-containing (N_2O, NH_2NH_2, $(CH_3)_2NNH_2$) ring cleavage intermediates. It was not clear from Kitts et al.[54] whether RDX was used as a carbon source or nitrogen source. Also, the authors did not elaborate on whether NO_2^-, NO_3^-, or NH_4^+ were formed or whether they were used as a nitrogen source by the degrading microorganisms. The reported increase in alkalinity of their system (pH 8 to 9) might indicate the formation of NH_3. Boopathy et al.[8] studied the metabolism of RDX and HMX using the sulfate-reducing bacterium *Desulfovibrio* sp. and observed the formation of ammonia, which was subsequently used as a nitrogen source for the degrading microorganism. Kitts et al.[54] concluded that most of the initial radioactivity from [$U^{14}C$]-HMX remained in the aqueous phase, but did not explain the nature of these polar metabolites. A degradation pathway similar to that reported by McCormick et al.[63] (Figure 11.4) was eventually suggested based only on the disappearance of RDX and the formation of its nitroso derivatives.[54]

Their claim, however, is not supported by a rigorous product analysis which is essential to the understanding of the degradation pathway. Likewise, Young et al.[106,107] used a mixed culture from horse manure and the isolate *Serratia marcescens* to degrade RDX and proposed a pathway that also resembles that of McCormick et al.[63] Huang[45] found that the thermophilic bacterium *Caldicellulosiruptor owensensis* can degrade HMX (30 mg/L) at 75°C under anaerobic conditions. They concluded that 94% of the initial radioactivity from [$U^{14}C$]-HMX remained in the aqueous phase with no mineralization. However, under thermophilic conditions, HMX is bound to undergo some hydrolysis in the aqueous phase.

The above discussion clearly illustrates a general trend among literature reports to relate any process of RDX disappearance to the pathway described by McCormick,[63] without critical evaluation of the intermediate degradation products or even their thermal and hydrolytic stabilities in water. This is a major concern that should be addressed in future research. For instance, none of the earlier reported studies[54,106,107] accounted for the potential formation of gaseous products (N_2O, N_2, and CH_4), which are very important in understanding the fate of cyclic nitramines.[41] Recently, Shen et al.[84] reported a high degree of mineralization from the degradation of RDX and HMX by microbial communities from municipal anaerobic sludge using molasses as a carbon source (pH 7). The authors used the sludge to try to create conditions close to those used in the study of McCormick et al.[63] Mineralization (liberated $^{14}CO_2$) from biodegradation of RDX (850 mg/L) and HMX (267 mg/L) by the sludge reached 66 and 55% in 105 days, respectively. There was a 10-day delay time for HMX mineralization in the presence of RDX, whereas there was no such delay for RDX, suggesting that RDX is preferentially degraded over HMX which is in line with their chemical stabilities.[2]

In a subsequent study (Hawari et al., unpublished data) involving the treatment of RDX (200 mg/L) by anaerobic sludge, the authors detected nitrous oxide (25% of total nitrogen content of the RDX that disappeared), methane (12% of total carbon content of the RDX that disappeared), CO_2 (60% of total carbon content of the RDX that disappeared), and traces of ammonium and nitrogen gas as end products. The sludge is expected to contain a complex microbial community with varied enzymes, including nitroreductases (reduction of the $-NO_2$ groups) and hydrolases (cleavage of the $-C-N-$ inner bonds in RDX) that would make it difficult to understand what is happening during biodegradation. Therefore, the physiology and molecular biology associated with the microbial degradation process will not be easy to address unless specific strains are isolated and their enzymes together with the biochemistry of the degradation process are understood. The previous argument demonstrates the importance of conducting an extensive product study, particularly for intermediate ring cleavage products, on the biodegradation of cyclic nitramines to help evaluate the fate of the chemicals during biodegradation by anaerobic sludge.

11.3.1.1 Product Analysis in Anaerobic Liquid Cultures: Novel Ring Cleavage Intermediates

Hawari et al.[41] studied the degradation of RDX and HMX with a municipal anaerobic sludge (pH 7) using LC/MS and found that the initial disappearance of the cyclic nitramine was accompanied by the formation of corresponding nitroso derivatives (MNX and DNX). Two other intermediate metabolites appearing at the same time as the nitroso compounds were tentatively identified using LC/MS and uniformly ring labeled [^{15}N]-RDX as methylenedinitramine, $O_2NNHCH_2NHNO_2$ (III), and bis(hydroxymethyl)nitramine, $(HOCH_2)_2NNO_2$ (V) (Figure 11.7). The metabolites did not accumulate, and they disappeared to produce the nitrogen containing products (N_2O, NH_4^+, and traces of N_2) and the carbon containing products ($HCHO$, $HCOOH$, CH_3OH, CH_4, and CO_2).[41] The formation of N_2O and N_2 as RDX metabolites was confirmed by preparing fresh microcosms under a blanket of argon with ring labeled [^{15}N-RDX], which resulted in the detection of $^{15}N^{14}NO$ and $^{15}N^{14}N$, respectively. Traces of N_2O, $HCHO$, and $HCOOH$ were also observed in the control containing RDX or HMX and the buffer without the sludge, emphasizing the potential occurrence of hydrolysis of the energetic chemicals even at mild pH (7 to 8). The detection of nitrous oxide and carbon dioxide in relatively high yield in the presence of the sludge (25 and 60%, respectively) and their presence only in trace amounts with autoclaved sludge emphasized that the two gases were formed biologically. Although the extent of hydrolysis was small (approximately 1%), it was significant and is expected to increase with pH. No RDX mineralization was observed in the control compared to the 60% mineralization level in the presence of the sludge.[41]

Increasing the pH from 7 to 8 during incubation led to an increase in the amount of metabolite V. The two metabolites III and V did not appear in the control containing RDX and the buffer (pH 8) alone. Interestingly, both RDX and HMX produced the two ring cleavage products III and V, suggesting that ring cleavage in both cyclic nitramines was catalyzed by the same enzyme. As discussed above, methylenedinitramine (III) has been detected or postulated as a ring cleavage product during photodegradation,[5,17] treatment with zero valent metals,[46] and biodegradation[63,64] of RDX. Although the anaerobic municipal sludge, which is rich in microbial diversity, allowed us to generate sufficient amounts of intermediates for direct detection and identification by LC/MS, specific microorganisms or the enzymes involved in the production of the above metabolites were not identified.

11.3.1.2 Metabolic Pathways of Biodegradation with Anaerobic Sludge in Liquid Cultures

As described above, since the early study of McCormick et al.[63] it appears that no additional information or insight has been gained into the postulated pathway for biodegradation of RDX by a municipal anaerobic sludge. The following summarizes the potential degradation routes proposed for RDX biodegradation in liquid cultures with anaerobic sludge.

Ring Cleavage via RDX or Its Initially Formed Nitroso and Hydroxylamino Derivatives — The formation of the nitroso derivatives was probably caused by reductive enzymes such as nitroreductases using a two-electron transfer process (type I). Such reduction of the nitro group in polynitroorganic compounds has been reported for TNT.[15,53,61] The resulting nitroso derivatives could either be cleaved or

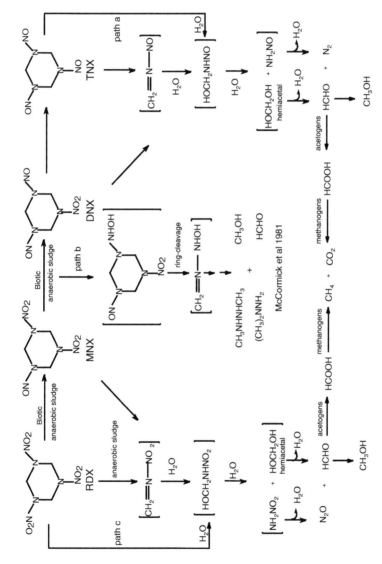

Figure 11.6 A constructed pathway for the biotransformation of RDX with anaerobic municipal sludge at pH 7 that includes a route via reduction to the nitroso derivatives and another via concerted decomposition. Dashed arrows represent direct hydrolytic ring cleavage routes. Intermediates in square brackets were not detected, but their presence was inferred from their products. Routes that were suspected of being biological are indicated on the arrows.

undergo further reduction (2e-transfer) to give the corresponding hydroxylamines (HOHN-RDX) for subsequent ring cleavage (Figure 11.6, *paths a,* and *b*). The chain of events that takes place following the ring cleavage, whether through RDX itself, its primary nitroso metabolites, or the hydroxylamine derivatives, might not be exclusively biotic. By analogy to the chemical decomposition patterns of cyclic nitramines (described above), the initial changes introduced to the RDX (or HMX) molecule by enzymes might produce unidentified initial intermediates (identified as HOHN-RDX by McCormick et al.[63]) that undergo spontaneous decomposition. The product distribution (Figure 11.6) is to some extent similar to that obtained by chemical means (Figure 11.2) because formaldehyde and nitrous oxide are the main products in both types of decomposition. The subsequent biotransformation of form-aldehyde to produce methanol, formic acid, methane, and carbon dioxide with the anaerobic sludge caused the difference observed in the product distribution between biological and abiotic decomposition processes.

Alternatively, RDX could undergo direct ring cleavage via a concerted elimi-nation of methylenenitramine, $CH_2=NNO_2$ (Figure 11.6, *path c*), which in turn undergoes spontaneous dissociation to N_2O and HCHO. Such a concerted elimi-nation mechanism is also feasible with the nitroso-RDX or HOHN-RDX deriva-tives, in which case methylenenitrosamine ($CH_2=N-NO$) (Figure 11.6, *path a*) or methylene(hydroxylamino)amine ($CH_2=N-NHOH$) (Figure 11.6, *path b*) would be produced. Ring cleavage via HOHN-RDX (Figure 11.6, *path b*) is the main decomposition route reported by McCormick et al.[63] The subsequent decomposi-tion of methylenenitrosamine is expected to produce N_2 in a pathway similar to that described for the elimination of nitrous oxide from methylenenitramine (Fig-ure 11.3). As described above, a trace amount of nitrogen gas was detected and confirmed by the formation of $^{15}N^{14}N$ following the use of ring labeled [^{15}N]-RDX as substrate in the degradation process. Whether the previous concerted ring cleavage routes are catalyzed by enzymes is an important question that awaits further investigation.

Hawari et al.[41] found that formaldehyde disappeared and produced formic acid, methanol, methane, and carbon dioxide. The anaerobic sludge is expected to contain acetogenic and methanogenic bacteria. The acetogenic bacteria biotransform HCHO to formic acid, which is then transformed by methanogenic bacteria to methane and carbon dioxide (Figure 11.6, *paths a* and *c*). The redox potential (E_h) and pH for the sludge before the addition of RDX were 100 mV and 7, respectively. Upon the start of RDX disappearance in the incubated culture medium, E_h and pH dropped to –220 mV and 6, respectively, emphasizing the reducing potential of the culture medium. On the other hand, the product distri-bution reported by McCormick et al.[63] is surprising because few components should withstand the highly reducing conditions expected in the sludge. As can be seen from Figure 11.6, the McCormick et al.[63] pathway described in Figure 11.4 constitutes only one route of the actual processes that might take place during biodegradation of RDX by anaerobic sludge. The quantitative contribution of each route in Figure 11.6 remains undefined due to a the lack of knowledge regarding the enzymes and insufficient stoichiometric data for each step of the degradation processes.

Degradation via Methylenedinitramine (III) and bis*(Hydroxymethylene) Nitramine (V)* — The previous detected RDX metabolites methylenedinitramine (III) and *bis*(hydroxymethylene)nitramine (V) accounted for the three nitro groups (-NO$_2$) of RDX. In addition both products were completely absent in controls containing RDX without the sludge, indicating that hydrolase enzymes catalyzed the cleavage of the cyclic nitramine (Figure 11.7). A time course study showed that neither intermediate accumulated in the system and that their disappearance produced HCHO and probably hydroxymethylnitramine (O$_2$NNHCH$_2$OH). Hydroxymethylnitramine was not detected, and it might be dissociated rapidly to HCHO and the nitramine NH$_2$NO$_2$. This hypothesis is supported by a previous report[56,57] that hydroxymethylnitramine can exist as an equilibrated mixture with HCHO and the nitramine NH$_2$NO$_2$. The latter decomposes in water to give N$_2$O. The transformation events described above are summarized in Figure 11.7. Once again, in the case of the anaerobic sludge, formaldehyde biotransformed further to formic acid, which in turn biotransformed to methane and carbon dioxide, whereas nitramine, NH$_2$NO$_2$, (bio)transformed to nitrous oxide and probably ammonia. It is very difficult at the present time to verify the extent of biodegradation that took place in Figure 11.7, because we did not identify the enzymes or the stoichiometric formation of all of the products in the degradation process. Furthermore, the source of ammonium

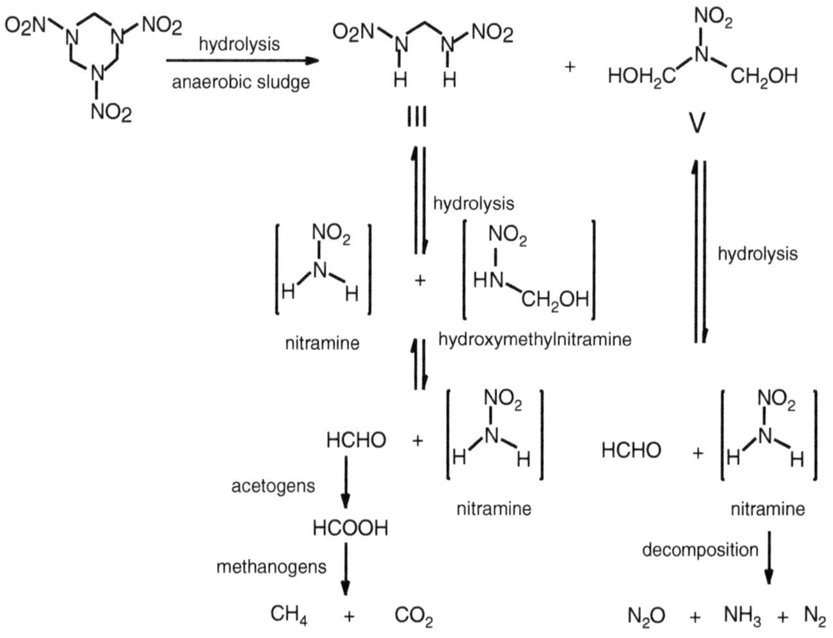

Figure 11.7 Proposed enzymatic induced hydrolysis of the cyclic nitramine RDX (or HMX) in the presence of anaerobic municipal sludge. Methlynedinitramine (III) and *bis*(hydroxymethyl)nitramine (V) have been observed only in the presence of the sludge and could not be observed in control experiments containing the cyclic nitramine and the buffer without the sludge. Intermediates in square brackets were not detected.

(NH_4^+) as an RDX degradation product remains unclear. A potential route might involve the reaction of the acidic metabolite HCOOH with the nitramine (NH_2NO_2) to form an ammonium salt or chemical reduction across the N-N or the N-O bond to give ammonia,[58,86] amines, and hydrazines.[58] The enzymatic reduction of nitramine to ammonia in the municipal sludge should not be ignored.

A Potential Abiotic Hydrolysis: Nitrite Formation — During RDX biodegradation by the anaerobic sludge at pH 8, a transient intermediate with a molecular formula of $C_2H_5N_3O_3$ was detected by LC/MS. A study using ring labeled [^{15}N]-RDX revealed that two of the nitrogen atoms of the original RDX ring were still in the product. A control experiment containing RDX and the carbonate buffer without the sludge also produced $C_2H_5N_3O_3$, at pH 8 but not at pH 7.[41] Under alkaline conditions, RDX loses HNO_2 through an E2 reaction to produce the cyclohex-1-ene derivative (II) which possibly undergoes rapid ring cleavage[21,38,44,51] to produce hydroxymethylnitramine (O_2NNHCH_2OH) and the intermediate VI which exists as an equilibrated mixture with its detected isomer VII (Figure 11.8, *path a*). Two potential decomposition routes are shown for the initially formed hydrolyzed product II. One passes through the elimination of HCN to produce hydroxymethylnitramine (O_2NNHCH_2OH) (Figure 11.8, *path a*) and the second through the elimination of O_2NNHCH_2OH and the generation of intermediate VII (Figure 11.8, *path b*). Both O_2NNHCH_2OH and VII can be degraded to eventually produce formaldehyde (HCHO) and nitrous oxide (N_2O).

As described below, the aerobic biodegradation of RDX produced a product with the same LC/MS characteristics as VII which accumulated in the system. When the aerobic hydrolysis mixture was treated with anaerobic sludge, VII disappeared to produce N_2O, NH_4^+, HCHO, HCOOH, CH_4, and CO_2. Although no quantitative product analysis was carried out, the previous observation might indicate that once RDX undergoes ring cleavage in water the subsequent (bio)transformations look basically similar (Figures 11.2, 11.6 to 11.8). The previous observation also indicates that abiotically formed RDX hydrolysis products can be degraded under anaerobic conditions. Therefore, one might conclude that mineralization of cyclic nitramines can be enhanced under slightly alkaline conditions.

Neither McCormick et al.[63,64] nor any of the subsequent studies[6,18,54,106,107] reported hydrolysis as a potential ring cleavage route for RDX because the experiments were conducted at neutral pH. The transformations summarized in Figure 11.8 become very critical for the degradation of cyclic nitramine in soils where some metal contamination is expected. Zero valent metals can function in place of the alkaline conditions and trigger abiotic reactions similar to those described earlier for the alkaline hydrolysis of RDX. This is in agreement with the suggestion made by Singh et al.[86] to remediate RDX by a joint process using zero metal valent reduction followed by biodegradation. Anaerobic bioremediation strategies for cyclic nitramines might be favored over the aerobic strategies, especially under slightly alkaline conditions which enhance ring cleavage for subsequent mineralization. As described below, the use of lime (alkaline) as a supplement in anaerobic flooded biopiles used to treat RDX and HMX in soil[31] could have contributed to enhanced mineralization by first causing an abiotic ring cleavage in both compounds. One critical question that has

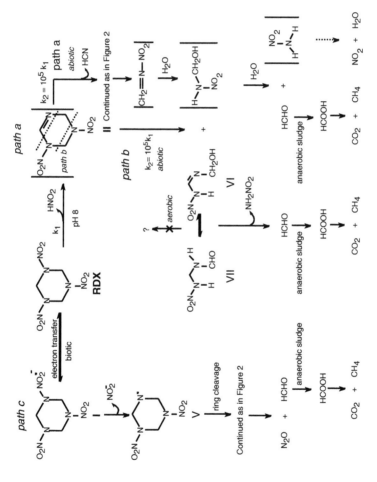

Figure 11.8 Other potential RDX degradation routes that involve denitration: *paths a* and *b*, abiotic denitration followed by ring cleavage due to hydrolysis of RDX in water; *path c*, a suspected biological denitration via a one electron transfer process followed by ring cleavage. Dashed arrows indicates abiotic or biological decomposition. Intermediates in square brackets were not detected.

not been addressed in the previous discussion is the effect of sterilized sludge on the degradation of cyclic nitramine explosives. Municipal sludge can contain several other ingredients, including metals, which might play a critical role in the overall degradation process. Future studies, therefore, should look into this point.

Enzymatic Cleavage of the N–NO$_2$ Bond in RDX: Biological Formation of Nitrite — In addition to the hydrolytic elimination route (E2) of NO$_2^-$ (Figure 11.8, *paths a* and *b*), another denitration pathway that involves bacterial enzymes might also be involved. As mentioned above, municipal sewage sludge is a rich source of microorganisms producing enzymes such as nitroreductases and cytochrome P-450. Type II nitroreductases are known to reduce nitroorganic compounds via a one-electron transfer process to produce nitro anion radicals.[62,69] Elimination of the nitrite anion (NO$_2^-$) from the parent anion radical (RDX$^-$) leaves a radical behind (RDX·), which as we described earlier would undergo a rapid ring cleavage (Figures 11.2 and 11.8, *path c*). This mechanism is potentially important for cases that involve the disappearance of RDX (or HMX) without the formation of RDX nitroso derivatives. The preceding discussion is highly speculative since neither type I nor type II nitroreductase enzymes were isolated from the sludge. Kitts et al.[55] recently reported that oxygen-insensitive type I nitroreductases in soil enterobacteria (*M. morganii* and *Enterobacter cloacae*) reduce RDX. However, a detailed analysis of products and their stoichiometry were not attempted. Unpublished work carried out in our laboratories showed that less than 5% of the total molar content of nitrogen in RDX was released as nitrite within the first 5 h of incubation with the sludge at pH 7 and without the formation of product VII. The ring cleavage product VII was only detected at pH 8 under both abiotic and biological conditions (Figure 11.8). Apparently, the observed denitration occurred via a biological process (Figure 11.8, *path c*). The earlier product distribution might indicate that in addition to the biological denitration for RDX (Figure 11.8, *path c*), other degradation routes such as reduction via the nitroso derivatives (Figure 11.6) or direct hydrolytic ring cleavage (Figure 11.7) are also important.

The intermediate products described thus far may not account for all the transformation routes experienced during the biodegradation of cyclic nitramines by the complex assemblage of microorganisms in the sludge. For instance, methylenedinitramine (III), a suggested metabolite in Figure 11.7, has been reported to undergo cleavage of its –N–C– linkage[56,57] and to react with HCHO and ammonia,[16,103] which are both degradation products themselves. This might explain the frequent observation of water soluble nonextractable RDX ring cleavage products after biodegradation.[34,54,63,83,106,107] No matter which degradation pattern is followed among Figures 11.6 to 11.8, nitramine, NH$_2$NO$_2$, and formaldehyde, HCHO, are obviously key elements in the degradation process. The spontaneous decomposition of NH$_2$NO$_2$ leads to N$_2$O,[56,57,99] and its presence in a biological system might produce NH$_4^+$. In addition, the potential presence of acetogenic and methanogenic bacteria in the sludge will lead to the biotransformation of HCHO to other carbon-containing products such as HCOOH, CH$_4$, and CO$_2$. Surprisingly, McCormick et al.[63] included nitramine and formaldehyde but did not report any of their expected biotransformed products in their pathway (Figure 11.4). The above discussion presents a controversy regarding the degradation patterns of cyclic nitramines[41,63] that warrants further investigation.

11.3.2   Mineralization in Soil Slurry Microcosms with Anaerobic Sludge

Young et al.[107] studied the biological breakdown of RDX in mixed soil slurry reactors under anaerobic conditions using either indigenous microorganisms in horse manure or the single isolate *M. morganii*. A complicated three-phase kinetic model was derived to postulate the degradation pathway of RDX based on the appearance and disappearance of the three nitroso metabolites (MNX, DNX, and TNX) of RDX. No other intermediates or end products were considered in the study in spite of the evidence that other important RDX products can have a detrimental effect on the kinetics of RDX degradation. Furthermore, the complexity of the microbial community in horse manure would yield different and sometimes unrelated degradation routes that would not fit a single kinetic model. The three-phase kinetic model of Young et al.[106,107] is thus an oversimplification of the actual events that take place during RDX biodegradation with manure. The disappearance of RDX without the formation of nitroso metabolites might indicate the occurrence of other degradation routes similar to those described in Figures 11.6 to 11.8. Shen et al.[84] reported poor (12%) mineralization of RDX in a soil slurry (40% w/w) using indigenous soil microorganism, but obtained 58% mineralization with anaerobic sludge. The presence of HMX slightly increased RDX mineralization (to 65%), whereas RDX seemed to decrease the extent of HMX mineralization. The mass balance for the UL [^{14}C-RDX] after 48 days of biodegradation in the soil slurry ($^{14}CO_2$ and acetonitrile extractable and bound residue, respectively) was distributed as follows: 59.1, 28.1, and 5.9% for RDX and 46.1, 34.3, and 14.9% for HMX.[84] Neither the bound residue nor the acetonitrile extractable fraction was further characterized. Although no extensive product analysis was performed for the biotransformation of the cyclic nitramines in soil slurries with anaerobic sludge, it was clear that the nitroso derivatives of both RDX (all three nitroso derivatives) and HMX (four nitroso derivatives) accumulated to a greater extent than in tests conducted in liquid cultures. Although the authors reported high mineralization amounts for the explosive in the aqueous phase, no discussion was provided on soil sorption and mass transfer and their effect on the mineralization process.[83,84] Also, no intermediate ring cleavage products were described. Thus, the stoichiometry and the kinetics of the actual steps involved in the biodegradation process are unknown. The above observation, however, will be useful in future field applications because both cyclic nitramines, expected to be present together at contaminated sites, can be mineralized together, leaving low amounts of the original explosive as bound residue.[84]

Young et al.[106] reported that RDX degradation in soils is not limited by mass transfer. This may be because RDX sorbed by soil can reportedly be biotransformed to nitroso derivatives by a nitroreductase enzyme,[14] which can either be membrane bound[53,69] or extracellular,[70] thus allowing biotransformation of sorbed RDX to take place. There are several unanswered questions in the study of Young et al.[106,107] The source of electrons and the mechanisms through which they reduced RDX were not explained. None of the previous studies[84,106,107] provided evidence of the actual involvement of enzymes during degradation. In addition, soil sorption–desorption data on RDX metabolites and information about ring cleavage products were not provided. With such uncertainties surrounding the nature of the degradation products of RDX and their routes of formation, conclusions regarding mass-transfer vs. mineralization-controlled limitations to biodegradation are incomplete and require further investigation.

11.3.3 Differences in the Reactivities of RDX and HMX

In general, RDX is more amenable to biodegradation than HMX.[54,63,64] RDX and HMX are structurally similar and are constructed of $CH_2=N-NO_2$ monomeric units. However, HMX is less water soluble (5 mg/L) than RDX (40 mg/L)[96,105] and is chemically more stable.[2] RDX, which exists preferentially in a chair conformation, and HMX, which exists in a crown-type conformation, resemble their corresponding cyclohexyl and cyclooctyl compounds in reactivity. For example, cyclohexyl halides are more reactive than the corresponding cyclooctyl halides toward elimination reactions.[108] In addition, molecular modeling based on the above conformations revealed that the transition state of HMX experiences more steric effects than RDX due to the crowding of atoms in the $-CH_2-N-NO_2$ reacting group of the nitramine. Consequently, the E2 reaction in HMX (loss of NO_2) demands more energy than the corresponding reaction of RDX.[21] This trend in the chemical reactivity of RDX and HMX is consistent with the observed differences in their biodegradability under both aerobic[50] and anaerobic conditions.[54,63,64]

11.4 AEROBIC DEGRADATION

11.4.1 Degradation by Indigenous Soil Microorganisms and Isolated Strains

The aerobic biodegradation of RDX and HMX has been a controversial issue for some time,[67,82,89] but evidence of their mineralization has been obtained recently.[6,18,19,31,36,50,97] For instance, RDX and HMX[82,83] were inert in activated sludge. In contrast, Osmon and Klausmeier[67] reported the aerobic degradation of RDX, whereas, Bell et al.[4] reported the partial removal of HMX and AcHMX in a bench- and pilot-scale semicontinous activated sludge system.

Jones et al.[50] employed *Rhodococcus* sp. strain A, isolated from a contaminated soil, to degrade RDX (used as a nitrogen source) in pure culture and observed 30% mineralization, 3% association with biomass, and 50% as nonextractable polar metabolites. The remaining 17% was unaccounted for. The use of RDX as a sole nitrogen source is in agreement with other reports that describe the use of energetic chemicals as the sole nitrogen source by pure cultures.[6,13,18,59,104] When soil contaminated with RDX was bioaugmented with the isolated *Rhodococcus* sp. strain A, 10% mineralization was observed. Increasing the concentration of the RDX gradually reduced mineralization to undetectable levels when the concentration reached 3000 mg/kg.[50] The reduction in mineralization with increasing RDX concentration was attributed to the toxic effects of RDX.[75] Interestingly, neither HMX nor TNT was mineralized by the *Rhodococcus* strain. Binks et al.[6] also showed that *Stenotrophomonas maltophilia* can degrade RDX, but not HMX when the cyclic nitramine is used as a nitrogen source. The reported variation in the biological response of cyclic nitramines under aerobic conditions is certainly a controversial one that requires explanation and further experimentation. It is not clear if these variations are caused by differences in the experimental conditions used by different authors, or as one reviewer commented, a timing factor in the evolution of the right degraders.

11.4.1.1 Intermediate Products from RDX and HMX Biodegradation under Aerobic Conditions

Degradation of RDX under aerobic conditions in soil and in water by specific isolates is receiving considerable attention among researchers interested in the degradation of cyclic nitramines.[6,18,19,50,97] Binks et al.[6] reported the formation of two products from the degradation of RDX with S. maltophilia. One product was identified as methylene-N-(hydroxymethyl)-hydroxylamine-N'-(hydroxymethyl) nitramine and the second as the chloride salt of methylene-N-nitroamino-N'-acetoxyammonium chloride (m/z 171 Da). As described above, the latter product could have been the chloride salt (IV) of the methylenedinitramine metabolite (III) detected earlier during biodegradation of RDX with anaerobic sludge (Figure 11.5).

Biodegradation of RDX under aerobic conditions using Rhodococcus sp. strain A produced what appeared to be a hydrolytic ring cleavage product VII which accumulated (see discussion given above, Figure 11.8, path b). Several authors[6,18,19,50,97] did not observe any of the nitroso derivatives under aerobic conditions. Recently, two independent communications reported the involvement of cytochrome P-450 in the degradation of RDX by Rhodococcus sp. strains YH11[97] and DN22.[19] Neither study identified metabolites other than nitrite. The previous discussion indicates that our understanding of the biodegradation of cyclic nitramines under aerobic conditions is incomplete. Therefore, future research should focus on areas such as the search for new isolates and the characterization of the enzymes (hydrolase, nitroreductase, and peroxidase) involved in the degradation of the cyclic nitramine.

11.4.1.2 Potential Degradation Routes with Isolated Strains Under Aerobic Conditions

There seems to be insufficient detail on the product distribution (stoichiometry and kinetics), the molecular biology, and the enzymes involved in the biodegradation of cyclic nitramines under aerobic conditions. Apart from the work published by Binks et al.,[6] who tentatively identified some RDX ring cleavage products, insufficient detail is known regarding the biochemistry of the aerobic biodegradation of cyclic nitramines. Regardless of these limitations, some potential pathways can be postulated based on the available information.

Some potential degradation routes may be ascribed to the reduction of the nitro groups in RDX by nitroreductases. Reduction by nitroreductases can take place either via a two electron transfer process to produce nitroso derivatives (Figure 11.6, path a) or via a one electron transfer process to produce the free anion radical of RDX (or HMX) (Figure 11.8, path c), although it has been reported that RDX nitro anion radicals revert back to the nitro compound in the presence of oxygen.[62,69] The denitration of the anion radical to give the corresponding RDX (or HMX) radical (Figure 11.8, *path c*), should not be neglected. This latter transient intermediate would then undergo rapid ring cleavage to produce the end product (Figure 11.8, *path c*). Furthermore, the superoxide anion radical, $O_2^{\cdot-}$, generated from the reaction of oxygen with RDX anion radical is extremely reactive and its presence may contribute to the degradation of the cyclic nitramine.

Because nitrite was initially detected[6,18,50] but did not accumulate, it was suspected that the microorganisms used this ion as their nitrogen source. Jones et al.[50] showed that the disappearance of RDX initially produced nitrite that later disappeared with the detection of traces of ammonium ions. A similar pattern of nitrite utilization has been reported for *R. erythropolis* growing on 2,4-dinitrophenol as a sole nitrogen source, where slow initial growth and nitrite accumulation is followed by exponential growth and nitrite disappearance.[59] More recently, Coleman et al.[18] isolated and identified *Rhodococcus* sp. strain D22 (from an RDX-contaminated site) that used nitrite liberated from RDX as its sole nitrogen source. The liberated nitrite was transformed to ammonium enzymatically prior to its use by the microorganism. In contrast, the presence of ammonium (NH_4^+) was shown to inhibit RDX mineralization.[6,18,104] Ammonium competes with RDX as a nitrogen source, and thus RDX degradation is repressed.

The microbial process for nitrogen incorporation (as nitrite, ammonium, or otherwise) during RDX degradation is not clearly understood and requires further investigation, as this may help to understand the stoichiometry and the kinetics of the degradation process. Coleman et al.[18] reported that three of the six atoms of N originally present in RDX ended up as biomass. Interestingly, Binks et al.[6] also reported a similar ratio for the incorporation of nitrogen from RDX into biomass. Unpublished work carried out in our laboratory showed that when RDX is degraded by a *Rhodococcus* strain, less than 5% of the theoretical nitrite accumulated. None of the previous studies explained which type of nitrogen in RDX (the exterior $-NO_2$, inner N atoms, or from both) ended up in biomass. The earlier nitrogen stoichiometry and the absence of nitroso RDX derivatives as degradation products in several other studies[6,18,41,50] might support the occurrence of a denitration mechanism (Figure 11.8, *path c*). As Figure 11.8, *path a* indicated, denitration is usually accompanied by a rapid ring cleavage. This conclusion is highly speculative and requires further investigation.

In view of the previous analysis, the claim provided by Coleman et al.[18] for the presence of oxidative denitration may necessitate further investigation with a special focus on the stoichiometry of the denitration process and the type of the catalytic enzymes involved in denitration. In their most recent report, Coleman and Duxbury[19] suggested the involvement of cytochrome P-450 in the degradation of RDX by *Rhodococcus* sp. strain DN22. No products were identified besides nitrite. In contrast, Binks et al.[6] concluded that denitration of RDX with strain *S. maltophilia* must have occurred under highly reducing conditions. It seems that the stoichiometry and the kinetics of nitrite and ammonium formation and disappearance together with characterization of the enzymes involved are important issues for future research. It has been reported that an important initial step in the biodegradation of polynitroaromatic compounds is the removal of the nitrite groups as either nitrite or ammonium.[87,88]

In another study, Harkins[36] employed mixed cultures from horse manure to degrade RDX and HMX separately using dextrose and alfalfa as supplementary carbon sources. Both Gram-negative bacteria (*Alcaligenes* sp., *Hydrogenophaga flava*, and *Xanthomonas oryzae*) and several facultative bacteria (*Escherichia coli, Kingella kingae,* and *Capnocytophaga canimorus*) were tested. The nitramines RDX and HMX disappeared in 9 days and produced the corresponding nitroso derivatives. Harkins[36] concluded that the nitramine explosives were neither used as a carbon nor as a nitrogen source, which is in contrast to the work described by Jones et al.,[50] Binks et al.[6] and Coleman et al.[18] who employed RDX as a nitrogen source and did not observe nitroso derivatives of RDX. In these last three cases, it was suggested that the microorganisms extracted their nitrogen from RDX.

The preceding discussion indicates that the metabolic pathways for the biodegradation of cyclic nitramines under aerobic conditions are poorly understood. Most of the available literature describes the disappearance of cyclic nitramines without providing rigorous information on metabolites or the catalytic enzymes that carry the sequence of the degradation process. Since the cyclic nitramine appears to undergo a rapid decomposition once one of its bonds, possibly N–NO_2, in the parent molecule is cleaved, an additional nitrogen source must become available to compensate for the losses caused by the formation of gaseous inorganic products such as N_2O.

11.4.2 Degradation by Fungi

The widespread distribution of fungi in the environment, particularly in soil, combined with their unique capabilities to produce several extracellular enzymes suggest their potential application in the degradation of cyclic nitramines. Despite these advantages of fungi, particularly the white rot fungus *Phanerochaete chrysosporium*, that is known to degrade a wide variety of organic pollutants including TNT, few reports can be found on their application to degrade the cyclic nitramines RDX and HMX. Fernando and Aust[25] studied [^{14}C]-RDX (0.028 mg/L) mineralization by *P. chrysosporium* in liquid culture and in soil slurries in the presence of pure oxygen. They reported that 66.6 ± 4.1 and 76.0 ± 3.9% mineralization occurred in liquid culture and soil slurries, respectively. In the liquid culture, 20.2% of the [^{14}C]-RDX was recovered as water soluble metabolites that were not subsequently identified. On the basis of the high levels of mineralization that were observed, the authors concluded that degradation of [^{14}C]-RDX occurred by an oxidative mechanism, presumably by the lignin peroxidases. However, lignolytic conditions were not demonstrated through enzyme assays for the systems studied. As discussed above, fungi can degrade chemicals using several enzymes such as peroxidases that are known to catalyze a number of free radical reactions.[91] The electronegative –NO_2 group in RDX readily accepts a free electron to form an anion radical. The elimination of the nitrite anion –NO_2^- would leave an unstable free radical intermediate behind to continue the decomposition process (Figures 11.2 and 11.8, *path c*). Other studies have used the same fungal strain to degrade mixtures of RDX and TNT in soil and water.[92] In the previous study neither the physiology nor the biochemistry of the degradation process is known. Furthermore, the actual fate of the nitramine explosive is not known since no product identification was reported. Further details on the role of fungi and their catalytic enzymes can be found in Chapter 9.

11.4.3 Biodegradation under Mixed Anaerobic and Aerobic Conditions

Aerobic conditions are difficult to ensure during composting and treatment of subsurface environments such as sediments and soil in flooded biopiles. For instance, although RDX can be biologically mineralized during composting[29,32,33,47,101] it is impossible to exclude the presence of protected microanaerobic niches in the compost. In addition, the relatively high temperatures (50 to 70°C) during composting are likely to enhance the elimination of oxygen, implying that a purportedly aerobic biological process may be anaerobic or even abiotic instead. Further details on composting can be found in Chapter 13. Likewise, degradation that takes place in sediments is more likely to be anaerobic since many subsurface environments are dominated by anoxic and anaerobic processes (processes using electron acceptors other than oxygen).

The above discussion emphasizes that proper knowledge of the type of degradation (aerobic, anaerobic, or both) is critically important since understanding degradation pathways, process optimization, and field application will all depend on the microbial communities and their habitats. In addition, degradation of cyclic nitramines under mixed conditions may have its advantages in future development of joint processes for the degradation of cyclic nitramines. For instance, Ronen et al.[76] suggested that a sequential combined anoxic/aerobic process is a suitable approach for the degradation of RDX in wastewater containing high levels of nitrate from munitions plants. The anoxic process first removes the nitrate and the aerobic process degrades the explosive. Earlier, Bell et al.[4] treated wastewater contaminated with RDX and HMX with an activated sludge in a sequencing batch reactor (anoxic, aerobic, settling, and decanting periods) that was operated on a semicontinuous pilot scale. Removal efficiencies for RDX ranged from 47 to 80%. Alternatively, the anaerobic process can be replaced with reduction by zero valent metals (Fe(0)) or by alkaline hydrolysis. Both abiotic processes lead to ring cleavage with subsequent production of formate and several other unidentified C–N intermediates.[46] Fe(0) reduction followed by aerobic bacterial treatment has been suggested as an effective method to mineralize nitramine explosives.[46,86]

11.4.4 Biodegradation of Mixed Contaminants

Experiments that describe the degradation behavior of more than one energetic chemical in a single system are important for future field applications because mixed contamination by several energetic chemicals is common in the field. Regan and Crawford[71] employed a strain of *Clostridium bifermentans* to degrade TNT and RDX under anaerobic conditions and found TNT to biotransform 2 h before RDX. No other details on the metabolites or their catalytic enzymes were described. Light et al.[60] studied the degradation of the three explosives RDX, HMX, and TNT in soil slurry reactors under aerobic and anoxic conditions (nitrate-reducing, methanogenic, and sulfate-reducing conditions). The three explosives disappeared preferentially under nitrate reducing conditions in the following order TNT > RDX > HMX, but were inhibitory to microbial activity under sulfate-reducing and methanogenic conditions. The above studies described neither metabolites nor the ultimate fate of the explosives and their degradation products.

More recently, Shen et al.[83] studied biodegradation of RDX and HMX in various soils with anaerobic sludge and observed high levels of mineralization (60%) for RDX. The presence of TNT reduced RDX mineralization to 47%, but TNT itself was not mineralized. Rather TNT (bio)transformed to mono-, di-, and triaminonitrotoluenes (ADNT, DANT, and TAT).[83] Boopathy et al.[9,10] showed that the three explosives RDX, HMX, and TNT can be degraded in a soil slurry and in semidefined liquid media under sulfate-reducing conditions. The previous discussion reveals that it is possible to treat soil containing more than one explosive under anaerobic (or anoxic) conditions, although more research is needed. This conclusion is important because most sites contaminated with cyclic nitramine explosives (manufacturing, training, OB/OD) contain a mixture of other contaminants.

The current understanding of the biodegradation of RDX and HMX, particularly under aerobic conditions, is far from complete and further fundamental research is required. New microbial strains together with the catalytic enzymes and metabolites they produce must be identified. The stoichiometry of metabolites and the kinetics of their formation must be carefully analyzed since such data carry powerful information for the construction of metabolic pathways. Aerobic degradation is sought as an environmentally acceptable technology because of the large volumes of soil and water that the technology can handle and also because it is more cost-effective. Furthermore, aerobic conditions do not normally produce metabolic intermediates as frequently as composting or anaerobic conditions, thus an aerobic-based technology is relatively cleaner and the toxicity associated with the production of metabolites is lower. On the other hand, as the above discussion revealed, cyclic nitramines biodegrade preferentially under anaerobic conditions. Thus, insight from laboratory work will provide the basis for the selection of the most effective field strategies.

11.5 FIELD STUDIES

Several field-scale remediation studies have been conducted, but the actual fate of explosives in the environment is generally unknown.[52,94,101] Site remediation necessitates extrapolating laboratory data to the field. However, laboratory data obtained through basic research under controlled conditions may not be applicable to the field where contaminant transport (sorption and desorption) and transformation (photodegradation and biodegradation) processes are not fully understood.[68,80,96] Field remediation thus requires that the type, concentration, breakdown products, and distribution of the explosive at the site are first fully characterized because variation in concentrations due to site heterogeneity can exceed analytical error by a factor of 10. For example, HMX concentrations were two orders of magnitude higher than those of TNT at a selected group of antitank firing ranges, even though the munition originally fired was composed of 30% HMX and 69% TNT.[49] Soil-water partition coefficients for RDX and HMX ($K_d < 1$) are less than those reported for TNT ($K_d > 4$),[20,68,80,85] indicating that the nitramines are more mobile than TNT and capable of migration through subsurface soil to groundwater supplies. Therefore, one must be aware of the actual fate of contaminants specific to the site in question. Extrapolation of laboratory data to the field also requires a gradual scale up to avoid creating logistical and technical problems. A gradual scale-up from microcosms (100 mL) to mesocosms (5 L) to pilot-scale soil biopiles (4 m^3) allowed us to develop a soil bioremediation technology for the treatment of RDX and HMX.[31] The process used indigenous soil microorganisms in the presence of an organic supplement and lime. Roughly 1 month was sufficient to reduce the concentration of RDX in soil leachates to zero, although HMX disappearance took close to 4 months. Appreciable amounts of the two nitramines degraded in the controls, which suggested the occurrence of other abiotic processes such as hydrolysis in the biopile due to the presence of lime. In addition, the irreversible sorption of polar hydrolyzed products by soil would drastically interfere with the biodegradation process, making the fate of the cyclic nitramine unclear. Also as described above, some of the major products of both biological and abiotic degradation of RDX are gaseous N_2O and CO_2 which are difficult to quantify in an open environment such as the biopiles.

In another recent study, a clay soil (40 to 44% clay) contaminated with RDX and HMX was successfully treated in drum-style reactors (8 L).[34] The disappearance rates for RDX and HMX were 213 and 141 mg/kg of soil per day, respectively. A further scale up of the soil slurry reactor to 250 L in a cement mixer was used to treat high levels of RDX (24,058 mg/kg) and HMX (7,855 mg/kg), where high rates of disappearance (513 and 160 mg/kg of soil per day, respectively) were achieved. The capability of the slurry reactors to treat high concentrations of RDX and HMX makes the technology suitable for treatment of ditches, process water, wastewater, and lagoons. A recent comparative study between three biological treatment technologies (composting, white rot fungi and anaerobic bioslurry) for explosives-contaminated soils has been reported.[94] Only the anaerobic bioslurry met the cleanup goals for RDX (5.8 mg/kg). Other *ex situ* remediation technologies based on composting[22,101] and anaerobic digesters[64,34] have been reported, but there is a lack of information on the *in situ* bioremediation of sites contaminated with energetic chemicals.

As mentioned above, knowledge of the type, concentration, and distribution of pollutants at a site is necessary to support accurate predictions of the fate of the explosive during the remediation process. In field testing and application, what is often planned as a biological process may in fact turn out to be an abiotic process. This is especially true for the case of the cyclic nitramines such as RDX and HMX, because the first change (whether chemical or biochemical) introduced in the molecule will result in its decomposition to produce several ring cleavage products (i.e., HCHO, HCOOH, N_2O, and NH_3). The presence of lime, zero valent metals, sunlight irradiation, and alkaline conditions are all potential causes for the disappearance of RDX and HMX.

Disappearance, however, should not imply safe removal or mineralization of the energetic chemical. Both energetic chemicals may (bio)transform to other polar products, polymerize, interact with biomass, or irreversibly bind to soil. Therefore, the fate of the energetic chemical supported by mass balances and toxicity assessments must be known to allow the evaluation of the bioremediation process.

11.6 CONCLUSIONS AND RECOMMENDATIONS

Current Status of Biodegradation — Contamination of soil and water by RDX and HMX is widespread and often caused by various production and military activities. Therefore, governments have an obligation to continue and further support research on the environmental fate and remediation of such contaminants. Consequently, most environmental research regarding explosives has been conducted in either governmental or private laboratories affiliated with governmental agencies. The research conducted thus far shows tentatively that biodegradation of RDX and HMX under both aerobic and anaerobic conditions is feasible. Unlike TNT which is subject to biotransformation rather than mineralization, the two cyclic nitramines RDX and HMX undergo ring cleavage following initial transformation of the molecules by anaerobic sludge. Under reducing conditions, small organic molecules (HCHO, HCOOH, CO_2) and small nitrogen-containing products (N_2O and NH_3) are produced. Mineralization of either RDX or HMX under anerobic conditions is extensive (60% with anaerobic sludge). Under aerobic conditions the biodegradation of the two explosives is clearly far from being understood and certainly further research is needed. In general, the current understanding of the biodegradation of cyclic nitramines, particularly with respect to the microbial populations and enzymes involved, is incomplete. The degradation pathway for cyclic nitramines remains controversial since no additional knowledge has been established or gained since the early work of McCormick et al.,[63] who postulated a very complicated pathway without extensive experimental evidence. Certainly, there is a serious need to conduct further research under both aerobic and anaerobic conditions. One clear conclusion from this chapter is that once the ring in cyclic nitramine cleaves, the resulting degradation products are expected to be thermally unstable and to hydrolyze readily in water. Such abiotic reactions would compete with other biological reactions during attempted biodegradation. If the biodegradation reactions are considerably slower, it would be impossible to evaluate the degradation process at the microbial level unless some innovative research is undertaken to either slow down the abiotic processes or enhance the enzymatic ones.

Fate and Behavior of Cyclic Nitramines in the Environment — The actual fate of the explosives in natural settings (phytoremediation and natural attenuation) constitutes a notable uncertainty for this family of compounds. Also, the scarcity of proper on-site monitoring tools (chemical detection, biosensors, and gene probes) makes it difficult to understand the type and extent of contamination and biodegradability of the explosive in the environment. Most studies in the biodegradation of RDX and HMX have been conducted in controlled laboratory experiments where the contaminated matrix and the type and level of contamination are known. This picture is completely different in the field where cyclic nitramines can be present with other explosives and other contaminants such as metals and petroleum-related products. Furthermore, the source of RDX and HMX varies from site to site and depends on the commercial and military activities at the site. For instance, RDX and HMX are frequently used in several applications including explosives formulations, civilian rockets, and nuclear warheads. The impact of these sources on the environment should be addressed, and processes for characterization and remediation should be developed in the future.

Environmental and Ecological Risk Assessment — The toxicity of energetic chemicals, particularly RDX and HMX, has been documented using different aquatic organisms[93,96] including bacteria, algae, invertebrates and fish, and terrestrial species such as mammals,[96] and earthworms.[73,74] Therefore, the fate and distribution of these energetic chemicals in the environment must be understood to allow the assessment of energetic chemicals for their environmental impact. Regulatory agencies are currently struggling with the cleanup criteria because fundamental information on the fate (migration and transformation) of explosive compounds is still lacking. Bioavailability and toxicity data would facilitate the estimation of maximum tolerable levels of the contaminant in soil and groundwater. Current risk assessment models tend to be more conservative (due to uncertainty and missing information) and may overestimate the environmental and ecological risk. Clearly, more research must be carried out to characterize the ecotoxicity of these energetic substances under natural and field conditions. Fundamental physical and biogeochemical interactions of toxic energetic chemicals with the soil matrix must be identified so one can accurately predict the fate and transport of the toxicants. Biological and toxicant–receptor interactions should also be considered to provide better estimates of the ecological risk at the contaminated sites.

ACKNOWLEDGMENTS

The author would like to thank the following people from the Biotechnology Research Institute of the National Research Council of Canada: Drs. S. Guiot and F. Shen for revising the discussion part of the bioslurry section, Dr. C. W. Greer for revising the discussion part of the aerobic section, Dr. G. Sunahara for insight about the current status of ecotoxicology of cyclic nitramines, Dr. S. Beaudet and Mme. A. Halasz for the library search, and Dr. T. Sheremata and Mr. C. Groom for editing the manuscript. Also, the author would like to thank Drs. S. Thiboutot and G. Ampleman from the Defence Research Establishment Valcartier, DOD Canada for providing the introduction on Canadian sites.

REFERENCES

1. Agrawal, A. and P. G. Tratnyek. 1996. Reduction of nitroaromatic compounds by zero-valent iron metal. *Environ. Sci. Technol.* 30:153–160.
2. Akhavan, J. 1998. *The Chemistry of Explosives.* The Royal Society of Chemistry, Cambridge, U.K.
3. Behrens, R., Jr. and S. Bulusu. 1991. Thermal decomposition of energetic materials. 2. Deuterium isotope effects and isotopic scrambling in condensed-phase decomposition of octahydro-1,3,5,7-tetranitro-1,3,5,7-tetrazocine. *J. Phys. Chem.* 95:5838–5845.

4. Bell, B. A., W. D. Burrows, and J. A. Carrazza. 1987. Pilot scale testing of a semi-contineous activated sludge treatment system for RDX/HMX wastewater. Report ARAED-CR-87018. U.S. Army Armament, Research, Development & Engineering Center, Picatinny Arsenal, New Jersey, AD-E401–740.

5. Bier, E. L., J. Singh, Z. Li, S. D. Comfort, and P. Shea. 1999. Remediating hexahydro-1,3,5-trinitro-1,3,5-triazine-contaminated water and soil by Fenton oxidation. *Environ. Toxicol. Chem.* 18:1078–1084.

6. Binks, P. R., S. Nicklin, and N. C. Bruce. 1995. Degradation of hexahydro-1,3,5-trinitro-1,3,5-triazine (RDX) by *Stenotrophomonas maltophilia* PB1. *Appl. Environ. Microbiol.* 61:1318–1322.

7. Boopathy R., J. Manning, and C. F. Kulpa. 1997. Anaerobic bioremediation of explosives-contaminated soil: a laboratory study, p. 463–474. In Wise, D. L. (ed.), *Global Environmental Biotechnology, Vol. 39.* Kluwer Academic Publishers, Dordrecht, Netherlands.

8. Boopathy, R., M. Gurgas, J. Ullian, and J. Manning. 1998a. Metabolism of explosive compounds by sulfate-reducing bacteria. *Curr. Microbiol.* 37:127–131.

9. Boopathy, R., C. F. Kulpa, and J. Manning. 1998b. Anaerobic biodegradation of explosives and related compounds by sulfate-reducing and methanogenic bacteria: a review. *Bioresource Technol.* 63:81–89.

10. Boopathy, R., J. Manning, and C. F. Kulpa. 1998c. Biotransformation of explosives by anaerobic consortia in liquid culture and in soil slurry. *Int. Biodeterior. Biodegrad.* 41:67–74.

11. Brill, B. T. 1990a. Structure-thermolysis relationships for energetic materials, p. 277–326. In Bulusu, S. N. (ed.), *Chemistry and Physics of Energetic Materials.* Kluwer Academic Publishers, Dordrecht, Netherlands.

12. Brill, B. T. 1990b. Thermochemical modeling. I. Applications to decomposition of energetic materials, p. 21–49, In Bulusu, S. N. (ed.), *Chemistry and Physics of Energetic Materials.* Kluwer Academic Publishers, Dordrecht, Netherlands.

13. Bruhn, C., H. Lenke, and H.-J. Knackmuss. 1987. Nitrosubstituted aromatic compounds as nitrogen source for bacteria. *Appl. Environ. Microbiol.* 53:208–210.

14. Brunsbach, F. R. and W. Reinkee. 1993. Degradation of chloroanilines in soil slurry by special organisms. *Appl. Microbiol. Biotechnol.* 40:402–407.

15. Bryant, D. J. and M. DeLuca. 1991. Purification and characterization of an oxygen insensitive NAD(P)H nitroreductase from *Enterobacter cloacae. J. Biol. Chem.* 266:4119–4125.

16. Chapman, F., P. G. Owston, and D. Woodcock. 1949. Studies of nitroamines. IV. The reaction of nitroamine with formaldehyde and primary or secondary amines. *J. Chem. Soc.* p. 1638–1641.

17. Chapman, R. D., R. A. O'Brien, and P. A. Kondracki. 1997. Novel Reagents for N - NO$_2$ scission. Technical Report AD-A328-022, IRN18052150, WL-TR-1997–7010. Wright Laboratory, Armament Directorate, TPL, Inc., Albuquerque, NM.

18. Coleman, N. V., D. R. Nelson, and T. Duxbury. 1998. Aerobic biodegradation of hexahydro-1,3,5-trinitro-1,3,5-triazine (RDX) as a nitrogen source by a *Rhodococcus* sp., Strain DN22. *Soil Biol. Biochem.* 30:1159–1167.

19. Coleman, N. and T. Duxbury. 1999. Biodegradation of RDX by *Rhodococcus* sp. strain DN22. Second Int. Symp. Biodegradation of Nitroaromatic Compounds and Explosives. Leesburg, VA, Sept. 8–9, Abstract p13.

20. Comfort, S. D., P. J. Shea, L. S. Hundal, Z. Li, B. L. Woodbury, J. L. Martin, and W. L. Powers. 1995. TNT transport and fate in contaminated soil. *J. Environ. Qual.* 24:1174–1182.

21. Croce, M. and Y. Okamoto. 1979. Cationic micellar catalysis of the aqueous alkaline hydrolyses of 1,3,5-triaza-1,3,5-trinitrocyclohexane and 1,3,5,7-tetraaza-1,3,5,7-tetranitrocyclooctane. *J. Org. Chem.* 44:2100–2103.

22. Emery, D. D. and P. C. Faessler. 1997. First production-level bioremediation of explosives-contaminated soil in the United States. *Ann. N.Y. Acad. Sci.* 829:326–340.

23. EPA document:http://www.epa.gov/swerffrr/doc/octscan.htm. 1995. Improving Federal Facility Cleanup. Report of the Federal Facilities Policy Group. U.S. Environmental Protection Agency.

24. Ermisch, V. and P. Schuldt. 1998. General environmental situation, risk assessment and remediation of former military sites in East Germany, p. 13–22. In Fonnum, F. et al. (eds.), *Environmental Contamination and Remediation Practices at Former and Present Military Bases.* Kluwer Academic Publishers, Dordrecht, Netherlands.

25. Fernando, T. and S. D. Aust. 1991. Biodegradation of munition waste, TNT (2,4,6-trinitrotoluene), and RDX (hexahydro-1,3,5-trinitro-1,3,5-triazine) by *Phanerochaete chrysosporium.* ACS Symposium Series. *Ind. Eng. Chem.* 486:214–232.

26. Fiorella, P. D. and J. C. Spain. 1997. Transformation of 2,4,6-trinitrotoluene by *Pseudomonas pseudoalcaligenes* JS52. *Appl. Environ. Microbiol.* 63:2007–2015.

27. Freedman, D. L. and K. W. Sutherland. 1998. Biodegradation of hexahydro-1,3,5-trinitro-1,3,5-triazine (RDX) under nitrate-reducing conditions. *Water Sci. Technol.* 38:33–40.

28. Funk, S. B., D. J. Roberts, D. L. Crawford, and R. L. Crawford. 1993. Initial-phase optimization for bioremediation of munition compound-contaminated soils. *Appl. Environ. Microbiol.* 59:2171–2177.

29. Garg, R., D. Geasso, and G. Hoag. 1991. Treatment of explosives contaminated lagoon sludge. *Hazardous Waste Hazardous Mater.* 8:319–340.

30. Gorontzy, T., O. Drzyzga, M. W. Kahl, D. Bruns-Nagel, J. Breitung, E. von Loew, and K.-H. Blotevogel. 1994. Microbial degradation of explosives and related compounds. *Crit. Rev. Microbiol.* 20:265–284.

31. Greer, C. W., J. Godbout, B. Zilber, S. Labelle, G. Sunahara, J. Hawari, G. Ampleman, S. Thiboutot, and C. Dubois. 1997. Bioremediation of RDX/HMX-contaminated soil: from flask to field, p. 393–398. In Hinchee, R., R. E. Hoeppel, and D. B. Anderson (ed.), *In-Situ and Ex-Situ Bioremediation*, Battelle Press, Columbus, OH.

32. Griest, W. H., A. J. Stewart, R. L. Tyndall, J. E. Caton, C.-H. Ho, K. S. Ironside, W. M. Caldwell, and E. Tan. 1993. Chemical and toxicological testing of composted explosives-contaminated soil. *Environ. Toxicol. Chem.* 12:1105–1116.

33. Griest, W. H., R. L. Tyndall, A. J. Stewart, J. E. Caton, A. A. Vass, C.-H. Ho, and W. M. Caldwell. 1995. Chemical characterization and toxicological testing of windrow composts from explosives-contaminated sediments. *Environ. Toxicol. Chem.* 14:51–59.

34. Guiot, S. R., C. F. Shen, L. Paquet, J. Breton, and J. Hawari. 1999. Pilot-scale anaerobic bioslurry remediation of RDX and HMX-contaminated soils. In Hinchee, R., R. E. Hoeppel, and D. B. Anderson (eds.), *In-Situ and Ex-Situ Bioremediation.* Battelle Press, Columbus, OH (in press).

35. Haas, R., I. Schreiber, E. v. Löw, and G. Stork. 1990. Conception for the investigation of contaminated munitions plants. 2. Investigation of former RDX-plants and filling stations. *Fresenius J. Anal. Chem.* 338:41–45.

36. Harkins, V. A. R. 1998. Aerobic Biodegradation of HMX (Octahydro-1,3,5,7-tetranitro-1,3,5,7-tetrazocine) with Supplemental Study of RDX (Hexahydro-1,3,5-trinitro-1,3,5-triazine). Ph.D. dissertation, Texas Tech. University, Lubbock.

37. Harvey, S. D., R. J. Fellows, D. A. Cataldo, and R. M. Bean. 1991. Fate of the explosive hexahydro-1,3,5-trinitro-1,3,5-triazine (RDX) in soil and bioaccumulation in bush bean hydroponic plants. *Environ. Toxicol. Chem.* 10:845–855.

38. Hawari, J., L. Paquet, E. Zhou, A. Halasz, and B. Zilber. 1996. Enhanced recovery of the explosive hexahydro-1,3,5-trinitro-1,3,5-triazine (RDX) from soil: cyclodextrin versus anionic surfactants. *Chemosphere.* 32/10:1929–1936.

39. Hawari, J., A. Halasz, L. Paquet, E. Zhou, B. Spencer, G. Ampleman, and S. Thiboutot. 1998. Characterization of metabolites in the biotransformation of 2,4,6-trinitrotoluene with anaerobic sludge: role of triaminotoluene. *Appl. Environ. Microbiol.* 64:2200–2206.

40. Hawari, J., A. Halasz, S. Beaudet, G. Ampleman, and S. Thiboutot. 1999a. Biotransformation of 2,4,6-trinitrotoluene (TNT) with *Phanerochaete chrysosporium* in agitated cultures at pH 4.5. *Appl. Environ. Microbiol.* 65:2977–2986.

41. Hawari, J., A. Halasz, S. Beaudet, L. Paquet, C. Groom, G. Ampleman, and S. Thiboutot. 1999b. Biotransformation Routes of Hexahydro-1,3,5-trinitro-1,3,5-triazine with Anaerobic Sludge. Unpublished.

42. Heilmann, H. M., M. K. Stenstrom, R. P. X. Hesselmann, and U. Wiesmann. 1994. Kinetics of the aqueous alkaline homogenous hydrolysis of high explosive 1,3,5,7-tetraaza-1,3,5,7-tetranitrocyclooctane (HMX). *Water Sci. Technol.* 30:53–61.

43. Heilmann, H. M., U. Wiesmann, and M. K. Stenstrom. 1996. Kinetics of alkaline hydrolysis of high explosives RDX and HMX in aqueous solution and adsorbed to activated carbon. *Environ. Sci. Technol.* 30:1485–1492.

44. Hoffsommer, J. C., D. A. Kubose, and D. J. Glover. 1977. Kinetic isotope effects and intermediate formation for the aqueous alkaline homogenous hydrolysis of 1,3,5-triaza-1,3,5-trinitrocyclohexane (RDX). *J. Phys. Chem.* 81:380–385.

45. Huang, C.-Y. 1998. The Anaerobic Biodegradation of the High Explosive Octahydro-1,3,5,7-tetranitro-1,3,5,7-tetrazocine (HMX) by an Extremely Thermophilic Anaerobe *Caldicellulosiruptor owensensis*, sp. Ph.D. dissertation, University of California, Los Angeles.

46. Hundal, L. S., J. Singh, E. L. Bier, P. J. Shea, S. D. Comfort, and W. L. Powers. 1997. Removal of TNT and RDX from water and soil using iron metal. *Environ. Pollut.* 97:55–64.

47. Isbister, J. D., G. L. Anspach, J. F. Kitchens, and R. C. Doyle. 1984. Composting for decontamination of soils containing explosives. *Microbiologica.* 7:47–73.

48. Jackson, M., J. M. Green, R. L. Hash, D. C. Lindsten, and A. F. Tatyrek. 1978. Nitramine (RDX and HMX) wastewater treatment at the Holston Army Ammunition Plant. Report ARLCD-77013. U.S. Army Armament Research and Development Command, Dover, NJ.

49. Jenkins, T. F., M. E. Walsh, P. G. Thorne, P. H. Miyares, T. A. Ranney, C. L. Grant, and J. R. Esparza. 1998. Site characterization for explosives contamination at a military firing range impact area. Special Report 98–9. U.S. Army Corps of Engineers, CRREL, Hanover, NH.

50. Jones, A. M., S. Labelle, L. Paquet, J. Hawari, D. Rho, R. Samson, C. W. Greer, J. Lavigne, S. Thiboutot, G. Ampleman, and R. Lavertu. 1995. Assessment of the aerobic biodegradation potential of RDX, TNT, GAP, and NC, p. 368–381. In Moo-Young, M., W. A. Anderson, and A. M. Chakrabarty (eds.), *Environmental Biotechnology: Principles and Applications*. Kluwer Academic Publishers, Dordrecht, Netherlands.

51. Jones, W. H. 1954. Mechanism of the homogeneous alkaline decomposition of cyclo-trimethylenetrinitramine: kinetics of consecutive second- and first-order reactions. A polarographic analysis of cyclotrimethylenetrinitramine. *J. Am. Chem. Soc.* 76:829–835.

52. Kaplan, D. L. 1998. Biotransformation and bioremediation of munitions and explosives, p. 549–575. In Sikdar, S.K. and R.L. Irvine (eds.), *Bioremediation: Principles and Practice Vol. II. Biodegradation Technology Developments.* Technomic Publishing Inc., Lancaster-Basel.

53. Kinouchi, T. and Y. Ohnishi. 1983. Purification and characterization of 1-nitropyrene reductases from *Bacteroides fragilis. Appl. Environ. Microbiol.* 46:596–604.

54. Kitts, C. L., D. P. Cunningham, and P. J. Unkefer. 1994. Isolation of three hexahydro-1,3,5-trinitro-1,3,5-triazine degrading species of the family Enterobacteriaceae from nitramine explosive-contaminated soil. *Appl. Environ. Microbiol.* 60:4608–4711.

55. Kitts, C. L., C. E. Green, R. A. Otley, M. A. Alvarez, and P. J. Unkefer. 2000. Type I nitroreductases in soil enterobacteria reduce TNT (2,4,6-trinitrotoluene) and RDX (hexahydro-1,3,5-trinitro-1,3,5-triazine). *Can. J. Microbiol.* 46:278–282.

56. Lamberton, A. H., C. Lindley, P. G. Owston, and J. C. Speakman. 1949a. Studies of nitroamines. V. Some properties of hydroxymethyl- and aminomethyl-nitroamines. *J. Chem. Soc.* p. 1641–1646.

57. Lamberton, A. H., C. Lindley, and J. C. Speakman. 1949b. Studies of nitroamines. VII. The decomposition of methylenedinitroamine in aqueous solutions. *J. Chem. Soc.* p. 1650–1656.

58. Larson, R. A. and E. J. Weber. 1994. Reduction, p. 169–215. In Larson, R. A. (ed.), *Reaction Mechanisms in Environmental Organic Chemistry.* Lewis Publishers, Boca Raton, FL.

59. Lenke, H., D. H. Pieper, C. Bruhn, and H. Knackmuss. 1992. Degradation of 2,4-dinitrophenol by two *Rhodococcus erythropolis* strain HL24–1 and HL24–2. *Appl. Environ. Microbiol.* 58:2928–2932.

60. Light,W. C., G. G. Wilber, and W. W. Clarkson. 1997. Biological treatability of RDX-contaminated soil. *52nd Purdue Ind. Waste Conf. Proc.* 15:135–148.

61. Liu, D., K. Thomson, and A. C. Anderson. 1984. Identification of nitroso compounds from biotransformation of 2,4-dinitrotoluene. *Appl. Environ. Microbiol.* 47:1295–1298.

62. McCalla, D. R., A. Reuvers, and C. Kaiser. 1970. Mode of action of nitrofurazone. *J. Bacteriol.* 104:1126–1134.

63. McCormick, N. G., J. H. Cornell, and A. M. Kaplan. 1981. Biodegradation of hexahydro-1,3,5-trinitro-1,3,5-triazine. *Appl. Environ. Microbiol.* 42:817–823.

64. McCormick, N. G., J. H. Cornell, and A. M. Kaplan. 1985. The anaerobic biotransformation of RDX, HMX and their acetylated derivatives. Technical Report AD Report A149464 (TR85-008). U.S. Army Natick Research & Development Center, Natick, MA.

65. Melius, C. F. 1990. Thermochemical modeling. I. Application to decomposition of energetic materials, p. 21–49. In Bulusu, S. N. (ed.), *Chemistry and Physics of Energetic Materials.* Kluwer Academic Publishers, Dordrecht, Netherlands.

66. Myler, C. A. and W. Sisk. 1991. Bioremediation of explosives contaminated soils (scientific questions/engineering realities), p. 137–146. In Sayler, G. S., R. Fox, and J. W. Blackburn (eds.), *Environmental Biotechnology for Waste Treatment.* Plenum Press, New York.

67. Osmon, J. L. and R. E. Klausmeier. 1972. The microbial degradation of explosives. *Dev. Ind. Microbiol.* 14:247–252.

68. Pennington, J. C. and W. H. Patrick. 1990. Adsorption and desorption of 2,4,6-trinitrotoluene by soils. *J. Environ. Qual.* 19:559–567.

69. Peterson, F. G., R. P. Mason, J. Hovesepian, and J. L. Holtzman. 1979. Oxygen-sensitive and -insensitive nitroreduction by *Escherichia coli* and rat hepatic micro-cosms. *J. Biol. Chem.* 254:4009–4014.

70. Rafii, F., W. Franklin, R. H. Heflich, and C. E Cerniglia. 1991. Reduction of nitroaromatic compounds by anaerobic bacteria isolated from the human gastrointestinal tract. *Appl. Environ. Microbiol.* 57:962–968.

71. Regan, K. M. and R. L. Crawford. 1994. Characterization of *Clostridium bifermentans* and its biotransformation of 2,4,6-trinitrotoluene (TNT) and 1,3,5-triaza-1,3,5-trini-trocyclohexane (RDX). *Biotechnol. Lett.* 16:1081–1086.

72. Rieger, P.-G. and H.-J. Knackmuss. 1995. Basic knowledge and perspectives on biodegradation of 2,4,6-trinitrotoluene and related nitroaromatic compounds in con-taminated soil, p. 1–18. In Spain, J. C. (ed.), *Biodegradation of Nitroaromatic Com-pounds.* Plenum Press, New York.

73. Robidoux, P. Y., C. Svendsen, J. Caumartin, J. Hawari, G. Ampleman, S. Thiboutot, J. M. Weeks, and G. Sunahara. 1999a. Chronic toxicity of energetic compounds in soil using the earthworm (*Eisenia andrei*) reproduction test. *Environ. Toxicol. Chem.* (in press).

74. Robidoux, P. Y., J. Hawari, S. Thiboutot, G. Ampleman, and G. Sunahara. 1999b. Chronic toxicity of octahydro-1,3,5,7-tetranitro-1,3,5,7-tetrazocine (HMX) in soil using the earthworm (*Eisenia andrei*) reproduction test. *Environ. Pollution.* (in press).

75. Rocheleau, S., R. Cimpoia, L. Paquet, I. van Koppen, S. R. Guiot, J. Hawari, S. Thiboutot, G. Ampleman, and G. I. Sunahara. 1999. Ecotoxicological evaluation of a bioslurry process treating TNT and RDX contaminated soil. *Bioremediation J.* 3:233–245.

76. Ronen, Z., A. Brenner, and A. Abeliovich. 1998. Biodegradation of RDX-contami-nated wastes in a nitrogen-deficient environment. *Water Sci. Technol.* 38:219–224.

77. Rosenblatt, D. H., E. P. Burrows, W. R. Mitchell, and D. L. Parmer. 1991. Organic explosives and related compounds, p. 195–234. In Hutzinger, O. (ed.), *The Handbook of Environmental Chemistry, Volume 3 Part G.* Springer-Verlag, Berlin-Heidelberg.

78. Schneider, N. R., S. L. Bradley, and M. E. Anderson. 1977. Toxicology of cyclotri-methylenetrinitramine: distribution and metabolism in the rat and the miniature swine. *Toxicol. Appl. Pharmacol.* 39:531–541.

79. Schneider, N. R., S. L. Bradley, and M. E. Anderson. 1978. The distribution and metabolism of cyclotrimethylenetrinitramine (RDX) in the rat and after subchronic administration. *Toxicol. Appl. Pharmacol.* 46:163–172.

80. Selim, H. M., S. K. Xue, and I. K. Iskandar. 1995. Transport of 2,4,6-trinitrotoluene and hexahydro-1,3,5-trinitro-1,3,5-triazine in soils. *Soil Sci.* 160:328–339.

81. Sewell, T. D. and D. L. Thompson. 1991. Classical dynamics study of unimolecular dissociation of hexahydro-1,3,5-trinitro-1,3,5-triazine. *J. Phys. Chem.* 95:6228–6242.

82. Shen, C. F., S. R. Guiot, S. Thiboutot, G. Ampleman, and J. A. Hawari. 1997. Bioremediation of energetic compounds-contaminated soil using bioslurry reactors. *Proceeding of the 4th International In Situ and On-Site Bioremediation Symposium (April 1997)*, Vol. 4, p. 171. Battelle Press, Columbus, OH.

83. Shen, C. F., S. R. Guiot, S. Thiboutot, G. Ampleman, and J. Hawari. 1998. Fate of explosives and their metabolites in bioslurry treatment processes. *Biodegradation.* 8:339–347.

84. Shen, C. F., S. R. Guiot, S. Thiboutot, G. Ampleman, and J. Hawari. 1998b. Complete degradation of RDX and HMX in anoxic soil slurry bioreactors: laboratory and pilot-scale experiments, p. 513–522. Proc. Sixth Int. FZK/TNO Conf. on Contaminated Soil. Edinburgh, U.K.

85. Sheremata, T. W., A. Halasz, G. Ampleman, S. Thiboutot, and J. Hawari. 1999. Fate of 2,4,6-trinitrotoluene and its metabolites in natural and model soil systems. *Environ. Sci. Technol.* 33:4002–4008.

86. Singh, J., S. D. Comfort, and P. J. Shea. 1998. Remediating RDX-contaminated water and soil using zero-valent iron. *J. Environ. Qual.* 27:1240–1245.

87. Spain, J. C. 1995a. Bacterial degradation of nitroaromatic compounds under aerobic conditions, p. 19–35. In Spain, J. C. (ed.), *Biodegradation of Nitroaromatic Compounds.* Plenum Press, New York.

88. Spain, J. C. 1995b. Biodegradation of nitroaromatic compounds. *Annu. Rev. Microbiol.* 49:523–555.

89. Spalding, R. F. and J. W. Fulton. 1988. Groundwater munition residues and nitrate near Grand Island, Nebraska, U.S.A. *J. Contam. Hydrol.* 2:139–153.

90. Spanggord, R. J., J. C. Spain, S. F. Nishino, and K. E. Mortelmans. 1991. Biodegradation of 2,4-dinitrotoluene by a *Pseudomonas* sp. *Appl. Environ. Microbiol.* 57:3200–3205.

91. Stahl, J. D. and S. D. Aust. 1995. Biodegradation of 2,4,6-trinitrotoluene by the white rot fungus *Phanerochaete chrysosporium*, p. 117–133. In Spain, J. C. (ed.), *Biodegradation of Nitroaromatic Compounds.* Plenum Press, New York.

92. Sublette, K. L., P. Cho, A. Maule, S. Schwartz, and D. Pak. 1991. Microbial and biomimetic degradation of nitrobodies, p. 756–763. In Nath, B. (ed.), *Proceedings of the International Environmental Pollution, 1-ICEP.1*, Interscience Enterprises Ltd.

93. Sunahara, G. I., S. Dodard, M. Sarrazin, L. Paquet, G. Ampleman, S. Thiboutot, J. Hawari, and A. Y. Renoux. 1999. Ecotoxicological characterization of energetic substances using a soil extraction procedure. *Ecotoxicol. Environ. Saf.* 43:138–148.

94. Sundquist, J. A. and S. S. Sisodia. 1996. Comparative treatability studies of three biological treatment technologies for explosives-contaminated soil. Hazardous and Industrial Wastes: *Proceedings of 28th Mid-Atlantic Industrial and Hazardous Waste Conference.* 28:93–100.

95. Suryanarayana, B., R. J. Graybush, and J. R. Autera. 1967. Thermal degradation of secondary nitramines: a nitrogen-15 tracer study of HMX (1,3,5,7-tetranitro-1,3,5,7-tetrazacyclooctanel). *Chem. Ind.* 2177–2178.

96. Talmage, S. S., D. M. Opresko, C. J. Maxwel, C. J. E. Welsh, F. M. Cretella, P. H. Reno, and F. B. Daniel. 1999. Nitroaromatic munition compounds: environmental effects and screening values. *Rev. Environ. Contam. Toxicol.* 161:1–156.

97. Tekoah, Y. and N. A. Abeliovich. 1999. Participation of cytochrome P-450 in the biodegradation of RDX by a *Rhodococcus* strain. Second Int. Symp. Biodegradation of Nitroaromatic Compounds and Explosives. Leesburg, VA, Sept. 8–9, Abstract p. 7.

98. Thiboutot, S., G. Ampleman, A. Gagnon, A. Marois, T. F. Jenkins, M. E. Walsh, P. G. Thorne, and T. A. Ranney. 1998. Characterization of anti-tank firing ranges at CFB Valcartier. Report DREV-R-9809. WATC Wainwright and CFAD Dundurn, Quebec.

99. Urbanski, T. 1967. *Chemistry and Technology of Explosives, Vol III.* p. 17–77. Jurecki, M. (trans), Laverton, S. (ed.). Pergamon Press, Oxford.

100. Wight, C. A. and T. R. Botcher. 1992. Thermal decomposition of solid RDX begins with N-N bond scission. *J. Am. Chem. Soc.* 114:8303–8304.

101. Williams, R. T., P. S. Ziegenfuss, and W. E. Sisk. 1992. Composting of explosives and propellant contaminated soils under thermophilic and mesophilic conditions. *J. Ind. Microbiol.* 9:137–144.

102. Won, W. D., L. H. Di Salvo, and J. Ng. 1976. Toxicity and mutagenicity of 2,4,6-trinitrotoluene and its microbial metabolites. *Appl. Environ. Microbiol.* 31:575–580.

103. Woodcock, D. 1949. Studies of nitroamines. III. The reactions of nitroamines with formaldehyde. *J. Chem. Soc.* 1635–1637.

104. Yang, Y. X., X. Wang, P. Yin, W. H. Li, and P. J. Zhou. 1983. Studies on three strains of *Corynebacterieum* degrading cyclotrimethylene-triamine (RDX). *Acta Microbiol. Sin.* 23:251–256.

105. Yinon, J. 1990. *Toxicity and Metabolism of Explosives.* CRC Press, Boca Raton, FL.

106. Young, D. M., P. J. Unkefer, and K. L. Ogden. 1997a. Biotransformation of hexahydro-1,3,5-trinitro-1,3,5-triazine (RDX) by a prospective consortium and its most effective isolate *Serratia marcescens. Biotechnol. Bioeng.* 53:515–522.

107. Young, D. M., C. L. Kitts, P. J. Unkefer, and K. L. Ogden. 1997b. Biological breakdown of RDX in slurry reactors proceeds with multiple kinetically distinguishable paths. *Biotechnol. Bioeng.* 56:258–267.

108. Zavada, J., J. Krupicka, and J. Sicher. 1967. Cycloalkene formation from cycloalkyl bromides: variation of rate with ring size as criterion of mechanism. *J. Chem. Soc. Chem. Commun.* 2:66–68.

109. Zhao, X., E. J. Hintsa, and Y. T. Lee. 1988. Infrared multiphoton dissociation of RDX in a molecular beam. *J. Chem. Phys.* 88:801–810.

CHAPTER 12

Subsurface Chemistry
of Nitroaromatic Compounds

Stefan B. Haderlein, Thomas B. Hofstetter, and René P. Schwarzenbach

CONTENTS

12.1 INTRODUCTION

Uptake and transformation of nitroaromatic compounds (NACs) by microorganisms, fungi, or plants can contribute to the natural attenuation of the compounds in the subsurface and provide the basis for various technologies to remediate contaminated sites (see Chapter 4). In soil and groundwater, however, NACs also interact with non-living matrix constituents, e.g., by sorption to minerals and organic matter or reactions with various dissolved, adsorbed, or solid chemical species. Abiotic and biotic processes that are relevant to the fate of NACs are often interconnected and occur simultaneously. For instance, slow desorption of NACs from the solid to the aqueous phase may limit the bioavailability and thus the rate of biotransformation in the subsurface. Conversely, microbial activity creates an anoxic milieu that favors various abiotic reduction processes of NACs. Abiotic reactions and sorption processes change the product distribution and bioavailability of NACs and thus affect their biodegradation. Therefore, the subsurface chemistry of NACs must be understood to allow assessment of natural or engineered attenuation processes at contaminated sites.

In this chapter we discuss important interactions and reactions of NACs with natural subsurface constituents. To this end, we will start with a brief summary of the properties and the chemistry of aromatic nitro compounds followed by a review of sorption and transport of NACs in porous media as well as of the major routes of abiotic transformation of NACs. Such knowledge will help to assess the relative importance and mutual dependence of biological and abiotic processes that act on NACs in a given system. Furthermore, it provides a scientific basis for designing appropriate remediation strategies as well as for evaluating their performance at contaminated sites.

We will deal with information on the interactions between NACs and the non-living natural matrix constituents obtained from very different scales of observation, ranging from spectroscopic characterization of sorption mechanisms to field scale observations on the fate of NACs in aquifers under complex geochemical conditions. Using illustrative case studies, we will also address the question of how to transfer and scale up concepts of basic theory and mechanistic information to complex natural systems. As a leading principle, we will address the major interactions and reactions of NACs with subsurface constituents from a structure reactivity point of view.

12.2 PROPERTIES AND CHEMICAL CHARACTERISTICS OF NITROAROMATIC COMPOUNDS

The physico-chemical properties of NACs and the chemistry of the aromatic nitro group are the keys to evaluating and understanding the distribution and reactivity of NACs in the subsurface. Due to the polarizing effect of the nitro group and its ability to form H-bonds with water, NACs exhibit high affinities for aqueous and polar environments in contrast to other "classic" organic pollutants such as halogenated or alkylated aromatic hydrocarbons (Table 12.1). Strong interactions of nitro substituents with the aromatic system in NACs determine the reactivity of these compounds. Nitro groups have a very strong negative inductive effect on the nitrogen–carbon bond (Hammett meta-value = 0.71). In contrast to most other electron-withdrawing substituents, nitro groups are also strong π-acceptors when in resonance with an aromatic system (Hammett para-value = 1.25). The concomitant action of inductive and delocalizing effects results in an accumulation of negative charge at the oxygen atoms of the nitro group. At the same time, the electron density of the aromatic ring system is strongly depleted, most notably in the case of multiple nitro substitution. As a consequence of the electron deficiency of the aromatic ring, NACs are good π-acceptors and are able to form coplanar complexes with suitable electron donors (Figure 12.1a). Such associations are referred to as charge transfer- or electron donor-acceptor (EDA) complexes and can involve π-π or n-π interactions.[11,31–33,100]

EDA complex formation between NACs and constituents of natural soils and aquifers affects the mobility, bioavailability, and reactivity of such compounds in a manner which is not seen for other priority contaminants. The electron deficiency of the aromatic system of NACs also impedes electrophilic attack on ring carbons, which is reflected by the low tendency of NACs to undergo chemical or microbial ring oxidation.[29,86] In contrast, the aromatic nitro group is quite susceptible to reductive transformation either by microorganisms or by abiotic reactions (see Figure 12.1b). Steric and electronic effects of NAC substituents strongly affect the stabilization of nitroaryl radicals and therefore determine the susceptibility of NACs for reductive transformation reactions. In the course of the reduction of one nitro group to an amino group, the reversible transfer of the first electron often involves the highest activation energy and, thus, is the rate-limiting step of the reaction.[43] One-electron reduction potentials ($E^1_h{}'$) are available for various NACs of environmental concern[49] and are very useful for estimating the (relative) rates of NAC reduction in a given system.

Table 12.1 Physico-Chemical Properties of Nitrobenzene (R = NO_2) Compared to Other Substituted Benzenes

	Symbol	Units	R = H	R = CH_3	R = Cl	R = NO_2	R = NH_2
Melting point	T_m	°C	5.5	−95	−46	**5.7**	−6.3
Boiling point	T_b	°C	80	111	132	**211**	184
Vapor pressure [a]	P^o	atm	0.13	0.038	0.016	**0.0004**	0.0013
Aqueous solubility [a]	$C_{w,sat}$	mol L^{-1}	0.023	0.0056	0.0045	**0.017**	0.39
Henry's Law constant [a]	K_H	(mol L^{-1} air)/(mol L^{-1} water)	0.23	0.28	0.14	**0.001**[b]	0.0013[b]
Octanol-water partitioning constant [a]	K_{ow}	(mol L^{-1} octanol)/(mol L^{-1} water)	135	490	832	**68**	8.0

Note: Data from Reference 95 and references cited therein.

[a] At 25°C.
[b] Estimated from P^o and $C_{w,sat}$.

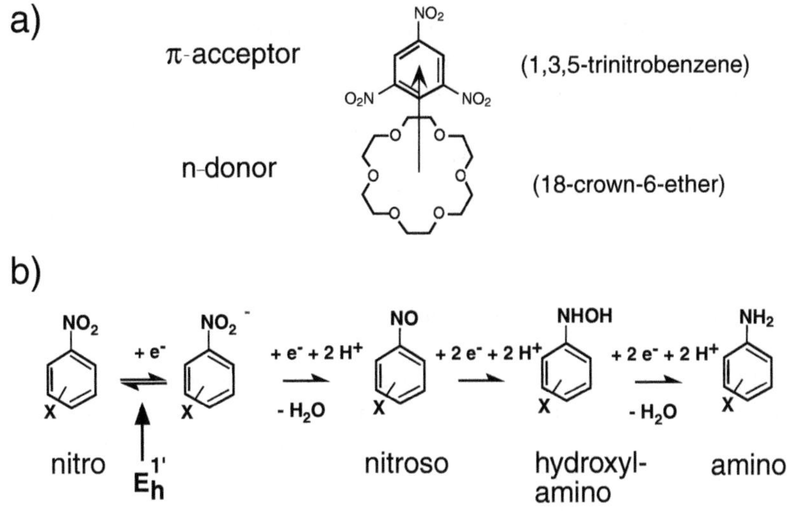

Figure 12.1 Types of reactions of NACs that are important for their behavior in natural subsurface environments: (a) Formation of co-planar electron donor-acceptor (EDA) complexes (here, n-π complex of trinitrobenzene with oxygens of a crown ether; after Spange and Heublein[100]); (b) Sequential reduction of nitro groups showing major intermediates and aromatic amines as the product of complete NAC reduction.

12.3 SORPTION AND SUBSURFACE TRANSPORT OF NITROAROMATIC COMPOUNDS

The distribution between the (mobile) aqueous phase and the various (stationary) sediment or soil phases is one of the key parameters determining both the transport and the transformation of NACs in the subsurface. Although the impact of adsorption/desorption on the transport and bioavailability of NACs in porous media is obvious, the potential effects of these processes on chemical transformation of such compounds have rarely been studied but can also be significant. For instance, the rate of abiotic reduction of *para*-substituted *n*-alkyl-nitrobenzenes in certain pond and river sediments appeared to correlate with the fraction of NACs present in the sorbed phase.[72,133] It is important to realize that not only the *extent,* but also the mechanism of sorption of a solute with solid phase(s) can affect the reactivity of NACs in porous media.

Attempts to identify the sorption mechanism(s) of NACs to natural surfaces from studies conducted with complex matrices such as soils, sediments, or aquifer material generally have had very limited success. Typically, the sorption of only one representative of NACs was studied. The results of such investigations are difficult to generalize since sorption of NACs is strongly affected by the structure of the compounds, i.e., by the type(s) and position(s) of substituents. Also, the large number of solid phases typically present in natural porous media offers a confusing variety of surfaces that NACs potentially might interact with. Furthermore, sorption studies of NACs in natural systems often lack good mass balances. Since NACs are very susceptible to chemical and microbial reduction reactions, disappearance from aqueous solution may be due not only to sorption, but often also to transformation processes, which complicates the interpretation of the results of such studies. Thus, it is necessary that we first deal with the sorption behavior of NACs in well-defined model systems using representatives of major classes of soil constituents as model sorbents. The knowledge of how functional groups of soil constituents interact with NACs will then enable us to assess their sorption behavior in complex matrices on more solid grounds.

12.3.1 Interactions of NACs with Organic Matter

Partitioning into the organic phase is commonly assumed to be a major sorption mechanism for uncharged organic compounds, provided that the natural matrix has an appreciable amount of organic matter.[15,53,97] In contrast to other important organic contaminants, very little reliable data on sorption of NACs to natural organic matter (NOM) are available. Recent studies revealed that 2,4,6-trinitrotoluene (TNT) and aminonitrotoluenes are removed from aqueous solution by colloidal NOM to a much higher extent than expected from their hydrophobicity.[19,65] Although the nature of specific NAC–NOM interactions is still the subject of current research, it is likely that in certain environmental settings such interactions may affect the sorption of NACs to the solid matrix. NOM can be a sorbent for NACs either in the aqueous or in the solid phase, and/or it can interfere with EDA complexes of NACs with clays (see below) by its tendency to form surface coatings on minerals.[108] Some authors speculated on the possibility of EDA complex formation or other specific interactions of NACs with certain functional groups of NOM, particularly with hydroquinone moieties.[62] Such constituents of reduced organic matter, however, are also effective reductants of NACs.[22,84,96,119] The reactivity of (partially) reduced organic matter makes it very difficult to distinguish between sorption and reductive transformation processes, which is probably the reason for the scarce data on sorption of NACs to natural organic phases. Weber and coworkers[25,118] studied the sorption of several munitions-related NACs to sediments rich in organic matter. Their results suggest that sorption of NACs to NOM under oxidizing conditions is consistent with hydrophobic, i.e., nonspecific, partitioning. Similar results were found for sorption of NACs to NOM coatings of minerals.[125] The NACs studied covered a wide range of substituents and about three orders of magnitude differences in hydrophobicity. Partitioning constants of uncharged NACs to soil organic carbon (K_{oc} values) were in the range of their octanol-water partitioning constants (K_{ow}), as generally expected for neutral organic solutes. Since many NACs of environmental concern exhibit fairly low K_{ow} values (in the order of 100 L kg^{-1},[42,44]) sorption due to hydrophobic partitioning should be negligible in most groundwaters (low fraction of organic carbon, f_{oc}) and significant only in porous media that are rich in organic carbon (e.g., peats, top layers of soils or sediments, etc.). This conclusion is corroborated by results of a screening study on TNT sorption to various soils,[82] that reported only a rather weak correlation between TNT adsorption constants and the organic matter contents of the soils.

Some transformation products of NACs such as hydroxylamino and aniline compounds can bind strongly to soil organic matter and form covalently bound residues.[25,86,121] One important pathway of such interactions is nucleophilic addition to carbonyl moieties present in the organic matrix.[79,110] Since the nucleophilicity of aromatic amines strongly decreases with the number of nitro or other electron-withdrawing substituents, this mechanism is expected to affect primarily the fate of more reduced intermediates of polynitroaromatic compounds such as diaminonitrotoluenes. Irreversible binding of such compounds to organic matter is the subject of ongoing research, particularly in the context of remediation of contaminated sites[85,86] (see also Chapter 4).

12.3.2   Interactions of NACs with Minerals

Several sorption studies using complex natural matrices point to the significance of specific rather than hydrophobic interactions of NACs with matrix components. Pennington and Patrick[82] found nonlinear sorption isotherms of TNT for various soils, which suggested specific interactions of TNT with soil constituents. TNT sorption correlated best with parameters related to the mineral fraction of the soils, such as the content of iron oxides and the cation exchange capacity. However, none of the bulk properties of solid matrices commonly used as predictors to describe the sorption of contaminants were very closely correlated with NAC sorption.

Weissmahr et al.[123] conducted an extensive survey of NAC sorption to major classes of minerals present in natural subsurface systems. Adsorption constants of NACs (exemplified in Figure 12.2 by TNT) to the various minerals varied by more than five orders of magnitude. The horizontal line in Figure 12.2 indicates the range of adsorption constants, K_d, expected for nonspecific interactions of neutral organic solutes with hydrophobicities similar to that of TNT (log K_{ow} = 1.86) with minerals containing negligible amounts of organic carbon.[42] As can be seen, phyllosilicates possess a unique potential for adsorbing NACs from aqueous solution. Adsorption constants of TNT and other NACs were very high for essentially all types of clay minerals, as well as for uncharged phyllosilicates such as pyrophyllite and talc (see Figure 12.2).

For most classes of soil minerals, however, sorption of TNT (and other NACs) was generally low and followed linear adsorption isotherms. Thus, minerals that form the bulk of most soil and aquifer matrices, including carbonates, iron and aluminum (hydr)oxides, feldspars, and quartz, are not expected to contribute significantly to the overall adsorption of NACs to natural matrices.

12.3.2.1 Phyllosilicates as Major Sorbents for NACs

Figure 12.3 shows typical examples of adsorption isotherms for strongly adsorbing NACs in aqueous suspensions of clay minerals. At higher surface concentrations of NACs, the isotherms are convex and converge to a saturation level. The principal shape of the isotherms can be approximated by a Langmuir equation. At low surface coverage the isotherms are quasi-linear (see insert in Figure 12.3), and an adsorption constant, K_d, can be used to describe the affinity of a given NAC to the clay mineral surface.

A particular clay mineral exhibits very different affinities (i.e., K_d values) for different NACs (see below), but its adsorption capacities for strongly sorbing NACs

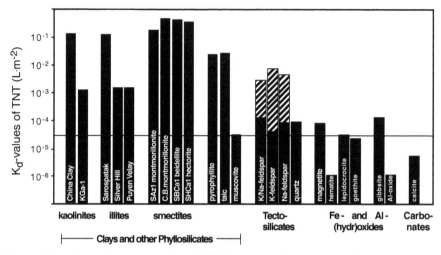

Figure 12.2   Adsorption constants, K_d, of TNT for some major groups of soil minerals normalized to the external surface areas of the sorbents. The horizontal line indicates
the range of adsorption expected for non-specific interactions of neutral organic
solutes with minerals. (Adapted from Weissmahr, K.W. et al. 1998. *Soil Sci. Soc.
Am. J.* 62(2):373.)

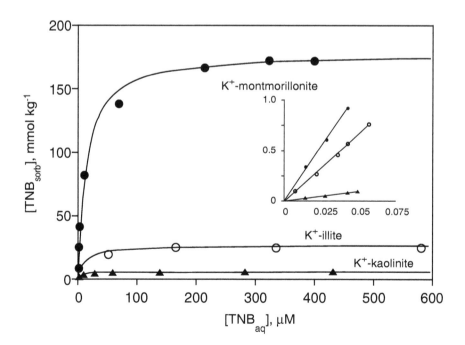

Figure 12.3   Sorption isotherms of 1,3,5-trinitrobenzene (TNB) illustrating the typical features
of NAC sorption to clay minerals. (Data from Reference 44.)

are similar. Sorption capacities of most clays closely corresponded to monolayer coverage of the external siloxane surface(s) of the clays, suggesting that such surfaces make up the majority of sorption sites for NACs. However, the interlayers of easily expandable three-layer clays host a small fraction of sorption sites with a particularly high affinity for NACs. Surface sites of clays that have a variable, pH-dependent charge (i.e., –SiOH and –AlOH at edge or gibbsite surfaces) do not contribute significantly to the strong adsorption of NACs.

12.3.2.2 Adsorption Mechanism of NACs at Phyllosilicates

Structure-adsorptivity data, measured adsorption enthalpies (approximately –40 kJ mol^{-1}), and other macroscopic observations suggest that NACs form EDA complexes with siloxane oxygens of phyllosilicates.[42,44] Weissmahr et al.[124] investigated the adsorption mechanism(s) of nitroaromatic compounds to clay minerals from aqueous and organic solvent solutions using an array of spectroscopic techniques (^{13}C-NMR, ATR-FTIR, UV/VIS, XRD). Their results proved that coplanar EDA complexes of NACs at siloxane surfaces are responsible for the strong sorption of NACs to such minerals in *aqueous* environments. Figure 12.4 illustrates schematically some of the principal features of such n–π EDA interactions of NACs at clay minerals in aqueous environments.

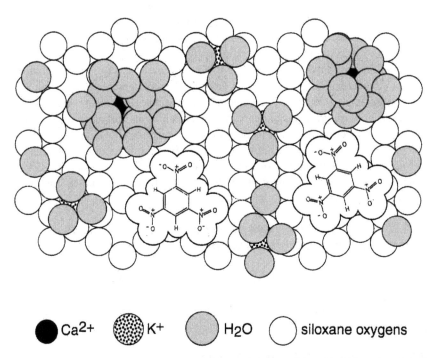

● Ca2+   ▦ K+   ◯ H2O   ◯ siloxane oxygens

Figure 12.4   Schematic representation of the basal siloxane layer of a clay mineral in the presence of hydrated exchangeable cations and adsorbed NACs. Adsorbed NACs (here, 1,3,5-trinitrobenzene) form co-planar EDA complexes with weakly hydrated siloxane oxygens. All structures are depicted using van der Waal dimensions.[123]

Planar NACs with a highly electron deficient π-electron system due to the presence of several electron-withdrawing substituents (e.g., 1,3,5-trinitrobenzene [TNB]) adsorb the most strongly. Such compounds adsorb coplanarly to the siloxane surface of the clay minerals to attain a maximum overlap of the orbitals of the non bonding electrons of the siloxane oxygens (n-donors) and the π-system of the NACs (π-acceptors).

Water molecules coordinated to strongly hydrated exchangeable cations such as Ca^{2+} interfere with such an intimate interaction of NACs and siloxane oxygens, whereas weakly hydrated exchangeable cations (e.g., K^+, Cs^+) shield the siloxane oxygens less efficiently due to their small hydration shell. Thus, exchangeable cations are a major environmental factor determining the affinity of clays for NACs (see below). Exchangeable cations on clays are located preferentially at siloxane oxygens that are highly polarized due to isomorphic substitution of metal centers in the clay lattice. The Lewis basicity and thus the e^--donor properties of siloxane oxygens not adjacent to isomorphic substitution are rather low, since the electron density of the non-bonding electrons is delocalized due to a d-π interaction over the vacant 3d orbitals of the silicon atoms.[24]

Weissmahr et al. demonstrated that adsorption of NACs to clays in aqueous compared to organic solvents involves very different mechanisms and sorption sites.[124] In apolar solvents, direct coordination of the nitro groups to surface sites and/or adsorbed cations contributed to or even dominated the sorption of NACs. Thus, only spectroscopic studies conducted in the presence of a bulk water phase provide reliable *in situ* information on the sorption mechanisms of NACs in aqueous systems. Unfortunately, different adsorption mechanisms of NACs can provide similar spectroscopic information depending on the spectroscopic technique used.[124] For example, adsorption of NACs due to H-bonding or EDA complex formation had similar effects on their UV spectra. In order to identify unambiguously such adsorption mechanism(s), an array of complementary spectroscopic methods accompanied by adsorption experiments under various conditions is necessary. The combined evidence of spectroscopic and macroscopic observations shows that hypotheses on the adsorption mechanisms of NACs involving H-bonding or direct coordination of nitro groups to metal centers or to adsorbed cations[62,90,91,131] are not appropriate in aqueous systems because they are based on spectroscopic studies conducted in organic solvent systems or at dry surfaces.

12.3.3   Structure-Adsorptivity Considerations for Assessing the Sorption of NACs to Natural Solids

To provide a better understanding of the factors that control interactions of NACs with clay mineral surfaces and to develop predictive models of NAC adsorption to phyllosilicates, the following sections address those parameters that determine the properties of NACs as sorbates and minerals as sorbents, respectively.

12.3.3.1 Effects of NAC Substituents

NACs exhibiting similar K_{ow} values show more than four orders of magnitude differences in K_d values for the same clay mineral,[42,44] confirming that other than hydrophobic effects play a pivotal role in the ability of NACs to form complexes with siloxane oxygens. The electronic and steric effects of substituents on the tendency of NACs to form n–π EDA complexes with phyllosilicates are illustrated by the data in Figure 12.5, and can be summarized as follows.

1. Planar NACs with an electron deficient π-system due to several electron-with-drawing and electron-delocalizing substituents show the highest adsorption. TNB has the highest K_d value of all NACs investigated so far. TNB can be considered as the optimal electron acceptor for EDA complex formation due to its three electron-withdrawing aromatic NO_2 groups and the lack of steric substituent effects. Poor electron acceptors, i.e., NACs with several electron-donating sub-stituents (e.g., $-CH_3$, $-NH_2$), as well as nonaromatic nitro compounds (e.g., nitro-cyclohexane or the explosive RDX [1,3,5-trinitro-hexahydro-1,3,5-triazine]) exhibit very low affinities.

2. *Ortho*-substituents such as alkyl, halogen, or even nitro groups diminish complex formation due to steric effects in two ways: they prevent complete coplanarity and thus optimal resonance of the nitro substituent(s) with the aromatic ring, which reduces the π-acceptor properties of NACs. In addition, *ortho*-substituents hinder a coplanar spatial arrangement of NACs and oxygen donors, which is necessary for optimal EDA complex formation.[34,124] A very similar effect is caused by bulky substituents, e.g., branched alkyl chains, at any position of the ring. This bulky substituent effect is illustrated by the various dinitrophenol herbicides shown in Figure 12.5a.

3. Compounds with substituents that are able to form intramolecular H-bonds with a neighboring NO_2 group (e.g., *ortho*-nitrophenols and *ortho*-nitroanilines) do not show a significant *ortho* effect.

Ortho- and *para*-substituted isomers of such NACs have very similar K_d values since coplanarity is not disturbed. Note that in the case of nitrophenols sorption of the anionic phenoxide species is negligible (see below).

The electronic and steric effects of substituents discussed earlier account for the huge differences in EDA complex formation of TNB, TNT, and some of its amino reduction products (Figure 12.5b). TNT adsorbs less than TNB due to the *ortho* effect of the methyl group. Substitution of an electron-withdrawing nitro group by an electron-donating amino group strongly decreases the tendency of NACs to form EDA complexes. The difference in K_d between TNT and 2,4-diamino-6-nitrotoluene is more than three orders of magnitude. Isomers of aminodinitro- and diaminonitro-toluenes have very different K_d values due to the *ortho* effect of the methyl groups next to a NO_2 group. The previous examples illustrate that the various compounds commonly present in munitions residues exhibit very different mobilities and bio-availabilities in subsurface environments when specific adsorption to clay minerals is the dominant sorption process.

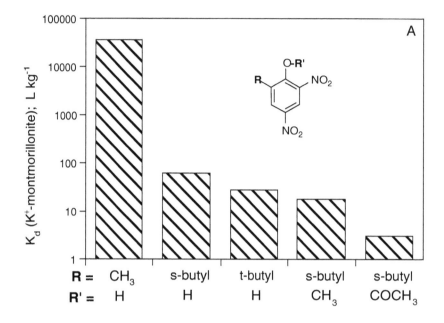

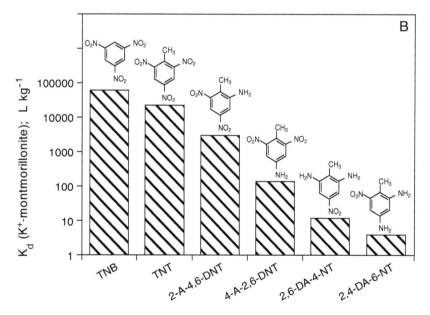

Figure 12.5 Illustration of substituent effects on EDA complex formation of NACs with clays: (A) effect of bulky substituents that hinder co-planar orientation of NACs and siloxane oxygens, exemplified by various alkyl dinitrophenol herbicides; (B) adsorption constants of some common explosives and some of their transformation products for homoionic K+-montmorillonite. (Adapted from Haderlein, S.B. et al. 1996. *Environ. Sci. Technol.* 30(2):618.)

The preceding qualitative discussion of substituent effects clearly reveals that electronic and steric factors control the adsorption of NACs and such factors can be interrelated, e.g., in the case of substituents *ortho* to nitro groups. In order to derive quantitative structure adsorptivity relationships for EDA complex formation of NACs with a given siloxane surface, not only must the π-acceptor properties of the NACs be quantified, but also the (micro)topology of both the donor and the acceptor reactants in aqueous solution must be known. This is a challenging task which has not yet been fully accomplished.[37,122]

12.3.3.2 Reactivity of Donor Sites at Phyllosilicates

Using several natural phyllosilicates as well as artificially modified clays, Weiss-mahr et al.[123] studied mineralogical properties and other factors that determine the tendency of siloxane oxygens to form EDA complexes with NACs. As is evident from the above discussion, a necessary prerequisite is a planar arrangement of NACs and such sites at a given mineral. In contrast to other minerals that have siloxane oxygens, phyllosilicates exhibit extended and planar networks of siloxane oxygens, which explain, in part, why minerals such as quartz and tectosilicates do not form strong EDA complexes with NACs. In addition, two major factors control the extent of complex formation: the electron donor properties of siloxane oxygens and their accessibility to the solutes. Both factors are related to the layer charge which is responsible for the net negative charge of clays and other types of phyllosilicates. This surplus of structural negative charge is compensated by the adsorption of variable amounts of exchangeable cations. The layer charge of phyllosilicates results from substitution of structural Al and Si centers by lower valent but isomorphic cations and varies between 0 and 2 negative charges per unit cell of the mineral.[39] In the vicinity of isomorphic substitution, especially if Si^{4+} is replaced by Al^{3+}, the Lewis basicity and, thus, the electron donor properties of siloxane oxygens are significantly enhanced.[101,130] The accessibility of such activated siloxane oxygens for NACs in aqueous environments is inhibited by the presence of exchangeable cations. Highly charged phyllosilicates such as muscovite not only have strongly polarized siloxane oxygens with good electron donor properties, but also bear a high density of exchangeable cations which restricts the access of NACs or other π-acceptors to the siloxane sites, resulting in very low adsorption constants (see Figure 12.2). Using modified montmorillonites, Weissmahr et al.[123] demonstrated that the efficiency of a given siloxane surface as n-donor in EDA complex formation is inversely correlated to the density of exchangeable cations at the surface.

The previous results are consistent with the interpretation that exchangeable cations potentially hinder but never enhance EDA complex formation of NACs at siloxane surfaces of phyllosilicates (see discussion of Figure 12.4). Siloxane oxygens of uncharged phyllosilicates (e.g., talc or pyrophyllite; Figure 12.2) are easily accessible due to the absence of exchangeable cations, but their donor properties are low due to the absence of activation by isomorphic substitution. In the case of natural clays, the sorption promoting effect of isomorphic substitution is to a large extent compensated by the shielding effects of adsorbed cations. This leads to a large variation in and a poor predictability of the ability of such minerals to form EDA complexes (see Figure 12.2). The variations in the affinities of siloxane sites of the various clays are consistent for *all* NACs that are able to form EDA complexes.[123] The constant relative affinities of the various clays for NACs confirm that the same mechanism (i.e., EDA complex formation) dominates the adsorption of NACs at clay minerals and phyllosilicates in general.

Figure 12.6 illustrates the unique pattern of relative K_d values for a given set of NACs. The NACs and phyllosilicates reflect the large range of e^--acceptor–e^--donor properties of the solutes and sorbents, respectively. Despite the large differences in the tendency of the NACs to form EDA complexes, their adsorption *relative* to each other is very similar at the various clays. At SAz-1 montmorillonite, for example, the K_d of 2,4-dinitrotoluene (2,4-DNT) was two orders of magnitude higher than that of 4-nitrotoluene (4-CH$_3$-NB). Ratios of the K_d values of the two compounds were very similar for other clays. The results demonstrate that similar substituent effects governed the EDA complex formation with natural clay minerals. In other words, the *relative* tendency of a given π-acceptor compound to adsorb to siloxane donors is independent of the structure of the phyllosilicates and types of exchangeable cations or other solutes present.

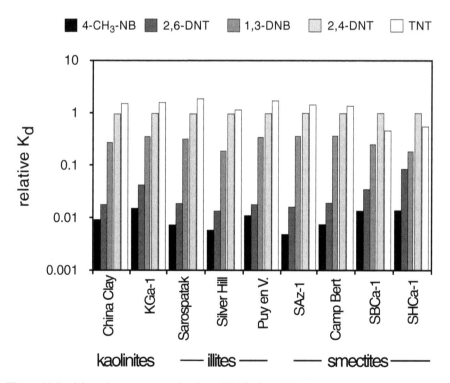

Figure 12.6 Adsorption constants of selected NACs for various homoionic K$^+$-clay minerals. The K$_d$ values are normalized to a reference NAC (2,4-DNT; arbitrarily chosen) and demonstrate the remarkably constant *relative* affinity of the different clays for EDA complex formation with NACs. (Adapted from Weissmahr, K.W. et al. 1998. *Soil Sci. Am. J.* 62(2):376.)

12.3.4 Effects of Solution Composition on Sorption of NACs

As was pointed out above, the structures of the sorbates are not the only variables that control EDA complex formation in aqueous systems. Other factors such as cation exchange processes and sorption or precipitation of co-solutes also affect the accessibility of surface sites and, thus, are expected to affect the extent of NAC sorption in natural porous media. The significance of such environmental factors on the sorption of NACs in the subsurface will be discussed in the following sections.

12.3.4.1 pH

A number of environmentally relevant NACs have acid or base functional groups, e.g., -OH, -COOH. At typical pH values of natural porous media ($5 < pH < 8$), mononitrophenols are partially and polynitrophenols (including dinitrophenol herbicides) as well as nitrobenzoic and nitrosulfonic acids are predominately present in their deprotonated, i.e., anionic, forms. Thus, the effects of speciation of both NACs and surface sites on the sorption process need to be evaluated. The pH effects on EDA interactions of NACs with phyllosilicates have been studied in detail.[42,44] K_d values of neutral NACs were constant between pH 3 and 9. Ionizable NACs showed a strong pH dependence of their apparent K_d values in the pH region corresponding to the pK_a of the compounds. Adsorption of nitroaromatic anions or ion pairs was insignificant compared to the adsorption of the corresponding neutral species (HA). This lack of anion sorption can be rationalized by two concomitant effects, namely, decreased π-acceptor properties and electrostatic repulsion of nitroaromatic anions by negatively charged siloxane oxygen donors.

The pH dependence of the apparent K_d values for clays of dissociable NACs can be modeled by considering the dissociation of the NACs and a pH-independent K_d value of the non-dissociated species, (K_d^{HA}):

$$K_d(pH) = \alpha_o \cdot K_d^{HA} \tag{1}$$

where α_o is the fraction of the non-dissociated species and is given by

$$\alpha_o = \frac{1}{1 + 10^{(pH - pK_a)}} \tag{2}$$

Since nitroaromatic anions also have a rather low tendency to sorb to organic matter,[50,51] the mobility of NACs such as alkylated dinitrophenol herbicides (e.g., DNOC, Dinoseb or Dinoterb, R = H; R' = methyl, sec-butyl, or tert-butyl, respectively; see Figure 12.5a) is generally quite high in the subsurface. This interpretation is consistent with the fact that picric acid did not accumulate in soils contaminated with munitions residues although this explosive was widely used during World War II.

12.3.4.2 Dissolved Organic Matter

Adsorption of "dissolved" organic matter (DOM) can have various effects on the sorption of NACs to minerals. DOM adsorbed at mineral surfaces can act as an organic sorbent for NACs, or it might reduce the accessibility of siloxane sites in natural porous media, thus decreasing sorption of NACs. Weissmahr et al.[125] studied the adsorption of NACs to clays in the presence of DOM. Adsorption of DOM at three-layer clays such as montmorillonite was quite low and had no noticeable effect on the sorption of the NACs. DOM, however, adsorbed significantly (up to 6 g kg^{-1}) to the two-layer clay kaolinite and caused a slight decrease in the K_d values of NACs as well as in the sorption capacity of the homoionic K$^+$-clays. The findings are consistent with that of other investigations showing that humic acids adsorb predominantly at edge aluminol groups rather than at siloxane surfaces of clay minerals.[113,134] Since cosolvent effects of humic acid were negligible, the minor effects on EDA complex formation at K$^+$-kaolinite were probably due to small amounts of humic acids adsorbed at the siloxane plane.

12.3.4.3 Competitive Sorption: Implications for the Mobility of NAC Mixtures

At many contaminated sites, e.g., in the vicinity of munitions plants, mixtures of dozens of nitroaromatic contaminants may be present.[43] Technical grade TNT including various byproducts (e.g., mono- and dinitrotoluenes) as well as transformation products of these compounds such as nitrotoluidines are major contaminants at those sites.[40,87,109]

Since all NACs interact by the same sorption mechanism with phyllosilicates, competition for sorption sites occurs among mixtures of sorbing NACs. When a strongly sorbing compound (e.g., TNT) is added to a clay suspension containing a more weakly sorbing NAC (e.g., 4-nitrotoluene (4-CH$_3$-NB)), the weaker sorbing NAC is desorbed from the clay.[44] The competition efficiency of NACs for adsorption sites parallels their relative affinities (K_d values) to the clay mineral(s). NACs sorb independently without noticeable competition effects only in the linear range of their adsorption isotherms, i.e., at trace concentrations.

Figure 12.7 illustrates the significance of competition effects for the transport of NACs in porous media. Shown are breakthrough curves of NACs from miscible displacement experiments carried out in laboratory columns. The solid matrix of the columns consisted of quartz sand coated with aggregated montmorillonite clay minerals.[28] This model sorbent mimics the typical distribution pattern of clays in natural porous media. Generally, NACs exhibit strongly tailing breakthrough curves due to their nonlinear sorption isotherms.[28] When mixtures of several NACs were present in the feeding solution, the mobility of all NACs present in the column was enhanced compared to single solute conditions. This is reflected by an earlier breakthrough as well as by less tailing of the breakthrough curves. However, the degree of mobilization of the individual NACs due to competitive sorption strongly depended on their relative affinities to the clay. The data in Figure 12.7 illustrate the effects of competing cosolutes with both higher (TNT) and lower (4-CH$_3$-NB) affinities than 1,3-dinitrobenzene (1,3-DNB), respectively (compare the relative strength of adsorption of the NACs shown in Figure 12.6). Competition efficiency of the NACs paralleled their sorption affinities in the column systems. Very similar results were obtained in batch experiments.[44] The breakthrough curves of mixtures of NACs exhibit other remarkable features in addition to the general trend of increased solute mobility. Weaker competitors (e.g., 1,3-DNB in the presence of TNT [Figure 12.7]) showed a distinct "overshooting" effect, i.e., the eluting concentrations partially exceeded the input concentrations of the solutes. Note that in the experiments shown, all NACs were fed simultaneously and in similar concentrations to the columns. The area under the overshooting peaks, which represents the amount of solute replaced from the clays by the stronger sorbing solute, corresponded to the decrease in tailing of the desorption fronts of the weaker sorbing solute. The competition effects of NACs in mixtures and their impact on solute transport can be predicted from the sorption isotherms of single solutes based on a competitive Langmuir approach.[27]

These above results have implications for assessing the mobility of NAC mixtures at contaminated sites. First, differences in the mobility of individual NACs will be even larger in mixtures than expected from their sorption behavior under single solute conditions. Cleanup of contaminated sites by flushing techniques may require a much longer time than anticipated from an eventually rapid depletion of mobile NACs such as nitrotoluenes. Second, nitroreduction, an important transformation pathway of NACs in the subsurface (see below), leads to products with lower K_d values than their (poly)nitrated precursors. Such compounds, e.g., (nitro)anilines, are also weaker competitors, which makes them even more mobile in the subsurface.

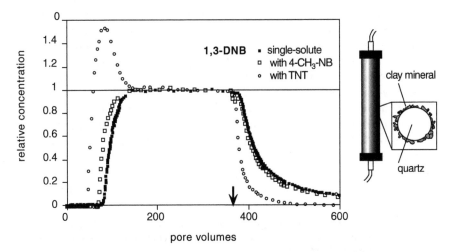

Figure 12.7 Illustration of the effects of competitive sorption of NACs at clays on their transport in porous media. Breakthrough curves are shown from column experiments of 1,3-dinitrobenzene (1,3-DNB) as single solute and in mixtures with other munitions-related NACs (4-CH$_3$-NB, TNT). 1,3-DNB was fed as a single solute pulse or in binary mixtures with either 4-nitrotoluene or TNT (c_0 of each component 10 μM) to the columns. The arrow indicates the duration of the contaminant pulses. The columns were packed with a mixture of clay-coated sand particles.[27]

Thus, competitive sorption must be considered as a potentially crucial process in evaluating the mobility and the exposure of NACs in the subsurface.

12.3.4.4 Type and Concentration of Electrolytes

Since exchangeable cations do not (fully) dehydrate upon adsorption at clays,[78,102] the density of exchangeable cations at siloxane sites and the size of their hydration spheres have a strong effect on the formation of complexes between NACs and clays.[42,44] Table 12.2 illustrates the effect of exchangeable cations on the ability of the clays to adsorb NACs. Homoionic K^+ clays have a very high tendency to adsorb TNT (and other planar NACs[123]), whereas the same clays are very poor sorbents when strongly hydrated Ca^{2+} is the exchangeable cation. The adsorption affinity of the clays is inversely correlated with the enthalpy of hydration of monovalent cations.[42] Strongly hydrated cations such as Li^+, Na^+, Mg^{2+}, or Ca^{2+} efficiently hinder the access of NACs to the siloxane donor sites and thus prevent EDA complex formation. More weakly hydrated cations such as NH_4^+, K^+, Rb^+, or Cs^+ shield siloxane oxygens by coordinated water molecules to a much lesser extent.[73,78]

Exchangeable cations also show a shielding effect in mixed ionic systems. Partial saturation of clays with weakly hydrated exchangeable cations also enhances the affinity of phyllosilicates for NACs. The affinity of clays for NACs increases steadily as the fraction of weakly hydrated exchangeable cations at clays increases.[42,125] Under environmentally relevant conditions, mixed ionic clays prevail and exchangeable potassium and ammonium cations largely determine the accessibility of siloxane sites for NACs. It should be noted that the accessibility of siloxane sites for EDA complex formation in mixed ionic systems may depend in a complex manner on factors such as ionic strength and the type of cations and clays present. With homoionic clay minerals, ionic strength had no measurable effect on the adsorption of NACs in the range of 10^{-4} to 10^{-1} M (values that are typical of freshwater systems). Thus, K_d values measured at I = 0.1 M are representative for lower ionic strength and vice versa.[42] However, in mixed ionic systems the ionic strength can affect the selectivity coefficients for cation exchange reactions with clays[7] and, thus indirectly, also the adsorption of NACs. Hence, a quantitative prediction of EDA complex formation constants in a given sediment is exceedingly complicated. The relative affinities of NACs to siloxane sites, however, remain unaffected by the composition of exchangeable cations and the type of clay minerals present in a given system. We conclude that the extent of EDA complex formation between clays and NACs is controlled by the composition and distribution of hydrated cations at the Stern layer

Table 12.2 Adsorption Constants (K_d Values) of TNT for Homoionic K^+ and Ca^{2+} Clays as an Example for the Effect of Exchangeable Cations on the Adsorption of NACs to Clay Minerals[44]

Clay Mineral	K_d (L kg^{-1})	
	Ca^{2+} Clay	K^+ Clay
Kaolinite	0.3	1,800
Illite	1.2	12,500
Montmorillonite	1.7	21,500

of siloxane surfaces. Although a quantitative prediction of the accessibility of siloxane sites for NACs at clays as a function of the type and composition of exchangeable
cations and of the ionic strength is still missing, changing the cation composition
on clays is a promising option to manipulate specifically the sorption and thus the
transport of NACs in the subsurface.

12.3.5   Groundwater Injection of Electrolytes to Control
Field Scale Retardation of NACs

Using 4-CH$_3$-NB as model compound, Weissmahr et al.[125] tested the applicability
of electrolyte injections to control the groundwater transport of NACs in a dual well
forced gradient field injection experiment. A 4-CH$_3$-NB pulse was injected along
with KCl, followed by the injection of pristine groundwater.

The transport of 4-CH$_3$-NB and electrolytes was monitored at a distance of 3.8
m through an extraction well. The breakthrough of 4-CH$_3$-NB occurred about 2
days after the breakthrough of the conductivity pulse (chloride) due to sorption of
4-nitrotoluene to the aquifer matrix. The apparent average retardation factor (R_f)
of 4-CH$_3$-NB relative to the chloride breakthrough was R_f = 1.8, corresponding to
an average sorption coefficient of 4-CH$_3$-NB to the aquifer matrix of about 0.2 L
kg^{-1}. The second part of the injection experiment was started at day 15 with the
injection of a CaCl$_2$ pulse (right-hand panels in Figure 12.8), after 4-CH$_3$-NB
concentrations in the extraction well dropped below the quantification limit of 1
μg L^{-1}. The goal of the second part of the field test was to evaluate if and to what
extent sorbed 4-nitrotoluene could be remobilized by exchanging weakly hydrated
cations (K$^+$) at clays by strongly hydrated cations (Ca^{2+}). Two days after the injection
of the CaCl$_2$ pulse, simultaneous breakthrough of conductivity and 4-CH$_3$-NB
occurred at the extraction well. Thus, the stimulated Ca^{2+}/K$^+$ ion exchange on clays
of the aquifer matrix resulted in desorption and quasi-conservative transport of 4-
nitrotoluene. This result suggests that sorption of 4-CH$_3$-NB to the aquifer matrix
was primarily due to EDA complex formation with clays, since other potential
sorption mechanisms of 4-CH$_3$-NB are not affected by Ca^{2+}/K$^+$ ion exchange. The
plateau of the remobilized 4-CH$_3$-NB pulse lasted for about 8 days. Thereafter, the
concentration of 4-CH$_3$-NB in the injection well dropped to values close to the
quantification limit. For practical reasons 4-CH$_3$-NB was used in this field study
as a relatively weakly sorbing NAC. Polynitroaromatic compounds such as TNT
or 1,3-DNB have a much higher affinity for clay minerals and are much more
susceptible than 4-CH$_3$-NB to mobilization due to induced cation exchange in
natural porous media.[125]

The above review of sorption of NACs demonstrates that the composition of
exchangeable cations at clays is crucial for assessing the sorption and mobility of
NACs under typical groundwater conditions. The observed effects of NAC substituents and the influence of exchangeable cations on the adsorption of NACs to aquifer
matrices are fully consistent with the adsorption features of NACs in suspensions
of pure clay minerals. Thus, mechanistic information obtained in clay mineral model
systems proved to be useful to interpret the sorption behavior and also to control
the mobility of NACs in natural porous media.

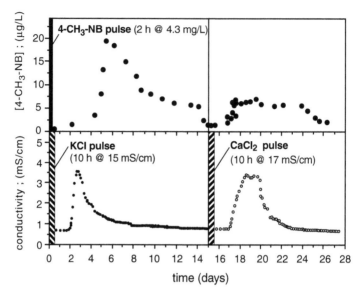

Figure 12.8   Dual well field injection test conducted to evaluate the effects of induced cation exchange at clays on the mobility of NACs (here, 4-CH₃-NB) in contaminated aquifers. Injection of KCl electrolyte retarded (left-hand panels) 4-CH₃-NB, whereas injection of CaCl₂ electrolyte mobilized 4-CH₃-NB (right-hand panels). Shown are solute breakthrough curves measured in the extraction well 3.8 m downstream of the injection well. (Adapted from Weissmahr, K.W. et al. 1999. *Environ. Sci. Technol.* 33(15):2598.)

12.4 ABIOTIC REDUCTION OF NITROAROMATIC COMPOUNDS

Nitroaromatic compounds as well as other classes of important subsurface pollutants (e.g., halogenated solvents, dyes, pesticides) are quite persistent in the presence of oxygen, but they may be susceptible to abiotic and/or biological reductive transformation under anaerobic conditions.[9,16,52,72,88,115,120] Several studies in natural systems or in systems mimicking the complex natural environment reported that reduction of NACs occurred fast and without a lag phase under anoxic conditions,[5,116,128] suggesting that reduction of NACs, at least partly, involves abiotic reactions. A survey of investigations in which abiotic reduction of NACs was postulated in anoxic groundwaters and sediments can be found in References 43, 72, and 129. However, many natural reductants are involved continuously in biogeochemical reactions, and it is somewhat arbitrary to classify NAC reductions into strictly abiotic and biological processes. Due to microbial activity in the pristine and contaminated subsurface, a variety of chemical reductants can be formed that are thermodynamically capable of reducing NACs under anoxic conditions. Since the reactivity as well as the product distribution for the reduction of NACs with the various reductants can vary considerably, it is important to know which reductant(s) is(are) predominantly involved in the reaction. Unfortunately, the information is not readily available from the various studies reporting the reduction of NACs in subsurface environments. It is the purpose of the following sections to evaluate the

various types of environmental- and compound-specific factors that may control the
rates of reduction of NACs in natural systems.

12.4.1 Thermodynamic and Kinetic Aspects of NAC Reduction

Whether a redox reaction can occur spontaneously under given conditions can
be seen by inspection of the reduction potentials (E_h) of the reactants involved.
Figure 12.9 summarizes standard reduction potentials (relative to the standard hydrogen electrode [SHE]) of a selection of environmentally relevant organic and inorganic
redox couples. The $E_h^o(w)$ values represent standard half-cell reduction potentials
at molar concentrations of the redox couples under typical environmental conditions
(for more details see References 95 and 107). Since the changes of the standard free
energy ($\Delta G_h^0(w)$) are negative for spontaneous reactions, a thermodynamically
favorable reaction requires the $E_h^o(w)$ values of oxidant $E_h^o(w)^{ox}$, e.g., NAC, to
exceed that of the reductant $E_h^o(w)^{red}$ (see Equations 3 and 4).

$$\Delta G_h^0(w) = -z \cdot F \cdot \Delta E_h^0(w) \tag{3}$$

$$\Delta E_h^0(w) = E_h^0(w)^{ox} - E_h^0(w)^{red} \tag{4}$$

where z is the number of electrons transferred and F represents Faraday's constant.
Thus, as can be seen from the relative positions of the redox couples on the $E_h^o(w)$
scale, under standard environmental conditions, nitrobenzene (and other NACs, data
not shown) can, in principle, be reduced by many of the reductants typically present
under strictly anoxic conditions in the subsurface. For instance, a variety of reduced
iron and sulfur species, certain moieties of natural organic matter (e.g., hydroquinone
structures as represented by juglone [Figure 12.9]), as well as complexes of other
transition metals (e.g., copper, cobalt; not shown in Figure 12.9) exhibit reduction
potentials that are low enough to reduce NACs fairly quantitatively to the corresponding anilines (for details, see Reference 43).

Such considerations, however, do not allow conclusions about the activation
energies of the various reaction steps involved and, thus, about the kinetics of the
overall NAC reduction. Indeed, the thermodynamically favorable reduction of NACs
to the corresponding anilines by major soluble bulk reductants such as H_2S can be
very slow.[22,49,55,96] Abiotic reduction of NACs requires the transfer of two, four, or
six electrons to yield stable intermediates (Figure 12.1b). In the majority of the
reactions, electrons are transferred in sequential steps.[23] According to Equation 5,
the transfer of the first electron from a reductant to the NAC leads to the formation
of a nitroaryl radical anion, which is much more reactive than the parent compound.
The formation of reactive nitroaryl radical intermediates, i.e., transfer of the first
electron, is an endergonic reaction, which typically represents the rate-determining
step in the overall reduction of NACs (Equation 5).

$$ArNO_2 + e^- = ArNO_2^{\bullet -} \qquad E_h^{1\prime}(ArNO_2) \tag{5}$$

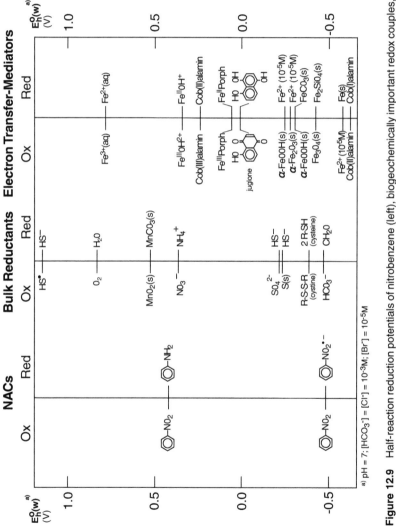

Figure 12.9   Half-reaction reduction potentials of nitrobenzene (left), biogeochemically important redox couples, and selected electron transfer mediators including various Fe(III)/Fe(II) couples (right) (aq = aqueous; porph = porphyrin). $E_h^o(w)$ indicates the standard reduction potential at standardized environmental conditions, i.e., T = 25°C, pH = 7.0, [Cl⁻] = [HCO₃⁻] = 10⁻³ M, [Br⁻] = 10⁻⁵ M. (Data from References 95 and 106 and references cited therein.)

$E_n^{1\prime}(ArNO_2)$ denotes the one-electron reduction potential of the half-reaction shown in equation 5 at pH 7.0. Comparison of the $E_h^{1\prime}$ values for Equation 5 (typically -0.6 to -0.3 V for various NACs, see later) with the potential for the overall NAC reduction to the corresponding anilines (Figure 12.9) reveals that the formation of $ArNO_2^-$ species involves a considerable energy barrier in the overall reduction of NACs. The reduction of NACs by HS^- illustrates that the transfer of the first electron to the NACs is highly endergonic in contrast to the standard reduction potentials for the overall reaction given in Figure 12.9.

$$ArNO_2 + HS^- = ArNO_2^{\bullet -} + HS^{\bullet} \tag{6}$$

$$\Delta G_1^0(w) = -F \cdot (E_h^{1\prime}(ArNO_2) - E_h^{1\prime}(HS^{\bullet})) = +154 \text{ kJ mol}^{-1} \tag{7}$$

In Equation 6, it is likely that the rate-determining step is the actual electron transfer. If one assumes that the free energy of activation is proportional to the standard free energy of the transfer of the first electron at pH 7 ($\Delta G_1^0(w)$), the high value calculated from Equation 7 ($+154$ kJ mol^{-1}) explains why the reduction of nitrobenzene by HS^- is very slow.

12.4.2 Linear Free Energy Relationships to Evaluate the Reactivity of NACs

Depending on the biogeochemical conditions and the reactant(s) involved in the reduction of NACs, various steps in the formation of nitroaromatic radical anions can be rate determining (e.g., biological uptake, adsorption to reactive surfaces, actual electron transfer) and can therefore contribute to the observable overall rate constant. Linear free energy relationships (LFERs of the type of Equation 9) are helpful tools to evaluate the factors that control the reactivity of different NACs in a given system and, under certain conditions, to predict the reactivity of NACs in a given system. The underlying principle of LFERs is that the free energy of activation of the first electron transfer to an NAC ($\Delta G^{\neq}$) is proportional to the standard free energy of the transfer of the first electron ($\Delta G_1^0(w)$). Measured rate constants, k_{obs}, are related to the free energy of activation of the reaction as follows:

$$k_{obs} = \alpha \cdot e^{(-\Delta G^{\neq}/RT)} \tag{8}$$

where α is a proportionality factor. Rate constants can be correlated with the $E_h^{1\prime}$ values to yield an LFER of the general form (for details, see References 23 and 95).

$$\log k_{obs} = a \cdot \frac{E_h^{1\prime}(ArNO_2)}{RT/nF} + b \tag{9}$$

where RT/nF has a value of 0.059 V at 25°C for a one-electron transfer. Equation 9 can be used to evaluate which steps (e.g., the actual transfer of the electron) are determining the rate of reduction. A significant correlation between log k_{obs} and $E_h^{1'}$ (ArNO$_2$) with a slope ($\underline{a}$) of close to unity indicates that the actual transfer of the first electron from the reductant to the NACs is rate determining (for details, see References 23 and 96). If a much weaker dependency of log k_{obs} from $E_h^{1'}$ is found, other reaction steps such as the formation of a precursor complex of NACs with the reductants and/or other processes are important. LFERs of the type in Equation 9 exist for reactions of NACs with various biogeochemically important reductants in model systems based on the rate constants summarized in Table 12.3. For such LFERs the availability of $E_h^{1'}$ values of NACs is a prerequisite. Since the formation of the nitroaryl radical anion is reversible (Figure 12.1b), $E_h^{1'}$ values can be measured (e.g., by pulse radiolysis[74–76]) and are published for a variety of such compounds.[95,117]

For NACs of environmental relevance such as TNT and its aminonitrotoluene reduction products directly determined $E_h^{1'}$ values are not available in the literature. Using an existing LFER established from a subset of five (P)NACs with known $E_h^{1'}$ values,[95,117] unknown $E_h^{1'}$ values of NACs can be estimated by measuring their rate of reduction. For this purpose, it is essential to choose an experimental system, where the actual electron transfer contributes significantly to the overall rate of reduction.[96] Figure 12.10 shows the measured second-order rate constants for the reduction of five reference NACs as well as TNT and related aminonitrotoluenes with a model hydroquinone [HJUG$^-$, i.e., reduced juglone (8-hydroxy-1,4-naphtho-quinone)] in the presence of hydrogen sulfide.[49] An excellent correlation of log k_{HJUG^-} and $E_h^{1'}$ values over a range of almost 250 mV of $E_h^{1'}$ (i.e., more than five orders of magnitude in k_{HJUG^-}) was obtained. The $E_h^{1'}$ data estimated for TNT and the aminonitrotoluenes cover a range of 215 mV, reflecting the changes of the electron acceptor properties of the compounds which occur upon sequential reduction of TNT. The sequence of decreasing $E_h^{1'}$ values (TNT $\gg$ 2-A-4,6-DNT $>$ 4-A-2,6-DNT $\gg$ 2,6-DA-4-NT $>$ 2,4-DA-6-NT) can be rationalized by the decreased stabilization of a nitroaryl radical anion with an decreasing number of electron-withdrawing nitro substituents. Generally, electronic, resonance, and steric effects of the substituents of NACs that favor the stabilization of nitroaryl radicals lead to less negative $E_h^{1'}$ values and thus favor NAC reduction.[95] Such is the case if one or more electron-withdrawing substituents are present (e.g., –Cl, –COCH$_3$, –NO$_2$). Electron-donating substituents (e.g., –CH$_3$, –NH$_2$), as well as substituents *ortho* to the nitro group, lower the one-electron reduction potential. Note that the substituent effects on $E_h^{1'}$ values are similar to the effects on the π-acceptor properties of NACs.

12.4.3 Reduction of NACs in Homogeneous Aqueous Solution

Reduction of most NACs by soluble reductants which are commonly present under anoxic conditions (e.g., H$_2$S, H$_2$) is very slow due to the high activation energies involved. In the presence of suitable electron transfer mediators such as dissolved NOM, NACs are reduced at significant rates in the subsurface. Compared to the reaction with aqueous H$_2$S/HS$^-$, the reduction of NACs is accelerated by orders of magnitude upon addition of dissolved NOM or other solutes that have reducible quinone moieties to an aqueous solution of H$_2$S/HS$^-$.[22,96,112] In the presence of H$_2$S, highly reactive species of reduced NOM or juglone are formed by Michael-addition

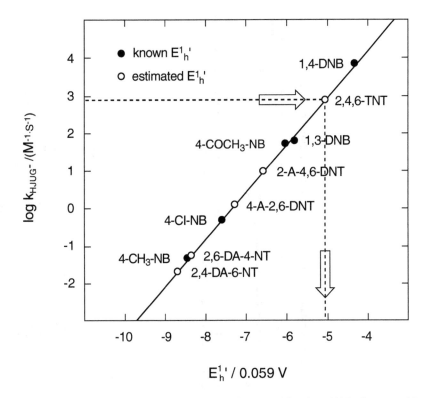

Figure 12.10 Plot of the second-order rate constants k_{HJUG^-} of various NACs (measured in 5 mM H_2S and variable concentrations of juglone (2–20 µM) at pH 6.60) vs. the $E_h^{1\prime}$ of the compounds. Filled circles (●) represent reference compounds with known $E_h^{1\prime}$, whereas $E_h^{1\prime}$ of the TNT and aminonitrotoluenes (○) are calculated from the measured second-order rate constants using a rearranged LFER: $E_h^{1\prime}$ = 0.0476 log k_{HJUG^-} − 0.436.[49]

of HS⁻.[84] Such reduced moieties of NOM react with many NACs at significant rates (k_{NOM} and k_{HJUG^-}, Table 12.3). Because such NOM species can be rereduced continuously and at high rates by bulk reductants such as H_2S, they can act as electron transfer mediators as illustrated in Scheme 1.

Scheme 1

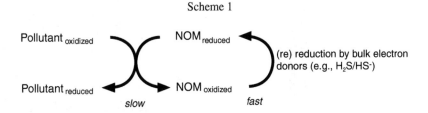

The pseudo-first-order reduction rate constants, k_{obs}, determined for several substituted nitrobenzenes using either various NOMs from very different environments or

Table 12.3 Names, Abbreviations, One-Electron Reduction Potentials ($E_h^{1\prime}$), and Normalized Rate Constants of the Reduction of NACs in Various Experimental Systems

| | | | Sulfate-Reducing Model Systems | | Iron-Reducing Model Systems | | | | | Grindsted Landfill Site | |
| | | | | | Batch | | | | Columns | Batch | |
	Abbreviation	$E_h^{1\prime}$ (mV)	NOM[a]	Juglone[b]	Fe(II)/ Goethite[c]	Fe(II)/ Magnetite[d]	Fe(II)/ Porphyrin[e]	GS-15/ FeOOH[f]	Aquifer Sediment[f]	Ground-water and Sediment[g]	Reducing Ground-water[h]
2,4,6-Trinitrotoluene	TNT	−300		1,540	21				50		
2-Amino-4,6-dinitrotoluene	2-A-4,6-DNT	−390		20	2.1				2.7		
4-Amino-2,6-dinitrotoluene	4-A-2,6-DNT	−430		2.5	1.6				3.2		
2,4-Diamino-6-nitrotoluene	2,4-DA-6-NT	−515		0.044	0.24				0.35		
2,6-Diamino-4-nitrotoluene	2,6-DA-4-NT	−495		0.12	0.28				0.14		
Nitrobenzene	NB	−486	0.13		0.82	0.82	0.36	0.34	0.33	1.0	
2-Nitrotoluene	2-CH₃-NB	−590	0.01		0.59	0.62	0.09	0.60	0.75	0.72	
3-Nitrotoluene	3-CH₃-NB	−475	0.18			0.92	0.41	0.40	0.55		
4-Nitrotoluene	4-CH₃-NB	−500	0.08	0.10	0.31	0.47	0.25	0.30	0.40	0.73	
2-Chloronitrobenzene	2-Cl-NB	−485	0.46			2.5	3.3	4.2	4.4		
3-Chloronitrobenzene	3-Cl-NB	−405	2.4			2.1	2.0	3.0	3.0		1.0

Compound	Abbrev.		a	b	c	d	e	f	g	h		
4-Chloronitrobenzene	4–Cl–NB	−450	1.0	1.0	1.0	1.0	1.0	1.0	1.0			
2-Acetylnitrobenzene	2–COCH$_3$–NB	−470	1.0			2.1	8.8	1.2	1.5			
3-Acetylnitrobenzene	3–COCH$_3$–NB	−405	3.0			1.7	2.3	2.1	2.5			
4-Acetylnitrobenzene	4–COCH$_3$–NB	−358	118	106		4.1	7.1	5.2	6.0			
1,2-Dinitrobenzene	1,2-DNB	−287			20				6.1	28	170	
1,3-Dinitrobenzene	1,3-DNB	−345		127	3.0				3.0	6.7	65	
1,4-Dinitrobenzene	1,4-DNB	−257	14,400									
2-Nitroaniline	2–NH$_2$–NB	<−560		0.0048	0.34				0.080			
3-Nitroaniline	3–NH$_2$–NB	−500		0.090	0.59				0.25			
Absolute value of the *reference compound* (4-Cl-*NB*, *NB*; units)			$5.0 \cdot 10^{-4}$ (M^{-1} s^{-1})	0.48 (M^{-1} s^{-1})	$2.2 \cdot 10^{-4}$ (s^{-1})	0.38 (s^{-1})	2.7 (M^{-1} s^{-1})	(−)	(−)	$1.0 \cdot 10^{-6}$ (s^{-1})	$5.3 \cdot 10^{-8}$ (s^{-1})	

Note: Experimental conditions (T = 25°C unless otherwise stated).

[a] pH 7.5; 5 mM H$_2$S; NOM Hyde County.[22]
[b] pH 6.6; 5 mM H$_2$S.[49,96]
[c] pH 7.2; 1.5 mM, Fe(II), 11.2 m^2 L^{-1} goehtite.[49]
[d] pH 7.0; 2.3 mM Fe(II), 17 m^2 L^{-1} magnetite.[55]
[e] pH 7.0; 5 mM cysteine.[96]
[f] pH 7.2; 20 mM NaHCO$_3$.[45,49] (See Equation 13.)
[g,h] T 10°C.[89]

reduced juglone were proportional to the concentration of the electron transfer mediator.[22,49,96] For a given compound at given experimental conditions (Table 12.3), a normalized second-order rate constant, $k_{Mediator}$, was calculated according to Equation 10.

$$k_{Mediator} = k_{obs}/[Mediator] \qquad (10)$$

where [Mediator] is the total concentration of the electron transfer mediator, i.e., dissolved natural organic matter (NOM) or juglone (HJUG⁻). Good linear correlations and slopes of close to unity were obtained between the reactivity (expressed by log $k_{Mediator}$) and the $E_h^{1\prime}$ values of the compounds for the reduction of several mono- and polynitroaromatic compounds, including TNT and related aminonitrotoluenes by hydroquinone model compounds[49,96] (Figure 12.10), by reduced NOM,[22] as well as in solutions of H_2S containing NOM derived from exudates of bacteria.[38] Note that on a carbon normalized basis, NOM derived from very different environments exhibited quite similar electron transfer mediator properties.[22] Thus, if NOM controls the reduction of NACs in a given system, a very strong dependence of the reactivity of NACs on their $E_h^{1\prime}$ can be expected. As becomes evident from the data shown in Table 12.3 and Figure 12.10, the rate of the complete reduction of TNT, i.e., formation of 2,4,6-triaminotoluene (TAT) by reaction of TNT with NOM, is about five orders of magnitude lower than the rate of TNT disappearance.

Besides NOM, some other soluble electron donors present in anaerobic environments might contribute to abiotic NAC reduction. It has been postulated that proteins and other biomolecules, including cofactors that contain transition metal complexes, might be responsible for NAC reduction.[126,127] However, the available data are too scarce to allow a sound assessment of the importance of these electron donors for NAC reductions in the environment.

12.4.4 Reduction of NACs in Heterogeneous Aqueous Systems

There is a fairly large body of evidence in the literature suggesting that abiotic reduction of many NACs proceeds much faster in the presence of solid matrices than in homogeneous aqueous solution under otherwise similar conditions.[25,72,89] An appropriate characterization of such systems in terms of active reductants is very difficult, since some highly reactive species may be present only as intermediates at low steady state concentrations and powerful analytical methods to characterize the reactants present at surfaces under *in situ* conditions are hardly established. Nevertheless, there is growing evidence that ferrous iron species associated with mineral surfaces play a pivotal role in the rapid reductive transformation of NACs and other contaminants in many anoxic subsurface matrices.[41,45,55,70,71,77,89]

12.4.4.1 Significance of Iron-Reducing Conditions for the Transformation of NACs

As illustrated in Scheme 2, the transformation of NACs by Fe(II) surface species in the anoxic subsurface can be coupled with biogeochemical processes such as microbial reduction of ferric iron or adsorption of Fe(II) from solution.

Scheme 2

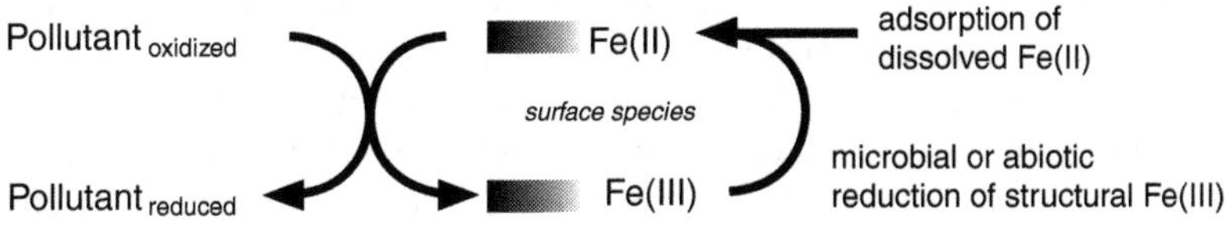

Pollutant $_{oxidized}$

Fe(II)
surface species

adsorption of
dissolved Fe(II)

Pollutant $_{reduced}$

Fe(III)

microbial or abiotic
reduction of structural Fe(III)

Iron-reducing conditions prevail in many sub- or anoxic aquifers and soils. In the following we will discuss the basic factors controlling the reaction of NACs and ferrous iron at mineral surfaces. For more details concerning the formation and reactivity of surface bound Fe(II) species, see Haderlein and Pecher.[41]

Figure 12.11 shows a typical example of the reactivity of ferrous iron at ambient pH in (anoxic) aqueous solutions or suspensions containing an NAC and Fe(III)-bearing minerals, respectively. In the presence of both Fe(II) and goethite, TNT was reduced by reactive Fe(II) species formed after adsorption of Fe(II) to the surface. Reactivity was not observed during the time scale of the experiments in assays containing solely Fe(II) or iron(hydr)oxides (Figure 12.11). Addition of Fe(III)(hydr)oxides or magnetite to anoxic solutions of Fe(II) creates highly reactive systems, whereas ferrous iron in aqueous solution was a poor reductant for NACs (Figures 12.9 and 12.11). Systems containing dissolved Fe(II) and iron(hydr)oxide minerals showed a high reactivity as long as sufficient aqueous Fe(II) was available

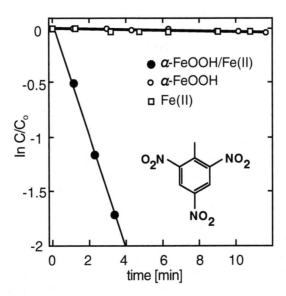

Figure 12.11 Reduction kinetics of TNT in the presence of Fe(II) and goethite in batch suspensions follow pseudo-first-order behavior (assays contained 1.5 mM of Fe(II) and 11.2 m^2 L^{-1} of goethite at pH 7.2.[49]). In suspensions of either dissolved Fe(II) or goethite, no reaction occurred.

to regenerate the reactive surface sites and the amount of electrons transferred to NACs matched very well the consumption of Fe(II).[55] Thus, reactive Fe(II) species formed at these surfaces can be regenerated rapidly by uptake of Fe(II) from solution once they have been consumed by oxidation of pollutants. Although thermodynamically feasible (Figure 12.9), the reduction of NACs by structural Fe(II) of minerals such as magnetite is a relatively slow processes.[55] Long-term reaction rates of other reducible pollutants with Fe(II)-bearing minerals such as iron sulfides or phyllosilicates were rather slow at near-neutral pH.[8,26,58,59,98] Therefore, it seems unlikely that such reactions can compete with the significant rates of formation and regeneration of reactive, surface bound Fe(II).

In addition to "abiotic" processes, reactive Fe(II) surface species can also be generated by microbial reduction of ferrous iron,[17,56,57,66,67,69] e.g., see Scheme 2). The role of microbially generated Fe(II) species as reductants of NACs was investigated in columns containing FeOOH-coated sand using *Geobacter metallireducens* strain GS-15, a naturally occurring dissimilatory iron-reducing bacterium[17] that is not able to reduce NACs directly.[49] In these systems NACs were transformed stoichiometrically to the corresponding anilines, whereas no reduction of the NACs was observed in the absence of either FeOOH coatings or *G. metallireducens*. Since reduction of NACs by dissolved Fe(II) was shown to be fairly slow,[55] reactive Fe(II) surface species must have been the reductants of NACs in this system. The coupling of microbial Fe(III) reduction with the transformation of NACs can be accelerated by the addition of acetate as a readily available electron donor for *G. metallireducens* to the influent of the columns[17] (see Scheme 2). The amount of NACs reduced to the corresponding anilines matched the consumption of acetate by *G. metallireducens* following the stoichiometric ratio of 1.33 given in equations 11 and 12.[68]

$$CH_3COO^- + 8\equiv Fe(III) + 3H_2O = 8\equiv Fe(II) + HCO_3^- + CO_2 + 8H^+ \qquad (11)$$

$$ArNO_2 + 6\equiv Fe(II) + 6H^+ = ArNH_2 + 6\equiv Fe(III) + 2H_2O \qquad (12)$$

The same influence of acetate on the reduction of NACs was found in columns filled with aquifer material from a river-groundwater infiltration site and operated under iron-reducing conditions.[45] The concentration of dissolved Fe(II) within the aquifer columns was very low, and, thus, reactive Fe(II) surface sites primarily were (re)generated by the respiratory activity of iron-reducing microorganisms   rather than by adsorption of Fe(II) from solution (Scheme 2). The studies show that in natural matrices the reduction of NACs can occur by an abiotic reaction with Fe(II) surface species that originate from the oxidation of organic material by dissimilatory iron-reducing bacteria. Therefore, surface bound iron acts as a mediator for the transfer of electrons from the organic substrate of bacteria to the NACs.

Redox reactions in heterogeneous systems often involve complex rate laws due to the change of the surface properties during the course of the reaction. Most studies on abiotic reduction reactions focused on such initial reaction rates, which may be very helpful to elucidate mechanistic aspects of reactions and to determine relative rates of reduction for a series of closely related compounds. It is obvious that such rate constants are not very useful for predicting reaction rates of NACs under typical environmental conditions, where a prolonged exposure of a given contaminant to the reactive sites has to be assumed. Under such conditions, processes that regenerate reactive sites at iron(hydr)oxide surfaces, as well as aging of such surfaces, can affect the overall rate of reaction (see Scheme 2). The biogeochemical conditions prevailing in the ferrogenic column experiments mentioned above can be regarded as an extreme example, since reducing conditions in natural porous media often involve the presence of significant amounts of dissolved ferrous iron due to chemical and microbial reduction of iron(hydr)oxide minerals or dissolution of Fe(II)-containing minerals. Under such conditions, readsorption of Fe(II) from solution, which is a fairly fast process,[41,55,81] may largely control the regeneration of reactive Fe(II) surface sites. The rate-limiting step in the overall reduction of NACs under such conditions is expected to be at least partly controlled by compound specific factors.

12.4.4.2 Kinetics of NAC Reduction by Fe(II) Surface Species

To evaluate the reactivity of NACs with various Fe(II) surface species, LFERs based on initial rate constants measured in various experimental systems and the $E_h^{1'}$ of the NAC may be useful. Figure 12.12 shows the pseudo-first-order rates of TNT and related aminonitrotoluenes measured in suspensions containing goethite (α-FeOOH) and dissolved and adsorbed Fe(II). From the rate constants, $k_{obs, Fe(II)/goethite}$, listed in Table 12.3, it is evident that under the prevailing experimental conditions, TNT, aminodinitrotoluenes, and diaminonitrotoluenes can be reduced by surface bound Fe(II) species within hours. Note that the differences in reactivities of the

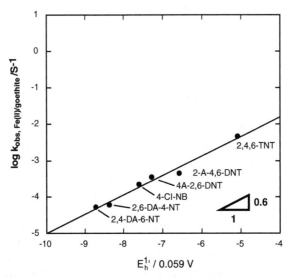

Figure 12.12 LFER of pseudo-first-order rate constants determined in Fe(II)/goethite batch suspensions (1.5 mM of Fe(II), 11.2 m^2 L^{-1} of goethite; pH 7.2) of TNT and related aminonitrotoluenes. The rate constants are normalized to the rate of 4-Cl-NB (arbitrarily chosen reference compound). (Adapted from Hofstetter, T.B. et al. 1999. *Environ. Sci. Technol.* 33(9):1484.)

(P)NACs exhibiting very different $E_h^{1\prime}$ values (e.g., TNT and 2,4-DA-6-NT, Figure 12.10) are much less pronounced in the Fe(II)/goethite system than with dissolved reductants in homogeneous solution (NOM and juglone). A similar picture of relative reactivities (as in Figure 12.12) was found for a series of ten other NACs in Fe(II)/magnetite suspensions[55] (Table 12.3). The results mentioned above suggest that in the two systems containing surface bound Fe(II) reductants, similar interactions of Fe(II) surface species with NACs were rate determining. Hence, under the experimental conditions studied, the relative reactivity of surface bound Fe(II) with respect to relative rates of the reduction of NACs in suspensions containing either goethite or magnetite seems not to be influenced by the underlying mineral surface.

One common feature in the reduction of NACs by Fe(II) species is the rather small slope of the linear relationship between log $k_{obs, rel}$ (relative rates of reduction) and $E_h^{1\prime}$ values of the compounds. For *meta-* and *para-*substituted nitrobenzenes the slopes of regression lines (Equation 9) are in the range of 0.5 to 0.6. In addition, the reactivity of some *ortho-*substituted nitrobenzenes deviated from the LFER, i.e., their relative reaction rates were higher than expected from their $E_h^{1\prime}$ values. The slopes of the correlations between log $k_{obs, rel}$ and $E_h^{1\prime}$ values of NACs, as well as the behavior of certain *ortho-*substituted compounds, indicate that in contrast to the reactions with reductants in homogeneous solution, the formation of a precursor complex contributes to the rate-limiting step in the formation of the nitroaryl radical by Fe(II) species. This interpretation is corroborated by the fact that strong competition for reactive Fe(II) sites was observed between different NACs present in mixtures.[49,55,89,96] The relative affinities of the NACs for reactive Fe(II) surface sites reflects the relative tendency of the NACs to form precursor complexes with the reductant(s)[55] and can be expressed as competition coefficient Q_c (see Equation 13).

$$Q_c = \frac{k_{obs}(NAC)}{k_{obs}(4 - Cl - NB)} \tag{13}$$

where k_{obs} is the rate constant of the reduction of a given NAC in the presence of a reference compound [4-chloronitrobenzene (4-Cl-NB); binary mixture experiments]. The Q_c values of the NACs paralleled their reactivity in single solute experiments (Table 12.3). These results demonstrate that the reduction of NACs by Fe(II) surface species is generally subject to very consistent and characteristic substituent effects. This pattern of relative reactivity of NACs can be visualized by a LFER and is therefore characteristic for the abiotic reduction of NACs by surface bound Fe(II) species.

If the reduction of NACs is coupled to biogeochemical processes providing reduced Fe(II) surface species, the rate-limiting step of the overall reaction (Scheme 2) is not necessarily the actual reduction of the NAC. In such situations, NAC reduction can be controlled by the rate of microbial Fe(III) reduction, i.e., the regeneration of reactive Fe(II) surface species (see Table 12.3). Nevertheless, relative affinities of the NACs for reactive sites, rationalized by their Q_c values (Equation 13), are indicative of the presence of similar Fe(II) surface species as active reductants. Figure 12.13 shows the results of kinetic studies of the reduction

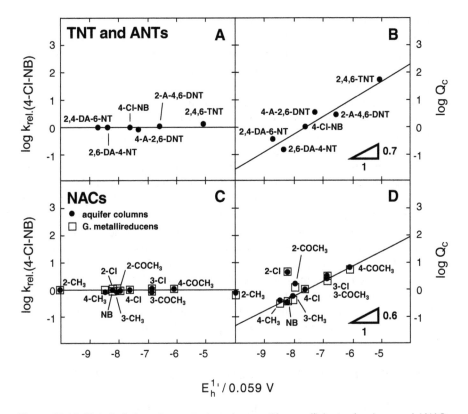

Figure 12.13 Plot of relative rate constants and competition coefficients of various model NACs as well as TNT and related aminonitrotoluenes (ANTs) measured in column systems vs. the $E_h^{1'}$ of the NACs. Panels A and C show the results of single solute experiments with aquifer sediment (●) and the *G. metallireducens*/FeOOH-coated sand system (□). Rate constants, k_{obs}, of an NAC were normalized to the k_{obs} of 4-Cl-NB. Panels B and D show competition experiments in binary mixtures (reference compound: 4-Cl-NB). (Adapted from Hofstetter, T.B. et al. 1999. *Environ. Sci. Technol.* 33(9):1484.)

of NACs in columns packed with an aquifer sediment matrix under iron-reducing conditions compared to the results obtained in systems containing iron oxyhydroxides and a pure culture of iron-reducing bacteria, i.e., *G. metallireducens* (see earlier). In both experimental systems, all NACs were transformed stoichiometrically to the corresponding anilines.[45,49] The rate constants of NAC reduction by *G. metallireducens* (Figure 12.13C [□]) as well as by anaerobic consortia of iron-reducing microorganisms in the sediment (Figures 12.13A and 12.13C (●)) were independent of solute concentration (pseudo-zero-order kinetics) and were very similar for all NACs (Table 12.3). These findings suggest that processes other than the actual electron transfer dominated or contributed to the rate-limiting step of the overall reaction. When binary mixtures of NACs were fed to the columns, reactivity of the solutes was competitive, which indicated that the number of reactive Fe(II) surface sites was limited. The Q_c values of NACs correlated well with the $E_h^{1'}$ values (Figures 12.13B and 12.13D), except for the *ortho*-substituted compounds whose reactivity was underestimated by the proposed LFER. The effects of NAC substituents on the Q_c values were very similar to those of $k_{obs,}$ $_{Fe(II)/goethite}$ values in sterile Fe(II)/goethite suspensions (Figure 12.12). Furthermore, very similar competition and reactivity patterns, i.e., slopes of about 0.6 in the linear correlations of log Q_c vs. $E_h^{1\prime}/0.059$ V, were found with a water soluble Fe(II)porphyrin, and Fe(II) bound to either goethite or magnetite (Table 12.3). The above results corroborate the interpretation that Fe(II) complexed to and/or associated with solid Fe(III) surfaces exhibit a very characteristic reactivity with respect to abiotic nitroaryl reduction.

The competition behavior of the NACs for biogenic Fe(II) surface sites strongly determines the relative reactivity of NACs in ferrogenic porous media. In mixtures of NACs the compounds are reduced sequentially in order of their Q_c values.[89] The reduction of NACs with low affinity for reactive Fe(II) sites is inhibited as long as stronger competitors are present. Such observations were made for the complete reduction of TNT to aminonitrotoluenes and 2,4,6-triaminotoluene (TAT, Figure 12.14). Aminonitrotoluenes, which have lower Q_c than TNT (Table 12.3), were not reduced before the complete disappearance of TNT. If the rate of regeneration and/or the number of reactive Fe(II) surface sites was increased by the addition of acetate in the feeding solution in order to increase the activity of iron-reducing bacteria (Scheme 2), TNT was rapidly and completely transformed to TAT without detectable intermediates (Figure 12.14B).

12.4.4.3 Significance of Other Solids as Reductants of NACs

Yu and Bailey[132] have shown that, in principle, NACs are reduced in aqueous solution by various metal sulfides (e.g., MnS, FeS, FeS$_2$, ZnS) which are often

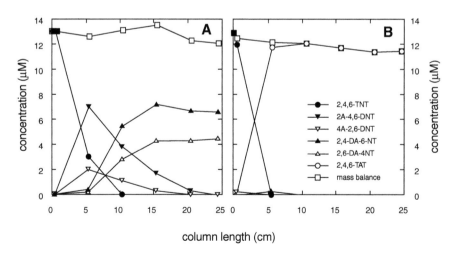

Figure 12.14 Reduction of TNT (13 µ*M* initial concentration) to aminonitrotoluenes and 2,4,6-triaminonitrotoluene (TAT) in a column (residence time of the compounds, τ = 53 h) filled with aquifer sediment and operated under iron-reducing conditions. The sequential reduction of the (P)NACs is shown in the absence (A) and presence (B) of 44 µ*M* of acetate in the influent.[49]

present in sulfate-reducing environments. Unfortunately, the conditions under which the reactions were studied were not very representative of natural environments. Lanz and Schwarzenbach[61] found that the reduction of 4-Cl–NB in aqueous suspensions of aged FeS and FeS_2 phases was very slow at circumneutral pH values. Because the oxidative dissolution of metal sulfides by NACs under environmental conditions is a relatively slow process, it seems rather unlikely that such reactions are of great environmental significance. Molecular hydrogen associated with solid surfaces might play a role for NAC reduction under methanogenic conditions or in the remediation measures involving metal iron.[6,111] In the presence of certain metal catalysts, H_2 is a very versatile reducing agent.[99] Hence, NACs could be reduced either directly by H_2 in the presence of catalytic surfaces or by electron transfer mediators that react with H_2. Although there is not enough experimental evidence to judge the role of H_2 as reductant in the subsurface, recent investigations in systems containing metal iron seem to contradict the hypothesis of NAC reduction by H_2.[92]

12.4.5 Reduction of NACs in a Geochemically Heterogeneous Anaerobic Aquifer

In pristine or polluted anaerobic environments several biogeochemical processes can take place in parallel,[10,16,46,47,63,71,89] giving rise to a variety of potential reductants for pollutants. In this matrix the characterization of the active redox species based on chemical analysis is very difficult since various potential reductants might be present simultaneously even at the grain scale.

The reduction of five nitroaromatic probe compounds (NB, 2-CH_3-NB, 4-nitrotoluene, 1,2-DNB, and 1,3-DNB; see Table 12.3) was studied in an anaerobic leachate plume of a landfill in Grindsted, Denmark.[89] The objective of the study was to identify reactants and processes that control the reductive transformation of pollutants in anaerobic aquifers, i.e., to evaluate the relative importance of biotic and abiotic processes, and to characterize the type of reductants involved. The set of model NACs had very different reactivities, as reflected by a range of 350 mV in their $E_h^{1\prime}$ values (Table 12.3). The patterns of reactivity of the NACs were studied in the leachate plume by means of an injection experiment in the plume area, by *in situ* microcosm experiments installed at five distances along a flow line in the plume, by laboratory batch experiments containing groundwater, and aquifer sediment from the same locations. The *in situ* reactivity of the compounds in the aquifer was compared to their reactivity in well defined systems as described previously in order to delineate the processes and reductants that were active in the leachate plume.

The characterization of the redox conditions in the Grindsted Landfill indicated a complex redox environment, where many of the observed redox processes occurred simultaneously. Closest to the landfill, methane production and sulfate reduction prevailed, followed by sulfate and iron reduction within the first 80 m of the plume.[10] Thus, the major reductants of NACs expected under these conditions include microorganisms, reduced NOM (50 to 120 mg C/L), and Fe(II) species at mineral surfaces (total dissolved Fe(II) as present up to 1.4 mM).

In experimental systems containing solid aquifer matrix, the added nitroaromatic probe compounds were completely reduced without the accumulation of hydroxyl-amino or aminonitro intermediates within a period of 2 to 70 days. Reductants associated with the aquifer sediment matrix strongly enhanced the reduction rates of NACs compared to those measured in filtered groundwater samples. Figure 12.15 compares the reactivity pattern of the NACs (for rates see Table 12.3) in the aquifer system with their reactivity in the presence of NOM and reactive Fe(II) sites, respectively. In the aquifer systems, DNBs, i.e., compounds exhibiting the highest $E_h^{1\prime}$, were generally reduced at very high rates and without any noticeable lag phase. Furthermore, these compounds were always reduced first within the mixture of the NACs studied. These results and the very similar pattern of NAC reduction under unchanged and microbially deactivated conditions suggest that abiotic reactions rather than biotransformation dominated NAC reduction.

The relative reactivities of NACs and their sequence of reduction in mixtures measured in the aquifer closely resembled the observations made in systems where surface bound Fe(II) species were the reductants (Table 12.3, Figure 12.15). There was competition among the NACs present in mixtures, and the relative affinity of NACs for reactive sites was consistent in both systems. The results provide strong evidence that in the anaerobic part of the Grindsted Landfill leachate plume ferrous iron adsorbed to Fe(III)(hydr)oxides was the dominant abiotic reductant and

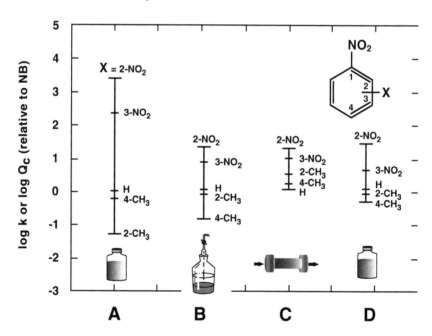

Figure 12.15 Comparison of the reactivity of the nitroaromatic tracers (1,2-DNB, 1,3-DNB, 4-CH$_3$-NB, 2-CH$_3$-NB, NB) measured in different laboratory and field systems (rates normalized to the value obtained for NB, Table 12.3). Note that similar reactivities were found in biologically active and deactivated experiments with homogeneous (groundwater) and heterogeneous (groundwater and sediment) matrix from the Grindsted aquifer (A, B). C and D represent the results measured in suspensions of Fe(II)/goethite and in aquifer sediment columns. (Data from Reference 89).

significantly contributed to the fast overall rates of reduction of the NACs in the plume. If NOM was the predominant reductant in the leachate plume, the NAC mixture would be expected to be reduced without competition effects, and dinitrobenzenes would be expected to react two to three orders of magnitude faster than the mono-NACs, which was not the case. Furthermore, 2-CH_3-NB, the compound with, by far, the lowest one-electron reduction potential, should have been reduced at the lowest rate (Table 12.3). Throughout the studied part of the plume, there was a very similar pattern of NAC reduction, which indicated that the same type of reactions occurred in the entire anaerobic part of the aquifer. Despite the presence of various potential reductants (e.g., H_2S/HS^-, $Fe(II)_{aq}$, reduced NOM, microorganisms), the patterns of relative reactivity of the probe compounds indicated that ferrous iron associated with Fe(III)(hydr)oxide surfaces was the dominant reductant of NACs throughout the entire anaerobic part of the plume. These findings are consistent with the high concentrations of aqueous Fe(II) present in this part of the aquifer, which provided an abundant reservoir of Fe(II) bulk reductants. Under such conditions, the (re)generation of reactive Fe(II) species at the mineral surface is not, or, to a much lesser extent, is controlled by the activity of iron-reducing bacteria, since regeneration of Fe(II) sites can proceed by adsorption of aqueous ferrous iron to the mineral surfaces.[41,55,80]

12.4.6 Products of the Reduction of NACs by Fe(II) Surface Species

Under iron-reducing subsurface conditions, NACs are reduced completely to the corresponding aromatic (poly)amines irrespective of their very different $E_h^{1'}$ values. Thus, in anoxic soils and aquifers, contamination with NACs, even in the absence of microbial transformation reactions, inevitably results in the formation of environmentally problematic compounds, i.e., aromatic (poly)amines.[30,79] The reduction of NACs leads to an increased mobility of NAC-related contaminants in porous media, especially in environments that are low in organic carbon. Nitroanilines and anilines are poor π-acceptors and thus exhibit a much lower tendency to form EDA complexes than their parent compounds. Several studies[19,22,25,45,49,55,89,96] pointed out that under anoxic conditions aromatic amines accumulate in the aqueous phase and can remain stable over considerable time periods (e.g., up to 250 days as found in the anaerobic aquifer).[89] The evidence suggests that processes that contribute to the transformation of aromatic (poly)amines (e.g., binding to the solid matrix[18,54,60,114]) or organic material,[19,25,110,121] biotransformation[83,94]), either proceed very slowly or are absent under iron-reducing conditions. In fact, the irreversible binding of aromatic (poly)amines to NOM through nucleophilic addition reactions is thought to be inhibited as long as quinone moieties of the NOM are reduced.[25,121]

12.5 PERSPECTIVES ON PRACTICAL APPLICATIONS AND OUTLOOK

Remediation strategies for contaminated sites require the removal of NACs at the source of the contamination as well as in the polluted groundwater plume. Soil contaminated with NACs can be treated microbiologically or, in the case of nonbio-degradable NACs, by using technical systems that apply a two-stage anaerobic/aerobic process. The procedure involves complete microbial and/or chemical reduction of NACs and subsequent binding of the products to organic matter under oxic conditions (see Chapter 4 and References 12, 13, 19, 21, and 64). Contaminated aquifers, however, have much lower concentrations of both NACs and organic material than contaminated soils, thus requiring different approaches for remediation. The present knowledge on abiotic reduction and sorption processes of NACs allows improved *ex situ* treatment of NAC contaminated soils as well as *in situ* remediation of contaminated groundwater.

The removal of di- and trinitro compounds from contaminated soils in two-stage anaerobic/aerobic treatments can be optimized by reduction of NACs under iron-reducing conditions. The complete reduction of NACs by Fe(II) surface species, as shown for aquifer matrices, yielded the corresponding aromatic amines without the accumulation of problematic intermediates such as arylhy-droxylamines.[49] Arylhydroxylamines can compromise the goal of complete NAC reduction since such intermediates can bind to organic matter.[2,4,25,86,121] Such "bound" nitro- and azoxy-functional groups are reduced at much slower rates than in solution.[1,3]

The remediation of contaminated aquifers can benefit from the effect of background electrolytes on the sorption of NACs. Enhanced release of adsorbed NACs from contaminated sediments might be desirable in the case of *ex situ* treatment of the sediments or for *in situ* pump and treat or funnel and gate remediation schemes (e.g., Reference 20). In such situations, the fraction of NACs in the mobile phase and thus their transport in groundwater as well as their availability for microbial and chemical transformations might be enhanced by the exchange of (naturally adsorbed) K^+ cations by strongly hydrated cations such as Na^+ or Ca^{2+}. Conversely, immobilization of NACs due to enhanced adsorption might be considered to secure groundwater wells and to gain time for further evaluation of the contaminated sites. In those situations, injection of KCl electrolytes into an aquifer might significantly decrease the mobility of NACs. The available results further suggest that the efficiency of electrolyte injections to mobilize or retard NACs depends on the ionic strength of the groundwater. Since changes in the K^+ saturation affect the sorption properties of clays to a higher degree at low and medium ionic strength, enhanced retardation of NACs by KCl injections would be favorable under such conditions. Conversely, NACs should be mobilized most efficiently at higher ionic strength by injection and/or in the presence of strongly hydrated cations. Due to the peculiar subsurface chemistry of NACs, common practices used to mobilize organic groundwater contaminants, such as the injection of surfactants, are probably much inappropriate.

The distribution of NACs between aqueous and solid phases affects their mobility, availability for biotransformation(s), and, possibly, their susceptibility to abiotic reduction reactions. Under anoxic conditions the reduction of structural iron of clay minerals[35,36,57,103,105] can have an impact on both the adsorption and the abiotic reduction of NACs. Certain physical and chemical properties of the phyllosilicates that are relevant to EDA complex formation depend on the oxidation state of structural iron,[103,104] and reduced iron in clays is a potential reductant of pollutants.[26] Preliminary results show that at the surfaces of reduced clays both EDA complex formation and reduction of NACs to aromatic amines take place.[48] Thus, enhancement of geochemical processes that lead to an immobilization of NACs and subsequent reduction at the same clay mineral surfaces could offer a future remediation perspective on treatment of contaminated groundwaters.

Biotransformation of NACs can also yield products such as substituted nitrobenzoic acids.[14,93] In contrast to neutral NACs, nitroaromatic anions neither form EDA complexes with phyllosilicates nor do they sorb significantly to organic matter. Such polar metabolites presumably are highly mobile in contaminated aquifers and may threaten drinking water facilities. The fate of such mobile metabolites during anaerobic/aerobic treatments as well as their importance compared to other degradation products of NACs such as aminonitrotoluenes is not yet known. Therefore, it is crucial to understand the processes that determine the formation and the fate of such metabolites in the subsurface as well as during treatment measures.

REFERENCES

1. Achtnich, C., E. Fernandes, J.-M. Bollag, H.-J. Knackmuss, and H. Lenke. 1999. Covalent binding of reduced metabolites of [^{15}N$_3$]TNT to soil organic matter during a bioremediation process analyzed by ^{15}N NMR spectroscopy. *Environ. Sci. Technol.* 33(24), 4448–4456.
2. Achtnich, C., H. Lenke, U. Klaus, M. Spiteller, and H.-J. Knackmuss. 1999. Influence of extensive anaerobic treatment on the immobilization of reduced TNT in soil. IGB Berichte Nr. 015 Fraunhofer Institut für Grenzflächen und Bioverfahrenstechnik Stuttgart, Germany.
3. Achtnich, C., P. Pfortner, M. G. Weller, H. Lenke, and H.-J. Knackmuss. 1999. Reductive transformation of bound trinitrophenyl residues and free TNT during a bioremediation process analyzed by immunoassay. *Environ. Sci. Technol.* 33(19):3421–3426.
4. Achtnich, C., U. Sieglen, H. J. Knackmuss, and H. Lenke. 1999. Irreversible binding of biologically reduced 2,4,6-trinitrotoluene to soil. *Environ. Toxicol. Chem.* 18(11):2416–2423.
5. Adhya, T. K., B. Sudhakar, and N. Sethunathan. 1981. Fate of fenitrothion, methylparathion, and parathion in anoxic, sulfur-containing soil systems. *Pestic. Biochem. Physiol.* 16:14–20.
6. Agrawal, A. and P. G. Tratnyek. 1996. Reduction of nitro aromatic compounds by zero-valent iron metal. *Environ. Sci. Technol.* 30(1):153–160.
7. Appelo, C. A. J. and D. Postma. 1993. Geochemistry, Groundwater and Pollution. A.A. Balkema, Rotterdam.
8. Assaf-Anid, N., K.-Y. Lin, and J. Mahony. 1997. Reductive dechlorination of carbon tetrachloride by iron(II) sulfide, p. 194–195. Presented at the 213th ACS National Meeting, Symposium on Redox Reactions in Natural and Engineered Aqueous Systems, San Francisco, CA, American Chemical Society, Washington, D.C.
9. Baughman, G. L. and E. J. Weber. 1994. Transformation of dyes and related compounds in anoxic sediment: kinetics and products. *Environ. Sci. Technol.* 28:267–276.

10. Bjerg, P. L., K. Rügge, J. Cortsen, P. H. Nielsen, and T. H. Christensen. 1999. Degradation of aromatic and chlorinated aliphatic hydrocarbons in the anaerobic part of the Grindsted Landfill leachate plume: in situ microcosm and laboratory batch experiments. *Ground Water.* 37(1):113–121.

11. Briegleb, G. and J. Czekalla. 1960. Elektronenüberführung durch Lichtabsorption und –emission in Elektronen-Donator-Acceptor-Komplexen. *Angew. Chem.* 72(12):401–413.

12. Bruns-Nagel, D., J. Breitung, E. von Löw, K. Steinbach, T. Gorontzy, M. Kahl, K. H. Blotevogel, and D. Gemsa. 1996. Microbial transformation of 2,4,6-trinitrotoluene in aerobic soil columns. *Appl. Environ. Microbiol.* 62(7):2651–2656.

13. Bruns-Nagel, D., O. Drzyzga, K. Steinbach, T. C. Schmidt, E. von Löw, T. Grontzy, K.-H. Blotevogel, and D. Gemsa. 1998. Anaerobic/aerobic composting of 2,4,6-trinitrotoluene-contaminated soil in a reactor system. *Environ. Sci. Technol.* 32(11):1676–1679.

14. Bruns-Nagel, D., T. C. Schmidt, O. Drzyzga, E. von Löw, and K. Steinbach. 1999. Identification of oxidized TNT metabolites in soil samples of a former ammunition plant. *Environ. Sci. Pollut. Res.* 6(1):7–10.

15. Chiou, C. T., L. J. Peters, and V. H. Freed. 1979. A physical concept of soil-water equilibria for nonionic compounds. *Science.* 206:831–832.

16. Christensen, T. H., P. Kjeldsen, H.-J. Albrechtsen, G. Heron, P. H. Nielsen, P. L. Bjerg, and P. E. Holm. 1994. Attenuation of landfill leachate pollutants in aquifers. *Crit. Rev. Environ. Sci. Technol.* 24(2):119–202.

17. Coates, J. D., E. J. P. Phillips, D. J. Lonergan, H. Jenter, and D. R. Lovley. 1996. Isolation of *Geobacter* species from diverse sedimentary environments. *Appl. Environ. Microbiol.* 62(5):1531–1536.

18. Cowen, W. F., A. M. Gastinger, C. E. Spanier, J. R. Buckel, and R. E. Bailey. 1998. Sorption and microbial degradation of toluenediamines and methylenedianiline in soil under aerobic and anaerobic conditions. *Environ. Sci. Technol.* 32(5):598–603.

19. Daun, G., H. Lenke, M. Reuss, and H.-J. Knackmuss. 1998. Biological treatment of TNT-contaminated soil. 1. Anaerobic cometabolic reduction and interaction of TNT and metabolites with soil components. *Environ. Sci. Technol.* 32(13):1956–1963.

20. Domenico, P. A. and F. W. Schwartz. 1998. *Physical and Chemical Hydrogeology*, 2nd ed. Wiley, New York.

21. Drzyzga, O., D. Bruns-Nagel, T. Gorontzy, K. H. Blotevogel, D. Gemsa, and E. von Löw. 1998. Incorporation of ^{14}C-labeled 2,4,6-trinitrotoluene metabolites into different soil fractions after anaerobic and anaerobic-aerobic treatment of soil/molasses mixtures. *Environ. Sci. Technol.* 32(22):3529–3535.

22. Dunnivant, F. M., R. P. Schwarzenbach, and D. L. Macalady. 1992. Reduction of substituted nitrobenzenes in aqueous solutions containing natural organic matter. *Environ. Sci. Technol.* 26(11):2133–2141.

23. Eberson, L. 1987. *Electron Transfer Reactions in Organic Chemistry.* Springer-Verlag, Berlin.

24. Egorochkin, A. N. and S. E. Skobeleva. 1979. Infrared spectroscopy of the hydrogen bond as a method for the investigation of intramolecular interactions. *Russ. Chem. Rev.* 48(12):1198–1211.

25. Elovitz, M. S. and E. J. Weber. 1999. Sediment-mediated reduction of 2,4,6-trinitrotoluene and fate of the resulting aromatic (poly)amines. *Environ. Sci. Technol.* 33(15):2617–2625.

26. Ernstsen, V. 1996. Reduction of nitrate by Fe^{2+} in clay minerals. *Clays Clay Miner.* 44(5):599–608.

27. Fesch, C. and S. B. Haderlein. 1998. Competitive sorption and multiple-species subsurface transport of nitroaromatic explosives: implications for their mobility at contaminated sites, p. 27–34. In M. Herbert and K. Kovar (eds.), *Groundwater Quality: Remediation and Protection*, IAHS, Tübingen, Germany.

28. Fesch, C., W. Simon, S. B. Haderlein, P. Reichert, and R. P. Schwarzenbach. 1998. Nonlinear sorption and nonequilibrium solute transport in aggregated porous media: experiments, process identification and modeling. *J. Contam. Hydrol.* 31(3–4):373–407.

29. Feuer, H. (ed.). 1981. *The Chemistry of the Nitro and Nitroso Groups*, 2nd ed. Robert E. Krieger Publishing, New York.

30. Fishbein, L. 1984. Aromatic amines, p. 1–40. In O. Hutzinger (ed.), *The Handbook of Environmental Chemistry–Anthropogenic Compounds*. Springer-Verlag, Berlin.

31. Foster, R. (ed.). 1975. *Molecular Association*. Academic Press, New York.

32. Foster, R., B. Dale, and D. L. Hammick. 1954. Interaction of polynitrocompounds with aromatic hydrocarbons and bases. XIII. The effect of changing the nitro-compound. *J. Chem. Soc.* (76) 3986.

33. Foster, R. and R. K. Mackie. 1962. Interaction of electron acceptors with bases. 5. The Janovsky and Zimmermann reactions. *Tetrahedron.* 18:1131–1135.

34. Foster, R. E. 1969. *Organic Charge Transfer Complexes*. Academic Press, New York.

35. Gates, W. P., A. M. Jaunet, D. Tessier, M. A. Cole, H. T. Wilkinson, and J. W. Stucki. 1998. Swelling and texture of iron-bearing smectites reduced by bacteria. *Clays Clay Miner.* 46(5):487–497.

36. Gates, W. P., H. T. Winkinson, and J. W. Stucki. 1993. Swelling properties of microbially reduced ferruginous smectite. *Clays Clay Miner.* 41(3):360–364.

37. Gelessus, A. 1997. Die Berechnung von Solvationsenergien mit dem COSMO-Verfahren bei semiempirischen Wellenfunktionen. Dissertation. University of Zurich.

38. Glaus, M. A., C. G. Heijman, R. P. Schwarzenbach, and J. Zeyer. 1992. Reduction of nitroaromatic compounds mediated by *Streptomyces* sp. exudates. *Appl. Environ. Microbiol.* 58(6):1945–1951.

39. Greenland, D. J. and M. H. B. Hayes (eds.). 1978. *The Chemistry of Soil Constituents*. Wiley, New York.

40. Haas, R. and E. von Löw. 1986. Grundwasserbelastung durch eine Altlast. Die Folgen einer ehemaligen Sprengstoffproduktion für die heutige Trinkwassergewinnung. *Forum Städte-Hygiene.* 37:33–43.

41. Haderlein, S. B. and K. Pecher. 1999. Pollutant reduction in heterogeneous Fe(II)/Fe(III)-systems, p. 342–356. In D. L. Sparks and T. Grundl (eds.), *Mineral/Water Interfacial Reactions: Kinetics and Mechanisms, Symposium Series 715*. American Chemical Society, Washington, D.C.

42. Haderlein, S. B. and R. P. Schwarzenbach. 1993. Adsorption of substituted nitrobenzenes and nitrophenols to mineral surfaces. *Environ. Sci. Technol.* 27(2):316–326.

43. Haderlein, S. B. and R. P. Schwarzenbach. 1995. Environmental processes influencing the rate of abiotic reduction of nitroaromatic compounds in the subsurface, p. 199–225. In J. Spain (ed.), *Biodegradation of Nitroaromatic Compounds*, Plenum Press, New York.

44. Haderlein, S. B., K. W. Weissmahr, and R. P. Schwarzenbach. 1996. Specific adsorption of nitroaromatic explosives and pesticides to clay minerals. *Environ. Sci. Technol.* 30(2):612–622.

45. Heijman, C. G., E. Grieder, C. Holliger, and R. P. Schwarzenbach. 1995. Reduction of nitroaromatic compounds coupled to microbial iron reduction in laboratory aquifer columns. *Environ. Sci. Technol.* 29:775–783.

46. Heron, G., P. L. Bjerg, P. Gravesen, L. Ludvigsen, and T. H. Christensen. 1998. Geology and sediment geochemistry of a landfill leachate contaminated aquifer (Grindsted, Denmark). *J. Contam. Hydrol.* 29(4):301–317.

47. Heron, G. and T. H. Christensen. 1995. Impact of sediment-bound iron on redox buffering in a landfill leachate polluted aquifer (Vejen, Denmark). *Environ. Sci. Technol.* 29(1):187–192.

48. Hofstetter, T. B. 1999. Reduction of polynitroaromatic compounds by reduced iron species–coupling biogeochemical processes with pollutant transformation. Dissertation ETH Nr. 13'140. ETH Zürich.

49. Hofstetter, T. B., C. G. Heijman, S. B. Haderlein, C. Holliger, and R. P. Schwarzenbach. 1999. Complete reduction of TNT and other (poly)nitroaromatic compounds under iron reducing subsurface conditions. *Environ. Sci. Technol.* 33(9):1479–1487.

50. Jafvert, C. T. 1990. Sorption of organic acids to sediments: initial model development. *Environ. Toxicol. Chem.* 9:1259–1268.

51. Jafvert, C. T., J. C. Westall, E. Grieder, and R. P. Schwarzenbach. 1990. Distribution of hydrophobic ionogenic organic compounds between octanol and water: organic acids. *Environ. Sci. Technol.* 24(12):1795–1803.

52. Jafvert, C. T. and N. L. Wolfe. 1987. Degradation of selected halogenated ethanes in anoxic sediment-water systems. *Environ. Toxicol. Chem.* 6:827–837.

53. Karickhoff, S. W. 1984. Organic pollutant sorption in aquatic systems. *J. Hydraul. Eng.* 110:704–735.

54. Klausen, J., S. B. Haderlein, and R. P. Schwarzenbach. 1997. Oxidation of substituted anilines by aqueous MnO_2: effect of co-solutes on initial and quasi steady state kinetics. *Environ. Sci. Technol.* 31(9):2642–2649.

55. Klausen, J., S. P. Tröber, S. B. Haderlein, and R. P. Schwarzenbach. 1995. Reduction of substituted nitrobenzenes by Fe(II) in aqueous mineral suspensions. *Environ. Sci. Technol.* 29(9):2396–2404.

56. Kostka, J. E. and K. H. Nealson. 1995. Dissolution and reduction of magnetite by bacteria. *Environ. Sci. Technol.* 29:2535–2540.

57. Kostka, J. E., J. W. Stucki, K. H. Nealson, and J. Wu. 1996. Reduction of structural Fe(III) in smectite by a pure culture of *Shewanella putrefaciens* strain MR 1. *Clays Clay Miner.* 44(4):522–529.

58. Kriegmann-King, M. R. and M. Reinhard. 1994. Transformation of carbon tetrachloride by pyrite in aqueous solution. *Environ. Sci. Technol.* 28(4):692–700.

59. Kriegmann-King, M. R. and M. Reinhard. 1992. Transformation of carbon tetrachloride in the presence of sulfide, biotite, and vermiculite. *Environ. Sci. Technol.* 26(11):2198–2206.

60. Laha, S. and R. G. Luthy. 1990. Oxidation of aniline and other primary aromatic amines by manganese dioxide. *Environ. Sci. Technol.* 24(3):363–373.

61. Lanz, K. and R. P. Schwarzenbach. 1988. Unpublished results. EAWAG (CH).

62. Leggett, D. C. 1991. Role of Donor-Acceptor Interactions in the Sorption of TNT and Other Nitroaromatics from Solution. Special Report 91–13. U.S. Army Cold Regions Research and Engineering Laboratory (CRRL), Hanover, NH.

63. Lendvay, J. M., S. M. Dean, and P. Adriaens. 1998. Temporal and spatial trends in biogeochemical conditions at a groundwater-surface water interface: implications for natural bioattenuation. *Environ. Sci. Technol.* 32(22):3472–3478.

64. Lenke, H., J. Warrelmann, G. Daun, K. Hund, U. Sieglen, U. Walter, and H.-J. Knackmuss. 1998. Biological treatment of TNT-contaminated soil. 2. Biologically induced immobilization of the contaminants and full-scale application. *Environ. Sci. Technol.* 32(13):1964–1971.

65. Li, A. Z., K. A. Marx, J. Walker, and D. L. Kaplan. 1997. Trinitrotoluene and metabolites binding to humic acids. *Environ. Sci. Technol.* 31(2):584–589.
66. Lovley, D. R. 1991. Dissimilatory Fe(III) and Mn(IV) reduction. *Microbiol. Rev.* 55(2):259–287.
67. Lovley, D. R., S. J. Giovannoni, D. C. White, J. E. Champine, E. J. P. Phillips, Y. A. Gorby, and S. Goodwin. 1993. *Geobacter metallireducens* gen. nov. sp. nov., a microorganism capable of coupling the complete oxidation of organic compounds to the reduction of iron and other metals. *Arch. Microbiol.* 159:336–344.
68. Lovley, D. R. and E. J. P. Phillips. 1988. Novel mode of microbial energy matabolism: organic oxidation coupled to dissimilatory reduction of iron or manganese. *Appl. Environ. Microbiol.* 54(6):1472–1480.
69. Lovley, D. R., J. F. Stolz, G. L. J. Nord, and E. J. P. Phillips. 1987. Anaerobic production of magnetite by a dissimilatory iron-reducing microorganism. *Nature.* 330:252–254.
70. Lyngkilde, J. and T. H. Christensen. 1992. Fate of organic contaminants in the redox zones of a landfill leachate pollution plume (Vejen, Denmark). *J. Contam. Hydrol.* 10:291–307.
71. Lyngkilde, J. and T. H. Christensen. 1992. Redox zones of a landfill leachate pollution plume (Vejen, Denmark). *J. Contam. Hydrol.* 10:273–289.
72. Macalady, D. L., P. G. Tratnyek, and T. J. Grundl. 1986. Abiotic reduction reactions of anthropogenic organic chemicals in anaerobic systems: a critical review. *J. Contam. Hydrol.* 1:1–28.
73. McBride, M. B. 1989. Surface chemistry of soil material, p. 35–88. In J. B. Dixon and B. B. Weed (eds.), *Minerals in Soil Environments*, 2nd ed. Soil Science of America, Madison, WI.
74. Meisel, D. and P. Neta. 1975. One-electron redox potentials of nitro compounds and radiosensitizers. *J. Am. Chem. Soc.* 97(18):5198–5203.
75. Neta, P. and D. Meisel. 1976. Substituent effects of nitroaromatic radical anions in aqueous solution. *J. Phys. Chem.* 80:519–524.
76. Neta, P., M. G. Simic, and M. Z. Hoffman. 1976. Pulse radiolysis and electron spin resonance studies of nitroaromatic radical anions. Optical absorption spectra, kinetics, and one-electron redox potentials. *J. Phys. Chem.* 80(18):2018–2023.
77. Nielsen, P. H., H. Bjarnadottir, and P. L. Winter. 1995. *In situ* and laboratory studies on the fate of specific organic compounds in an anaerobic landfill lechate plume. II. Fate of aromatic and chlorinated aliphatic compounds. *J. Contam. Hydrol.* 20:51–66.
78. Parker, J. C. 1986. Hydrostatistics in porous media, Chapter 6. In D. L. Sparks (ed.), *Soil Physical Chemistry.* CRC Press, Boca Raton, FL.
79. Parris, G. E. 1980. Environmental and metabolic transformations of primary aromatic amines and related compounds. *Residue Rev.* 76:1–24.
80. Pecher, K., S. B. Haderlein, and R. P. Schwarzenbach. 1997. Unpublished results. EAWAG (CH).
81. Pecher, K. and R. P. Schwarzenbach. 1997. Unpublished results. EAWAG (CH).
82. Pennington, J. C. and W. H. Patrick, Jr. 1990. Adsorption and desorption of 2,4,6-trinitrotoluene by soils. *J. Environ. Qual.* 19:559–567.
83. Peres, C. M., H. Naveau, and S. N. Agathos. 1998. Biodegradation of nitrobenzene by its simultaneous reduction into aniline and mineralization of the aniline formed. *Appl. Microbiol. Biotechnol.* 49(3):343–349.
84. Perlinger, J. A., W. Angst, and R. P. Schwarzenbach. 1996. Kinetics of the reduction of hexachloroethane by juglone in solutions containing hydrogen sulfide. *Environ. Sci. Technol.* 30(12):3408–3417.

85. Renner, R. 1997. New research reveals how contaminants can be "locked" into soil. *Environ. Sci. Technol.* 31(6):270A–271A.

86. Rieger, P.-G., and H.-J. Knackmuss. 1995. Basic knowledge and perspectives on biodegradation of 2,4,6-trinitrotoluene and related nitroaromatic compounds in contaminated soil, p. 1–18. In J. Spain (ed.), *Biodegradation of Nitroaromatic Compounds*, Plenum Press, New York.

87. Rippen, G. (ed.). 1996. *Handbuch Umwelt-Chemikalien: Stoffdaten, Prüfverfahren, Vorschriften*, Vol. 3, Chapter II-2.6 Rüstungsaltlasten. ecomed Publisher, Landsberg/Lech, Germany.

88. Roberts, L. A. and P. M. Gschwend. 1994. Interaction of abiotic and microbial processes in hexachloroethane reduction in groundwater. *J. Contam. Hydrol.* 16:157–174.

89. Rügge, K., T. B. Hofstetter, S. B. Haderlein, P. L. Bjerg, S. Knudsen, C. Zraunig, H. Mosbæk, and T. H. Christensen. 1998. Characterization of predominant reductants in an anaerobic leachate-contaminated aquifer by nitroaromatic probe compounds. *Environ. Sci. Technol.* 32(1):23–31.

90. Saltzman, S. and S. Yariv. 1976. Infrared and X-ray study of parathion-montmorillonite sorption complexes. *Soil Sci. Soc. Am. Proc.* 40:34–38.

91. Saltzman, S. and S. Yariv. 1975. Infrared study of the sorption of phenol and p-nitrophenol by montmorillonite. *Soil Sci. Soc. Am. Proc.* 39:474–479.

92. Scherer, M. M., B. A. Balko, and P. G. Tratnyek. 1999. The role of oxides in reduction reactions at the metal-water interface, p. 301–322. In D. L. Sparks and T. Grundl (eds.), *Mineral/Water Interfacial Reactions: Kinetics and Mechanisms, Symposium Series 715.* American Chemical Society, Washington, D.C.

93. Schmidt, T. C., K. Steinbach, E. von Löw, and G. Stork. 1998. Highly polar metabolites of nitroaromatic compounds in ammunition wastewater. *Chemosphere.* 37(6):1079–1090.

94. Schnell, S. and B. Schink. 1991. Anaerobic aniline degradation via reductive deamination of 4-amino-benzoyl-CoA in Desulfobacterium anilini. *Arch. Microbiol.* 155:183–190.

95. Schwarzenbach, R. P., P. M. Gschwend, and D. M. Imboden. 1993. *Environmental Organic Chemistry.* John Wiley & Sons, New York.

96. Schwarzenbach, R. P., R. Stierli, K. Lanz, and J. Zeyer. 1990. Quinone and iron porphyrin mediated reduction of nitroaromatic compounds in homogeneous aqueous solution. *Environ. Sci. Technol.* 24(10):1566–1574.

97. Schwarzenbach, R. P. and J. C. Westall. 1981. Transport of nonpolar organic compounds from surface water to groundwater. Laboratory sorption studies. *Environ. Sci. Technol.* 15:1360–1367.

98. Sivavec, T. M. and D. P. Horney. 1997. Reduction of chlorinated solvents by Fe(II) minerals, p. 115–117. Presented at the 213th ACS National Meeting, Symposium on Redox Reactions in Natural and Engineered Aqueous Systems, San Francisco, CA, American Chemical Society, Washington, D.C.

99. Smith, J. G. and M. Fieser. 1990. *Collective Index for Volumes 1–12 of Fieser and Fieser's Reagents for Organic Synthesis.* Wiley, New York.

100. Spange, S. and G. Heublein. 1983. Formation of complexes between aminobenzene derivatives and 18-crown-6 in the presence of π-acceptors. *Z. Chem.* 23(10):378–379.

101. Sposito, G. 1984. *The Surface Chemistry of Soils.* Oxford University Press, New York.

102. Sposito, G. and R. Prost. 1982. Structure of water adsorbed on smectites. *Chem. Rev.* 82(6):553–573.

103. Stucki, J. W. 1988. Structural iron in smectites, p. 625–675. In J. W. Stucki, B. A. Goodman, and U. Schwertmann (eds.), *Iron in Soils and Clay Minerals*. D. Reidel Publishing, Dordrecht.

104. Stucki, J. W., G. W. Bailey, and H. Gan. 1995. Redox reactions in phyllosilicates and their effects on metal transport, p. 113–182. In H. E. Allen, C. P. Huang, G. W. Bailey, and A. R. Bowers (eds.), *Metal Speciation and Contamination of Soil*. Lewis Publishers, Boca Raton, FL.

105. Stucki, J. W., P. Komadel, and H. T. Winkinson. 1987. Microbial reduction of structural iron(III) in smectites. *Soil Sci. Soc. Am. J.* 51:1663–1665.

106. Stumm, W. 1992. *Chemistry of the Solid-Water Interface*. Wiley, New York.

107. Stumm, W. and J. J. Morgan. 1996. *Aquatic Chemistry*, 3rd ed. John Wiley & Sons, New York.

108. Theng, B. K. G. 1974. *The Chemistry of Clay-Organic Reactions*. John Wiley & Sons, New York.

109. Thieme, J., R. Haas, and P. Kopecz. 1996. Bestandsaufnahme von Rüstungsaltlastverdachtsstandorten in der Bundesrepublik Deutschland. Texte 25/96–30/96. Umweltbundesamt, Berlin, Germany.

110. Thorn, K. A., P. J. Pettigrew, W. S. Goldenberg, and E. J. Weber. 1996. Covalent binding of aniline to humic substances. 2. ^{15}N NMR studies of nucleophilic addition reactions. *Environ. Sci. Technol.* 30(9):2764–2775.

111. Tratnyek, P. G. 1996. Putting corrosion to use: remediation of contaminated groundwater with zero-valent metals. *Chem. Ind.* 1(13):499–503.

112. Tratnyek, P. G. and D. L. Macalady. 1989. Abiotic reduction of nitroaromatic pesticides in anaerobic laboratory systems. *J. Agric. Food Chem.* 37(1):248–254.

113. Varadachari, C., A. H. Mondal, and K. Ghosh. 1995. The influence of crystal edges on clay humus complexation. *Soil Sci.* 159(3):185–190.

114. Vasudevan, D. and A. T. Stone. 1996. Adsorption of catechols, 2-aminophenols, and 1,2- phenylenediamines at the metal (hydr)oxide/water interface: effect of ring substituents on the adsorption onto TiO_2. *Environ. Sci. Technol.* 30(5):1604–1613.

115. Vogel, T. M., C. S. Criddle, and P. L. McCarty. 1987. Transformations of halogenated aliphatic compounds. *Environ. Sci. Technol.* 21(8):722–735.

116. Wahid, P. A., C. Ramakrishna, and N. Sethunathan. 1980. Instantaneous degradation of parathion in anaerobic soils. *J. Environ. Qual.* 9:127–130.

117. Wardman, P. 1989. Reduction potentials of one-electron couples involving free radicals in aqueous solution. *J. Phys. Chem. Ref. Data.* 18(4):1637–1755.

118. Weber, E. 1997. Unpublished results. U.S. EPA, Athens, GA.

119. Weber, E. J. 1995. Iron-mediated reductive transformations: Investigation of reaction mechanism. *Environ. Sci. Technol.* 30(2):716–719.

120. Weber, E. J. and R. L. Adams. 1995. Chemical- and sediment-mediated reduction of the azo dye Disperse Blue 79. *Environ. Sci. Technol.* 29:1163–1170.

121. Weber, E. J., D. L. Spidle, and K. A. Thorn. 1996. Covalent binding of aniline to humic substances. 1. Kinetik studies. *Environ. Sci. Technol.* 30(9):2755–2763.

122. Weissmahr, K. W. 1996. Mechanism and Environmental Significance of Electron Donor Acceptor Interactions of Nitroaromatic Compounds with Clay Minerals. Dissertation ETH Nr. 11'631. ETH Zürich.

123. Weissmahr, K. W., S. B. Haderlein, and R. P. Schwarzenbach. 1998. Complex formation of soil minerals with nitroaromatic explosives and other π-acceptors. *Soil Sci. Soc. Am. J.* 62(2):369–378.

124. Weissmahr, K. W., S. B. Haderlein, R. P. Schwarzenbach, R. Hany, and R. Nüesch. 1997. *In situ* spectroscopic investigations of adsorption mechanisms of nitroaromatic compounds at clay minerals. *Environ. Sci. Technol.* 31(1):240–247.

125. Weissmahr, K. W., M. Hildenbrand, R. P. Schwarzenbach, and S. B. Haderlein. 1999. Laboratory and field scale evaluation of geochemical controls on groundwater transport of nitroaromatic ammunitions residues. *Environ. Sci. Technol.* 33(15):2593–2600.

126. Wolfe, N. L. 1992. Abiotic transformations of pesticides in natural waters and sediments, p. 93–104. In J. L. Schnoor (ed.), *Fate of Pesticides and Chemicals in the Environment*, Wiley, New York.

127. Wolfe, N. L. and D. R. Burris. 1992. Abiotic redox transformations of organic contaminants in aquifer materials, p. 2–16. Presented at the AWMA, 85th Annual meeting. Air & Waste Management Association, Kansas City, MO.

128. Wolfe, N. L., B. E. Kitchens, D. L. Macalady, and T. J. Grundl. 1986. Physical and chemical factors that influence the anaerobic degradation of methyl parathion in sediment systems. *Environ. Toxicol. Chem.* 5:1019–1026.

129. Wolfe, N. L. and D. L. Macalady. 1992. New perspectives in aquatic redox chemistry: abiotic transformations of pollutants in groundwater and sediments. *J. Contam. Hydrol.* 9:17–34.

130. Yariv, S. 1992. The effect of tetrahedral substitution of Si by Al on the surface acidity of the oxygen plane of clay minerals. *Int. Rev. Phys. Chem.* 11(2):345–375.

131. Yariv, S., J. D. Russell, and V. C. Farmer. 1966. Infrared study of the adsorption of benzoic acid and nitrobenzene in montmorillonite. *Israel J. Chem.* 4:201–213.

132. Yu, S. Y. and G. W. Bailey. 1992. Reduction of nitrobenzene by four sulfide minerals: kinetics, products, and solubility. *J. Environ. Qual.* 21:86–94.

133. Zepp, R. G. and L. N. Wolfe. 1987. Abiotic transformation of organic chemicals at the particle-water interface, p. 423–455. In W. Stumm (ed.), *Aquatic Surface Chemistry*, Wiley, New York.

134. Zhou, J. L., S. Rowland, R. Mantoura, and J. Braven. 1994. The formation of humic coatings on mineral particles under simulated estuarine conditions — a mechanistic study. *Water Res.* 28(3):571–579.

CHAPTER 13

Composting (Humification) of Nitroaromatic Compounds

Dirk Bruns-Nagel, Klaus Steinbach, Diethard Gemsa, and Eberhard von Löw

CONTENTS

13.1 INTRODUCTION

Composting is a complex and not very well understood biochemical process of decomposition and conversion of organic substances to various macromolecules such as humic constituents (humification). The technique is of high interest for municipal waste management mainly because it causes a reduction of weight and volume. In the case of hazardous waste management, the high biological diversity and activity of compost leads to the efficient destruction of organic contaminants which resist biodegradation by other technologies.

In this chapter, the use of composting as a bioremediation technique for soil contaminated with explosives and related compounds will be discussed. The focus will be directed toward 2,4,6-trinitrotoluene (TNT)-contaminated soil. TNT is highly resistant to microbial mineralization, but is susceptible to transformation by a variety of microbes.[14,25,35,62] At present, composting is widely accepted in the U.S. (see Chapter 14). Incineration is the remediation technology accepted for TNT-contaminated soil by most authorities worldwide. Because incineration is very costly and destructive to the soil, biological remediation techniques are desirable for soil treatment. Numerous studies show that composting can be an efficient bioremediation technology for soils contaminated with xenobiotics. In addition to studies on explosives, which will be discussed in this chapter, composting has been applied to a number of other environmental contaminants. Kirchmann and Ewnetu[48] found that composting of petroleum-based oil wastes with horse manure can decompose 79 to 93% of the oil residues within 4.5 months. O'Reilly et al.[56] performed a pilot-scale composting treatment with three different oil-contaminated materials over a period of 6 months. The treatment reduced the diesel-range organic contaminants by 55 to 85%. Valo[78] reduced the polycyclic aromatic hydrocarbon (PAH) concentration of cresote-contaminated soil by 76% over a period of 6 months via composting. Composting was also investigated for chlorophenol-[69] and herbicide-contaminated soils.[23] An effective pilot-scale treatment for soil contaminated with pentachlorophenol (PCP) and total petroleum hydrocarbons (TPH) was conducted by Diehl et al.[20] In a 1-year pilot-scale study they achieved a 97% reduction of the PCP concentration and the concentration of TPH decreased by 93%. The composting treatment was much less effective with a different soil type and could not be recommended as a remediation technique. These selected examples are provided to illustrate the broad range of hazardous chemicals that can be bioremediated via composting.

To give an impression of the relevance of explosives and nitroaromatic compounds as environmetal pollutants, a short review about the structural diversity, chemical properties, and toxicological significance will follow. In further sections of this chapter, composting of explosives will be discussed in detail. Mass balance studies, different pilot-scale operations in Germany and the U.S., and toxicological investigations of composted soil are presented. Furthermore, the mechanisms of transformation of TNT under composting conditions will be described along with a qualitative description of the chemistry of humified TNT residues.

13.2 EXPLOSIVES AND NITROAROMATIC COMPOUNDS AS CONTAMINANTS

The manufacture and handling of explosives and the demolition of the manufacturing facilities have resulted in soils and groundwater contaminated with munition residues. In Germany, 2340 sites are known where ammunition was handled and, therefore, are potentially contaminated with residues of explosives.[73] Major contaminants of military sites are listed in Table 13.1. The nitroaromatic compound TNT was the most abundantly produced explosive manufactured worldwide during the past century.[45] The reduction products 4-amino-2,6-dinitrotoluene (4-ADNT) and 2-amino-4,6-dinitrotoluene (2-ADNT) can be found in soils contaminated with TNT. Furthermore, 2,4-dinitrotoluene (2,4-DNT) and 2,6-dinitrotoluene (2,6-DNT) contaminate production sites. DNTs are precursor substances in the production process of TNT and were also used as modifiers in explosives and propellants. 1,3,5-Trinitrobenzene (TNB) is an ubiquitous co-contaminant of TNT-contaminated soil. The TNB concentration exceeds the TNT concentration at some sites.[16] One possible source is the nitration of benzene instead of toluene when toluene was scarce during World War II. Furthermore, the oxidation of the methyl function of TNT and a subsequent decarboxylation has been proposed as a source for TNB.[52] In our studies we found that TNT metabolites oxidized at the methyl group are present in soils from former ammunition plants in a concentration of up to 500 mg/kg dry soil.[5,13] In all cases, 2-amino-4,6-dinitrobenzoic acid (2-ADNBA) represented the main oxidized TNT metabolite. 2,4,6-Trinitrobenzoic acid (TNBA) and 4-amino-2,6-dinitrobenzoic acid (4-ADNBA) could be detected in low amounts (0.5 to 10 mg/kg dry soil). Oxidized TNT metabolites have also been detected in groundwater from ammunition plants.[68,71] They have been neglected in previous surveillance programs of contaminated sites, in spite of their importance, because of their very high water solubility when compared to TNT and its reduction products. Their origin is uncertain at this time. No oxidized nitroaromatic compounds could be detected in TNT which was produced at sites where such compounds can be found as contaminations (unpublished data). TNT metabolites oxidized at the methyl group are known to be photochemical degradation products and might be present as impurities in low amounts in TNT.[50,52] The fact that major amounts of this substance can still be found in soil after the TNT production was terminated over 50 years ago implies that a biotic or abiotic generation still takes place. A photochemical catalysis seems very unlikely since soil samples were taken in about 30 to 100 cm depth and no TNT as a possible source could be found at the soil surface in former production sites.

In addition to TNT, other frequently used explosives such as 2,4,6-tetranitro-*N*-methylaniline (tetryl), 1,3,5-hexahydro-1,3,5-trinitrohydrazine (RDX), and 1,3,5,7-hexahydro-1,3,5,7-tetranitrotriazine (HMX) (Table 13.1) are also environmental pollutants. The toxicity of the explosives requires the remediation of contaminated soil and groundwater. The nitroaromatic compounds especially are acutely toxic, mutagenic, and potential or known carcinogens.[2,21,27,33,38,58,70,72,74]

Table 13.1 Explosives, Some Metabolites, and Other Nitroaromatic Compounds with Considerable Environmental Impact

Explosives and Related Compounds:

2,4,6-Trinitrotoluene

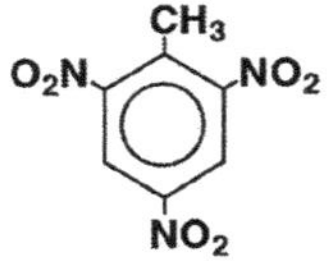

4-Amino-2,6-dinitrotoluene

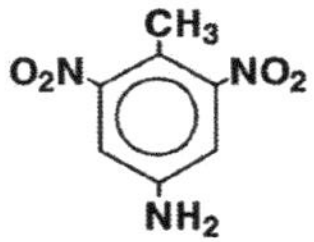

2-Amino-4,6-dinitrotoluene

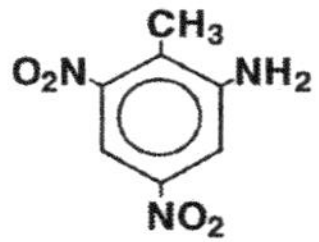

2,4,6-Trinitrobenzoic Acid

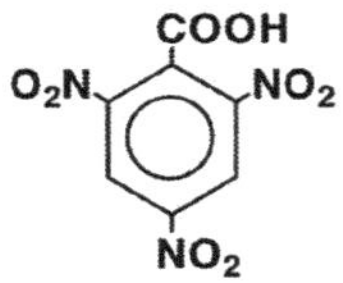

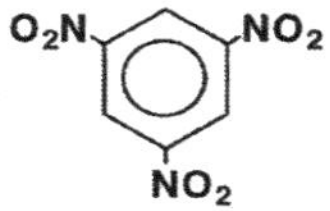

4-Amino-2,6-dinitrobenzoic Acid

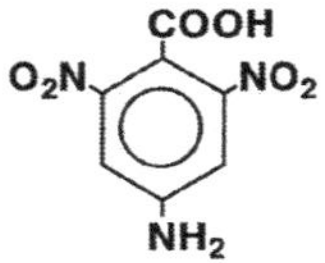

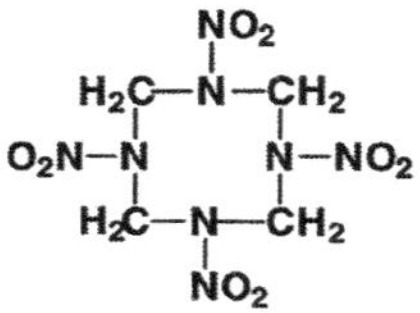

2-Amino-4,6-dinitrobenzoic Acid

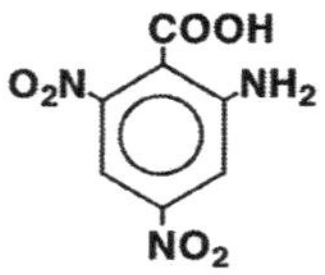

2,4-Dinitrotoluene

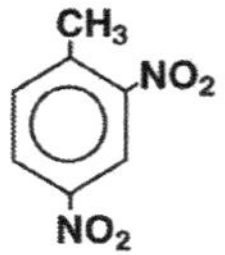

2,6-Dinitrotoluene

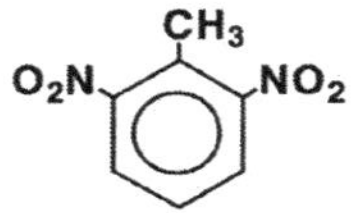

1,3,5-Trinitrobenzene

RDX

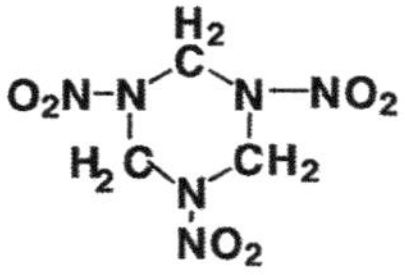

HMX

Fragrances:

Musk Xylene

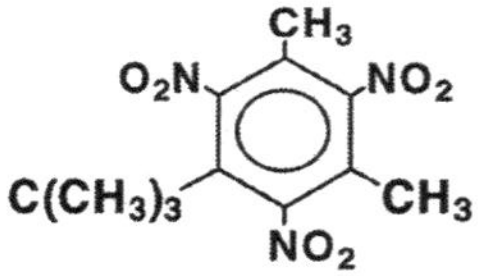

Musk Ketone

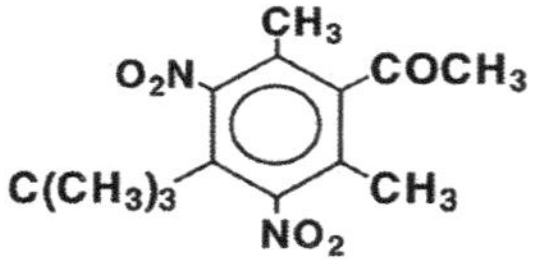

Another group of nitroaromatic compounds with considerable environmental impact are the synthetic musk fragrances: 1-t-butyl-3,5-dimethyl-2,4,6-trinitrobenzene (musk xylene) is structurally related to TNT and 1-t-butyl-3,5-dimethyl-2,6-dinitro-4-acetylbenzene (musk ketone) is structurally related to 2,4-DNT (Table 13.1). Both compounds are commonly used worldwide as fragrances in cosmetics, soaps, lotions, washing powder and other scented products. They are classified as persistant organic pollutants (POPs).[63] Because of the high environmental impact, musk xylene and musk ketone were added to the Third Priority List of the European Union,[28] and musk xylene was enlisted in the OSPAR (Oslo and Paris Commission, organization of 16 European countries and the Commission of the European Communities) List of Chemicals for Priority Action.[57] The worldwide production of musk xylene is about 1000 tons/year.[84] The environmental importance of these chemicals was previously demonstrated by Käfferlein et al.[44] who found musk xylene in 66 of 72 human blood samples. An accumulation of these compounds in breast milk, human adipose tissue, and food has also been reported.[3] Little is known about the toxicity of the musk fragrances. Musk xylene, however, is a suspected carcinogen.[51] To date, nearly nothing is known about the biodegradability of such compounds. However, the information is urgently needed because the compounds accumulate in sewage sludge which is often released into agricultural systems. Controlled composting of the sludge would be a possible treatment to detoxify such compounds.

13.3 PRINCIPLES OF SOIL COMPOSTING

Contaminated soil is mixed with degradable organic material, and the appropriate amount of moisture is added to start the composting process.[29] Indigenous bacteria and fungi break down organic substances via extracellular hydrolysis and a subsequent catabolic biodegradation under aerobic conditions.[64] Self-heating is caused by a mass proliferation of microorganisms. The process is divided into the following four phases: mesophilic phase (<45°C), thermophilic phase (45 to 60°C, up to 80°C is possible), cooling phase (45 to 20°C), and maturation phase (20°C).[32] The temperature changes lead to sequential changes of the microbial community. During the first phase, a highly diverse, but not well-described microbial community is active. During the thermophilic phase, spore formers (*Bacillus* spp.) are predominant and thermophilic fungi have been isolated.[32] During the cooling phase, fungi become predominant because the spores can withstand high temperatures and the fungi are able to degrade lignin and waxy fractions as energy sources. The highest decomposition activity takes place during the thermophilic phase, even though it usually lasts only a few days. If the temperature rises near 80°C, the microbial biomass declines and the composting process is inhibited.[29] A composting process can also be maintained under mesophilic conditions. The technique is more efficient, however, with a short phase at an elevated (>45°C) temperature.

Major factors affecting the compost process are compost size and/or insulation, structure, aeration, pH, moisture, and C/N ratio of the substrate.[64,77] A further important factor is the technical preparation of the contaminated soil, especially to prevent inhomogeneity of the distribution of TNT or agglomerations of non bioavailable crystalline TNT.

Aeration of the systems is necessary because composting is primarily an aerobic process. Anaerobic/aerobic composting systems are also highly efficient for treatment of TNT-contaminated soil.[4,7,10,11,12,80,81,92] In such systems, an anaerobic phase precedes the classical aerated composting treatment. To achieve aerobic conditions, the system can be aerated by injecting air or by turning the material regularly. The structure of the compost material significantly influences the aeration and may even create anaerobic micropockets. Wood chips, straw, or mineral material may be added to inhibit compression of the compost. In addition to supplying oxygen, the aeration can provide cooling to prevent an overheating of the compost.

The pH of the compost should be in the range of 5.5 to 8.0.[32] If an anaerobic pre-phase is performed, the pH can drop to values below 4.0. Usually it reaches the optimum value without the addition of any regulating agents during a subsequent aerated treatment. The optimum moisture content for bacteria in compost is 50 to 60%. A moisture content above 60% generates anaerobic conditions. The C/N ratio for an efficient composting process is 26 to 35.[32] If the nitrogen content is too high, ammonia liberation increases the pH and disturbs the composting process. On the other hand, a nitrogen deficiency impedes microbial growth and slows the composting process. Furthermore, it can cause an extensive formation of organic acids which lowers the pH and retards the microbial growth.

The goals of composting of environmentally hazardous chemicals depend on the chemicals. Some contaminants such as petroleum hydrocarbons can be mineralized. Compounds that are not subject to mineralization can be transformed cometabolically to less toxic compounds and incorporated into humic material (humification) under appropriate conditions. Decomposition and leaching of those complexes must be prevented and the long-term stability must be guaranteed[1,36,60] (see Chapter 4).

13.4 COMPOSTING (HUMIFICATION) STRATEGIES FOR EXPLOSIVES-CONTAMINATED SOIL

The use of composting as the bioremediation technology for soils and sediments contaminated with explosives has been the subject of a number of investigations. Composting of explosives is an established off-the-shelf commercial technology that is widely accepted by the U.S. Army and U.S. regulatory agencies. In Germany, at present, different composting strategies for TNT-contaminated soil are under investigation in full-scale experiments.

13.4.1 Laboratory Experiments

Laboratory experiments revealed that composting can cause a fast decline of the extractable amount of explosives and related compounds. Mass balance studies with ^{14}C labeled test chemicals were of great importance in understanding the process.

Table 13.2 summarizes some of the investigations performed with TNT, RDX, and HMX. The experiments differ in the compost composition, treatment (e.g., aeration, agitation), incubation time, temperature, and ratio of soil to auxiliary substrate. In the early 1980s, Kaplan and Kaplan[46] reported that composting caused a transformation of TNT to various reduction products and azoxy compounds. The disappearance of the explosive was not caused by mineralization, but by binding of ^{14}C-labeled products to humic soil components. Extension of the treatment time increased the amount of bound residues that were formed. Experiments performed by Isbister et al.[42] confirmed the previous findings with the exception that they identified only minor amounts of TNT reduction products. They used a different compost mixture and a slightly elevated incubation temperature. Up to 66.5% of ^{14}C-labeled compounds were bound to the solid phase. The degradation of RDX was also investigated in the study. In contrast to TNT, a major amount of the RDX was mineralized in the compost material.

In a very complex study, Doyle et al.[24] used two different compost mixtures and performed composting experiments with three different amounts of soil (sediment). They tested the system with ^{14}C-TNT, ^{14}C-RDX, and ^{14}C-HMX. When "hay-horse feed" compost was used, 100% of the applied radiolabled TNT was bound to the soil if the compost contained 10% soil. An increase of the amount of soil caused a decrease of the binding of ^{14}C to the compost. However, when 25% soil was used, 77.8% of the radioactivity originally applied as TNT became unextractable, whereas in sewage sludge compost the unextractable residues reached 97.6%.

A considerable portion of RDX was mineralized (46.5% was recovered as [14]CO_2) in hay-horse feed compost containing 10% soil. Higher proportions of soil drastically decreased the mineralization rate to 4% (18% soil) and 0.8% (25% soil). Sewage sludge compost mineralized RDX to 37.2% (10% soil) and even to 26.0% (25% soil). However, more bound residues were formed, and an increase of the soil ratio had less effect on the mineralization rate. HMX was not mineralized in hay-horse feed composts and only poorly bound to soil constituents. A sewage sludge compost containing 10% soil mineralized 31.1% of the added ^{14}C-HMX. Almost 43% of the initially added radioactivity was unextractable. Again, higher ratios of soil negatively affected the mineralization.

Recently, Pennington et al.[60] investigated the fate of ^{14}C-TNT in an aerated compost. They used similar conditions to those in previous studies in which the compost material contained more easily degradable materials such as apples and potatoes. About 80% of the radioactivity was bound to different humic compounds of the soil and to residues of cellulose. Again, no mineralization of the explosive was documented.

Drzyzga et al.[25] performed investigations with an anerobic/aerobic composting system using radiolabeled TNT. During the anaerobic treatment of the compressed compost the extractable amount of ^{14}C-labeled compounds was reduced to about 40%. A subsequent aerated phase decreased the extractable fraction to 9.4% and enhanced the amount of bound ^{14}C-labeled compounds to about 83%.

In the extract of the anaerobic/aerobic compost, the labeled, most highly polar compounds could not be assigned to known TNT metabolites by chemical analyses including HPLC. The results corresponded with those of Achtnich et al.[1] who found

Table 13.2 Examples of Investigations About Composting Explosives Using ¹⁴C-Labeled Compounds

Compost Composition, Treatment, and Temperature	Treatment Time (days)	Amount of Soil (dry wt%)	Concentration of Explosive (mg/kg dry wt)	Radio-Labeled Compound	$^{14}CO_2$ Liberation (%)	^{14}C Distribution in Different Soil Fractions	Ref.
Horse Manure, alfalfa hay, grass clippings, dead hardwood, leaves, garden soil	24	No data	TNT: 15,000 (spiked)	TNT	Not significant	E:86.6 Humin: 1.3 HA: 4.0 FA: 0.4	46
Continuous aeration, 55°C	91					E: 61.5 Humin: 13.9 HA: 7.8 FA: 0.4	
Alfalfa hay/horse feed	35	20	TNT: 10,000 RDX: 10,000 (spiked)	TNT RDX	TNT: 0.5 RDX: 54.5	TNT E: 19.7 R: 66.5 RDX E: 21.6 R: 16.1	42
Continuous aeration, 60°C							
Hay/horse feed compost	36		LAAP soil TNT: 114,600 RDX: 64,205 HMX: 7,043	TNT RDX HMX	TNT: 0.7 RDX: 46.5 HMX: 0.4 TNT: 0.3 RDX: 4.0 HMX: 0.6 TNT: 0.9 RDX: 0.8	TNT E: 7.8 R: 100.3 RDX E: 5.6 R: 29.1 HMX E: 76.8 R: 19.0 TNT E: 30.8 R: 75.2 RDX E: 70.9 R: 24.4 HMX E: 74.8 R: 20.6 TNT:E: 21.6 R: 77.8 RDX E: 69.5	24
Continuous aeration, 60°C		10 18 25					

Substrate					
Sewage sludge compost Continuous aeration, 60°C	55	10	HMX: 0.6		R: 25.8 HMX E: 71.6 R: 27.6
		18	TNT: 2.4 RDX: 37.2 HMX: 31.1		TNT E: 1.4 R: 94.4 RDX E: 11.0 R: 54.6 HMX E: 27.1 R: 42.8
		25	TNT: 1.9 RDX; 24.2 HMX: 14.6		TNT E: 1.6 R: 97.4 RDX E: 15.5 R: 66.8 HMX E: 34.2 R: 50.6
			TNT: 2.2 RDX: 26.0 HMX: 13.4		TNT E: 2.3 R: 97.6 RDX E: 25.4 R: 46.6 HMX E: 43.3 R: 43.6
Green cow manure, sawdust, apples, potatoes, alfalfa Continuous aeration, 55°C	55	10	UMDA soil TNT: 34,600 RDX: 8,500 HMX: 2,140	TNT	0.17	E: 15.1 Cellulose: 31.98 Humin: 21.94 HA: 8.84 FA: 17.54	60

continued

Table 13.2 (continued) Examples of Investigations About Composting Explosives Using ^{14}C-Labeled Compounds

Compost Composition, Treatement, and Temperature	Treatment Time (days)	Amount of Soil (dry wt%)	Concentration of Explosive (mg/kg dry wt)	Radio Labeled Compound	$^{14}CO_2$ Liberation (%)	^{14}C Distribution in Different Soil Fractions	Ref.
Molasses compost		77.5	Soil from "Tanne" TNT: 1,180	TNT	Not significant	E: 40.2 Humin: 47 HA: 4.5 FA: 5.3	26
35°C Anaerobic	35						
Anaerobic/aerobic	98				Not significant	E: 9.4 Humin: 62 HA: 11.1 FA: 9.5	

Note: UMDA: Umatilla Military Depot Activity, Oregon; LAAP: Louisiana Army Ammunition Plant, Louisiana; "Tanne": former ammunition plant in Germany; E: solvent extractable; HA: humic acid; FA: fulvic acid; and R: rest, not solvent extractable.

that the radioactivity associated with small amounts of fulvic and humic acid released with N,N'-1,2-ethanediy(bis[N-(carboxymethyl)glycine] (EDTA) did not contain any TNT or reduced metabolites. Most of the of the radioactivity was bound to humic components of the compost. Even boiling the compost material with concentrated NaOH did not lead to liberation of the labeled compounds. No mineralization was reported. When this study was compared to previous investigations (Table 13.2), the high binding rate of TNT residues to soil was striking despite a high soil content of 77.5%. Higher binding rates were only reported in other studies if the soil content did not exceed 25%. These results indicate that the anaerobic/aerobic treatment is an efficient technique to bioremediate TNT-contaminated soil. A disadvantage of the treatment is that minor amounts of TNT and transformation products are still extractable from the composted material after the treatment and the fact that a long incubation time is necessary.

Information on the bioremediation of RDX and HMX via anaerobic/aerobic composting is not available yet. In general, the various laboratory investigations demonstrate that composting of contaminated soil leads to a significant reduction of the extractable amount of explosives. HMX and RDX can be mineralized to a considerable extent along with binding to soil compounds. The elimination of TNT could only be explained by a binding of TNT transformation products to different soil compounds such as humic and fulvic acids. No significant mineralization of TNT was reported by any investigators.

Laboratory-scale composting systems are important for determining the mass balances of biodegradation of xenobiotic compounds because it is difficult to handle radiolabeled substances in a large-scale system. The critical points dealing with the transfer of results from laboratory-scale systems to pilot-scale composting are the following.

1. Small-scale laboratory composting experiments require a forced external heating which is not needed in large, self-heating compost systems. In other words, large and small compost systems are not always comparable in each aspect and to a certain degree a "scale down" effect is observed in laboratory models (unpublished data).
2. The effect of oxygen supply has not been sufficiently studied, especially as a consequence of its inhomogeneous distribution within a compost of different density and structure. Until now, most composting trials were treated in a more or less aerobic manner or they remained unaerated, except for the experiments performed by Achtnich et al.,[1] Breitung et al.,[4] Bruns-Nagel et al.,[6,11] and Drzyzga et al.[25]
3. With ^{14}C experiments, a mass balance of the metabolism of explosives and, in addition, evidence for mineralization can be achieved by detecting labeled CO_2. However, no information can be gathered on the chemistry of residues bound to humic structures that evade chemical analysis. Until now, it has been a common experience that a lot of extractable labeled metabolites also remain unidentified.
4. In addition, insufficient information is available on long-term behavior and possible remobilization of bound metabolites.

13.4.2 Pilot-Scale Composting

The numerous successful laboratory-scale experiments involving composting of explosives led to different pilot-scale treatments in both the U.S. and in Germany.[17,77]

Various composting methods have been employed, however, which were not always comparable because these practical approaches were often undertaken without standardization and sufficient scientific background. Composting methods can be divided into five different categories: (1) in-vessel static piles, (2) static piles, (3) mechanically agitated in-vessel composting (MAIV), (4) windrow composting, and (5) anaerobic/aerobic composting. Examples of composting explosives-contaminated soil with various treatments are listed in Table 13.3. Using in-vessel static piles, sewage sludge-wood chip compost was the most effective treatment if the composting process lasted 56 days. However, hay-horse feed compost also caused disappearance of 96% for TNT, 83% for RDX and 79% for HMX. About 11 to 16% contaminated soil was added in these studies. Static piles showed similar results. If only 3% soil was used, the degradation rates were between 96.5 and 99.9%. When up to 25% soil was added, MAIV composting resulted in degradation rates of 68 to 99.7%. The efficiency of windrow composting was even higher. In a recent study performed in Germany, Fischer and Walter[31] reported that nearly 100% of the TNT present in the soil was degraded in a biphasic (anaerobic/aerobic) windrow composting system containing 64% contaminated soil. In various previous studies conducted in the U.S., 30% contaminated soil was used for windrow composting.[17,79] The degradation rates for TNT in aerated compost reached 99.8%, and in unaerated compost it reached up to 99.7%.[79] The status review of Craig et al.[17] also showed that windrow composting was well suited to treat RDX and HMX. Finally, a pilot-scale anaerobic/aerobic composting treatment (dynamic pile) was performed in Germany.[22,80,82] For this treatment, about 80% (wt/wt) soil and only 20% (wt/wt) auxiliary substrate were used. Despite the high amount of soil, 99.9% of the TNT was degraded. This high amount of soil resulted in a material with only a slightly elevated organic carbon content of about 3% after the treatment. The quality of the composted material, particularly the content of organic carbon and nitrogen, is important for subsequent use.

The most important parameters for the evaluation of different composting treatments are the rate of disappearance of the contaminants and the amount of soil that can be included. Accordingly, windrow composting and anaerobic/aerobic composting are highly efficient strategies for the bioremediation of TNT-contaminated soil. Windrow composting results in a complete degradation of TNT and also efficiently remediates RDX and HMX. In the studies described above, 30 to 64% of the composting mixture was soil. Anaerobic/aerobic composting allowed the use of about 80% contaminated soil and resulted in a degradation of 99.9%. Unfortunately, the technology has not yet been applied to RDX and HMX.

13.5 TOXICOLOGICAL TESTING OF COMPOSTED SOIL CONTAMINATED WITH EXPLOSIVES

Although chemical analysis can be used to follow the disappearance of the parent contaminants, toxicological tests are essential to evaluate toxicity of the products. The tests are integrative methods that are designed to detect toxic compounds that cannot be analyzed chemically. TNT and DNT and some of their metabolites are

Table 13.3 Examples of Pilot-Scale Composting Treatments of Soil Contaminated with Explosives

Treatment	Treatment Time (days)	Temperature Range °C	Amount of Soil (dry wt%) and origin	Concentration of Explosive Initial/Final (mg/kg dry wt)	Reduction of Extractable Explosives (%)	Quantitity of Treated Soil	Ref.
In-vessel static pile							
Hay/horse feed compost	47	39–67	11.5 LAAP sediment	TNT: 46,178/1,716 ADNT: 27.5/170 RDX: 3,580/600 HMX: 575/161.5	96 +618 83 79	470 kg	24
Sewage sludge-wood chips	47	40–53	15.7 LAAP sediment	TNT: 39,236/10,446 ADNT: 1,153/1,005 RDX: 6,698/2,471 HMX: 962/914	73 13 63 5	504 kg	
Horse manure	56	26–71	11.8 LAAP sediment	TNT: 31,021/138 ADNT: 3,000/0 RDX: 7,400/445 HMX: 2,800/532	99.6 100 94 81	260 kg	
Static pile							
Alfalfa/manure/horse feed	153	35 (mesophilic)	3 LAAP sediment	TNT: 11,187/50 RDX: 4,630/242 HMX: 643/84	99.6 94.8 86.8	4,400 kg	17

continued

Table 13.3 (continued)   Examples of Pilot-Scale Composting Treatments of Soil Contaminated with Explosives

Treatment	Treatment Time (days)	Temperature Range °C	Amount of Soil (dry wt%) and origin	Concentration of Explosive Initial/Final (mg/kg dry wt)	Reduction of Extractable Explosives (%)	Quantitity of Treated Soil	Ref.
Alfalfa/manure/horse feed	153	55 (thermophilic)	3 (wt/wt) LAAP sediment	TNT: 11,840/3.0 RDX: 5,293/45 HMX: 739/26	99.9 99.1 96.5		17, 47, 77
Sawdust/apple/chicken manure/potatoes	90	50 (thermophilic)	10 UMDA soil	TNT: 4,984/200 RDX: 1,008/542 HMX: 180/12	95.9 46.2 21.3	2.3 m³	17, 47, 77
Mechanically agitated in-vessel composting (MAIV)							
Horse manure/buffalo manure/alfalfa/horse feed	44	No data	10 UMDA soil	TNT: 3,452/90 RDX: 1,011/104 HMX: 169/120	97.4 89.7 28.8	2.3 m³	17, 47, 77
Horse manure/buffalo manure/alfalfa/horse feed	44	No data	10 UMDA soil	TNT: 3,126/5 RDX: 574/3 HMX: 199/6	99.8 99.3 94.9	2.3 m³	
Sawdust/apple/potato/ cow manure	44	No data	25 UMDA soil	TNT: 5,208/14 RDX: 597/18 HMX: 161/51	99.7 97.0 68.0	2.3 m³	

Windrow composting							
Sawdust/apple/potato/ cow manure with forced aeration	40	Thermophilic	30 UMDA soil	TNT: 1,869/4 RDX: 1,069/8 HMX: 175/47	99.8 99.3 73.1	23.0 m³	17, 47, 77
Unaerated				TNT: 1,038/20 RDX: 944/2 HMX: 159/5	99.7 99.8 96.9	23.0 m³	
Full scale				TNT: 296/2.3 RDX: 290/1.2 HMX: 26.9/11.0	99.2 99.6 59.1		
Composition unknown	86	Thermophilic	64 soil from "Tanne"	TNT: 699/>0.5 No metabolites	100	70 tons	31
Dynamic pile (anaerobic/aerobic composting)							
Byproduct of food production	180	20–65	80 soil from "Tanne"	TNT: 917/1.3 Negligible amounts of transformation products.	99.9	64 tons	22, 80, 81

Note: UMDA: Umatilla Military Depot Activity, Oregon; LAAP: Louisiana Army Ammunition Plant, Louisiana; and "Tanne": former ammunition plant in Germany.

known to be mutagenic and carcinogenic. Furthermore, they are toxic to a variety of prokaryotes and eukaryotes (see Section 13.2). Consequently, soil contaminated with explosives shows a considerable toxic and mutagenic potential.[7,44] Two separate studies compared the mutagenicity (Ames test) and acute toxicity (*Ceriodaphnia dubia*) of soil treated by windrow composting, composting in static piles, and MAIV composting.[36,37] The contaminated soil contained TNT, RDX, and HMX. The different treatments reduced the extractable amount of explosives to 52.7 to 99.9%. The toxicological data (Table 13.4) indicate the highest detoxification for windrow composting. For example, the percent reduction from start to end of non-aerated composting was 99.7% in the Ames test (TA-98), and the reduction of acute toxicity (survival) toward *C. dubia* was 92%. Other compost treatments also showed high detoxification rates (Table 13.4).[37] Toxicological data for a pilot scale windrow composting performed in Germany also showed promising results. After a 12 week treatment, the nitroaromatics were reduced from 1.126 mg/kg dry weight to <0.5mg/kg (not detectable), and the soil extracts were determined to be non-genotoxic (G_{EA} 3) in the Ames test. At the beginning of the treatment the soil was genotoxic (G_{EA} 24). The composted material was not toxic to plants.[31]

Jarvis et al.[43] performed earthworm tests and investigations with the Mutatox assay with TNT-, RDX-, and HMX- contaminated soil which was composted for 30 days with cow manure and alfalfa in a 40-L adiabatic reactor. During the test, the extractable amount of explosives declined by 99.8 to 99.9% for TNT, 82 to 88% for RDX, and 25 to 28% for HMX. The treatment caused a significant detoxification as assayed by the earthworm test. The initial 14-day LC_{50} value of 4.4% (contaminated soil diluted in uncontaminated soil) was increased to 35.3 and 100% for the two prepared composts. Tests with Mutatox bacteria revealed a similar decline of toxicity. In contrast to the investigations performed by Griest et al.[36,37] the mutagenicity of the composted soil increased during the study. The authors suggest that the mutagenic metabolites were generated from the explosives during the composting treatment.

Soil remediated by anaerobic/aerobic composting was also subjected to various toxicological tests. A luminescence inhibition test with *Vibrio fischeri* provided

Table 13.4 Comparison of Toxicological Data of Different Compost Treatments of Explosive-Contaminated Soil[36,37]

Toxicological Test	Percent Reduction from Start to End of Composting				
	Windrow		Static Pile		MAIV[a]
	Non-Aerated 30% Soil	Aerated 30% Soil	7% Soil	10% Soil	25% Soil
Mutagenicity Ames test					
TA-98	99.7	99.2	88.2	83.6	97.1
TA-100	97.9	97.5	99.0	87.2	97.8
Acute toxicity toward *C. dubia*					
Survival	92	91	>80	93	>80
Reproduction	87	91	>86	>93	>97

[a] MAIV: mechanically agitated in-vessel composting.

preliminary evidence of detoxification.[81] A more systematic toxicological study of soil material treated by this process has not been performed as yet and would be desirable.

The major disadvantage of toxicological tests is that they measure the effects on animals, plants, and microorganisms, and it remains questionable whether the results can be extrapolated to humans. To overcome this drawback, a new *in vitro* lucigenin dependent chemoluminescence (CL) test with human monocytes as test cells was recently developed and applied to TNT and its metabolites, as well as to leachates of TNT-contaminated soil and anaerobic/aerobic composted soil.[12] TNT and the reduction product 2,4-diamino-6-nitrotoluene were considerably more toxic than 4- and 2-ADNT. Leachates of anaerobic/aerobic composted TNT-contaminated soil showed about the same toxicity as a compost free of any nitroaromatic compounds. The results imply that the remediated soil material is not detrimental to human monocytes. For the future, this *in vitro* test might be a valuable addition to established test systems for the risk assessment of contaminated and remediated soil because it provides information about immunotoxic effects on human cells.

Palmer and coworkers investigated the inhalation of dust from TNT composts.[59] These investigations are of interest because airborne particles of composted TNT-contaminated soil might be generated by handling the compost. They composted ^{14}C-labeled TNT-spiked soil and susequently exposed rats to the dust. Rats treated with neat ^{14}C-TNT excreted about 35% of the radioactivity with their urine within 3 days. Uncomposted ^{14}C-TNT-spiked soil showed similar results. In contrast, only 2.3% of the ^{14}C from composted material was excreted during the first 3 days. [14]C accumulated in the kidneys of rats treated with compost, but not in those treated with neat ^{14}C-TNT and ^{14}C-TNT spiked soil. From these results the authors deduced that TNT residues in compost are not stable when introduced into the lungs. Furthermore, they point out that a proper risk assessment of TNT compost is not possible without the knowledge of the chemical structure of bound TNT residues formed during composting.

There is a substantial body of literature that clearly establishes the effectiveness of composting in decreasing the toxicity of contaminated soil. In some instances the detoxification has been reported to be complete.[37] On the other hand, composting increased the mutagenicity of one TNT-contaminated soil.[43] It seems clear that there is a dearth of knowledge about the chemical structures of bound residues formed from TNT under composting conditions. Such knowledge is essential for an accurate risk assessment of composted soil.

While a large body of evidence exists on the toxicity of pure TNT and other nitroaromatic compounds and on the (eco-)toxicity of TNT-contaminated soil that was subject to a bioremediation process, insufficient data are available about the long-term stability and possible release of toxic metabolites during the turnover process of compost and toxic effects of composted nitroaromatic compounds on humans.

13.6 TRANSFORMATION OF TNT UNDER COMPOSTING CONDITIONS

To date, the knowledge about the processes involved in the remediation and detoxification of TNT-contaminated soil by composting is rather fragmentary. From a scientific point of view, a better understanding of the transformation and reaction pathways participating in this process is essential.

In the following sections, the transformation of TNT under composting conditions will be described in detail. We will focus on the anaerobic/aerobic composting process which was investigated in our laboratory.

Early investigations demonstrated that TNT is reduced stepwise to aromatic amines in biological systems.[15,18] McCormick et al.[53] showed in experiments with enzyme preparations from *Veillonella alkalescens* that the reduction pathway proceeds via nitroso and hydroxylamino intermediates. The reductive microbial transformation takes place under aerobic as well as anaerobic conditions.[30,35,61] However, TNT is only completely reduced to triaminonitrotoluene (TAT) under strictly anaerobic conditions.[19,62] Kaplan and Kaplan[46] were the first to describe the transformation of TNT under composting conditions. They could identify 2- and 4-ADNT, 2,4- and 2,6-diaminonitrotoluene (DANT), and several azoxy compounds as TNT metabolites in compost material. The formation of mono- and diaminonitrotoluenes was also affirmed for different composting strategies.[36,37] Bruns-Nagel et al.[11] identified the same reduction products under anaerobic/aerobic composting conditions (Figure 13.1A). During the anaerobic/aerobic composting process the explosive was first transformed to ADNT and subsequently to a DANT isomer; 2,6-DANT was not detected during the experiment. The amount of ADNT was strikingly lower than TNT, and the detected amount of DANT represented only a fraction of the ADNT. The lack of stoichiometric conversion indicates that TNT metabolites undergo further reactions. They might be transformed to unknown metabolites or react with soil constituents and, therefore, cannot be extracted any longer. The reduction of TNT took place during the anaerobic phase, whereas the amount of extractable TNT residues continuously decreased during the aerobic phase. The sudden increase of TNT at day 48 was caused by an inhomogeneous distribution of the chemical in the sample. The results of the investigations conducted with ^{14}C-TNT (Table 13.2) and the time course of the TNT transformation and disappearance under composting conditions indicate that the disappearance of the explosive is most likely caused by a binding of TNT reduction products to humic compounds of the soil.

Subsequent detailed analyses revealed that, in addition to a reduction of the nitro groups, azoxy compounds and several acylated metabolites were generated during the anaerobic/aerobic composting process (Figure 13.1B). The following acylated metabolites have been identified in compost systems to date: 4-*N*-acetyl-amino-2-hydroxylamino-6-nitrotoluene (4-*N*-AcOHANT), 4-*N*-formylamino-2-amino-6-nitrotoluene (4-*N*-FAmANT), 4-*N*-acetylamino-2-amino-6-nitrotoluene (4-*N*-AcANT),[11] and 4-*N*-acetylamino-2,6-dinitrotoluene (4-*N*-AcDNT) (Bruns-Nagel, unpublished data). A time course of the generation and degradation of 4-*N*-FAmANT and 4-*N*-AcANT under anaerobic/aerobic composting conditions (Figure 13.1B) reveals that the compounds do not represent deadend products; however, their relevance during biotransformation of TNT is still unknown. The appearance of acylated TNT metabolites is not restricted to anaerobic systems or composting. Gilcrease and Murphy[34] described the formation of 4-*N*-AcANT by a *Pseudomonas fluorescens* strain under anaerobic as well as aerobic conditions.

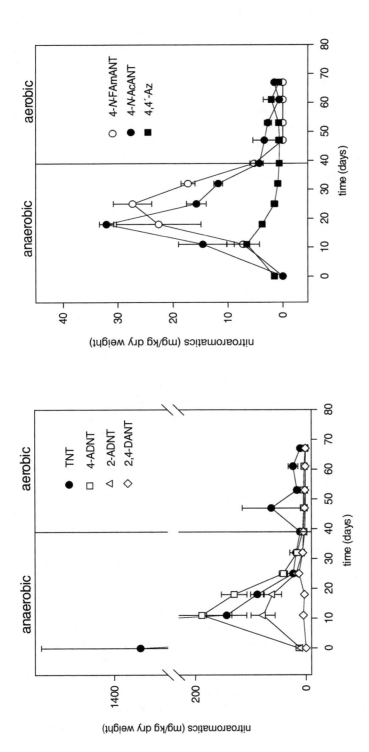

Figure 13.1 Anaerobic/aerobic composting of TNT-contaminated soil in a laboratory experiment. The contaminated soil was mixed with about 20% (wt/wt) substrate and composted at 30°C. During the anaerobic phase the material was compressed. It was blended weekly and compressed again. During the aerobic phase the material was blended daily. The soil was extracted with methanol by ultrasonic treatment. Part A shows concentrations of TNT and reduction products, and Part B shows concentrations of two acylated TNT metabolites and one azoxy compound.

A *Serratia* strain recently used by Drzyzga et al.[25] in a study on the aerobic metabolism of ^{14}C-TNT generated 4-*N*-AcANT in aerobic liquid culture experiments (Bruns-Nagel, unpublished data). Acylated TNT metabolites are also produced by the fungus *Phanerochaete chrysosporium*. Formylated metabolites were first detected as intermediates in the generation of DANT by the fungus.[54] Hawari et al.[39] recently identified nine different acylated TNT derivatives produced by the fungus. Besides the four metabolites previously identified in composts, they found 2-*N*-acetylamino-4,6-dinitrotoluene, 2-*N*-formylamino-4,6-dinitrotoluene and its *p*-isomer, 4-*N*-acetylhydroxy-2,6-dinitrotoluene, and 4-*N*-acetoxy-2,6-dinitrotoluene. The compounds did not accumulate in the culture media. No *o*-acylated TNT reduction products have been detected in composts to date. A reasonable explanation might be that in most cases the reduction of the nitro group in the *para* position is favored and therefore the amount of *p*-acylated compounds is detectable, whereas the *o*-acylated metabolites are below the detection limit. One possible reason for the generation of the compounds might be a detoxification reaction. Hawari et al.[39] speculated that acylated compounds might be a key to the TNT bioremediation process.

Unfortunately, to date the knowledge about the toxicity of acylated TNT metabolites is limited. Acetylated ADNT derivatives were subjected to an "immunotox" test in which primary human monocytes were used as test cells. The EC_{50} values of 4-*N*-AcDNT and 2-ADNT were 50 µg/mL and that of 2-*N*-acetylamino-4,6-dinitrotoluene was >50 µg/mL. The highest toxicities were those of 2,4-DANT (EC_{50}: 5 µg/mL) and TNT (8 µg/mL), respectively.[12]

The *o*-acetylated metabolite of ADNT was considerably genotoxic toward the *Salmonella typhimurium* strain TA98 in the Ames test, whereas the *p*-acetylated product was not.[7] The fact that toxic effects of acylated TNT metabolites cannot be excluded demonstrates the importance of including them in analytical surveillance programs of compost processes.

A summary of the present understanding of composting TNT-contaminated soil, with emphasis on anaerobic/aerobic systems, is compiled in Figure 13.2. The explosive is transformed by microorganisms to several reduction products. Condensation reactions can lead to the formation of azoxy compounds (see Chapter 4). The reduction products can also be acylated. Neither the conditions of the reaction nor the microorganisms involved are known. All these condensation and acylation reactions can take place under aerobic and, in contrast to prior knowledge, anaerobic conditions (Figure 13.1B),[11] whereas the reduction of the nitro groups is more effective under anaerobic conditions. In all investigations we conducted, it was never possible to identify TAT in compost systems. It is very likely that TAT is generated during the anaerobic treatment of the compost, if a redox potential is less than −250 mV, but it is highly reactive and polymerizes fast if oxygen is introduced into the system. TAT also probably reacts very rapidly with humic soil constituents and becomes unextractable. All TNT transformation products bearing amino and hydroxylamino functions can be incorporated into humic compounds by humification reactions (see Chapter 4). The resulting humus and the bound residues are mineralized slowly as a result of natural conversion. TNT residues that are part of the

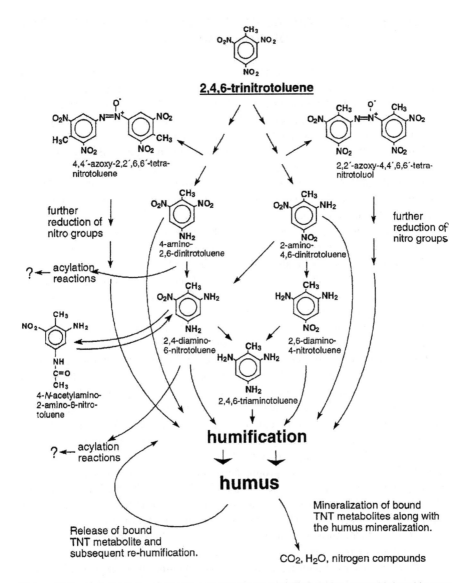

Figure 13.2 Transformation pathway and humification of TNT under anaerobic/aerobic composting conditions. All the depicted chemical structures were identified in compost except triaminotoluene, because it is produced under strictly anaerobic conditions. (Adapted from Bruns-Nagel et al.[6])

complex structure might be completely degraded to carbon dioxide, water, and nitrogen compounds in the course of the humus turnover process.[40] On the other hand, it is well accepted that humus is a dynamic structure.[85] Parts of the macromolecule are reconstructed constantly. For this reason, a release of bound TNT residues cannot be excluded. If this happens, however, a fast rehumification of the xenobiotic is very likely.

13.7 NUCLEAR MAGNETIC RESONANCE (NMR) INVESTIGATIONS OF TNT COMPOSTS

Chemical analyses of soil extracts with different chromatographic techniques and investigations with radioactive labeled compounds are only partly suitable to gain insight into humification processes. Solid-state NMR investigations are a common tool for the characterization of soils. The cross-polarization magic angle spinning (CPMAS) ^{15}N solid-state NMR technology is well suited to characterize bound TNT residues that were formed under composting conditions.[49] Because the sensitivity of solid-state NMR for ^{15}N is considerably low, it was necessary to use ^{15}N-TNT-spiked soils to investigate bound TNT residues by this technique. The low sensitivity of this technique has the advantage that the interference of naturally occurring ^{15}N can be excluded.

In our experiments, soil was spiked with 20,000 mg of ^{15}N$_3$-TNT per kilogram of soil to provide high resolution NMR spectra. The soil was obtained from a site originally contaminated with about 1000–2000 mg of TNT per kilogram of dry soil. This spiked soil was subjected to an anaerobic/aerobic composting process. As a result of the high contamination, about 200 mg of TNT and about 2500 mg of ADNT were still extractable after the composting process.

Figure 13.3 shows the solid-state ^{15}N NMR spectrum of a bulk soil after composting in comparison to the same soil which was extracted in 15 steps in a 1:500 ratio (wt/vol) with acetonitrile (Figure 13.3b). For an interpretation of the NMR spectra, the tentative assignments of chemical shift regions to ^{15}N functional groups was compiled in Table 13.5. Besides a conspicuous peak at –8 ppm, which represents nitro functions as proved by an untreated spiked control,[49] major signals appeared in the chemical shift region of –165 to –350 ppm. The signals between –270 and –350 ppm most likely represent amino nitrogen of aniline derivatives. The sharp peak at –320 ppm (Figure 13.3) is indicative of non-bound aromatic amines. The presence of nitro groups shifts the signal to lower fields. Therefore, the downfield shoulder of the peak was probably caused by ADNTs and DANTs. The sharp signal at –320 ppm of the soil prior to extraction was flattened after extraction (Figure 13.3b), which supports the interpretation that the signal was caused by aromatic amines which were extractable. Furthermore, substituted aromatic amines such as secondary amines and tertiary amines, phenoxazines, anilinohydroquinones, hydrazines, and aminodiphenylamines were expected in this area from –270 to –310 ppm. Obviously, it was not possible to extract these structures because a comparable signal was visible in both unextracted and extracted soils. This indicates a binding of anilines to soil constituents. The signals in the chemical shift region of –165 to –270 ppm further support the assumption that TNT reduction products bound covalently to soil constituents. Resonance lines of nitrogen in heterocyclic structures such as pyrroles, imidazoles, indoles, quinolones, and carbazoles are expected in the region in addition to enaminones and amides (Table 13.5). The structures causing signals in the region from –165 to 270 ppm were also not extractable.

Most strikingly, the structures causing the nitro peak (–8 ppm) resisted acetonitrile extraction. Thus, it can be concluded that partially reduced TNT formed

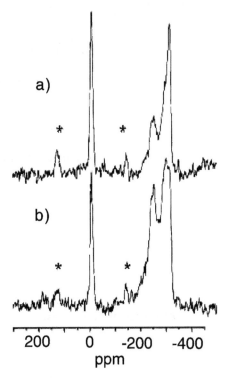

Figure 13.3 Solid-state ^{15}N NMR spectra of (a) non-extracted, anaerobic/aerobic composted ^{15}N-TNT-spiked soil and (b) the same soil after a 1:500 ratio acetonitrile extraction in 15 steps.

non-extractable residues attached to soil constituents. The remaining nitro functions of the bound residues would be responsible for the signals at –8 ppm in the extracted soil.

Figure 13.4 represents the solid-state ^{15}N NMR spectra of bulk soil (Figure 13.4b) treated anaerobically (Figure 13.4b) and, subsequently, aerobically (anaerobically/aerobically) (Figure 13.4a). Signals were visible in the same chemical shift regions as in Figure 13.3. Striking is the fact that the relative signal intensity from –165 to –270 ppm of the anaerobically/aerobically treated soil was about twice as high as that of the anaerobically treated soil. It might be concluded that the aerobic treatment promotes the generation of more stable residues, in which the nitrogen function of TNT metabolized to aniline compounds during the anaerobic phase has become part of heterocyclic structures bound covalently to humic material. A similar multistep binding reaction was previously proposed for the binding of dichloroaniline (DCA) with humic compounds.[67] The authors proposed that DCA first reacts with quinoid groups of humic acid compounds and subsequent addition and oxidation reactions may lead to the formation of heterocyclic structures. Recently, Thorne and Leggett[76] investigated the time course of TNT metabolism and binding of the transformation products. They proposed a two-stage covalent binding of TNT reduction products to compost matrix during both aerated and unaerated treatment.

Table 13.5 Tentative Assignment of Chemical Shift Regions to [15]N Functional Groups[47]

Chemical Shift Region (ppm)	Assignment
148 to 50	Azo compounds
50 to −25	Nitrate, nitrite, nitro groups
−25 to −120	Imines, phenoxazinones, pyridines, quinolines
−120 to −165	Nitriles, oxazoles
−165 to −270	Imidazoles, indoles, pyrroles, carbazoles, quinolone, anilide, amide, enaminones

−270 to −310	Aniline derivatives, phenoxazine, bisphenylhydrazine

−310 to −350	Aniline (primary amines), phenylamines

−358	Ammonium

During the first stage, mainly bonds hydrolyzable by acid and/or base are formed. During the second phase more stable bonds are generated. Our findings support their hypothesis. Therefore, it is important during the application of composting as a remediation technique not to stop the process when the solvent-extractable amount of nitroaromatics is still detectable. A prolonged treatment seems to be required and must be monitored by chemical analysis.

13.8 DISTRIBUTION OF BOUND ^{15}N RESIDUES IN HUMIC FRACTIONS

We analyzed bulk soil and different humic fractions of the anaerobic/aerobic composted soil via solid-state ^{15}N NMR.[6,49] Prior to the extraction of humic

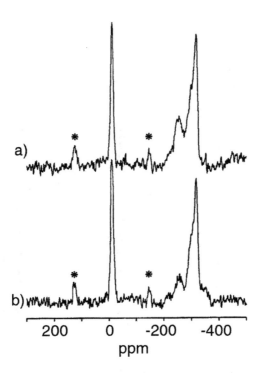

Figure 13.4   Solid-state ^{15}N NMR spectra of ^{15}N-TNT-spiked soil. **(a)** anaerobic/aerobic composted and **(b)** anaerobic treated. (From Bruns-Nagel et al.[6] With permission.)

compounds, the soil was extracted with acetonitrile to remove releasable TNT and
residues. The humic compounds were then extracted in two steps. First, a mild
extraction with 0.5 N NaOH was carried out. Thereafter, the solid residue was
extracted with NaOH for 3 h at 95°C in a reflux unit. The highest ^{15}N enrichment
was in humic acid (HA) fractions (Figure 13.5). It was sufficient to achieve high
resolution spectra of humic acid (Figures 13.5A and 13.5B). In the case of fulvic
acid (FA) and humin, far more accumulations of scans were necessary. All spectra
of humic fractions extracted under mild conditions show distinct nitro peaks (–8
ppm), which support the hypothesis that partially reduced TNT residues are bound
to humic compounds during the humification process. The spectrum of humic
acids shows signals of similar intensity for heterocyclic and/or condensed structures and aniline derivatives. Whereas the heterocyclic and condensed structures
were predominant in the fulvic acid fraction, the aniline derivates (substituted
aromatic amines) were found in the humin fraction. The spectra of the humic
compounds extracted under harsh conditions were similar, with the major difference being that the peak associated with the nitro groups disappeared. The results
indicated that all humic compounds containing nitro groups were extracted during
the mild extraction procedure, but released TNT was not detected by GC and
HPLC. Another possible explanation is a denitration, as has been described for

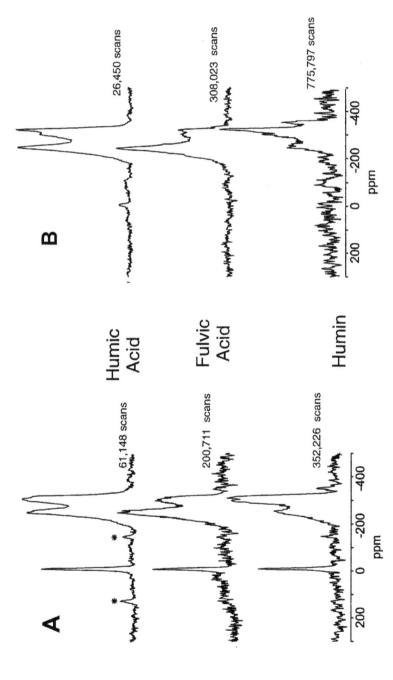

Figure 13.5   Solid-state ^{15}N NMR spectra of different soil fractions of ^{15}N-TNT-spiked soil after an anaerobic/aerobic composting process: (A) soil fractions obtained by mild extraction, and (B) fractions obtained by drastic extraction of the humin fraction. (Adapted from Bruns-Nagel et al.[6])

TNT under basic hydrolysis.[66] The bound TNT transformation products, however, were obviously not affected by the extraction under harsh conditions, which indicates the stability of the linkages.

The solid-state ^{15}N NMR is a predominantly descriptive method. We have used this method to obtain data about the quality of bound TNT residues formed under anaerobic/aerobic composting conditions. We can only speculate at this time about possible chemical or biochemical pathways leading to the formation of the structures found by solid-state ^{15}N NMR. Thorne et al.[75] have extensively discussed different chemical reactions that may be involved during TNT composting. Known biochemical reaction mechanisms are insufficient to explain the formation of the different bound residues that seem to be present in the composted soil.

A recently proposed pathway involving the conversion of 3,4-DCA to a succinimide catalyzed by the fungus *Phanerochaete chrysosporium*[65] shows great potential to explain the generation of heterocyclic structures. Soil microbes might be able to generate a similar reaction sequence with TNT reduction products to generate heterocyclic compounds.

13.9 ^{15}N MASS BALANCE

[15]N-TNT can also be used for mass balance studies under composting conditions. For this purpose, we used pyrolysis capillary gas chromatography-atomic emission detection to determine the ^{15}N/^{14}N ratio. Furthermore, we analyzed the CHN (carbon, hydrogen, nitrogen) content of soil and humic constituents. On the basis of these data, the ^{15}N content of the different samples originating from ^{15}N-TNT in relation to the total N of the fractions was calculated.[6] The results of such a calculation are given in Figure 13.6. The amount of ^{15}N detectable in humic acid and fulvic acid was slightly higher than the ^{14}C value obtained in an analogous experiment performed with ^{14}C-TNT (Table 13.2).[26] Altogether, 56% of the labeled compound could be recovered in various humic fractions (Figure 13.6). About one third (33%) of the labeled compound was extractable with acetonitrile and 11% was not recovered.[6] A explanation for the missing material is a possible denitration of TNT residues during the drastic extraction and a loss of ^{15}N nitro functions during a dialysis step which was performed to enrich the humic compounds. The integration of the solid-state ^{15}N NMR spectra and the mass balance allowed an assignment of the ^{15}N found in the different humic fractions to chemical structures (Figure 13.6). According to the calculation described above, 23.3% of the ^{15}N residues were bound as heterocyclic and/or condensed structures and 15.2% as aniline derivatives. Only 1.9% were present as nitro functions. Furthermore, 15.2% was attributed to unbound amino functions of anilines. The reliability of the assignment was proved with isotope GC/MS. The result indicates that the mass of the non-extractable ^{15}N was bound covalently to humic structures. Free nitro and amino groups probably belong to bound TNT residues in which not all the nitro functions were reduced or not all amino functions were bound.

The use of ^{15}N-TNT-spiked soil and solid-state ^{15}N NMR allowed for the first time a qualitative description of TNT residues formed during a composting process. The approach revealed a rather stable binding of a significant amount of the explosive

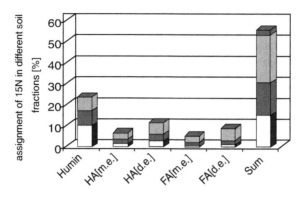

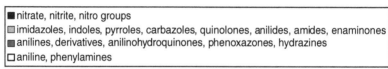

■ nitrate, nitrite, nitro groups
▨ imidazoles, indoles, pyrroles, carbazoles, quinolones, anilides, amides, enaminones
▨ anilines, derivatives, anilinohydroquinones, phenoxazones, hydrazines
☐ aniline, phenylamines

Figure 13.6 ^{15}N mass balance and assignment of ^{15}N to chemical structures in different soil fractions of ^{15}N-TNT-spiked soil after an anaerobic/aerobic composting process. (From Bruns-Nagel et al.[6] With permission.)

to humic constituents. A major drawback of all previous investigations about composting was that it was only possible to speculate about the fate of the explosive. In the future the approach applied here will provide important information about the composting processes. Furthermore, it will be useful to demonstrate to regulatory authorities the role of composting as an important bioremediation technology. Such investigations, however, cannot substitute for studies about the long-term stability of bound residues of explosives.

13.10 CONCLUSIONS AND AREAS OF FUTURE WORK

A variety of composting strategies have proven to be highly effective to bioremediate soil contaminated with the explosives TNT, RDX, and HMX. It is clear that assessment of the different pilot-scale composting treatments should focus mainly on the degradation rate and the optimum ratio of soil to organic supplement. Both windrow composting and anaerobic/aerobic composting seem to be highly effective. Experiments with radiolabled tracers demonstrate that HMX and RDX can be mineralized to a certain extent under composting conditions. That was not the case for TNT. The elimination of TNT was caused by the formation of non-extractable residues in the soil. Most toxicological studies substantiated a detoxification of explosives-contaminated soil by composting. In one case, composting increased the mutagenicity of the contaminated material. In another case, TNT transformation products accumulated in the kidneys of rats after inhalation of composted TNT-contaminated soil.

Composting of explosives is an established method to remediate explosive-contaminated soil. The technology is also applicable for smaller sites and widespread contamination by relatively low concentrations of the explosives. On the other hand, there are still many unsolved questions, and it will be of great importance for the acceptance of composting to continue the scientific work on this subject and to transform the results into technical realization.

The scientific work should focus on further improvement of composting techniques and elucidation of possible risks arising from humified explosives residues. Furthermore, the applicability of composting toward other POPs, in particular nitroaromatics, should be evaluated. The main areas where future research could have a substantial impact are introduced in the following sections.

13.11 TNT BIOTRANSFORMATION: METABOLITES AND METABOLIC PATHWAYS

The transformation of TNT under composting conditions is still only partially understood. Acylated TNT metabolites are generated in a variety of biological systems. Their relevance and also their toxicological impact are not known and require further research. The question of understanding the transformation processes is not only of academic interest. Such knowledge will provide the basis to improve the established systems, and it will serve to make composting a more reliable remediation technique. A greater understanding of the microbiology and biochemistry will be required because strategies based solely on analytical chemistry are limited in their ability to elucidate mechanisms of microbial metabolism in soil and compost. An understanding of which bacteria and fungi in the compost are involved in the transformation process would allow stagies to promote such organisms.

The presence of polar TNT metabolites was previously reported for composting.[5] These metabolites that were analyzed by photometrical detection of the sum of aromatic amines after diazotization and coupling to azo dyes are still not identified. TNT metabolites oxidized at the methyl group were also previously identified in TNT-contaminated soil.[13] The origin of these substances is uncertain at this time. To date, no evidence for oxidation of the methyl function of TNT under composting conditions exists. However, it seems to be very important to intensify investigations about an oxidative attack of TNT by microbes. It will probably be necessary to investigate the biochemistry of the process using defined microbial cultures. At present, nothing is known about the fate of the oxidized metabolites under composting conditions.

13.12 TOXICOLOGICAL EVALUATION

Since the available toxicological data about composting are not uniform it seems to be necessary to evaluate carefully each composting application with a number of different toxicological tests. For this reason, the development of a test battery which covers different aspects such as mutagenicty, acute toxicity, immunotoxicity, ecotoxicity, etc. would be of great value. At this time, such a test battery is in development in a project funded by the German Federal Ministry of Education,

Science, Research and Technology (BMBF). For the future, it will also be important to develop test systems that are more closely related to humans. A test that uses human monocytes as test cells has recently been published.[12] A further improvement of such a test system and the development of specific cellular *in vitro* tests covering other functional aspects would be helpful for a solid toxicological evaluation of composted soil and (bio)remediated soil in general. The requirement of investigations of human toxicity is dependent on the subsequent use of the compost. It is also important for the assessment of risks resulting from handling the compost. In the future, tests of human toxicity will be of increasing interest for the application of *in situ* remediation techniques.

13.13 QUALITATIVE INVESTIGATIONS ON BOUND RESIDUES IN COMPOSTED SOIL

The acceptance of humification as a cleanup technique by scientists, authorities, and the public depends to a major extent on the evaluation and understanding of reaction pathways of contaminants and a useful description of the resultant soil material. All previous work about TNT composting points toward a covalent binding to humic compounds of aromatic amines generated by soil microorganisms. The character of the bonds, however, remains speculative. The use of ^{15}N-TNT and solid-state ^{15}N NMR can be a valuable tool to gain knowledge about the chemical character of non-extractable TNT residues formed under composting conditions. For the future, similar investigations about other bioremediation techniques applied to TNT-contaminated soil are desirable. The same reasoning also applies to the treatment of RDX, HMX, and other nitrogen-containing explosives that are not completely mineralized by microbes. Certainly, the use of the non-radioactive tracer ^{15}N-TNT and NMR is in its infancy. For the future it will be a great challenge to optimize the technology to achieve an insight into the humification process.

13.14 BIOZOENOSIS IN COMPOSTS

To date, research about composting has focused on analytical and toxicological aspects. Little or no attention has been given to the microorganisms responsible for the process. For further optimization of composting processes, knowledge about the bacteria and fungi involved and their biochemical properties will be of major importance. The latter might, for example, help to complete the understanding of the transformation pathway of TNT under composting conditions. Investigations of the microbiology require modern methods of biochemistry and molecular biology. The intention is to identify microorganisms which are involved in the TNT metabolism, to distinguish and analyze specific microbial populations and activities, to explain microbial interactions, and to describe the microbial ecology in contaminated soil and compost. The methods that should be performed include physiological (substrate utilization), biochemical (determination of protein and membrane composition, immunological assays, enzyme biochemistry), and molecular genetic (16S rDNA, 16S RNA, hybridization, PCR, and others).

13.15 FURTHER BIOREMEDIATION TECHNIQUES BASED ON HUMIFICATION

The infiltration of aqueous solutions amended with nutrients causes the elimination of TNT and DNT from contaminated soil.[8,9] The disappearance of the nitroaromatic compounds seems to be caused by humification of the transformation products. This infiltration technique can be used in both on-site and *in situ* remediation methods. Held et al.[40] also proposed an "enhanced humification" method as *in situ* technology to clean up TNT-contaminated soils. The application of an infiltration technique would complement composting methods since it might be more feasible for soils and sediments contaminated with low amounts of explosives (<100 mg/kg dry soil). Since only limited knowledge about the technique is available at this time, research and development work is necessary prior to pilot-scale and full-scale application. Aspects to be investigated are optimization of nutrient cocktails to be infiltrated, mass balance studies, and characterization of bound residues (see Chapter 4).

ACKNOWLEDGMENTS

We thank Dr. Jürgen Breitung, Dr. Oliver Drzyzga, Dr. Torsten C. Schmidt, Dr. Jocher Michels, Dr. Heike Knicker, and Dr. Ulf Bütehorn for their excellent cooperation in different projects about the bioremediation of TNT-contaminated soil which were carried out at the Institute of Umwelthygiene, Marburg (Germany). We also thank the technical staff of the institute for their valuable support. Furthermore, we thank the German Federal Ministry of Education, Science, Research and Technology (BMBF); the state Lower Saxony; and the Industrieverwaltungsgesellschaft AG (IVG), Bonn, Germany for supporting our research.

REFERENCES

1. Achtnich, C., U. Sieglein, H.J. Knackmuss, and H. Lenke. 1999. Irreversible binding of biologically reduced 2,4,6-trinitrotoluene to soil. *Environ. Toxicol. Chem.* 18: 2416–2423.
2. Bailey, H.C., R.J. Spanggord, H.S. Javitz, and D.H.W. Liu. 1985. Toxicity of TNT wastwaters to aquatic organisms. Vol. III. Chronic toxicity of LAP wastewater and 2,4,6-trinitrotoluene. Final Report. SRI International, Menlo Park, CA. DAMD 17-75-C-5056.
3. Behechti, A., K.-W. Schramm, A. Attar, J. Niederfellner, and A. Kettrup. 1998. Acute aquatic toxicities of four musk xylene derivatives on "Daphina magna." *Water Res.* 32: 1704–1707.

4. Breitung, J., D. Bruns-Nagel, K. Steinbach, L. Kaminski, D. Gemsa, and E. v. Löw. 1996. Bioremediation of 2,4,6-trinitrotoluene-contaminated soils by two different areated compost systems. *Appl. Microbiol. Biotechnol.* 44: 795–800.

5. Bruns-Nagel, D. 1999. Analyses of oxidized TNT metabolites and dinitrophenol in soil samples of the former ammunition plant "Tanne" (Germany), study for the Industrieanlagen-Betriebsgesellschaft mbH, Ottobrunn, Gernmany, unpublished.

6. Bruns-Nagel, D., H. Knicker, O. Drzyzga, E. v. Löw, and K. Steinbach. 2000. Characterization of ^{15}N-TNT residues after an anaerobic/aerobic treatment of soil/molasses mixtures by solid-state ^{15}N NMR spectroscopy. II. Systematic investigation of whole soil and different humic fractions. *Environ. Sci. Technol.* 34: 1549–1556.

7. Bruns-Nagel, D., J. Breitung, E. v. Löw, and K. Steinbach. 1996. Mikrobiologische Sanierung von Rüstungsaltlasten am Beispiel des ehemaligen Sprengstoffwerkes "Tanne" bei Clausthal-Zellerfeld, Niedersachsen. Final Report. (BMBF) German Federal Ministry of Education, Science, Research and Technology fund no. 1450791.

8. Bruns-Nagel, D., J. Breitung, E. v. Löw, K. Steinbach, T. Gorontzy, M. Kahl, K.-H. Blotevogel, and D. Gemsa. 1996. Microbial transformation of 2,4,6-trinitrotoluene (TNT) in aerobic soil columns. *Appl. Environ. Microbiol.* 62: 2651–2656.

9. Bruns-Nagel, D., J. Breitung, K. Steinbach, D. Gemsa, K.-H. Blotevogel, and E. v. Löw. 1995. Untersuchungen zum mikrobiellen Abbau sprengstoffspezifischer Nitroaromaten in aeroben Bodenperkolationsanlagen. *Wasser/Abwasser* 137: 192–198.

10. Bruns-Nagel, D., J. Breitung, T. Goronzty, K.-H. Blotevogel, D. Gemsa, E. v. Löw, K. Steinbach, and Th. Gorontzy. 1997. Bioremediation of 2,4,6-trinitrotoluene-contaminated soil by anaerobic/aerobic and aerobic methods. p. 9–14. In Allwman B.C. and A. Leeson (eds.), *In Situ and On-Site Bioremediation: Volume 2. Fourth International In Situ and On-Site Bioremediation Symposium.* Battelle Press, Columbus, OH.

11. Bruns-Nagel, D., O. Drzyzga, K. Steinbach, T.C. Schmidt, E. v. Löw, T. Gorontzy, K.-H. Blotevogel, and D. Gemsa. 1998. Anaerobic/aerobic composting of 2,4,6-trinitrotoluene-contaminated soil in a reactor system. *Environ. Sci. Technol.* 32: 1676–1679.

12. Bruns-Nagel, D., S. Scheffer, B. Casper, H. Garn, O. Drzyzga, E. v. Löw, and D. Gemsa. 1999. Effect of 2,4,6-trinitrotoluene and its metabolites on human monocytes. *Environ. Sci. Technol.* 33: 2566–2570.

13. Bruns-Nagel, D., T. C. Schmidt, O. Drzyzga, E. v. Löw, and K. Steinbach. 1999. Identification of oxidized TNT metabolites in soil samples of a former ammunition plant. *Environ. Sci. Pollut. Res.* 6: 7–10.

14. Carpenter, D. F., N. G. McCormick, J. H. Cornell, and A. M. Kaplan. 1978. Microbial transformation of ^{14}C-labeled 2,4,6-trinitrotoluene in an activated-sludge system. *Appl. Environ. Microbiol.* 35: 949–954

15. Channon, H. J., G. T. Mills, and R.T. Williams. 1944. The metabolism of 2,4,6-trinitrotoluene (α-TNT). *Biochem. J.* 38: 70–85.

16. Checkai, R. T., M.A. Major, R. O. Nwanguma, J. C. Amos, C. T. Phillips, R. S. Wentsel, and M. C. Sadusky. 1993. Transport and fate of nitroaromatic and nitramine explosives in soils from open burning open detonation operations at Anniston Army Depot. ERDEC Technical Report 135. Cited in Major, M.A., W.H. Griest, J.C. Amos, and W.G. Palmer. 1996. Evidence for the chemical reduction and binding of TNT during the composting of contaminated soils. Toxicological Study No. 87–3012–95.

17. Craig, H. D., W. E. Sisk, M. D. Nelson, and W. H. Dana. 1995. Bioremediation of explosives-contaminated soils: a status review. In 10th Annual Conference on Hazardous Waste Research, pp. 168–179, Manhatten, KS.

18. Dale, H. H. 1921. The fate of TNT in the animal body. *Med. Res. Counc. (G.B.) Spec. Rep. Ser.* 58: 53–61.

19. Daun G., H. Lenke, M. Reuss, and H.J. Knackmuss. 1998. Biological treatment of TNT-contaminated soil. I. Anaerobic cometabolic reduction and interaction of TNT and metabolites with soil components. *Environ. Sci. Technol.* 32: 1956–1963.

20. Diehl, S.V., M.S. Lybrand, A. Borazjani, and R.K. Penmetsa. 1997. Bioremediation of PCP- and TPH-contaminated soil in a pilot-scale field study. p. 97–102. In Allwman B.C. and A. Leeson (eds.), *In Situ and On-Site Bioremediation: Volume 2. Fourth International In Situ and On-Site Bioremediation Symposium.* Battelle Press, Columbus, OH.

21. Dilley, J.V., C.A. Tyson, R.J. Spanggord, D.P. Sasmore, G.W. Newell, and J.C. Dacre. 1982. Short-term oral toxicity of 2,4,6-trinitrotoluene in mice, rats and dogs. *J. Toxocol. Environ. Health.* 9: 565–586.

22. Dohnalik-Droste, A. and R. Winterberg. 1999. Das Plambeck ContraCon-Verfahren zur Sanierung TNT-kontaminierter Böden. p. 281. In Heiden, S., R. Erb, J. Warrelmann, and R. Dierstein (eds.), *Biotechnologie im Umweltschutz: Bioremediation: Entwiclung — Anwendung — Perspektiven.* Erich Schmidt Verlag, Berlin.

23. Dooley, M. A., K. Taylor, and B. Allen. 1995. Composting of herbicide-contaminated soil. p. 199–207. In Hinchee, R.E., R.E. Hoeppel, and D.B. Anderson (eds.), *Bioremediation of Recalcitrant Organics.* Battele Press, Columbus, OH.

24. Doyle, R.C., J.D. Isbister, G.L. Anspach, and J.F. Kitchens. 1986. Composting explosives/organics contaminated soils. U.S. Army Toxic and Hazardous Materials Agency, Aberdeen Proving Ground, MD, NTIS-Publication No. ADA 169994.

25. Drzyzga, O., D. Bruns-Nagel, T. Gorontzy, K.-H. Blotevogel, D. Gemsa, and E. v. Löw. 1989. Mass balance studies with ^{14}C-labeled 2,4,6-trinitrotoluene (TNT) mediated by an anaerobic *Desulfovibrio* species and an aerobic *Serratia* species. *Curr. Microbiol.* 37: 380–386.

26. Drzyzga, O., D. Bruns-Nagel, T. Gorontzy, K.-H. Blotevogel, D. Gemsa, and E. v. Löw. 1998. Incorporation of ^{14}C-labeled 2,4,6-trinitrotoluene (TNT) metabolites into different soil fractions after anaerobic and anaerobic/aerobic treatment of soil/molasses mixtures. *Environ. Sci. Technol.* 32: 3529–3535.

27. Drzyzga, O., T. Gorontzy, A. Schmidt, and K.-H. Blotevogel. 1995. Toxicity of explosives and related compounds to the luminescent bacterium *Vibrio fischeri* NRRL-B-11177. *Environ. Contam. Toxicol.* 28: 229–235.

28. EG-Verordnung Nr. 143/97 vom 27.1.1997, Abl. Nr. L 25/13 vom 28.1.1997. Cited in Rimkus, G. 1998. Synthetische Moschusverbindungen in der Umwelt. *Umweltmed. Forsch. Prax.* 3: 341–346.

29. Finstein, M.S. 1992. Composting in the context of municipal solid waste management. p. 355–374. In Mitchell, R. (ed.), *Environmental Microbiology.* John Wiley & Sons, New York.

30. Fiorella, P.D. and J.C. Spain. 1997. Transformation of 2,4,6-trinitrotoluene by *Pseudomonas pseudoalcaligenes* JS52. *Appl. Environ. Microbiol.* 63: 2007–2015.

31. Fischer, D. and U. Walter. 1999. *TNT-Humifizierung — Das Umweltschutz Nord-Verfahren, Werk Tanne — Erste Ergebnisse der Bodenbehandlung.* Umweltschutz Nord GmbH & Co, Germany.

32. Fogarty, A.M. and O.H. Tuovinen. 1991. Microbial degradation of pesticides in yard waste composting. *Microbiol. Rev.* 55: 225–233.

33. Furedi, E.M., B.S. Levine, J.W. Sagartz, V.S. Rac, and P.M. Lish. 1984. Determination of the chronic mammalian toxicological effects of TNT (twenty-four month chronic toxicity/carcinogenicity study of trinitrotoluene (TNT) in the B6C3F1 hybrid mouse). IITRJ Project No. L6116, Study No. 11, ADA 168754. IIT Research Institute, Chicago. DAMD 17–79-C-9120.

34. Gilcrease, P.C. and V.G. Murphy. 1995. Bioconversion of 2,4-diamino-6-nitrotoluene to a novel metabolite under anoxic and aerobic conditions. *Appl. Environ. Microbiol.* 61: 4209–4214.

35. Gorontzy, T., O. Drzyzga, M.W. Kahl, D. Bruns-Nagel, J. Breitung, E. v. Löw, and K.-H. Blotevogel. 1994. Microbial degradation of explosives and related compounds. *Crit. Rev. Microbiol.* 20: 265–284.

36. Griest, W.H., A.J. Stewart, R.L. Tyndall, J.E. Caton, C.-H. Ho, K.S. Ironside, W.M. Caldwell, and E. Tan. 1993. Chemical and toxicological testing of composted explosives-contaminated soil. *Environ. Toxicol. Chem.* 12: 1105–1116.

37. Griest, W.H., R.L. Tyndall, A.J. Stewart, J.E. Caton, A.A. Vass, C.-H. Ho, and W.M. Caldwell. 1995. Chemical characterization and toxicological testing of windrow composts from explosives-contaminated sediments. *Environ. Toxicol. Chem.* 14: 51–59.

38. Hankenson, K. and D.J. Schaeffer. 1991. Microtox assay of trinitrotoluene, diaminonitrotoluene, and dinitromethylaniline mixtures. *Bull. Environ. Contam. Toxicol.* 46: 550–553.

39. Hawari, J., A. Halasz, S. Beaudet, L. Paquet, G. Ampelman, and S. Thiboutot. 1999. Biotransformation of 2,4,6-trinitrotoluene with *Phanerochaete chrysosporium. Appl. Environ. Microbiol.* 65: 2977–2986.

40. Held, T., G. Draude, F.R.J. Schmidt, A. Brokamp, and K.H. Reis. 1996. Enhanced humification as an in-situ bioremediation technique for 2,4,6-trinitrotoluene (TNT) contaminated soils. *Environ. Technol.* 18: 479–487.

41. Honeycutt, M.E., A.S. Jarvis, and V.A. McFarland. 1996. Cytotoxixity and mutagenicity of 2,4,6-trinitrotoluene and its metabolites. *Ecotoxicol. Environ. Saf.* 35: 282–287.

42. Isbister, J.D., G.L. Anspach, J.F. Kitchens, and R.C. Doyle. 1984. Composting for decontamination of soils containing explosives. *Microbiologica.* 7: 47–73.

43. Jarvis, A.S., V.A. McFarland, and M.E. Honeycutt. 1998. Assessment of the effectiveness of composting for the reduction of toxicity and mutagenicity of explosive-contaminated soil. *Ecotoxicol. Environ. Saf.* 39: 131–135.

44. Käfferlein, H.U., T. Göen, and J. Angerer. 1997. Belastung der Allgemeinbevölkerung durch Moschus-Xylol. *Umweltmed. Forsch. Prax.* 2: 169–170.

45. Kaplan, D.L. and A.M. Kaplan. 1983. Reactivity of TNT and TNT-microbial reduction products with soil components. Final Report U.S. Army Natick R&D Laboratories, Natick, MA. No. Natick TR-83/041.

46. Kaplan, L.A. and A.M. Kaplan. 1982. Thermophilic biotransformation of 2,4,6-trinitrotoluene under simulated composting conditions. *Appl. Environ. Microbiol.* 44: 757–760.

47. Keehan, K. Erfahrungen in den Vereinigten Staaten bei der Sanierung von sprengstoffkontaminierten Standorten. 1992. Oral presentation at Expertengespräch "Rüstungssaltlasten aus der Produktion und Verarbeitung von Sprengstoff," 1–2 September, Marburg, Germany.

48. Kirchmann, H. and W. Ewnetu. 1998. Biodegradation of petroleum-based oil wastes through composting. *Biodegradation* 9: 151–156.

49. Knicker, H., D. Bruns-Nagel, O. Drzyzga, E. v. Löw, and K. Steinbach. 1999. Characterization of ^{15}N-TNT residues after an anaerobic/aerobic treatment of soil/molasses mixtures by solid-state ^{15}N NMR spectroscopy. I. Determination and optimization of relevant NMR spectroscopic parameters. *Environ. Sci. Technol.* 33: 343–349.

50. Layton, D., B. Mallon, W. Mitchell, L. Hall, R. Fish, L. Perry, G. Snyder, K. Bogen, W. Malloch, C. Ham, and P. Dowd. 1987. Conventional weapons demilitarization: a health and environmental effects data-base assessment. Explosives and their co-contaminants. Final Report, Phase II. Environmental Science Division, Lawrence Livermore National Laboratory, University of California.
51. Maekawa, A., Y. Matsushima, H. Onodera, M. Shibutani, H. Ogasawara, Y. Kodama, Y. Kurokawa, and Y. Hayashi. 1990. Long-term toxicity/carcinogenicity of musk xylol in B6C3F1 mice. *Food Chem. Toxicol.* 28: 581–586; cited in Käfferlein, H.U., T. Göen, and J. Angerer. 1997. Belastung der Allgemeinbevölkerung durch Moschus-Xylol. *Umweltmed. Forsch. Prax.* 2: 169–170.
52. Major, M.A., W.H. Griest, J.C. Amos, and W.G. Palmer. 1996. Evidence for the chemical reduction and binding of TNT during the composting of contaminated soils. Toxicological Study No. 87–3012–95.
53. McCormick, N.G., F.E. Feeherry, and H.S. Levinson. 1976. Microbial transformation of 2,4,6-trinitrotoluene and other nitroaromatic compounds. *Appl. Environ. Microbiol.* 31: 949–958.
54. Michels, J. and G. Gottschalk. 1995. Pathway of 2,4,6-trinitrotoluene (TNT) degradation by *Phanerochaete chrysosporium*. p. 135–149. In J.C. Spain (ed.), *Biodegradation of Nitroaromatic Compounds*. Plenum Press, New York.
55. Neumann, H.-G. 1996. Toxic equivalence factors, problems and limitations. *Food Chem. Toxicol.* 34: 1045–1051.
56. O'Reilly, K., T. Simpkin, K. Sobczak, and T. Moreno. 1997. Efficacy and economics of composting aged materials at a refinery. p. 73–84. In Allwman B. C. and A. Leeson (eds.), *In Situ and On-Site Bioremediation: Volume 2. Fourth International In Situ and On-Site Bioremediation Symposium*. Battelle Press, Columbus, OH.
57. OSPAR Strategy with Regard to Hazardous Substances, DYNAMEC 98/2/2-E. Cited in Rimkus, G. 1998. Synthetische Moschusverbindungen in der Umwelt. *Umweltmed. Forsch. Prax.* 3: 341–346.
58. Palazzo, A.J. and D.C. Leggett. 1986. Effect and disposition of TNT in terrestrial plants. *J. Environ. Qual.* 15: 49–52.
59. Palmer, W.G., J.R. Beaman, D.M. Walters, and D.A. Cresia. 1997. Bioavailability of TNT residues in composts of TNT-contaminated soil. *J. Toxicol. Environ. Health* 51: 97–108.
60. Pennington, J.C., C.A. Hayes, K.F. Myers, M. Ochman, D. Gummison, D.R. Felt, and E.F. McCormick. 1995. Fate of 2,4,6-trinitrotoluene in a simulated compost system. *Chemosphere* 3: 429–438.
61. Preuss, A., J. Fimpel, and G. Diekert. 1993. Anaerobic transformation of 2,4,6-trinitrotoluene (TNT). *Arch. Microbiol.* 159: 345–353.
62. Rieger, P.-G. and H.-J. Knackmuss. 1995. Basic knowledge and perspectives on biodegradation of 2,4,6-trinitrotoluene and related nitroaromatic compounds in contaminated soil, p. 1–18. In J.C. Spain (ed.), *Biodegradation of Nitroaromatic Compounds*. Plenum Press, New York.
63. Rimkus, G. 1998. Synthetische Moschusverbindungen in der Umwelt. *Umweltmed. Forsch. Prax.* 3: 341–346.
64. Ro, K.S., K.T. Preston, S. Seiden, and M.A. Bergs. 1998. Remediation composting process principles: focus on soils contaminated with explosive compounds. *Crit. Rev. Environ. Sci. Technol.* 28: 253–282.
65. Sandermann, H., Jr., W. Heller, N. Hertkorn, E. Hoque, D. Pieper, and R. Winkler. 1998. A new intermediate in the mineralization of 3,4-dichloraniline by the white rot fungus *Phanerochaete chrysosporium*. *Appl. Environ. Microbiol.* 64: 3305–3312.

66. Saupe, A., H.J. Garvens, and L. Heinze. 1998. Alkaline hydrolysis of TNT and TNT in soil followed by thermal treatment of the hydrolysates. *Chemosphere* 36: 1725–1744.

67. Saxena, A. and R. Bartha. 1983. Binding of 3,4-dichloraniline by humic acid and soil: mechanism and exchangeability. *Soil Sci.* 136: 111–116.

68. Schmidt, T.C., K. Steinbach, E. v. Löw, and G. Stork. 1998. Highly polar metabolites of nitroaromatic compounds in ammunition wastewater. *Chemosphere* 37: 1079–1091.

69. Semple, K.T. and T.R. Fermor. 1995. Composting systems for the bioremediation of chlorophenol-contaminated land. p. 93–100. In Hinchee, R.E., R.E. Hoeppel, and D.B. Anderson (eds.), *Bioremediation of Recalcitrant Organics*, Battele Press, Columbus, OH.

70. Smock, L.A., D.L. Stoneburner, and J.R. Clark. 1976. The toxic effects of trinitrotoluene (TNT) and its primary degradation products on two species of algae and the fathead minnow. *Water Res.* 10: 537–543.

71. Spanggord, R.J., W. Maybey, T. Mill, T.W. Chou, J.H. Smith, S. Lee, and D. Roberts. 1983. Environmental fate studies on certain munition wast-water constituents. Phase IV — lagoon model studies, Menlo Park, California: SRI International. ADA 138550. Cited in Walsh, M.E. 1990. Environmental transformation products of nitroaromatics and nitramines. Literature review and recommendations for analytical method development. U.S. Army Corps of Engineers, Cold Regions Research & Engineering Laboratory. Special Report 90-2.

72. Tan, E.L., C.H. Ho, W.H. Griest, and R.L. Tyndall. 1992. Mutagenicity of trinitrotoluene and its matabolites formed during composting. *J. Toxicol. Environ. Health* 36: 165–175.

73. Thieme, J., R. Haas, and P. Kopecz. 1996. Stock-taking of armament residual loads of the Federal Republic of Germany, Part II, UFOPLAN-Ref.-No. 103 40 102 (Report No. UBA-FB 90-030/1), p. 17.

74. Thierfelder, W., W.H. Mehnert, and H. Höring. 1996. Rüstungsaltlasten und Leukämierisiko. *Bundesgesundhbl.* 1: 9–12.

75. Thorne, K.A., P.J. Pettigrew, W.S. Goldenberg, and E.J. Weber. 1996. Covalent binding of aniline to humic substances. 2. ^{15}N NMR studies of nucleophilic addition reactions. *Environ. Sci. Technol.* 30: 2764–2775.

76. Thorne, P.G. and D.C. Leggett. 1997. Hydrolytic release of bound residues from composted soil contaminated with 2,4,6-trinitrotoluene. *Environ. Toxicol. Chem.* 16: 1132–1134.

77. Track, T. and D. Schenk. 1998. Recherche über das Abbau-/Transformationsverhalten von 2,4,6-Trinitrotoluol, 2,4- & 2,6-Dinitrotoluol sowie den Entwicklungsstand von Sanierungstechnologien für Sprengstoffkontaminationen. Johannes Gutenberg-Universität, Mainz, Germany, unpublished.

78. Valo, R.J. 1997. Bioremediation of cresote contaminated soil: effect of hydrogen peroxide pretretment and inoculation. p. 91–96. In Alleman B.C., A. Leeson (ed.), *In Situ and On-Site Bioremediation: Volume 2. Fourth International In Situ and On-Site Bioremediation Symposium*. Battelle Press, Columbus, OH.

79. Weston, R.F., Inc. 1993. Windrow composting demonstration for explosives-contaminated soils at the Umatilla Depot Activity, Oregon. Prepared for the U.S. Army Environmental Center (AEC), Aberdeen Proving Ground, MD. Cited in Craig, H.D., W.E. Sisk, M.D. Nelson, and W.H. Dana. 1995. Bioremediation of explosives-contaminated soils: a status review. In 10th Annual Conference on Hazardous Waste Research, pp. 164–179, Manhattan, KS.

80. Winterberg, R. 1998. Maßstabsgerechte Erprobung biologischer Verfahren zur Sanierung von Rüstungsaltlasten: Dynamisches Beetverfahren. p. L1–L12. In Verbundvorhaben Langzeit- und Remobilisierungsverfahren von Schadstoffen, Statusseminar 22. und 23. September 1998 in Bremen. Umweltbundesamt, Germany.

81. Winterberg, R. and A. Dohnalik-Droste. 1998. Technikumsversuche zur Optimierung und zum scale up der Sanierungsverfahren. In Verbundvorhaben, Biologische Sanierung von Rüstungsaltlasten. Tagungsband zum 4. Statusseminar am 07 Mai 1998 in Clausthal-Zellerfeld. CD edited by the German Federal Ministry of Education, Science, Research and Technology (BMBF).

82. Winterberg, W., E. v. Löw, and T. Held. 1998. Dynamisches Mietenverfahren zur Sanierung von Rüstungsaltlasten. *TerraTech.* 3: 39–41.

83. Won, W.D., L.H. DiSalvo, and J. Ng. 1976. Toxicity and mutagenicity of 2,4,6-trinitrotoluene and its microbial metabolites. *Appl. Environ. Microbiol.* 31: 576–580.

84. Zahao, Q. 1993. Chinas perfumery industry. *Perfum. Flavorist* 18: 47–48. Cited in Käfferlein, H.U., T. Göen, and J. Angerer. 1997. Belastung der Allgemeinbevölkerung durch Moschus-Xylol. *Umweltmed. Forsch. Prax.* 2: 169–170.

85. Ziechmann, W. 1996. *Huminstoffe und ihre Wirkung.* Spektrum, Akad. Verl., Heidelberg, Berlin, Oxford.

CHAPTER 14

Applications and Costs for Biological Treatment of Explosives-Contaminated Soils in the U.S.

Douglas E. Jerger and Patrick Woodhull

CONTENTS

14.1 INTRODUCTION

The manufacturing and handling of explosives and propellants have resulted in soil and sediment contamination at numerous locations. In the U.S. the sites are primarily Department of Defense (DOD) facilities. The facilities include former manufacturing facilities, load-assembly-pack operations, demilitarization activities, disposal and burn pit operations, and commercial production facilities. Whereas past environmental practices allowed for the contamination of soil, sediments, and groundwater at these facilities, current regulations and owner environmental awareness require the remediation of soils contaminated with explosive compounds.

Explosives can be categorized into three general categories of compounds: nitroaromatic compounds, nitrate esters, and nitramines. Trinitrotoluene (TNT), dinitrotoluene (DNT), and trinitrophenylmethylnitramine (tetryl) are the most common nitroaromatic compounds. The most common nitroamines are cyclotetramethylenetetranitramine (HMX) and cyclotrimethylenetrinitramine (RDX). Nitrate esters (which include nitrocellulose) are rarely encountered at concentrations requiring treatment at U.S. DOD sites.

The published literature contains numerous references to the microbial degradation of nitroaromatic explosive compounds.[28,19,8] The most common mechanism encountered during environmental remediation of nitroaromatic explosives compounds is reduction of the nitro groups. After reduction of the nitro groups, the resulting mono-, di-, and triaminotoluene compounds are either further degraded or irreversibly bound to humic fraction of the soil matrix.[19]

The primary objective of this chapter is to provide information on biological treatment processes currently employed in the field for the remediation of nitroaromatic compounds in soils. These processes have been developed over the past 10+ years and have met the criteria for successful application in the field, i.e., implementability, performance, and cost. In an effort to provide sufficient information on ongoing or completed full-scale soil treatment projects, the authors have limited their discussion to processes that have treated soil quantities greater than 1000 yd^3

in the U.S. The basic work that was used to develop these processes is contained in the references provided for field application.

14.2 PROCESS APPLICATION

Protection of human health and the environment is usually the regulatory goal for site cleanup. Source removal and treatment are typically the first steps in site remediation. Therefore, process application for full-scale treatment has focused on the treatment of excavated soils in the U.S. To the authors' knowledge, full scale site remediation of nitroaromatic-contaminated soils has not been conducted internationally. Basic research and process development for treatment of nitroaromatic compounds in soil has been discussed in other chapters in this book. Biological processes for the *in situ* treatment of groundwater contaminated with nitroaromatic compounds are under development and have not been applied full scale. The authors direct the reader to a series of documents recently published for a discussion and application of natural attenuation of nitroaromatic compounds in groundwater.[22,23] Full scale, aboveground groundwater treatment typically involves physical or chemical treatment (i.e., carbon adsorption or a combination of ozone and UV oxidation).

Successful field application of each process requires a project management plan and a detailed work plan which includes a mobilization plan, a field quality control plan, a sampling and analysis plan, a soil excavation plan, a plan for the processing of contaminated materials, a specific process plan for actual soil treatment, a site closure plan, and a quality assurance plan. A site-specific health and safety plan is also required. These plans are typically required and approved internally by the client and by the federal, state, and local regulatory agencies prior to initiation of work. In general these documents or project summaries are not published in peer-reviewed journals or are not readily accessible outside of the individual project teams. Publications and/or presentations on the results of the individual projects are typically prepared for general trade publications or presented at national conferences.

Site-specific cleanup goals are developed for each project based on risk-based action levels of constituents of concern. The risk-based action levels are derived to reflect the potential risk from exposure to the constituents of concern based on specific land use. Risk-based action levels integrate current U.S. Environmental Protection Agency (EPA) toxicity and carcinogenicity values with health protective exposure assumptions to evaluate chemical concentrations in environmental media. Therefore, a discussion of the final products and toxicity of these products encountered following treatment will not be presented here. Such information is reflected in the site-specific cleanup goals. Further information on the chemical and toxicological characterization of treated soils can be found in studies performed at the Oak Ridge National Laboratory (ORNL).[13,14,19,29]

Five bioremediation technologies were recently evaluated for the treatment of explosives-contaminated soils at the bench and pilot scale.[24] These technologies were selected on the basis of their potential to treat explosives-contaminated soils, were not similar to conventional composting or bioslurry technology, and were "new and innovative." Following the screening process, the technologies selected were Formula One, developed by Midwest Microbial, L.L.C.; BTS developed by Bioremediation Technology Services; Chemical–Biological Treatment developed by the Institute of Gas Technology; DARAMEND® developed by GRACE Bioremediation Technologies; and Fungal-Based Bioremediation developed by Intech One-Eighty Corporation and licensed to EarthFax Engineering. The DARAMEND and the Fungal-Based Bioremediation processes are presented in this chapter since they have met the requirements for full-scale treatment of explosives-contaminated soils. The other technologies were not able to achieve a substantial reduction in the nitroaromatic compounds in soils and/or exhibited high process costs.

14.3 ASSESSMENT OF THE SCOPE OF THE PROBLEM

14.3.1 Regulatory Framework

The DOD in the U.S. is the primary purchaser of services for the cleanup of explosives-contaminated soils. The DOD was responsible for the installations at which explosives were manufactured and munitions were loaded, stored, and tested.

The investigation and cleanup of DOD sites are governed by the Comprehensive Environmental Response, Compensation, and Liability Act (CERCLA), as amended by the Superfund Amendments and Reauthorization Act (SARA), the Resource Conservation and Recovery Act (RCRA), and state and local environmental laws and regulations. The DOD Installation Restoration Program (IRP) is equivalent to the EPA Superfund Program and is carried out in accordance with CERCLA and RCRA requirements with the DOD as the lead agency. The IRP is authorized by the Defense Environmental Restoration Program (DERP) to carry out the investigation and cleanup of past contamination at operational and closed installations, including Formerly Utilized Defense Sites (FUDS).[32]

In addition to cleanup activities managed under the IRP, the DOD also has environmental responsibilities related to Base Realignment and Closure (BRAC). The goals of the BRAC Program investigate and clean up bases that are to be closed in order to accelerate property transfers and to evaluate environmental conditions at installations that will receive additional personnel. Therefore, the cleanup activities at DOD installations may involve both IRP and BRAC Program requirements and may be funded through one or both programs.[32]

While there are significant differences among the regulatory requirements that apply to the investigation and cleanup processes at DOD installations, the sequence of events required in each case can be generalized to consist of four major components: investigation, interim action, design, and cleanup.

In the U.S., RCRA regulations apply primarily to active sites and hazardous wastes disposed of after November 19, 1980 (40 Code of Federal Regulations [CFR] 261.23).[30] Soils contaminated with explosives can be considered hazardous by RCRA guidelines under two conditions: due to the characteristics of the material (i.e., ignitable, corrosive, reactive, or toxic) or the specific process that generated the material (i.e., K047-Pink/Red water from TNT operations).

The most common RCRA waste codes encountered during the remediation of explosives-contaminated soils include[30]

- D003 — Characteristic for reactivity (capable of detonation or explosive reaction if subject to a strong initiating source or if heated under confinement; readily capable of detonation or explosive decomposition or reaction at standard temperature and pressure)
- D030 — Characteristic for toxicity (concentrations of 2,4-DNT in the leachate using the toxicity characteristic leaching procedure [TCLP] greater than 0.13 mg/L)
- K044 — Process code for wastewater treatment sludges from the manufacturing and processing of explosives
- K046 — Process code for wastewater treatment sludges from the manufacturing, formulation, and loading of lead-based initiating compounds
- K047 — Process code for material contaminated with Pink/Red water from TNT operations
- K111 — Process code for material contaminated through the contact of washwaters from the production of DNT via the nitration of toluene

Once the contaminated soil, sludge, or sediment is classified as a hazardous material, RCRA has established minimum guidelines for the management of the material. The guidelines govern the procedures for all facilities that treat, store, or dispose of hazardous waste, including provisions for groundwater monitoring; corrective actions program; records keeping; and the design, installation, and operation of the treatment systems. In general, for the onsite biological treatment of explosives-contaminated soils the guidelines will include containment liners and/or buildings for the treatment process, sampling and analytical requirements, and the ultimate disposition of the treated material.

14.3.2 Sequence of Cleanup Activities

The DOD has a goal of focusing restoration activities and DERP resources on sites that pose the greatest risk to human health and the environment. One way to assess which DOD sites pose the greatest risk and thus the priority for cleanup is to determine which sites have been placed on the National Priority List (NPL). In accordance with the National Contingency Plan, DOD installations are evaluated by the Hazard Ranking System (HRS) and when appropriate placed on the NPL. While the HRS score is an indicator of the relative potential hazard to human health and the environment posed by an installation, it does not address the degree of cleanup required nor does it indicate if explosives contamination is present at the site.[32]

In 1993 the DOD began implementing the use of Relative Risk Site Evaluation to categorize DERP sites into relative risk groups based on an evaluation of contaminants, pathways, and receptors in groundwater, surface water, sediment, and surface soil. The evaluation places a site into an overall category of "high," "medium," or "low" risk. The cleanup of high risk sites is to be completed by FY2002, medium risk sites by FY2008, and low risk sites by FY2015. The Relative Risk Site Evaluation scoring information is contained in the Defense Site Environmental Tracking System (DSERTS) database which is not available to the public.[32]

14.3.3   DOD Installations with Explosives-Contaminated Soils

Explosives-contaminated soils and groundwater exist at many federal installations. The majority of these installations are operational or former U.S. Army installations, including Army Ammunition Plants (AAP), Army Depots (AD), and Army Arsenals. Currently, an estimate of the total scope of the problem is unavailable because the DOD has not completed investigation activities at all installations. Even at installations where investigations have been completed, the full extent of the cleanup efforts is unknown until the remedial action has started.

The U.S. Army Environmental Center (USAEC) has identified at least 50 installations that have explosives-contaminated soils and are in various stages of remedial activities. The assessment was based upon a review of the DSERTS database[2] and personal communications. The USAEC has segregated these installations into three categories: Cleanup Completed, Cleanup in Progress, and Sites with Known or Suspected Contamination (Tables 14.1 to 14.3). The contaminants of concern at these sites are primarily TNT, RDX, HMX, and tetryl. There is a considerable amount of contaminated groundwater at some of the facilities, but a discussion of the extent of the problem is beyond the scope of this chapter.

Table 14.1 Installations Where Soil Remediation Is Complete

Installation	NPL	BRAC	Comment
Alabama AAP	x	x	Incinerated 32,000 yd³ in 1994
Cornhusker AAP	x		Incinerated 46,000 tons in 1988
Camp Mead		FUD	Incinerated 16,000 yd³ in 1998
Weldon Spring		FUD	Incinerated 58,000 yd³ in 1998
Louisiana AAP	x	x	Incinerated 119,000 tons in 1989–90
Savanna AD	x	x	Incinerated 42,000 tons in 1992
Camp Navaho	National	Guard	4,000 yd³ composted in 1999 Natural Guard
Umatilla AD	x	x	15,000 yd³ composted in 1996

Modified from Broder, M.F. and R.F. Westmoreland. 1998. Report No. SFIM-A56-ET-CR-98002. U.S. Army Environmental Center.

Table 14.2 Installations Where Cleanup Activity Is in Progress

Installation	NPL	BRAC	Comment
Badger			Pilot-Scale *In Situ* Treatment
NWS Crane			Composting over 100,000 yd³
Hawthorne AD			Composting over 64,000 yd³
Iowa AAP	x		10,000 yd low temp thermal stripping
Joliet AAP	x		Composting over 200,000 yd³
Milan AAP	x		Composting over 58,000 yd³
Newport AAP			Composting pilot testing completed — 9,000 yd³ to treat
Pueblo AD		x	Composting over 21,000 yd³
Sierra AD		x	Composting 2,000 yd³
Tooele AD	x	x	Composting pilot testing completed — 15,000 yd to treat

Modified from Broder, M.F. and R.F. Westmoreland. 1998. Report No. SFIM-A56-ET-CR-98002. U.S. Army Environmental Center.

Table 14.3 Installations with Known or Suspected Explosives Soil Contamination

Installation	NPL	BRAC	USAEC Soil Volume Estimate (yd³)
Aberdeen Proving Ground	x		10,000
Anniston AD	x		1,000
Picatinney Arsenal	x		27,000
Bluegrass AD			2,000
Dugway Proving Ground			5,000
Fort Carson			
Fort Drum			
Fort Gordon			
Fort Irwin			4,000
Fort Riley	x		
Fort Sill			
Fort Wingate		x	27,000
Holston AAP			
Indiana AAP		x	
Kansas AAP		x	24,000
Lake City AAP	x		
Letterkenny AD	x		
Lone Star AAP	x		
Longhorn AAP	x		4,000
McAlester AAP		x	
Radford AAP			10,000
Ravenna AAP			55,000
Red River AD		x	
Redstone Arsenal	x		
Riverbank AAP	x		
Seneca AD	x		10,000
Sunflower AD	x	x	17,000
Tobyhanna AD	x	x	
Twin Cities AD	x		
West Virginia Ordnance		FUD	14,000
Volunteer AAP			7,000

Modified from Broder, M.F. and R.F. Westmoreland. 1998. Report No. SFIM-A56-ET-CR-98002. U.S. Army Environmental Center.

The ten installations listed in Table 14.2 represent between 600,000 and 1,000,000 yd³ of soils containing explosives compounds. The majority of these facilities have completed the Remedial Investigation/Feasibility Study (RIFS) Process, the Engineering Evaluation/Cost Analysis (EE/CA), and Remedial Design (RD) Stages and are involved in the Remedial Action for Cleanup Stage. The installations listed in Table 14.3 represent approximately 400,000 yd³ of soil based on the USAEC estimates. The exact quantity of soil at any site is not known until the remediation is complete. The quantities are estimates based on available information. The majority of these sites contain less than 10,000 yd³ of contaminated soils and are in the RIFS Process or the EE/CA Stage. A few of the sites are in the RD or Remedial Action (RA) Stages in which specific cleanup technologies are being evaluated for technical and economic merit, or plans for implementation are in progress.

Table 14.4 Annual Explosives-Contaminated Soil Remediation Activity

Year	1998	1999	2000	2001	2002	2003	2004+	Total
Volume of soil to be remediated 1000 yd^3	7	5	10	15	3	9	107	155
Budget for RA, $M	5.5	2.4	4.5	5.9	1.1	3.6	45.0	68.1
Number of sites, RA completed	2	0	5	8	2	1	30	48
Projected sites with new Record of Decision	9	4	10	2	0	11	12	48

Modified from Broder, M.F. and R.F. Westmoreland. 1998. Report No. SFIM-A56-ET-CR-98002. U.S. Army Environmental Center.

Of the 155,000 yd^3 volume of soils to be remediated from Table 14.3, approximately 30% of the soil will be treated prior to FY 2004. The schedule for remediation of these sites and the estimated volumes are presented in Table 14.4. The cost of the remedial actions average $439/yd^3, which is a reasonable estimate for bioremediation at sites with varying quantities of soil to be treated.[17] The sites average 3,200 yd^3 of secondary explosives-contaminated soil each.

14.4 SOLID PHASE TREATMENT PROCESSES

14.4.1 Windrow Composting

14.4.1.1 Process Description

The USAEC conducted pilot-scale composting demonstrations at the Louisiana Army Ammunition Plant, the Badger Army Ammunition Plant in Wisconsin, and at the Umatilla Army Depot Activity in Oregon.[25] The tests were conducted to develop a process for the treatment of explosives-contaminated soils that offered the potential for substantial cost savings over more conventional hazardous waste treatment technologies such as incineration. In addition, composting was a simpler process to operate and offered higher public acceptability. The pilot demonstrations tested aerated static pile composting (compost is formed into piles and aerated with blowers), mechanically agitated in-vessel composting (compost is placed in a reactor vessel where it is mixed and aerated), and windrow composting (compost is placed in long piles known as windrows and periodically mixed using mobile equipment). The windrow composting process was selected for further process development because of better performance, ease of operation, and economics.

Implementation of the windrow composting process involves amendment delivery and stockpiling, soil preparation and windrow construction, windrow operations, and treated residual management. Amendment delivery, stockpiling, and management are critical factors in the success of the composting process. Based on pilot-scale amendment optimization testing, the required amendments are procured from local sources to minimize transportation costs and reduce odor and handling issues associated with the onsite amendment stockpile. The amendments are stored in individual "bins" constructed near the treatment area. Amendment bins are covered to minimize precipitation infiltration and runoff from the stockpiles.

A screen plant is used to remove debris, rocks, and concrete from the excavated and stockpiled soil. Depending on the quantity and type of material encountered, the debris can be powerwashed manually or a mechanical debris-washing system is used to remove contaminants below the remedial goals. All washwater is collected and used for moisture addition in the windrows, as necessary. After screening, the soil and amendments are blended through a mixer specifically designed for compost operations.

Proper selection of amendments for windrow composting is critical to the success of the process. The amendments can be divided into the following general classifications: bulking agents, nutrient sources, and inoculum. The addition of bulking agents allows air movement through the compost and provides a measure of insulation for self-heating in the windrow. Bulking agents may include wood chips, yard waste (a combination of grass clippings and wood chips), or other locally available products. Potential nutrient sources include alfalfa, starch (potato or corn), hay, silage, food processing wastes (potato waste, corn mash, etc.), or green waste (grass clippings). Animal wastes, especially cow manure, are an excellent source of inoculum for windrow composting of nitroaromatic compounds. Selection of the amendment mixture entails a correct balance of bulk density, carbon/nitrogen/phosphorus ratio, pH, moisture content, and inoculum for rapid self-heating after blending. Pilot-scale windrows are typically used to optimize amendment ratios and confirm process performance.[26,35]

Windrow operation includes aeration and mixing, process monitoring, and moisture addition. In general, the windrows are turned three to seven times per week with a commercial windrow turner. Process monitoring includes daily temperature readings throughout the windrow. Properly blended compost will achieve a thermophilic temperature within 1 to 3 days after construction. After the windrow has been verified to meet project cleanup goals, the windrow is removed from the treatment area.

14.4.1.2 Technology Applications

Umatilla Army Depot Activity, Oregon — The first production level bioremediation of explosives-contaminated soil in the U.S. was conducted at Umatilla Army Depot Activity (UMDA) near Hermiston, OR.[7,27] Soil from munitions washout lagoons which contained high concentrations of TNT and RDX as well as HMX was treated by windrow composting. The cleanup goal established by the Record of Decision (ROD) was 30 mg/kg each for TNT and RDX. Treatment time on the initial 2700 yd^3 batch (810 yd^3 of soil) was 10 to 12 days.

The amendments added to the soil were cow manure, chicken manure, potato waste, sawdust, and alfalfa. The tabulated data from the trial test and the full-scale production were presented.[7] The remediation costs of $256 per ton of soil were reasonable for the first production-level application. Based on recent discussions with the authors, current costs for application of windrow composting are similar to those presented in Section 14.6.3. Approximately 15,000 yd^3 of soil was successfully treated, and more than 70% of all analyses indicate non-detectable levels of both TNT and RDX. A savings of over $2.6 million was claimed compared to treatment using more conventional technologies such as incineration; however, a cost comparison was not presented.

U.S. Navy Submarine Base (SUBBASE) Bangor, WA — In September 1994, the Final ROD was signed which established bioremediation as the remedy for treating ordnance-contaminated soil at SUBBASE Bangor in Washington. This project was one of the first large-scale applications of composting for explosives-contaminated soils in the Northwest. Pilot-scale testing was conducted to determine scale up parameters for full-scale operation.[31] Compost temperatures ranging from 110° to 120°F were maintained for 14 days. These low temperatures were attributed to the low ambient temperatures. However, the moisture content of the windrow was above 40% which probably was also a factor.

The composting facility for full-scale treatment included a 30,000-ft² temporary building which contained four windrows. Each windrow was 250 ft long, 24 ft wide, and 6 ft high. The amendment compost mix included by volume 35% soil, 20% cow manure, 15% wood chips, 15% alfalfa, and 15% apple pomace/potato culls/potato waste.

Approximately 880 yd³ of soil contaminated with TNT were excavated from a former ordnance disposal area and treated for an average time of 53 days. Approximately 2300 yd³ of TNT- and RDX-contaminated soil from a former evaporation/infiltration lagoon that received wastewater from a munitions demilitarization facility were treated for an average of 30 days. An additional 1400 yd³ of soils from a third site were treated for an average of 33 days.[9] Insufficient information was provided to evaluate the reasons that the treatment times were longer than those reported for the windrow composting of explosives-contaminated soils at other sites.

Hawthorne Army Depot, Nevada — A full-scale production pilot study was conducted at Hawthorne Army Depot to determine the feasibility of windrow composting for treatment of explosives-contaminated soil from several settling pits.[3] The major goal of the test was to determine the effectiveness of windrow composting under specific site conditions using locally available amendments. In addition to TNT, RDX, and HMX, the effectiveness of windrow composting to treat soils containing ammonium picrate was evaluated because no information was available from previous work.

A major process modification tested was the construction and operation of the windrows on an outdoor compacted soil pad rather than on impermeable pads inside an enclosed structure. Another modification from the process used at Umatilla Army Depot was the elimination of the premixing of contaminated soil with the amendments prior to windrow construction. Instead appropriate amounts of amendments were added directly to the soil on the treatment pad during windrow construction, and the windrow was continuously mixed during the first day. This procedure was also used to construct the windrows at Crane Naval Surface Warfare Center in Indiana. Wood chips generated by the base from old ammunition crates and pallets were used as an inexpensive amendment for the compost blend. The wood chips were considered solid waste because they had been treated with pentachlorophenol and were scheduled for landfill disposal at considerable cost.

The explosives contaminants were degraded rapidly using a variety of ratios of amendments including hay, wood chips, manure, and potato waste. Because there was some scattering in the data and no indication of the statistics was provided, it is not possible to distinguish differences among the various mixtures. In each case the treatment goals of 233 mg of TNT per kilogram of soil and 64 mg of RDX per kilogram of soil were achieved in less than 28 days.

The site-specific conditions which allowed a departure from the process operations at Umatilla should result in lower unit costs for full-scale operation. The compacted soil used as the treatment pad was not contaminated at the end of the compost treatment; therefore, decontamination or disposal would not be required. Due to the arid climate, leachate was not generated during process operation from moisture addition or rainfall. Process operation outdoors would not be possible in an area that receives more rainfall. In addition, the base is located in a sparsely populated area so dust control and odor containment may not be as critical as in a more densely populated area. Moisture control was a major concern. It is important to provide enough moisture for the process and for dust control without creating leachate, and it is not clear from the published information how such control was accomplished.

The use of the readily available wood chips not only supported the composting process, but also provided a solution to the disposal of the contaminated wood. Pentachlorophenol was not detected at the end of the treatment, which indicated that the composting process also degraded the pentachlorophenol.

The test established the feasibility of windrow composting at the site, and the authors indicated that the costs were reduced substantially by the modifications cited previously. The costs cited ($250/yd^3) were higher than those reported later in this chapter, but no details were provided on the calculations.

The issue of dust control from mixing the windrows outside is important and must be resolved before the strategy described here can be widely implemented. Protection of the windrows from excessive wind and rainfall are also engineering challenges that must be resolved at each site. It seems likely that appropriate measures were employed during the Hawthorne demonstration, but no details were provided on the specific strategies.

Tooele Army Depot, Utah — Pilot-scale windrow composting tests on 6.5 yd^3 of soil were performed at Tooele Army Depot (TEAD) in Tooele, UT to determine the potential to treat explosives-contaminated soils under site-specific conditions.[15] Windrow composting was one of the corrective measures under consideration for treatment of TNT-, DNT-, RDX-, and HMX-contaminated soils from operation of a washout facility that was used to decommission explosive munitions. Bench-scale tests were performed prior to the pilot tests to determine the suitability of locally available amendments. Potato waste, a component of the amendment blend used at Umatilla, was not available locally. The amendments tested were lettuce, onions, barley, and molasses in different blends. The local supplies of wood chips, alfalfa, and manure were not tested in these blends. The test beds were 1.5 ft^3 and sufficient heating was only observed with the lettuce/barley mixture with the addition of molasses. However, the conditions for the tests were not given. There may have been insufficient volume to maintain self-heating conditions and overcome rapid heat loss with the other blends tested. The addition of molasses may not be necessary to maintain thermophilic conditions when composting is performed with larger volumes of material.

The amendment mixture used in the pilot-scale test included alfalfa, cow and chicken manure, lettuce, barley, manure, and wood chips. A mixture of molasses and water was used to add moisture and provide readily available nutrients. The actual blend of amendments or the amount of molasses was not provided. Sufficient nutrients and carbon should be available, however, without the addition of molasses. The use of molasses at full scale may not provide enough benefit to the process performance to overcome both capital and operational costs.

The TNT and RDX concentrations of 1890 and 724 mg/kg, respectively, were reduced to below cleanup goals of 94 mg of TNT per kilogram of soil and 34 mg of RDX per kilogram of soil within 10 days. The sample concentrations were the average of five samples; however, the individual sample results were not provided. Total composting costs from this study were approximately $230/yd^3 ($300/m^3), which is equivalent to $153/ton ($168/metric ton) of treated soil. A cost breakdown was not provided so it is not possible to evaluate the costs associated with the purchase, transportation, and management of the amendments at the project site. The authors recommended 1 year of curing before attempting to establish a vegetative cover on the composted material.

Pueblo Chemical Depot, Colorado — A pilot-scale demonstration of windrow composting technology was conducted on explosives-contaminated soils from a TNT washout facility at the Pueblo Chemical Depot (PCD) in Colorado.[36] The testing was conducted to determine whether the site-specific cleanup goals could be achieved using locally available amendments and to develop accurate costs for full-scale treatment. The amendments added to the soil at a final volume ratio of 70% amendments to 30% soil included cow manure, chicken manure, chopped alfalfa, potato waste, and yard waste (wood chips and grass clippings). Yard wastes had not been tested previously for treatment of explosives-contaminated soils and were considered an inexpensive bulking agent. The initial concentration of contaminants in the soil was 3000 mg/kg of TNT, 10 mg/kg of DNT, and 40 mg/kg of trinitrobenzene (TNB). After 15 days of treatment, DNT and TNB concentrations were degraded below 0.20 mg/kg and the average TNT concentration was 1.5 mg/kg.

Full-scale treatment is in progress in a 180-ft wide by 500-ft long building which contains ten 200-ft long windrows. Full-scale process optimization tests were conducted on ten windrows to determine the final amendment blend. These tests resulted in the selection of an amendment mixture that contained only cow manure, alfalfa hay, ground organic yard waste, and soil and a process that required only 16 days on average to achieve all required treatment standards. Under current operations, TNT concentrations as high as 1000 mg/kg are typically reduced to less than the risk-based treatment standard of 3.8 mg/kg within 16 days of treatment. The treated compost is cured outdoors and used onsite as a soil amendment and conditioner.[5]

Crane Naval Surface Warfare Center (NSWC Crane), Indiana — At the Crane Naval Surface Warfare Center several areas have soil heavily contaminated with TNT, RDX, and HMX. A full-scale bioremediation facility including three compost buildings has been constructed at the site. Explosives-contaminated soil is sampled, excavated, screened, and transported to the facility where it is deposited on previously chopped and blended amendments in a windrow 20 ft wide, 7 ft high, and 270 ft long. The soil/amendment mixture contains 25% explosives-contaminated soil, 15% chicken manure, and 60% straw. As of July 1999, a full-scale operation has been in progress for 15 months. A total of 17,560 tons of soil have been treated. Test kits are used to estimate end of cycle and when to take final samples to confirm that cleanup goals are met. The treatment cycle is typically 7 to 10 days. The treated compost is then returned to its site of origin for backfill after confirmation sampling of the excavated area.

The estimated quantity of soil requiring remediation is approximately 110,000 yd^3 (150,000 tons) which will yield close to 200,000 yd^3 of compost. Once acceptable reduction of explosives has been achieved, the finished compost will be stockpiled for use as daily cover in the onsite solid waste landfill (cost savings vs. obtaining offsite landfill cover). If possible, after full-scale operations are successful, the compost may be used as backfill at the original excavation site. The unit costs are dependent upon the final selected recipe, but are expected to be approximately $175/ton. Based on current estimated quantities and the cost of the final selected recipe, remediation of explosives-contaminated soils at NSWC Crane will take approximately 7 years to complete at a total cost of $26 million.[21]

14.4.2 Soil Pile — Two Stage

14.4.2.1 Process Description

A two-stage soil (TOSS) process was developed and tested by Waste Management, Inc. (WMI), Cincinnati, OH to provide a commercial service for explosives-contaminated soils at two of its landfill sites in the U.S. In the first stage of treatment, explosives-contaminated soil is combined with a microbial inoculum consisting of animal manure(s) and/or anaerobic digester sludge, an electron-donating substrate such as simple or complex carbohydrates, and water to achieve anaerobic conditions. The volume of amendments used during this stage is approximately 20%, which is considerably less than the volumes used for windrow composting. The soil and amendments are blended together using an excavator or windrow turner, and water is added to between 70 and 95% of the water holding capacity of the soil. The resulting mixture is constructed into a biopile or placed in a bermed construction and maintained under anaerobic conditions. In the second stage of the process, the anaerobically treated soil is combined with previously composted yard wastes and aerated. Waste Management, Inc. has applied for a patent for the TOSS process. The TOSS process is that it is the only technology currently offered as a commercial service for the offsite treatment of explosives-contaminated soil. Waste Management, Inc. began offering this capability at two of its U.S. landfills in 1999.

14.4.2.2 Technology Applications

Bench- and pilot-scale testing of the TOSS process has been completed on explosives-contaminated soils from two U.S. Army installations, the Joliet Army

Ammunition Plant (JOAAP) in Wilmington, IL and the Pueblo Chemical Depot (PCD) in Pueblo, CO, and a commercial explosives manufacturing facility in Sweden. The tests were performed using soils impacted with the following contaminants: TNT, 2,4-DNT, RDX, HMX, 1,3,5-TNB, and tetryl.

Results from bench-scale testing of soil from PCD demonstrated greater than 99% removal of TNT.[11] A pilot-scale demonstration at PCD on approximately 15 yd^3 of contaminated soil yielded a TNT removal efficiency of 75% in 113 days of anaerobic treatment at an operating temperature of 12°C.[17] The pilot test was performed in the winter, and ambient temperature probably had a impact on the TOSS process kinetics. The TOSS process is not a self-heating process because of the lower volume of amendments. Windrow composting of the same soil was not impacted by ambient temperatures.[17]

Green and coauthors have reported on the results of a pilot demonstration of the TOSS process on 100 tons of explosives-contaminated soil from the JOAAP.[12] The demonstration, conducted at a Waste Management, Inc. landfill in Louisville, KY, permitted for explosives bioremediation, achieved removals of greater than 98% for TNT, RDX, and HMX after 39 weeks of anaerobic–aerobic treatment.

Bench-scale testing of the technique on soil contaminated with approximately 1100 mg/kg of 2,4-DNT from an active explosives manufacturing facility in Sweden evaluated the influence of moisture content and carbohydrate concentration of removal efficiency.[6] The results indicated 2,4-DNT removal of greater than 96% in 104 days. The technology is being further evaluated in a 200-ton pilot study currently underway at the site.

Early seedling growth studies were conducted in conjunction with the JOAPP pilot test to evaluate residual toxicity of treated soil to plants.[1] The authors concluded that soil treatment resulted in statistically significant reductions in plant toxicity as measured by biomass yield and seedling length.

14.4.3 Soil Pile — Fungal-Based Remediation

14.4.3.1 Process Description

Fungal-based soil remediation is an *ex situ* treatment process that is based on the pollutant-degrading abilities of wood decay fungi[8] and consists of amending the contaminated soil with a fungal inoculum at a predetermined application rate. For full-scale treatment the fungal inoculum can either be prepared onsite, a process that requires from 3 to 6 weeks depending on ambient temperatures, or offsite and delivered in a pelletized form. The inoculum consists of a pure culture of the selected fungus grown on a lignocellulosic substrate. A variety of inexpensive substrates (e.g., cottonseed hulls, sawdust, straw, etc.) can be used depending on the fungal species. The most effective combination of fungal species, inoculum application rate, and type of inoculum are predetermined for each soil via bench- and pilot-scale tests. The inoculum/soil mixture is treated in a forced-aeration biopile. Aeration of the pile controls the temperature via removal of metabolic heat and ensures an adequate supply of oxygen to the fungus.[4]

14.4.3.2 Technology Application

EarthFax Engineering, Logan, VT conducted a treatability study on TNT-contaminated soil from the Mead Army Ammunition facility to evaluate the comparative abilities of several different species of white rot fungi to decrease the TNT concentrations. All the fungi were able to cause rapid and extensive reduction of TNT in the soils. However, there were significant differences among the organisms used in the rates and extents of treatment.[4]

EarthFax Engineering also performed bench tests with tetryl-contaminated soil from the Joliet Army Ammunition Facility, in Joliet, IL.[4] The initial concentration in the soil was 12,000 ppm. After 4 weeks of treatment with the white rot fungus *Trametes versicolor*, the tetryl concentration had decreased 59% to 4900 ppm. The rate of degradation is very similar to that observed for degradation of TNT by *Phanerochaete chrysosporium* in soil with an initial TNT concentration of 10,000 ppm. Longer treatment would be expected to result in more extensive degradation of tetryl. Work in liquid culture with *P. chrysosporium* suggests that as with TNT, rapid decreases in the tetryl concentration (i.e., complete disappearance in 24 h) were due to reduction of one of the nitro groups.

Application of the fungal-based remediation technology to soils with elevated concentrations of TNT and tetryl at the JOAAP was initiated in July 1998 and consisted of two pilot-scale evaluations, each involving treatment of 10 tons of soil.[24] The initial TNT concentration in the soil was approximately 3500 mg/kg. The inoculum application rate was 60% on a dry weight basis. Typical inoculum application rates, however, ranged from 10 to 30% on a dry weight basis. Following incorporation of the EarthFax fungal inoculum into the soil, the TNT concentration measured was 628 mg/kg on day 0. Since dilution would account for only a 60% reduction the remaining reduction was due to treatment by the active fungal inoculum upon mixing with the soil. The final TNT concentration was 332 mg/kg on day 60 of the pilot test. In the laboratory phase of the test the fungal technology reduced concentrations of TNT in the soil from 3190 to 460 mg/kg after 60 days. The fungal inoculum accounted for only 10% on a dry weight basis of the soil. The differences in the results between the bench and pilot tests may have been due to the difficulty of performing the field tests by non-EarthFax personnel.

Tetryl concentrations were reduced from 8410 to 1220 mg/kg in the laboratory phase of the tests and from 131 to 82 mg/kg in the pilot test. Very little radiolabeled TNT or tetryl was mineralized during the laboratory test. Both the fungal-treated TNT and tetryl soils showed lower toxicity on day 120 than the respective controls.

14.4.4 Land Treatment — Grace Daramend®

14.4.4.1 Process Description

DARAMEND® is a bioremediation technology developed by Grace Bioremediation Technologies, Mississanga, Ontario, an operating unit of W.R. Grace & Co. The patented technology (U.S. Patents No. 5,411,664; 5,480,579; and

5,618,427) has been applied to a wide variety of soils and sediments containing organic explosive compounds. The sites where DARAMEND has been applied at field scale are the U.S. Army's Raritan Arsenal site in New Jersey and JOAAP, and the U.S. Navy's Yorktown Naval Weapons Station in Virginia. Bench-scale tests conducted during the process development phase included explosives-contaminated soils from Pantex Plant in Amarillo, TX; Hawthorne Army Ammunition Plant in Nevada; Weldon Spring Ammunition Depot in Missouri; and the Naval Weapons Station in Crane, IN in addition to the sites where the pilot-scale tests were performed. The technology was evaluated and proven to be effective for treatment of soils contaminated with wood preserving wastes (pentachlorophenol and polycyclic aromatic hydrocarbons) in full-scale application during an independent audit conducted by the EPA as part of its Superfund Innovative Technology Evaluation (SITE) program (EPA No. 540/R-95/536.[34]

The DARAMEND process is essentially a modification of conventional land farming technology that incorporates the addition of proprietary amendments into the soil. The technology also includes operation of the soils through anaerobic, anoxic, and aerobic phases of treatment. The number and frequency of the cycles varies with the contaminant type, concentrations, and the treatment criteria. Mixing of the amendments and soil and the aeration of the soil for the aerobic phase is accomplished with the use of an agricultural tractor and a rototiller attachment. The soil is placed in a treatment cell, and a building is constructed over the cell to allow year round operation of the process. The proprietary DARAMEND organic amendments are manufactured from plant materials and may be supplemented with agricultural-grade or food-grade nutrients. The inorganic amendment most commonly applied is powdered metallic iron. Typical application rates for the organic and inorganic amendments are 0.5 to 10% (w/w) and 0.2 to 2.0% (w/w), respectively.

14.4.4.2 Technology Applications

U.S. Army, Joliet Army Ammunition Plant, Wilmington, IL — Application of DARAMEND technology to soils with elevated concentrations of TNT and tetryl at the JOAAP was initiated in July 1998 and consisted of two pilot-scale evaluations, each involving treatment of 10 tons of soil. The technology evaluations were conducted as part of an evaluation of innovative biotechnologies for remediation of soils containing organic explosive compounds.[24]

Soils with elevated levels of TNT and tetryl were excavated, screened, and homogenized prior to the onset of treatment. The demonstration ran for 120 days and involved the application of ten DARAMEND treatment cycles. One cycle consisted of an anaerobic and aerobic treatment phase. Toxicity testing including *Ceriodaphnia dubia* reproduction, Ames mutagenicity testing with *Salmonella typhimurium*, and MicroTox tests with *Vibrio fischeri* were conducted with treated soil from the pilot tests.

Over the 120-day course of the demonstration, TNT concentration in the soil was reduced from 3600 to 114 mg/kg, and the tetryl concentration was reduced from an initial value of almost 6000 to 16 mg/kg. Samples collected from both the TNT and tetryl treatment cells were less toxic than the respective controls. Results from laboratory tests showed a tetryl reduction of 26% after 60 days of treatment. TNT concentrations were reduced from 3500 to 446 mg/kg in the laboratory tests. Laboratory tests were conducted with radiolabeled TNT and tetryl prior to the pilot-scale tests to collect information on the pathways and mechanisms of degradation. Very little radiolabeled TNT or tetryl was mineralized using the DARAMEND process.

U.S. Navy, Naval Weapons Station Yorktown, Yorktown, VA — Full-scale DARAMEND bioremediation was recently initiated at the Naval Weapons Station Yorktown near Yorktown, VA.[10] The soil was excavated from Yorktown Site 6, a former munitions manufacturing facility, and contained a mixture of organic explosive compounds including TNT, ADNT, and RDX. Processwater and washwater from this facility were responsible for the deposition of TNT, RDX, and HMX to the soil.

Prior to initiation of treatment, the soil contained TNT at 10,200 mg/kg, ADNT at 1000 mg/kg, RDX at 210 mg/kg, and HMX at 5 mg/kg. Remedial goals for the site were 14 mg/kg for TNT and 5 mg/kg for RDX. Results following 35 days of active treatment showed concentrations of 2440 mg/kg for TNT and <1 mg/kg for RDX.

14.5 SLURRY-PHASE PROCESSES

14.5.1 SABRE™ (Simplot Anaerobic Bioremediation) Process

14.5.1.1 Process Description

Through a joint research effort with the University of Idaho, Moscow, ID, the J.R. Simplot Company (Simplot), Pocatella, ID developed a technology for remediating nitroaromatic compounds in soil. The bioremediation technology, the SABRE™ (Simplot Anaerobic Bioremediation) Process, was developed from studies on the biodegradation of dinoseb (2-*sec*-butyl-4,6-dinitrophenol) performed at the University of Idaho. Further research demonstrated the application of the technology on TNT, RDX, and HMX.[33] Simplot previously held the exclusive worldwide license to the technology. The University of Idaho Research Foundation currently holds the license to the technology. U.S. patents were issued in February 1995 and April 1997.

The SABRE bioslurry process is a proprietary technology that requires materials handling, slurry preparation, biological treatment in an engineered treatment lagoon (biocell) or above-ground reactor, and slurry disposal. Slurry preparation may include soil washing, while slurry disposal may include slurry dewatering. A significant process engineering design effort may be required prior to process implementation.

Operating parameters that influence the performance of the SABRE process include pH, redox, and temperature. The preferred pH range is between 6 and 7.

Anaerobic conditions suitable for the treatment of the nitroaromatic compounds occur when the oxidation/reduction potential is lower than -200 mV. During periodic mixing of the biocell, the solid phase is recontacted with the liquid in a manner to prevent aeration of the liquid. The semistatic conditions will achieve acceptable results when soil, water, and the carbon source are well mixed during loading of the biocell.

14.5.1.2 Process Applications

The most comprehensive publication on the application of the SABRE process for treatment of explosives-contaminated soils was the EPA SITE Program Report based on the results of the tests performed at Weldon Spring Ordnance Works.[33] The goal of the SITE program is to "... develop scientific and engineering information needed by EPA to support regulatory and policy decisions; and provide technical support and information transfer to ensure effective implementation of environmental regulations and strategies." This report is one of the most comprehensive evaluations for any of the technologies discussed in this chapter. The conclusions indicated that the SABRE process achieved a 99% treatment efficiency based on an average initial TNT concentration of 1500 mg/kg and a final concentration of 8.7 mg/kg. The treatment time was approximately 9 months, and 5 months was necessary to achieve 95% treatment efficiency. The common TNT intermediate transformation products increased during the initial phases of treatment then decreased to below the analytical detection limits. Relative toxicity tests including early seedling growth, root elongation, and earthworm reproduction showed the process had successfully reduced the toxicity of the soil. Problems were encountered with the loading and mixing of the soil due to the high clay content. The process costs, not including excavation, for the treatment of 5000 yd^3 of soil was approximately \$112/yd^3. Additional costs of up to \$100/yd^3 have been assessed by Simplot for technical assistance, soil nutrients, and other process enhancements.

Tests performed on explosives-contaminated soils from Nebraska Ordnance Plant, Mead, NE achieved the site treatment goals of 17.2 mg/kg for TNT and 5.8 mg/kg for RDX. The initial soil concentrations of TNT and RDX were 1700 and 2400 mg/kg, respectively. Even though the SABRE process successfully treated the explosives-contaminated soils, incineration was the treatment technology selected and applied for full-scale remediation because of the EPA ROD.

At Bangor Naval Submarine Base in Washington State a pilot-scale treatment of soils contaminated with TNT, RDX, and HMX was performed.[31] The demonstration involved the use of an innovative hydromixing system, and the authors tested a heating process under adverse weather conditions. Treatment goals for both soils were achieved. The full-scale remediation of soils at this site was performed by windrow composting, however, as selected by the U.S. Army Corp of Engineers.[9]

A field demonstration was conducted at the IAAP in 1996 under the direction of the USAEC. The objective of the demonstration was to accelerate the field application of the bioslurry technology. In addition to the SABRE process, the aerobic/anoxic process developed by the USAEC was also tested.[16]

The initial concentration of approximately 800 mg/kg of TNT was treated to the cleanup goal of 196 mg/kg in less than 8 weeks. RDX concentrations were reduced from 300 mg/kg to the cleanup goal of 53 mg/kg in approximately 10 weeks. The majority of the RDX disappearance occurred after disappearance of TNT. The treated slurry underwent a variety of tests: radiolabeled TNT metabolic fate studies, leaching stability tests, plant growth analysis, and toxicity testing including Ames assay and earthworm testing. Results from previous trials indicated that the slurry is suitable for direct land application by direct discharge.

Simplot conducted a treatability study to evaluate the potential to remediate explosives-contaminated soils at the Yorktown Naval Weapons Station using the SABRE process. The soils contained the explosives TNT, RDX, and HMX. Following the bench-scale study, a full-scale demonstration was performed on 650 yd^3 of soils at the Yorktown Naval Weapons Station.[18]

An initial TNT soil concentration of 1050 mg/kg was treated to the proposed goals of less than 30 mg/kg. The treatment goal of 100 mg/kg for RDX was also achieved starting with an initial soil concentration of 380 mg/kg. Difficulties loading the engineered treatment cell were encountered due to the nature of the excavated soil. The soil consisted of dense clays full of organic matter. The organic material was a combination of tree roots, decaying plant material, and root systems from marsh flora. This caused the loading of 650 yd^3 of soil to take 49 days. Full-scale treatment of 1200 yd^3 of soil was performed in 1997.[17] The treatment cell was operated for a period of 30 days to achieve treatment goals of 30, 8.0, 50, and 3900 mg/kg for TNT, DNT, RDX, and HMX, respectively.

14.5.2 U.S. Army Environmental Center Aerobic/Anoxic Process

The USAEC developed a bioslurry system to treat explosives-contaminated material to provide a non-proprietary process without a technology transfer fee for treatment of explosives-contaminated soils at DOD sites.[20]

The USAEC conducted a field demonstration at the JAAP with the aerobic/anoxic bioslurry process. The strategy was based on the successful performance of an aerobic/anoxic laboratory-scale soil slurry reactor which showed that the contaminated soil could be treated successfully in batches or semicontinuously.[20] The overall goal of these tests was to determine the effectiveness and cost of bioslurry systems for degrading explosives in soil. The bioslurry system was reported to be effective in the treatment of TNT. Explosives were not degraded in the control reactor without cosubstrates. The rate of TNT degradation was substantially impacted at operating temperatures less than 20°C, and a significant accumulation of the monoaminodinitrotoluene transformation product was observed. Overall, the important process parameters, as determined in this field demonstration, were the need for an organic cosubstrate (molasses), the operation of the reactors in an aerobic-anoxic sequence, and temperature.

The USAEC field tested both aerobic/anoxic and anaerobic soil slurry biotreatment processes at the IAAP to evaluate their performance and cost in remediation of explosives-contamination in soils.[6] The anaerobic process was similar to the SABRE process. The trials were conducted in an RCRA quality lagoon and a concrete trench reactor, employing a variety of commercial impeller and hydraulic mixing systems. Test results described both degradation kinetics and metabolic fate, as well as the capabilities and limitations of different reactor configurations and mixing strategies. The results will be used by environmental managers in their remedy selection.

TNT was reduced from 1500 mg/kg to below treatment levels in approximately 8 weeks, while limiting the accumulation of metabolic intermediates to less than 40 mg/kg. The temperature remained above 22°C, the dissolved oxygen fluctuated between 1.0 and 0 mg/L, and the redox potential of this system varied in the range of –50 to +50 mV, indicating the aerobic/anoxic nature of the process. The degradation of HMX and RDX followed that of TNT.

The demonstration also evaluated options for disposing of the treated soil and process water, if land application is not an option. Slurry was gravity dewatered to 40% soil moisture in 2 to 3 weeks. Options for disposing of the process water depend on discharge standards at the site. Residual explosives in the treated water were below the local National Pollutant Discharge Elimination System standards of 2 µg/L TNT, but the biotreatment processes in both reactors increased the biological oxygen demand (BOD) in the water above 10,000 mg/L. The results of analytical and toxicity tests can support land application of the process water. Otherwise, high BOD levels can require additional aerobic treatment or processing through activated carbon to bring it down to discharge levels. A report including cost data for the aerobic/anoxic process will be available in 2000.

14.6 PROCESS COMPARISON AND ECONOMIC EVALUATION

14.6.1 Introduction

In order to compare costs for the different treatment technologies, existing economic data for the various treatment technologies were reviewed. Economic data from the following technologies were evaluated.

- Windrow composting
- Simplot-SABRE slurry
- WR Grace-DARAMEND
- WMI-TOSS (offsite)
- EarthFax — Fungal-Based Remediation

Only economic and operating data from field-scale projects (greater than 1000 tons of treated soil) were used in the economic evaluation for each technology. The economics for two of the technologies, EarthFax — Fungal-Based Remediation and the WMI-TOSS, are prepared from similar projects that did not include treatment of explosives. The economic data was provided by each technology supplier and was evaluated for reasonableness based on past field experience. Existing data from the different technology suppliers were standardized for two "typical" sites encountered in the remediation market: 25,000 tons of RCRA hazardous soil and 100,000 tons of non-hazardous soil. The major cost contributors (i.e., labor, specialized equipment, proprietary amendments, licensing fees, etc.) for each technology are highlighted and briefly discussed in the following sections.

14.6.2 Site Descriptions

In order to simplify economic comparisons, several site characteristics and cost factors were considered common across the technologies. For all scenarios, the contaminated soil has been excavated and stockpiled near the treatment area and additional screening and segregation of oversized material (greater than approximately 2 to 4 in.) is not required. All treatment will be conducted in an engineered building constructed onsite. Ambient temperatures will average above 50 to 60°F for 8 months per year. If the treatment technology is not self-heating, an additional building heat source is required. After successful treatment, the material will remain onsite for use as excavation backfill or as a soil conditioner; therefore, additional transportation or offsite disposal is not required. Sampling and analytical activities include the collection of process monitoring and confirmation samples and analyses by a U.S. Army Corp of Engineers certified laboratory for explosives following USEPA SW-846 Method 8330 guidelines. Applicable health and safety considerations such as personal protective equipment, respirators, air monitoring, etc. are included in each of the treatment costs. In addition, a 100 yd^3 pilot-scale demonstration of each technology will be completed prior to full-scale treatment.

The first site evaluated (Site Condition 1) is based on the treatment of 25,000 tons of a RCRA-hazardous soil. The soil is classified as hazardous due to reactivity (D003) and toxicity for the potential to leach DNT (D030). A more detailed discussion of the RCRA classification was presented earlier. The RCRA classification impacts both analytical requirements and construction of the treatment building. Additional TCLP analyses are required in the confirmation sampling, and the treatment building must meet RCRA guidelines for construction, containment, and monitoring for treatment of hazardous soils. Treatment operations are conducted 5 days per week for 12 months per year and the total project duration is 2 to 3 years.

The other site evaluated (Site Condition 2) is based on the treatment of 100,000 tons of soil contaminated with TNT. After excavation, the soil contains less than 10% TNT by weight and is considered non-reactive and therefore is not managed as a RCRA hazardous material. In addition, toxicity is not an issue since DNT is not present in the soil and the material does not carry a RCRA process code. A 5-acre (approximately 220,000 ft^2) treatment facility is constructed onsite, and a 5 day per week, 12 month per year operating schedule produces a total project duration of 3 to 4 years.

14.6.3 Treatment Economics

For all of the five technologies and both site conditions, treatment costs were divided into five categories: preconstruction activities, site setup and mobilization, onsite pilot demonstration (this category includes transportation costs for the offsite WMI-TOSS process), treatment, and demobilization and site closeout. The treatment category was further divided into labor, equipment, materials and supplies, and analytical costs. In general, preconstruction activities include design, permits, and workplan preparation. Site setup include construction of treatment facilities, site preparation, and mobilization of personnel and equipment.

The estimated treatment costs for the five technologies at a typical Site Condition 1 are summarized in Table 14.5. The total costs range from $165/ton for the WR Grace-DARAMEND process to $309/ton for the Simplot-SABRE process. The treatment costs for the five technologies at a typical Site Condition 2 are summarized in Table 14.6. The total project costs range from $115/ton for WR-Grace-DARA-MEND process to $162/ton for the Simplot-SABRE technology. The major cost contributors for each technology for both sites are summarized in the following paragraphs. In addition, the overall process advantages and disadvantages are discussed for each technology.

14.6.4 Windrow Composting

As discussed previously, windrow composting requires the addition of a large proportion of organic amendments (approximately 70% by volume) with the contaminated soil (approximately 30% by volume). In addition to the cost of amendments, both equipment and labor costs are impacted by the requirements to manage the relatively large volume of both amendments and the resulting treated soil. As indicated in Table 14.5, the highest cost categories are labor, equipment, and materials and supplies, which include the cost of amendments. Site Condition 1 requires approximately six equipment operators and four technicians to operate, maintain, and monitor four rubber-tired loaders, a windrow turner, a batch mixer (to provide the initial blend of amendments and soil), and two 20-yd^3 dump trucks. Approximately 12,500 yd^3 of cow manure, 12,500 yd^3 of alfalfa, and 33,000 yd^3 of wood chips are required for the treatment of 25,000 tons of soil. Approximately 50,000 yd^3 of compost will remain after treatment for use as an onsite soil conditioner. This increase in volume also impacts analytical costs due since a large number of samples are required for confirmation monitoring.

Site Condition 2 requires eight equipment operators and four technicians to operate, maintain, and monitor four rubber-tire loaders, two portable batch mixers, and a windrow turner. Approximately 35,000 yd^3 of manure and 200,000 yd^3 of blended yard waste and wood chips are required and will produce approximately 200,000 yd^3 of treated material. Based on economies of scale, both labor and equipment unit costs are much lower for the larger volume of soil (Table 14.6). Due to the large volume of amendments required, the relative amendment unit costs are increased due to the logistics of procurement and transportation. Analytical costs are highest for windrow composting among the evaluated technologies due to the volume increase of the treatment material and the resulting increase in the number of confirmation samples.

Windrow composting offers several advantages over other treatment technologies. Among the technologies reviewed, windrow composting has been, by far, the most successfully implemented biological treatment technology for explosives-contaminated soils. Depending on contaminant type and concentration and amendment optimization, relatively rapid treatment can be achieved in individual windrows. The compost process is self heating which allows for year round treatment with minimal

Table 14.5 Comparison of Treatment Costs for Explosives-Contaminated Soils[a]

Task	Windrow Composting	Simplot - SABRE	WR Grace- DARAMEND	WMI - TOSS (Off Site)	EarthFax - WRF
Preconstruction activities	$20	$20	$20	$10	$15
Site setup and mobilization	$30	$55	$30	$10	$20
Onsite pilot test/transportation	$10	$12	$10	$80	$5
Treatment	$150	$204	$90	$205	$150
Labor	$41	$35	$31	NA	$40
Equipment	$55	$65	$18	NA	$25
Materials and supplies	$36	$95	$29	NA	$75
Analytical	$18	$9	$12	NA	$10
Demobilization and site closeout	$15	$17.5	$15	$10	$15
Total ($/ton):	$225	$309	$165	$315	$205

[a] Approximately 25,000 tons of RCRA hazardous soil.
[b] NA: Data not available.

Table 14.6 Comparison of Treatment Costs for Explosive-Contaminated Soils[a]

Task	Windrow Composting	Simplot- SABRE	WR Grace – DARAMEND	WMI-TOSS (Offsite)	EarthFax- WRF
Preconstruction activities	$5	$5	$5	$2.5	$5
Site setup and mobilization	$10	$25	$13	$2.5	$5
Onsite pilot test/transportation	$5	$8	$3	$32	$2.5
Treatment	$108	$115	$89	$90	$107
Labor	$21	$27	$20	NA	$20
Equipment	$14	$29	$33	NA	$15
Materials and supplies	$66	$56	$31	NA	$65
Analytical	$8	$4	$5	NA	$7
Demobilization and site closout	$5	$7.5	$5	$2.5	$5
Total ($/ton):	$133	$162	$115	$130	$125

[a] Approximately 100,000 tons of RCRA non-hazardous soil.
[b] NA: Data not available.

capital expense for heating of the treatment facilities. Due to the physical properties and high nutrient (nitrogen and phosphorous) value of the finished compost, the material could be used as a soil conditioner for revegetation activities. Drawbacks to the technology include the significant volume increase (up to twice the original volume) and the availability of the proper amendments. Site location can yield difficulties in locating and procuring a consistent, large volume of amendments over multiple years. Local weather conditions can impact amendment availability and cost from one growing season to the next. In addition to directly impacting treatment cost, difficulties with amendment selection and availability may produce ammonia gas during the composting process. The production of ammonia results in an increased treatment cost for air monitoring and treatment (if necessary) and increased cost for personnel protective equipment (PPE).

14.6.5   WR Grace-DARAMEND

For Site Condition 1, a 1.5-acre treatment cell will treat approximately 3500 tons of soil, 360 yd^3 of organic amendments, and 72 yd^3 of inorganic amendments per batch. With an average batch treatment time of approximately 3 months plus an additional 1 month per batch to load and unload the soil from the treatment cell, approximately 2.5 years are required to treat 25,000 tons of soil. Two equipment operators and one technician are assigned to the project on a part time basis to provide tilling and monitoring activities. Treatment equipment includes two agricultural tractors with tilling attachments, an irrigation system, and a building. During treatment cell loading and unloading, loaders, dozers, and trucks are brought to the site as needed, minimizing both labor and equipment costs. Although the overall quantity of amendments is less than composting, the unit cost for the required amendment is much higher, and, therefore, the overall amendment costs are similar to windrow composting (Table 14.6).

Minimal economy of scale is shown for the operation of the DARAMEND process under Site Condition 2. A 4-acre treatment cell enclosed in a heated building can treat approximately 9700 tons of soil per batch for an overall project duration of approximately 3.5 years. Six equipment operators and technicians are required to provide tilling, amendment addition, moisture management, and process monitoring. Higher personnel utilization reduces the unit cost for labor. However, this is offset by the need for larger equipment for soil loading and unloading and the resulting increase in equipment costs. Since the cost of the proprietary amendments is unchanged for increased quantities, the unit cost remains relatively constant (see Table 14.6).

The DARAMEND process offers two advantages over windrow composting relating to amendment addition: a smaller volume increase in the treated soil and less local and seasonal variability with amendment supply. The smaller final volume reduces the cost for confirmation sampling and handling of the treated material. Also, the equipment required for operation is typically available in the agricultural market as compared to the specialty equipment required for windrow composting. Since the DARAMEND process is not self-heating, the technology's disadvantages include the need for an external heat source for year-round operation. Although the smaller amendment volumes have some advantage, the proprietary nature of the amendment does increase unit costs for implementation. Also, due to the limited depth of operation (a manageable depth of 18 in. or less) and longer treatment times (90 to 120 days per batch), either a larger treatment cell or a longer project duration is required.

14.6.6 Simplot-SABRE Slurry

Site Condition 1 requires two biocells approximately 86 ft wide by 150 ft long by 7 ft deep each. The biocells will be double lined with an 80-mil high density polyethylene liner. Each biocell will hold approximately 500 tons of soil as a 40 to 50% slurry approximately 2.5 ft deep with 2.5 ft of water above the slurry. The treatment time for each batch will be approximately 40 days. Based on a soil quantity of 25,000 tons, 25 treatment cycles over a period of 3 to 4 years (depending on site location and winter operating temperatures) are necessary. Site Condition 2 requires four biocells. Based on a soil quantity of 100,000 tons, 50 treatment cycles over a period of 6 to 8 years are required.

In both site conditions, the soil is first processed through a series of vibrating bar screens to remove debris and rock greater than 2 in. in size. The screened material is then slurried with water to produce a final solids concentration of 40 to 50% by weight and then pumped to the treatment cell. Simplot has developed a proprietary gantry mixing system that bridges the treatment cell and allows for periodic mixing of the soil slurry and blending the inoculum, nutrients, and carbon source while maintaining anaerobic conditions. After treatment is complete, high solids slurry pumps on the gantry transfer the slurry from the cell to a filter press or a drying bed for separation of the solids and the water. Approximately three to four equipment operators or field technicians and a site supervisor are required to operate the system.

The SABRE bioslurry process is a proprietary technology that requires payment of royalties or a technology transfer fee. The technology, however, can treat high concentrations of explosives and has been successfully implemented in the field. However, a significant engineering effort is required due to the number and complexity of the process components, especially if the operation is conducted in engineered reactors or heated for year round operation.

14.6.7 EarthFax — Fungal-Based Remediation

Site Condition 1 requires treatment in aerated biopiles. Each biopile would be approximately 40 ft wide by 200 ft long by 8 ft deep and would treat approximately 1000 tons of soil based on a 50% by volume addition of an inoculum. Three biopiles would be constructed and operated concurrently in an enclosed facility. The total treatment cycle time for the biopiles would be four months which includes 1 month for pile construction and teardown; therefore up to nine biopiles would be constructed and operated during 1 year. Approximaely 3 to 4 years of operation is necessry to treat 25,000 tons of soil. For Site Condition 2, six biopiles would be operated concurrently in an enclosed facility during a 4-month period. Therefore, up to 18 biopiles would be constructed and operated during 1 year. Approximatley 5 to 7 years is necessary to treat 100,000 tons of contaminated soil.

14.6.8 WMI-TOSS

As shown in Table 14.7, a significant cost for Site Condition 1 and hazardous soils is the excavation and transportation of large quantities of soil to the Lake Charles Facility. The reduced costs for support facilities, including mobilization and demobilization, are offset by the transportation costs, even when considering shipments by rail. Even with Site Condition 2, where actual treatment costs for the TOSS process are comparable to the other technologies, transportation costs are significant disadvantages. The major advantage of the offsite WMI-TOSS process is realized when volumes of material requiring treatment at a single site are less than 5,000 to 10,000 tons. Under this scenario, the fixed costs (i.e., mobilization, treatment facilities, administration, etc.) for the onsite technologies would result in increased unit costs, whereas the offsite costs at a fixed facility can be spread over larger volumes of soil from other facilities and normal landfill operations, reducing the unit costs for treatment.

Table 14.7 Process Comparison for Treatment of Explosives-Contaminated Soils

Process	Advantages	Disadvantages
Windrow composting	Fast, proven	Amendments, volume bulking, curing
WMI TOSS	Offsite, accept liability	Transportation, temperature
EarthFax — Fungal-Based Remediation	Low operating costs	Temperature, inoculum
WR Grace-DARAMEND	Amendments	Space requirements, proprietary process
Simplot-SABRE bioslurry	Proven, treat high concentrations	Cost, engineering

14.7 SUMMARY

In summary, the processes presented in Table 14.7 have been applied at the demonstration or full scale for treatment of explosives-contaminated soils. Windrow composting has been used successfully to treat soils at numerous Army and Navy facilities. The process is fast, proven, and does not produce any persistent toxic transformation products. Windrow composting, however, requires large volumes of amendments on a year round basis that produce a substantial increase (typically 50%) in volume of material. Final curing of the compost is necessary after treatment of the nitroaromatic compounds prior to disposal or land application. A unique feature of the TOSS process is that it is the only technology currently offered as a commercial service for the offsite treatment of explosives-contaminated soil. WMI will accept liability for the contaminated soils. Transportation cost for the contaminated soils, however, may be significant. This commercial service may be very desirable for small soil volumes. The Earth Fax fungal process is a forced-aeration soil pile process providing effective treatment of explosives with low operating costs. Process temperature control, however, is critical for performance and a substantial volume (>10%) of fungal inoculum is required. The WR Grace DARAMEND process is a land treatment process which requires <10% by volume amendments. Therefore, increased soil volumes are not a concern. The process is proprietary and requires more space to treat a given volume of soil because the soil depths must be less than 18 to 24 in. for effective treatment.

The Simplot-SABRE process is a bioslurry treatment that can effectively treat high explosives concentrations in soil. The cost, however, is higher than that of the other commercial processes, and significant process engineering effort is required.

ACKNOWLEDGMENTS

The authors would like to thank Wayne E. Sisk with the U.S. Army Environmental Center for his contributions to Tables 14.1 to 14.3 and the companies who provided information through personal communications.

REFERENCES

1. Barnes, P.W., R.B. Green, and G.R. Hater. 1999. Phytotoxicity reduction by anaerobic-aerobic treatment of explosives-contaminated soil, pp. 39–44. In B.C. Alleman and A. Leeson (eds.), *Bioremediation of Nitroaromatic and Haloaromatic Compounds*. Battelle Press, Columbus, OH.
2. Broder, M.F. and R.F. Westmoreland. 1998. An estimate of soils contaminated with secondary explosives. Report No. SFIM-AEC-ET-CR-98002. U.S. Army Environmental Center, Aberdeen Proving Ground, MD.
3. Brunner, R. and D. Potter. 1999. Cost effective explosive compound bioremediation by windrow composting Hawthorne Army Depot, pp 57–61. In B. Alleman and A. Leeson (eds.), *Bioremediation of Nitroaromatic and Haloaromatic Compounds*. Battelle Press, Columbus, OH.
4. Earth Fax Development Corporation, Logan, UT. 1999. Personal communication.
5. Earth Tech Corporation, Pueblo, CO. 1999. Personal communication.
6. Eckman, G., T. Lundgren, A. Handel, and R. Green. 1999. Sequential anaerobic-aerobic treatment of explosives-contaminated soil. Presented at the XIII SAFEX International Congress, May 26–29. Dublin, Ireland.
7. Emery, D. and P. Faessler. 1997. First production-level bioremediation of explosives-contaminated soil in the United States. *Biomed. Surf. Subsurf. Contam.* 829:326–340.
8. Fritsche, W., K. Scheibner, A. Herre, and M. Hofrichter. 2000. Fungal degradation of explosives: tnt and related nitroaromatic compounds. Chap. 9. In J.S. Spam, J.B. Hughes, and H.-J. Knackmuss (eds.), *Biodegradation of Nitroaromatic Compounds and Explosives*. Lewis Publishers, Boca Raton, FL.
9. Fukutomi, E. and W. Zbitnoff. 1999. Composting ordnance contaminated soil pilot study to full scale reality. In Poster Abstracts. Proceedings from the Fifth International Symposium In Situ and On Site Bioremediation, San Diego, CA.
10. Grace Bioremediation. 1999. Personal communication.
11. Green, R.B., P.E. Flathman, G.R. Hater, D.E. Jerger, and P.M. Woodhull. 1998. Sequential anaerobic-aerobic treatment of TNT-contaminated soil, pp. 271–276. In G.B. Wickramanayake and R.E Hinchee (eds.), *Designing and Applying Treatment Technologies: Remediation of Chlorinated and Recalcitrant Compounds*. Battelle Press, Columbus, OH.

12. Green, R.B., P.W. Barnes, and G.R. Hater. 1999. Pilot demonstration of the sequential anaerobic-aerobic bioremediation of explosive-contaminated soil, pp. 45–50. In B.C. Alleman and A. Leeson (eds.), *Bioremediation of Nitroaromatic and Haloaromatic Compounds*. Battelle Press, Columbus, OH.

13. Griest, W.H., A.A. Vass, A.J. Stewart, and C. Ho. 1998. Chemical and toxicological characterization of slurry reactor biotreatment of explosives-contaminated soils. Report No. SFIM-AEC-ET-CR-96186. U.S. Army Environmental Center, Aberdeen Proving Ground, MD.

14. Griest, W.H., R.L. Tyndall, A.J. Stewart, C. Ho, K.S. Ironside, J.E. Canton, W.M. Caldwell, and E. Tan. 1991. Characterization of explosives processing waste decompostion due to composting. Report No. DOE IAG 1016-B1213-A1. U.S. Army Medical Research and Development Command, Ft. Detrick, Frederick, MD.

15. Gwinn, R. 1999. Composting treatment of explosives-contaminated soil, Tooele, Utah, pp. 51–56. In B. Alleman and A. Leeson (eds.), *Bioremediation of Nitroaromatic and Haloaromatic Compounds*. Battelle Press, Columbus, OH.

16. Hampton, M. and J. Manning. 1998. Field demonstration of multiple bioslurry treatment technologies for explosive-contaminated soils, pp. 259–266. In G. Wickramanayake and R. Hinchee (eds.), *Designing and Applying Treatment Technologies: Remediation of Chlorinated and Recalcitrant Compounds, Vol. 6*. Battelle Press, Columbus, OH.

17. Jerger, D. and P. Woodhull. 1999. Technology development for the biological treatment of explosives-contaminated soils, pp 33–38. In B. Alleman and A. Leeson (eds.), *Bioremediation of Nitroaromatic and Haloaromatic Compounds, Vol. 5*. Battelle Press, Columbus, OH.

18. Kaake, R., J. Bono, and T. Jergovich. 1997. Full-scale remediation of an explosives-contaminated site at Yorktown Naval Weapons Station using the SABRE process. In Proceedings from the Air and Waste Management Association 90th Annual Meeting. Toronto, Ontario.

19. Lenke, H., C. Achtnich, and H.-J. Knackmuss. 2000. Perspectives of bioelimination of polynitroaromatic compounds. Chap. 4. In J.S. Spain, J.B. Hughes, and H.-J. Knackmuss (eds.), *Biodegradation of Nitroaromatic Compounds and Explosives*. Lewis Publishers, Boca Raton, FL.

20. Manning, J.F., R. Boopathy, and E. Breyfogle. 1996. Field demonstration of slurry reactor biotreatment of explosives-contaminated soils. Report SFIM-AEC-ET-CR-96178. U.S. Army Environmental Center, Aberdeen Proving Ground, MD.

21. Naval Surface Warefare Center. 1997. Bioremediation facility — a technology for explosives-contaminated soils. Department of Navy, Crane, IN.

22. Pennnington, J.C. 1999. Natural attenuation of explosives in soil and water systems at Department of Defense sites. Interim Report. Technical Report El-99-8. U.S. Army Engineer Waterways Experiment Station, Vicksburg, MS.

23. Pennington, J.C. 1999. Natural attenuation of explosives in soil and water systems at department of defense sites. Final Report. Technical Report SERDP-99-1. U.S. Army Engineer Waterways Experiment Station, Vicksburg, MS.

24. Plexus Scientific Corporation. 1999. U.S. Army Environmental Center biotechnology demonstration. Final Report, SFIM-AEC-ET-CR-99012. U.S. Army Environmental Center, Aberdeen, MD.

25. Roy F. Weston, Inc. 1988. Field demonstration-composting of explosives-contaminated sediments at the Louisiana Army Ammunition Plant. Technical Report AMXTH-IR-TE-88242. U.S. Army Toxic and Hazardous Materials Agency, Aberdeen Proving Ground, MD.

26. Roy F. Weston, Inc. 1993. Windrow composting demonstration for explosives-contaminated soils at the Umatilla Depot Activity Hermiston, Oregon. Technical Report CETHA-TS-CR93043. U.S. Army Environmental Center, Aberdeen Proving Ground, MD.

27. Roy F. Weston, Inc. 1993. Windrow composting engineering/economic evaluation. Technical Report CET-HA-TD-5. U.S. Army Environmental Center, Aberdeen Proving Ground, MD.

28. Spain, J.C. (ed.). 1995. *Biodegradation of Nitroaromatic Compounds*. Plenum Publishing, New York, 232 pp.

29. Tan, E.L., C.H. Griest, and R.L. Tyndall. 1992. Mutagenicity of trinitrotoluene and its metabolites formed during composting. *J. Toxicol. Environ. Health.* 36:165–172.

30. Title 40, Code of Federal Regulations (40 CFR), Chapter I, Subchapter I, Part 261. 1980.

31. Tuomi, E., M. Coover, and H. Stroo. 1997. Bioremediation using composting or anaerobic treatment for ordnance contaminated soils. *Bioremed. Surf. Subsurf. Contam.* 829:160–178.

32. U.S. Army Environmental Restoration Programs Guidance Manual. April 1998. Defense Environmental Network and Information Exchange (online). www.denix.osd.mil/denix/Public/Policy/Army/ERPerptoc.html.

33. U.S. Environmental Protection Agency. 1995. J.R. Simplot *ex situ* bioremediation technology for treatment of TNT-contaminated soils. Report EPA/540/R-95/529. U.S. Environmental Protection Agency, Washinton, D.C.

34. U.S. Environmental Protection Agency. 1997. GRACE Bioremediation Technologies' DARAMEND™ Bioremediation Technology. Report EPA/540/R-95/536a. U.S. Environmental Protection Agency, Washington, D.C.

35. Williams, R.T., P.S. Ziegenfuss, and W.E. Sisk. 1992. Composting of explosives and propellant contaminated soils under thermophilic and mesophilic conditions. *J. Ind. Microbiol.* 9:137–144.

36. Woodhull, P., D. Jerger, P. Barnes, R. Staponski, and S. Wharry. 1999. Composting of explosives-contaminated soils: a pilot and full-scale case history, pp. 63–67. In B. Alleman and A. Leeson (eds.), *Bioremediation of Nitroaromatic and Haloaromatic Compounds*. Battelle Press, Columbus, OH.

Index

N

9 780367 398491